MW00534794

Note to the Student

Dear Student,

If you winced when you learned the price of this textbook, you are experiencing what is known as "sticker shock" in today's economy. Yes, textbooks are expensive, and we don't like it any more than you do. Many of us here at PWS-KENT have sons and daughters of our own attending college, or we are attending school part-time ourselves. However, the prices of our books are dictated by cost factors involved in producing them. The costs of typesetting, paper, printing, and binding have risen significantly each year along with everything else in our economy.

The prices of college textbooks have increased less than most other items over the past fifteen years. Compare your texts sometime to a general trade book, i.e., a novel or nonfiction book, and you will easily see substantial differences in the quality of design, paper, and binding. These quality features of college textbooks cost money.

Textbooks should not be considered only as an expense. Other than your professors, your textbooks are your most important source for what you learn in college. What's more, the textbooks you keep can be valuable resources in your future career and life. They are the foundation of your professional library. Like your education, your textbooks are one of your most important investments.

We are concerned, and we care. We pledge to do everything in our power to keep our textbook prices under control, while maintaining the same high standards of quality you and your professors require.

Wayne A. Barcomb
President
PWS-KENT Publishing Company

Computer Science Series List

Advanced Structured Basic: File Processing with the IBM PC
James Payne

Advanced Structured COBOL, Second Edition
Gary S. Popkin

Assembly Language for the PDP-11: RT-RSX-UNIX, Second Edition
Charles Kapps and Robert Stafford

Comprehensive Structured COBOL, Third Edition
Gary S. Popkin

Data Structures, Algorithms, and Program Style
James F. Korsh

Data Structures, Algorithms, and Program Style Using C
James F. Korsh and
Leonard J. Garrett

Design and Analysis of Algorithms
Jeffrey D. Smith

Introductory Structured COBOL Programming, Second Edition
Gary S. Popkin

Logic and Structured Design for Computer Programmers
Harold J. Rood

Problem Solving Using Pascal: Algorithm Development and Programming Concepts
Romualdas Skvarcius

Problem Solving with Pascal
Kolman W. Brand

Programming Using Turbo Pascal
John H. Riley

Structured BASIC for the IBM PC with Business Applications
James Payne

Structured Programming in Assembly Language for the IBM PC
William C. Runnion

Turbo Pascal for the IBM PC
Loren Radford and
Roger W. Haigh

Using BASIC: An Introduction to Computer Programming, Third Edition
Julien Hennefeld

Using Microsoft and IBM Basic: An Introduction to Computer Programming
Julien Hennefeld

Using Turbo Pascal
Julien Hennefeld

VAX Assembly Language and Architecture
Charles Kapps and Robert Stafford

Design and Analysis of Algorithms

Jeffrey D. Smith
San Jose State University

PWS-KENT Publishing Company
Boston

Acquisitions Editor: J. Donald Childress
Production Editor: Wanda K. Wilking
Production: Marjorie Sanders, Bookman Productions
Interior Design: Image House, Inc.
Cover Design: Nancy Lindgren
Artwork: Lisa C. Sparks
Manufacturing Coordinator: Marcia A. Locke
Composition: Bi-Comp, Inc.
Cover Printer: Phoenix Color Corporation
Text Printer and Binder: Book Press, Inc.

PWS-KENT Publishing Company is a division of Wadsworth, Inc.

Printed in the United States of America

1 2 3 4 5 6 7 8 9—93 92 91 90 89

Library of Congress Cataloging-in-Publication Data

Smith, Jeffrey Dean, 1951–
Design and analysis of algorithms/Jeffrey D. Smith

p. cm
Bibliography: p.
Includes index.
ISBN 0-534-91572-8
1. Algorithms. I. Title.
QA9.58.S57 1989
511′.8—dc19

88-25542
CIP

Preface

This text is intended to present the design and analysis of algorithms to students who do not necessarily have a thorough background in data structures and data types. We emphasize design techniques; we do not merely present a series of unrelated algorithms.

Most prominent among these techniques is the *divide-and-conquer* approach. This technique is emphasized because of its traits: naturalness, usefulness, relationship to other techniques such as dynamic programming, and relationship to recursion and induction. In turn, induction relates additional topics such as greedy methods and iteration to the main thrust of the text. Finally, algorithms based on a divide-and-conquer strategy are usually easy to implement once the design is finished.

A brief synopsis of the text is useful here. Chapter 1 is an introductory chapter that deals largely with asymptotic analysis, O-notation, and mathematical background. Chapter 2 introduces induction, recursion, and most of this text's design techniques. Chapter 3 covers iteration, relating it to the approach of the previous chapter. Chapter 4 applies the techniques of Chapters 1 through 3 typically to problems related to a particular data type. It also provides a brief introduction to the theory of computability. Chapter 5 introduces the notion of NP-completeness, suggests why NP-complete problems are likely to be intractable, and touches on several techniques for dealing with intractable problems.

One goal of this text is to make students as comfortable with recursion and induction as they are with iteration. No claim is made that iteration is better than recursion or that learning one approach first will spoil a student for the other. Chapter 2, dealing primarily with recursion, is longer than Chapter 3, dealing with iteration, largely because students are expected to be more familiar with iteration. Induction, recursion, and the related topics of recurrence relations and O-notation are emphasized more heavily than in other texts.

Ideally, students will have seen arrays, stacks, queues, and linked lists and will be familiar with the notions of abstract data types and structured programming. For those who are not, a brief introduction to these data types and their implementation is given in Appendix D. Students who have seen these topics before are unlikely to become bored when seeing them again, since the approach of this text is new.

Although students would greatly profit from having had a course in discrete mathematics before attempting this text, it is not absolutely necessary. A more detailed discussion of assumed material and useful definitions appears in Chapter 1.

Solutions to a large percentage of the exercises are presented in Appendix C. The exercises are of a wide range of difficulty; in general, exercises proceed from less to more difficult within a subject heading. For students with relatively little background in data structures, the solutions to the exercises serve as additional examples. Students with more background are likely to benefit from the solutions to exercises that require proof. Instructors may thus tailor this text to their classes, providing additional examples from the exercises.

A number of the solutions are actually references to other sources (the corresponding exercises are marked with double asterisks). Typically these references are other textbooks rather than original sources. Throughout this text the assumption is that students following references are seeking an alternate presentation of new material and not fodder for a dissertation. Many texts cited have substantial bibliographies. Credit has been given where appropriate and where known to the original discoverers of particular algorithms; no original algorithms or significant theorems are claimed here. Some original sources are presented in the sections on further reading at the end of each chapter.

In some cases (e.g., balanced search trees, partition), details found inappropriate for early coverage are presented later. Since the topics of an algorithms course aren't structured in a linear way, any attempt to linearize them will inevitably lead to compromises. In addition, a number of sections contain references to later sections. These generally point to applications of the material currently being presented and reassure the student that the current material really does have applications. Presumably, mentioning an application and later elaborating on it are preferable to simply mentioning it.

Since specifications of divide-and-conquer algorithms and recursive definitions of functions are so close to executable programs in many languages, frequently only the specification or the definition is given in this text. In particular, the same numbering system is used for definitions and algorithms.

As is common in texts on algorithms, but perhaps less common in texts on data structures and algorithms, a pseudolanguage has been used in preference to any existing programming language. In this way students avoid confusion over irrelevant details of syntax of a particular language and are better prepared to work in new languages. They also learn more thoroughly the skill of proceeding from a specification to an executable program; seeing a fully-fleshed out program in a common programming language is probably not the best way to learn this skill.

This text does not aim to be a text on structured programming or programming style. Most algorithms presented here would correspond to low-level operations in structured programming. However, the algorithms are consistent with structured programming techniques, and are of good style.

Students should not be intimidated by the formal statement of definitions and theorems and their proofs. This formalism is not intended to increase but to reduce complexity by isolating the formal material from the less formal discussion.

Some algorithms for which correctness proofs have been provided may seem trivially correct. Part of the reason the proofs have been provided is simply to give students practice in simple cases before they are expected to handle difficult cases. Ideally, students using this text will be able to construct at least simple proofs themselves.

Throughout the text, emphasis is placed on how to arrive at solutions, not merely on the solutions themselves. This emphasis applies to analysis of algorithms, constructions of invariants, and other proofs, as well as to the design of algorithms. In particular, the guess-and-verify strategy is emphasized so that students do not come to believe that proofs are natural only to an arcane priesthood.

Finally, the individuals who reviewed this text must be thanked for their helpful and constructive comments and criticisms:

Martin Fraser, Georgia State University
Robert Holloway, University of Wisconsin
James Korsh, Temple University
Jeffrey Leon, University of Illinois at Chicago
Steven Minsker, Camden College of Rutgers University
Ronald Peterson, Weber State College
Dean Sanders, Illinois State University
Richard Schlichting, The University of Arizona
Janice Stone, The Interfactor, Inc.
Helen Takacs, Mississippi State University
Alan Tharp, North Carolina State University

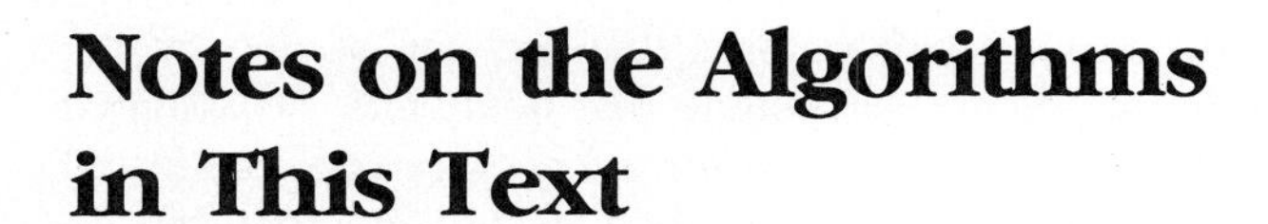

Notes on the Algorithms in This Text

The algorithms and recursive definitions in this text have been written in a structured pseudocode. Its syntax does not correspond exactly to that of any common high-level language. For the most part the algorithms can be easily read without bothering with the following description of the syntax, but a few points are worth making.

A "return" procedure is assumed. This procedure exits the algorithm, possibly with a value and possibly "unsuccessfully" as in the case of an unsuccessful search. This procedure is documented in bold face. An unspecified error-handling procedure is assumed; one syntax for exiting the algorithm to enter this procedure is **return error.**

Standard control structures of the form "if... then... else," "for...," "while ...," and "repeat..." are assumed. These are terminated by an explicit "end" statement (e.g., **end for** or **end else**), unless only a single statement is to be executed under control of one of the keywords **then, else,** or **for** and the statement appears on the same line as the keyword. The full form of the "for" statement is "for j from k to r step m," indicating that j is incremented by m or decremented by $-m$ if m is negative. A "for" loop will not be executed if $k > r$ and $m > 0$ or if $k < r$ and $m < 0$. The "end" of the "repeat" statement also includes the condition that terminates the loop (e.g., **end repeat if** $i = 0$). A statement **end repeat** without a condition indicates the corresponding loop is repeated until the execution of a "return" statement.

Assignment statements are indicated by **:=**. Comments are enclosed in braces in definitions as well as algorithms. Carriage returns and line feeds do not terminate comments. Function arguments are enclosed by parentheses and array indices by brackets.

A **swap** operation is assumed. It exchanges the values of its two arguments.

A **dummy** value is assumed in several algorithms. Its implementation has not been specified.

Parameter passing uses call by value unless documented otherwise. When using call by value, parameters may be used as ordinary local variables without affecting their values in the calling routine. We assume that data structures (e.g., arrays) can be passed as parameters with changes made locally reflected in the calling subprogram's copy.

In a conditional of the form "if A and B," if A is false then B is not tested. In a conditional of the form "if A or B," if A is true then B is not tested.

The field "fld" of the record to which the pointer "ptr" points is indicated by "fld(ptr);" we will not be dealing with records that are not labeled by the names of pointers pointing to them. We assume that fields may be assigned values by a statement of the form "fld(ptr) := *q*." In a few cases a field will itself have an array structure. If a field is an array, then its *k*th element can be accessed by "fld(ptr)[*k*]."

In a few cases we assume that a function is defined on objects of a given data type and can have its value accessed using similar syntax; for example, "size(a)" where A is an array. The implementation and initialization of these functions is language dependent, so we do not discuss their implementation.

A function **Get** is assumed for the creation of new storage. The size, type, and structure of the new storage are language dependent, so we have not provided a formal mechanism for specifying them. The documentation of an algorithm should clearly show how to implement the function in a given programming language.

Line numbers have been included in each algorithm. They are for convenience only and have no syntactic function.

The distinction between upper- and lowercase is not significant.

Contents

3 Iteration 159

4 Data Structures and Applications 191

Introduction and Mathematical Preliminaries

1.1 Introduction

A natural first question for students approaching a textbook is "Why bother with this material?" In this case, the short answer is that this text will help the reader learn the art of problem solving.

More precisely, this text is concerned with the design, analysis, and correctness of algorithms, especially those arising naturally from operations on data structures. Intuitively, an algorithm is a recipe for solving a problem. It is what remains of a program when details of particular languages, machines, and so on, have been abstracted away. That is why we have not written our algorithms in any common programming language in this text.

The *design* process begins with a description, perhaps informal, of the problem to be solved. It proceeds to a more formal specification of the problem and then to an algorithm expressed as a program in a programming language or in a pseudolanguage. The *analysis* of an algorithm determines how efficient the algorithm is in its use of resources such as time and space. *Correctness* arguments are used to show that the algorithm intended to solve a problem really does solve the problem (i.e., meets the specifications of the problem). Another important consideration is whether or not a program *terminates*. This may be considered a special case of either analysis or proving correctness.

Chapters 2 and 3 cover two families of techniques for dealing with the above issues. Chapter 2 treats divide-and-conquer techniques, induction, and recursion. Chapter 3 treats iteration, or the use of loops. Chapter 4 applies these techniques to various data structures, and Chapter 5 deals with hard problems and how to get around them. Throughout the text, the emphasis is on algorithms for operations on data structures.

Section 1.2 describes the mathematical and computer science background you will need for this text. This section also presents a few definitions. The rest of the chapter covers the material needed to understand data structures and the analysis of algorithms.

You should have the computer science background needed for this text if you have had the equivalent of one year or more of programming courses in which arrays, stacks, queues, and linked lists were discussed. Ideally, you will have seen how data structures, such as stacks, can be implemented in terms of other data structures, such as arrays or linked lists. You will also be able to write programs so structured that a change in implementation does not require substantial rewriting of the code. If you are unfamiliar with the above data types, a very brief discussion of them and their implementation is given in Appendix D. A more detailed list of assumed concepts is given at the beginning of Section 1.2.

These data types and others are covered in this text from a more mathematical point of view. You may still want to look through Section 1.3 even if you have seen the notions of data structures and data types before.

Other courses that would be helpful but are not absolutely necessary for understanding the material in this book include probability and statistics, computer organization, and especially discrete mathematics. Proofs are an important part of this text, and you will have an advantage if you have done, or at least followed, a number of mathematical proofs. However, the material on induction and recursion with which we begin Chapter 2 is probably the most painless place to begin the study of proofs.

If you are not used to texts containing many explicit theorems and proofs, parts of this one may look formidable. Don't worry. The use of explicit theorems, like the use of footnotes, is a means of isolating detail away from the main presentation. Most of the text is easier to read than use of the word "theorem" might imply.

1.2 Mathematical Background

Concepts assumed but not discussed in this text include real and rational numbers; functions (especially the exponential and logarithmic functions); sums of functions; products of functions; compositions of functions; sets, set notation, set operations (union, intersection, complementation and the universal set, De Morgan's laws, Cartesian products, membership); the mean and the median; intervals and interval notation; binary numbers and bits; and in general the use of a base, or *radix*, other than 10.

Concepts assumed from computer science are arrays, input and output, loops, and iteration conditional statements, subprograms or procedures and their parameters (or arguments), methods of parameter passing (e.g., value and variable parameters in Pascal), local and global variables, characters and character strings, compilation and interpretation, documentation and debugging, as well as the data types of the previous section.

We use some mathematical terms and notation you may not have seen:

The size of a set S will be denoted $|S|$.

If $y > 0$, x *modulo* y (usually abbreviated $x \bmod y$) is the remainder (also called the *residue*) when the integer x is divided by the integer y. The possible values run therefore from 0 through $y - 1$. If $x < 0$, it may not be clear exactly what is meant by the remainder. In this case $x \bmod y$ is defined to be $x - m$, where m is the largest multiple of y that is less than or equal to x. If $x \bmod y$ equals $z \bmod y$ then x and z are said to be *congruent* mod y. This is the same as saying that $x - z$ is a multiple of y.

If $x \bmod y$ is zero (i.e., if y divides x), we write $y|x$.

The *floor* of a number x is the largest integer that is less than or equal to x. The notation is $\lfloor x \rfloor$. The *ceiling* of a number x is the smallest integer that is greater than or equal to x. The notation here is $\lceil x \rceil$. For example, $\lfloor 3.2 \rfloor = 3$, $\lceil 3.2 \rceil = 4$, and $\lfloor 3 \rfloor = \lceil 3 \rceil = 3$. Notice $\lfloor -3.2 \rfloor = -4$, $\lceil -3.2 \rceil = -3$. Also, $n = \lceil n/2 \rceil + \lfloor n/2 \rfloor$ and $n = \lceil n \rceil = \lfloor n \rfloor$ for all integers n.

A function f is *bounded* if and only if there exists a number M such that for all x in the domain of f, $|f(x)| \leq M$.

Two sets that do not intersect are called *disjoint*. A set is the *disjoint union* of two sets A and B if 1) it is the union of A and B, and 2) A and B are disjoint sets. A *family* is another word for a set.

An *operator* is for our purposes the same as a function. The input values to which it is applied are called *operands* or *arguments*. The application of an operator is called an *operation*.

A *Boolean operator* is one of the operators **and, or** (both taking two operands), or **not** (taking one operand), or an operator that can be defined in terms of these (e.g., the following "exclusive or" function). They take as operands the *Boolean values* "true," symbolized below by T, and "false," symbolized below by F. Sometimes "false" is represented by 0 and "true" by 1 or some other nonzero value.

The operator **and** is often symbolized by $\wedge$, **or** by $\vee$, and **not** by $\sim$. Sometimes $\sim x$ is written as x'. The definitions of these three operators, together with the *exclusive or* operator (symbolized by $\oplus$), are given by the tables in Figure 1.1.

x	y	$x \vee y$	$x \wedge y$	$x \oplus y$
T	T	T	T	F
T	F	T	F	T
F	T	T	F	T
F	F	F	F	F

x	$\sim x$
T	F
F	T

Figure 1.1 Truth Tables Defining Four Boolean Functions

Tables that give the values of Boolean functions for all possible values of the operands are called *truth tables.*

The notation $p \rightarrow q$, read "p *implies* q," is synonymous with $p' \vee q$. The notation $p \leftrightarrow q$, read "p is *equivalent* to q," is synonymous with $(p \rightarrow q) \wedge (q \rightarrow p)$. The equivalence is true if and only if p and q have the same truth value. Sometimes we express $\leftrightarrow$ by saying "p iff q," where *iff* is an abbreviation of the phrase "if and only if." The $\leftrightarrow$ operator is just the negation of $\oplus$, as can be checked by constructing their truth tables.

The expression $(a \vee b)'$ is equivalent to $a' \wedge b'$. The expression $(a \wedge b)'$ is equivalent to $a' \vee b'$.

We can apply the **and** and **or** operators to binary numbers of the same length by applying them to corresponding pairs of bits. Here we use 1 to stand for T and 0 for F. For example, $110 \wedge 101 = 100$, $100 \vee 010 = 110$, $\sim 101 = 010$.

The *expected value,* or *average value,* of a function f is the sum over all possible function values v of $vp(v)$, where $p(v)$ is the probability that f has value v. As usual, we assume that the sum of all the probabilities is 1.

An undirected *graph* consists of a set V of vertices and a set E of edges, each of the form $\{v, w\}$ for v and w in V. The vertices v and w are said to be *adjacent.* A *directed graph,* or *digraph,* consists of sets V and E such that each element of E is of the form (v, w) rather than $\{v, w\}$ (i.e., is ordered) for v and w in V. We think of the edge (v, w) as going from v to w. A *subgraph* of a graph with vertices V and edges E is a graph whose vertices are all from V and whose edges are all in E.

Graphs will be considered to be undirected unless otherwise specified. Also, *loops,* edges connecting a vertex to itself, and multiple edges with the same direction between two vertices will not be allowed unless otherwise specified.

A *complete* graph is one such that every pair of distinct vertices is connected by an edge. In other words, $E = \{\{x, y\} | x, y \text{ are in } V \text{ and } x \neq y\}$ for undirected graphs. For directed graphs $E = \{(x, y) | x, y \text{ are in } V \text{ and } x \neq y\}$.

In Figure 1.2, the graph of (a) is a complete, undirected graph. The graph of (b) is a directed graph. The graph of (c) is a subgraph of the graph of (a). In it, vertices 2 and 4 are adjacent, as are vertices 1 and 4. Vertices 1 and 2 are not adjacent.

We generally use the word *increment* to mean "add one to" and *decrement* to mean "subtract one from." Sometimes we will say "increment by two" to mean "add two to," and so on. The word *increment* is also used to mean the number added; the word *decrement* is also used to mean the number subtracted.

A repetition of the statements in a loop is called an *iteration.* A procedure that calls itself is said to be *recursive.* This will be true even if the procedure calls itself indirectly (e.g., P calls Q, which calls P). We treat recursion more formally beginning with Section 2.1.

When parameters are passed by *value-result,* an assignment to a parameter variable in the called subprogram also changes the value of the corresponding variable in the calling subprogram.

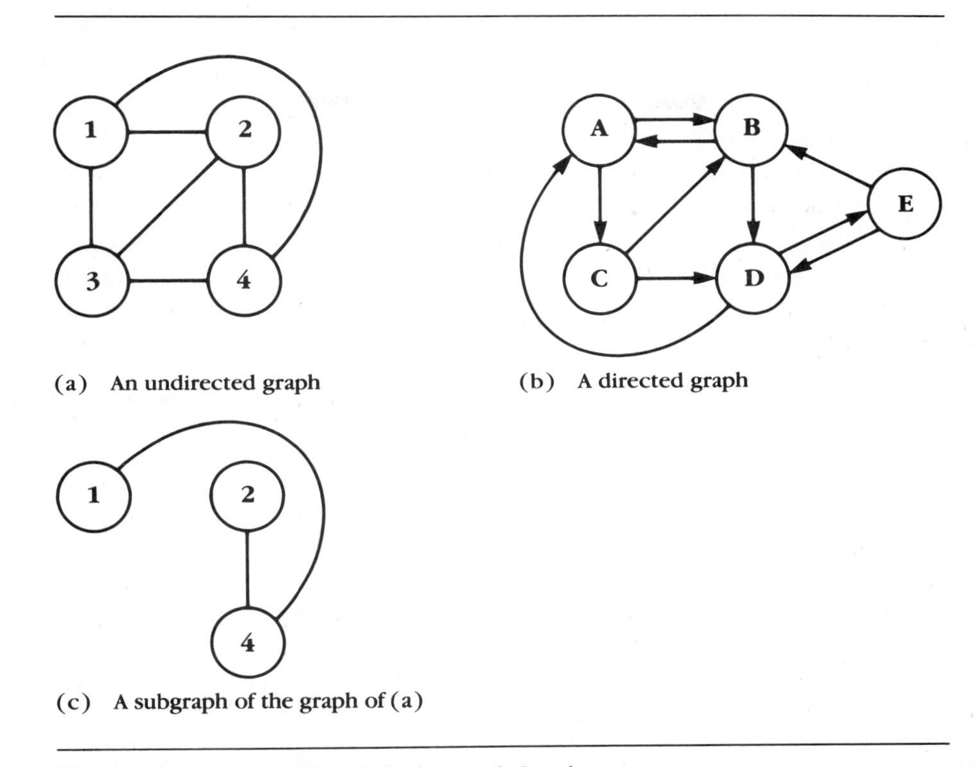

(a) An undirected graph

(b) A directed graph

(c) A subgraph of the graph of (a)

Figure 1.2 Directed and Undirected Graphs

An *instance* of a problem is a set of actual parameters corresponding to the formal parameters of the problem.

To *trace* a procedure means to mimic "by hand" the action of the procedure.

A *pointer* is a data item that gives the location of another data item. An example is an array index. Some languages provide a special data type or types for pointers. In these languages an array index would be of type "integer" instead of a pointer type.

A *dummy* value or variable is one whose only function is to occupy space.

The term *high-level language* refers to a programming language that is closer to English or other human languages than is the language of a particular machine. Examples include Pascal, FORTRAN, BASIC, COBOL, Lisp, Ada, and APL.

An *algorithm,* a procedure so specifically defined that in principle a human being could do it, consists of a finite set of instructions and terminates after a finite number of steps for all choices of input. Sometimes when we are formulating an algorithm, we will first describe a less formal or precise version called a *high-level* description of the algorithm.

External memory of a machine is that which is located outside of the machine proper, such as on disk or on tape.

1.3 Data Structures and Data Types

Data items may be divisible into smaller data items or may be indivisible. Examples of the second kind include the integer "3," the character "x," and the Boolean "false." Such indivisible data items are called *atomic,* or *elementary.* Objects of the first kind—such as arrays, fractions, or character strings—are called *composite* items, or *data structures.*

Data items of similar structure may be grouped into data types. Data types may thus be either elementary or composite. Because of the similarity of structure among data items of the same type, they can be processed in the same way (i.e., by using the same operations). In other words, operations are generally defined on all objects of a given data type. The result of a given operation is of a particular type. Operations may take multiple arguments of different types.

For example, the operations of multiplication and division are defined on integers and not on character strings.[1] For multiplication of integers, the result type is integer. Finding the numerator makes sense for a fraction but not for a character string. An operation may take an integer k and a character string and return the kth character of the string.

We are led to the following definition:

Def 1.1 A *data type* is a class of data items together with a class of operations defined on these items.

For convenience, however, we refer to "an integer" rather than "a data item of integer type," and do likewise for other types.

Programming languages differ in their treatment of data types. In some languages, like Lisp and APL, variables or data items have no data type of their own, but take on the type of their current value. This is called *dynamic typing.* In others, like Pascal and FORTRAN, the data type of a variable is fixed throughout execution. This is called *static typing.*

In languages with static typing, the permissible values for variables of a given data type are those in the class of values defining the type. Also in these languages, it can usually be determined by an examination of the program whether or not an operation in the program will have arguments of a wrong type, and the resulting error can be detected during compilation. In fact, Lisp and APL are not usually compiled at all. In assembly languages, the concept of data type is all but ignored, and virtually all operations are assumed to be legal.

[1] We do allow the possibility that some operations may result in an error for some objects of the specified argument type (e.g., division by 0).

A language in which type restrictions are strongly enforced is said to be *strongly typed.* Ada and Pascal are examples of strongly typed languages, although Pascal is not quite so strongly typed as Ada.

Some operations are defined for several data types; for example, addition may be defined for both integers and real numbers. If an integer is to be added to a real, most languages will treat the integer temporarily as a real and calculate the sum using the addition operation defined for reals. The result is a real number. This is an example of *type coercion,* the automatic conversion of a data value from one type to another. Another example occurs when a real variable is assigned an integer value. Explicit type conversion operations may also be available to the programmer.

Some languages (e.g., Pascal and Ada) allow user-defined data types. A programmer may then define operations on items of this new type, and the language will enforce the type restrictions for their arguments.[2] For example, a data type "fraction" may be defined as a pair of components of type "real," and then addition, subtraction, equality, and so on, may be defined for fractions. In Ada it is possible to restrict manipulation of fractions to the use of these operations, while in Pascal it is possible to access the numerator and denominator directly without using the operations defined for fractions.

Good programming style requires that if the representation of user-defined data types is changed, only the operations defined as part of the data type need be changed. For example, the operations "numerator" and "denominator" would be part of the definition of the "fraction" data type. The function that adds two fractions should be defined in terms of "numerator" and "denominator" so that if the definitions of these functions were changed, the addition function would not need to be rewritten.

Since this text is designed to be language independent, we do not specify in our algorithms a way in which type and other language-dependent issues are to be handled. If you are using the algorithms in this book as a reference, you should first read Section 0.3.

This text is not intended as a text in structured programming or programming style. Most algorithms in this text are too short to admit of much structure, but they are consistent with the philosophy of structured programming.

In the rest of this section, we discuss briefly the relationship between data structures and the creation of computer programs that model real-world objects and events. In our examples the way the data items are related or structured is as important as their individual values, providing a justification for use of the term "data structure."

For example, consider the family tree—either the pedigree that gives ancestors of one person or the genealogy that gives descendants of one person. This is a data structure according to the definition above and also in the informal sense that the structure of the data, such as who is the child or descendant of

---

[2] Data structures can be represented in other languages, but type constraints need not be enforced by the language.

whom, is as important as the data values, such as the name or birth date of an individual. Typical operations include merging two pedigrees to give the pedigree of a newborn and finding the mother of a given ancestor.

A corporate organization chart is similar to a genealogy. Here the concern is with who reports to whom and who supervises whom, as much as it is with the names of departments or employees. Again, the relations among data items may be thought of as defining the structure of the data. Typical operations include adding a new department and giving an existing department a new manager.

The multiplication table is another familiar data structure. In its usual form, the rows and columns represent factors, and the entries in the body of the table represent products. The numbers in the body of the table are not particularly important in themselves; their importance is in the relation between these numbers and the rows and columns in which they lie. A mileage chart giving the mileage between pairs of cities has the same property. In both cases, the structure of the data is again as important as the data items, and table lookup is the typical operation.

Another type of table for which table lookup is the dominant operation includes tables of functions such as the sine and square root functions. Once again, the fact that a number like 1.414 appears in a square root table is not so significant as the fact that it is associated with the number 2.

Even something as simple as a line in front of a theater may be thought of as a data structure. If you are in such a line, you are more concerned with your position in line and with the person whom you follow than you are with the names of the people in line. In this case the operations add someone to the end of the line and remove the person in front.

In general, only explicitly defined operations are permissible on data structures of a given type. For example, the movie line can change only if someone joins at the rear or exits from the front. Under normal circumstances no one joins the line in the middle or leaves from the middle of the line. Thus, if a data type (usually called a *queue*) were defined to represent such a line, the operations defined for a queue would allow joining the line at the end and leaving from the beginning, but not joining in or leaving from the middle.

One can imagine another type of queue in which people are served not in the order of how long they have been in the queue but in order of their importance. Such a data structure is called a *priority queue*.

Operations on objects of type "priority queue" would include insertion, deletion of the element with the highest priority, and an operation to test whether a priority queue is empty.

Different hierarchical data structures may have different operations. In a genealogy, people may be added only as children of people already in the tree. No one leaves a family tree. By contrast, in a corporation people and sometimes entire departments disappear, departments may divide in two, or a department head may be told to report to a different officer.

From the point of view of data types, when considering the representation of a real-world data object, we must consider the set of allowable operations as part of the definition. For example, a queue may be defined in terms of operations equivalent to Enter and Leave, and the result of the Leave operation will be the item (of those items remaining in the queue) that first Entered.

When trying to solve a problem involving large amounts of data, in general there are many ways to organize the data, some more appropriate than others. Generally the most profitable approach is to ask what type of operations—such as creations, insertions, deletions, *traversals* (the usual name for finding and processing all elements of a data structure),[3] copyings, combinations, searches, and sorts—are most likely to be performed. One also likes to know properties of a structure such as its size and whether or not it is empty. The following terms are useful:

Def 1.2 A data structure whose size may change is called *dynamic*. Otherwise the data structure is called *static*.

Knowing which operations will be most common, we can employ the data structures that most efficiently represent these operations. We call this the *operational point of view*.

For example, suppose you frequently must find and delete the largest of a certain set of numbers. Occasionally, you also must add new elements to this set, with other operations being less common. Then you might consider using a priority queue to represent the data, which has the disadvantage of making it difficult to delete an arbitrary element, since abstract priority queues do not have such an operation. On the other hand, if you used an ordinary queue, it would be difficult to perform the common operation of finding the largest element. You would have to delete all the elements one at a time from the front, keeping track of which was the largest, and then insert them back again.

The actual method of creation of a data structure depends on the data types provided by the language in use; in any case, complicated data objects are represented in terms of more fundamental ones. In constructing such a representation, the main problem is how to define operations on the data structure in terms of those operations defined for the old structure. Any programming language will provide at least one fundamental structured data type (and operations for structures of that type) in terms of which all new structures can be represented.

In most languages, including assembly languages, the fundamental data type of the *array* is present. In assembly language the entire computer memory is treated as a 1-dimensional array. There are a few exceptions; for example, the language Lisp has the *list* as its underlying data type.

---

[3] In general, one may define different sorts of traversals depending on what sort of processing is being done on each element. When we speak about traversals in general, we use the word *visiting* for the processing of individual elements.

1.4 Behavior of Algorithms

In a typical computing problem, there is a choice of data types to use; the question is which is most appropriate? The operational point of view suggests reformulating this question: which data type supports the most efficient versions of the operations to be performed on the data? This makes sense only if there is a way to measure the efficiency of the algorithms for these operations. Generally, these operations will be written as subprograms inside larger programs.

Algorithms may be efficient in any of several ways. They may be economical in terms of time or space (i.e., of computer memory), or they may be easy for humans to read, write, or understand. Of these virtues only the first two can be easily quantified.[4]

The main problem in measuring the time and space required by a computer program or subprogram lies in isolating the properties of the algorithm from the properties of the particular machine and language in use. For example, the same sorting algorithm will run much more slowly when implemented in BASIC on your home microcomputer than on a large mainframe in the machine's own language. Additionally, more space is needed to sort 10-digit telephone numbers than the same number of 5-digit ZIP codes.

Most of the following analysis defines and measures the running time of an algorithm, as opposed to a program implementing the algorithm and running on a particular machine. The arguments can be applied to space requirements as well. One point not emphasized but nevertheless important is that programs require space just for their own storage. Thus a fast and large program may be less useful than a slow but short program if saving space is more important than saving time.

Since machines and languages differ greatly in the time they take, it may not be clear what may be said about a particular algorithm without knowing the circumstances under which it is to be run. However, comparing two algorithms and concluding one is in some sense faster than the other is often possible.

Assume that for any algorithm we can count in advance the number of steps required for its completion. Although this assumption is not true in general,[5] we can often count the number of steps required in the worst case or best case or average case. Usually this number of steps varies with the input size.

The parameter n will be used throughout this text for the size of a data structure. For example, in a sorting algorithm n would be the number of items

[4] Another important property an algorithm may or may not have is correctness. This is not a joke. Correctness is a property sometimes sacrificed to obtain other advantages. Algorithms that find approximate values of functions and algorithms that are only usually correct can be valuable. These issues are considered further in Sections 5.4 and 5.5.

[5] We can't even predict in general cases whether or not a program will terminate for all inputs (i.e., when it represents an algorithm).

Function (representing running time)	Ratio of increase in running time (as input size n increases from 5 to 25)
c_1	1
$c_2 \log n$	2
$c_3 n$	5
$c_4 n \log n$	10
$c_5 n^2$	25
$c_6 n^3$	125
$c_7 2^n$	1,048,576

Figure 1.3 Growth of Various Functions of n As n Increases from 5 to 25

to be sorted; it seems reasonable that more instructions would have to be executed to sort a large file than a small one. In a graph algorithm, n may be the number of vertices, or n may be the number of edges. In a text algorithm, n may be the length of a character string.

Suppose that the number of steps taken by a particular algorithm is a function $f(n)$ of the input size n. In Figure 1.3, we account for the fact that algorithm steps may translate into different numbers of machine language instructions, which may require different amounts of time by multiplying $f(n)$ by an unspecified constant, denoted c_i for various i. This constant will be different for different machines.[6]

In Figure 1.3 seven of the most frequently encountered possibilities for $f(n)$ are listed. The value of $c_i f(n)$ represents the running time of a program that implements the algorithm. Of course, we cannot evaluate this expression without knowing the value of c_i. However, we can use the table to answer the following question: If the input size increases from 5 to 25, how much more time will the program take? We can generalize this question for other changes in input size; you are asked to do this in the exercises.

The calculations needed to construct the table entries are simple. For example, if $f(n) = c_3 n^2$, we want to compare $f(5) = c_3 5^2$ and $f(25) = c_3(25)^2$. Simple division shows that we get a running time increase of $(c_3(25)^2)/(c_3 5^2) = 25$ when n changes from 5 to 25. The other table entries may be found in the same way. In particular, a similar table may be constructed for any given change in n.

As n gets large, the table indicates it is more important to have an efficient algorithm than a fast machine. For example, suppose that the algorithms with running times $c_3 n$ and $c_2 \log n$ take the same time T when implemented on machine A when $n = 5$. If n increases from 5 to 25, then machine A will take

[6] Actually we don't assume and don't need to assume that each machine instruction takes a constant amount of time. It is enough that each instruction takes a bounded amount of time. (See Exercise 16.)

time $5T$ for the first algorithm. Even if the algorithm is run on machine B, which is twice as fast as A, the running time will still increase to $2.5T$. On the other hand, if the second algorithm is run on the slower machine A, only $2T$ will be required. This relative advantage of the second algorithm will continue to increase as n grows larger.

The functions in the table have been listed from slowest growing to fastest growing. Since the functions represent running times, the slowest growing ones are the most desirable. For example, an algorithm corresponding to our first function c_1 requires no more time for $n = 25$ than for $n = 5$. At the other extreme, an algorithm with running time $c_7 2^n$ would take over a million times longer for $n = 25$ than for $n = 5$. This is the difference between 1 second and 12 days!

1.5 O-Notation

Note that the conclusions of the previous paragraph and of the entire previous section hold regardless of the exact values of the constants. This suggests that if the numbers of steps (as functions of n) required by two algorithms differ by only a constant factor (i.e., correspond to two different values of c_i), then we may consider that the algorithms take essentially the same time. The term *asymptotically* is used in this context to describe behavior of functions for sufficiently large values of n.

More formally, we want to use the insights of the previous section to compare running times of algorithms rather than programs. To that end we make the following definition:

Def 1.3 If f and g are functions defined on the positive integers, then f is $O(g)$ iff there is a $c > 0$ and an $N > 0$ such that $|f(n)| \leq c|g(n)|$ for all $n > N$. We say, "f is big oh of g" or simply "f is oh of g."

Historically this notation is "$f = O(g)$," although the idea that f is equal to something called $O(g)$ is misleading. Sometimes set terminology is used, and f is said to be a member of the set $O(g)$ of functions. The function g is a kind of upper bound for f.

If f and g are positive-valued functions defined on the positive integers, then the definition can be simplified as follows (see Exercise 17 and its solution):

Def 1.4 If f and g are positive-valued functions defined on the positive integers, then f is $O(g)$ iff there is a $c > 0$ such that $f(n) < cg(n)$ for all positive integers n.

The definitions above apply to arbitrary functions f and g. Most of our applications will be to functions f and g measuring running times of algorithms F and G, so the second definition will apply. The following term is useful:

Def 1.5 A function f such that the running time of a given algorithm is $O(f)$ is said to measure the *time complexity* of the algorithm.

If f is $O(g)$, then F runs at least as fast as G (within a constant factor) for large n. Among the functions of Figure 1.3, f is $O(g)$ iff f is listed before g.

It is useful to accompany the "O" definition with two related definitions. The first generalizes the concept of "lower bound" in the same way that the O-relation generalizes the concept of "upper bound." The second corresponds to the idea that if g is both an upper and a lower bound for f, then f and g are in some sense equal. These definitions are also included in the concept of "O-notation."

Def 1.6 f is $\Omega(g)$ if and only if g is $O(f)$.

Def 1.7 f is $\Theta(g)$ if and only if f is $O(g)$ and g is $O(f)$ [i.e., if and only if f is $O(g)$ and f is $\Omega(g)$].

In other words, f is $\Omega(g)$ if and only if f grows at least as fast as g (within a constant factor) and f is $\Theta(g)$ if and only if f and g grow (within a constant factor) at the same rate.

In conclusion, given two algorithms, it is impossible to decide which requires less time for a given input without knowing the language in which or machine on which it will be run. It does often happen, however, that one algorithm will be asymptotically faster than another, independent of the language or machine used. That is, for sufficiently large n, we are guaranteed that an implementation of the first algorithm will run faster. We define the O-notation in order to help recognize such situations. If the second algorithm has time complexity $f(n)$ and the first has time complexity $O(f)$ but not $\Theta(f)$, then we may expect the first to be faster for sufficiently large n.[7]

The O-notation does not tell us how large "sufficiently large" is. In particular, it does not tell us which algorithm might be faster for small values of n. Typically, one algorithm is asymptotically faster than another because it does a lot of preliminary work. This preliminary work, called *overhead,* takes a certain amount of time to perform but will allow the rest of the algorithm to run faster. In general, the input size must be above a certain size to justify the expense of the overhead. It is the algorithm without the overhead that is likely to run faster for a small input size.

One special case of the O-notation is worth looking at. A positive-valued function $f(n)$ is $O(1)$ iff $f(n) \le c \times 1 = c$ for some c and for all n. In other words, a function is $O(1)$ if and only if it is bounded.

In the rest of this chapter we cover two topics. First, we describe properties of the O-notation that will allow us to simplify complicated functions of n. For example, the properties of O-notation enable us to show that if $f(n)$ is $O(n^3 + n^2 \log n + 5 \log n - 2)$, then f is $O(n^3)$. In Section 1.6 these properties are merely stated and examples are given. They are proved formally in Appendix A.

Second, in Section 1.7 we show how, given the description of an algorithm or listing of a program, to calculate the number of steps required as a function

---

[7] Actually this need not happen for general functions f, but it is decidedly the norm for functions measuring time complexity.

of n. Fortunately, this will be easy once we have seen the properties of O-notation. This simplicity is one of the reasons for defining O-notation in the first place.

1.6 Properties of O-Notation

In this section we first describe informally the properties of O-notation that we will need. Then we state the properties more formally, with examples and interpretations dealing with running times of algorithms. The proofs of the properties are deferred to Appendix A.

If you are familiar with calculus, the development of the properties should remind you of the derivation of the rules for derivatives, although our results are considerably simpler. We need to process complicated functions by thinking of them as sums, products, and constant multiples of elementary functions.

The rules as described informally should be believable:

1. Constant factors can be ignored in O-notation calculations.
2. The growth rate of a sum is given by the growth rate of its fastest-growing term.
3. If f grows faster than g, which grows faster than h, then f grows faster than h.
4. Higher powers of n grow faster than lower powers.
5. The growth rate of a polynomial is given by the growth rate of its leading term (ignoring the leading coefficient by rule 1).
6. The product of upper bounds for functions gives an upper bound for the product of the functions.
7. Exponential functions grow faster than powers.
8. Logarithm functions grow more slowly than powers.
9. All logarithm functions grow at the same rate.
10. The sum of the first n rth powers grows at the same rate as the $(r + 1)$st power of n.

Now we present the rules more precisely:

1. For all $k > 0$, kf is $O(f)$. In particular, f is $O(f)$.
   Of course, ignoring constant factors is a large part of the motivation for O-notation in the first place.
2. If f is $O(g)$ and h is $O(g)$, then $(f + h)$ is $O(g)$. In particular,

2′. If f is $O(g)$, then $(f + g)$ is $O(g)$.
   This corollary implies that if an algorithm has a fast part F and a slow part G, then the time required by G dominates the time required to implement the algorithm.

3. If f is $O(g)$ and g is $O(h)$, then f is $O(h)$.
   This says that if F is as fast as G and G is as fast as H, then F is as fast as H.
4. n^r is $O(n^s)$ if $0 \leq r \leq s$.

Now we can look at some examples of specific functions. For example, suppose we want to compare $5n^2$ and n^3. We know that n^2 is $O(n^3)$ by rule 4. By rule 1, $5n^2$ is $O(n^2)$. So by rule 3 and the previous two statements, $5n^2$ is $O(n^3)$. And then by rule 2′, $n^3 + 5n^2$ is $O(n^3)$. Actually, we can replace the last O by Θ, since n^3 is $O(n^3 + 5n^2)$. This example can be generalized as follows:

5. If p is a polynomial of degree d, then p is $\Theta(n^d)$.
   So we have for example that $6n^5 + 3n^3 + 7n + 1$ is $\Theta(n^5)$.

A couple of useful definitions can now be made.

Def 1.8 An algorithm has *polynomial* time complexity iff its time complexity is $O(n^d)$ for some integer d. A problem is *intractable* iff no algorithm with polynomial time complexity is known for it.

Rule 5 says that any algorithm with time complexity $O(p(n))$ for some polynomial p has polynomial time complexity. Of course, an algorithm may be both intractable and of polynomial time complexity if it has a polynomial time algorithm that nobody knows about. A large number of intractable problems are not known to be nonpolynomial.

6. If f is $O(g)$ and h is $O(r)$, then $f \times h$ is $O(g \times r)$. In particular, if f is $O(g)$ then $f \times h$ is $O(g \times h)$.
   Sometimes we will abbreviate this fact by saying "$f \times O(g)$ is $O(f \times g)$." Compare this fact to the corresponding theorem for positive numbers: if $f < g$ and $h < r$ then $f \times h < g \times r$.
7. n^k is $O(b^n)$, for all $b > 1$, $k \geq 0$.
   For example, n^3 is $O(2^n)$. The restrictions on b and k merely insure that both functions are actually increasing.
8. $\log_b n$ is $O(n^k)$ for all $b > 1$, $k > 0$.
   This is the inverse of rule 7. One important consequence is that $\log n$ is $O(n)$. So for example, since n is $O(n)$, rule 6 implies that $n \log_b n$ is $O(n \times n)$ is $O(n^2)$ for any base b.
9. $\log_b n$ is $\Theta(\log_d n)$, for all $b, d > 1$.
   The proof of this result is trivial; by the formula for changing the base of a logarithm, the two logarithmic expressions differ by a factor of $\log_b d$.

Although we have not formally defined the Σ notation for sums—we do so in Section 2.1—we include the following rule for completeness. The expression on the left represents the sum of the first n rth powers.

10. $\sum_{k=1}^{n} (k^r)$ is $\Theta(n^{r+1})$.
    Occasionally we must add the rth powers of consecutive integers when analyzing algorithms, so having a simple formula for handling this problem

is useful. Rule 10 should suggest the behavior of the integral from calculus; in fact, its proof is a straightforward application of the integral test.

The main use of these rules for O-notation is in finding a simple function $f(n)$ to describe the behavior of an algorithm. In the most common case, we have an algorithm with time complexity $T(n)$, and we want a function $f(n)$ such that $T(n)$ is $O(f(n))$. In general, many different functions will work; for example, the function $6n^2$ is $O(n^2)$, $O(n^3)$, and $O(n^4)$. Finding the smallest possible f corresponds to finding an f such that $T(n)$ is $\Theta(f)$. This may be done for polynomials by rule 5; many other types of functions are also easy to handle.

As an example, if $T(n)$ is $6n^2 + \log n$, then $T(n)$ is $O(n^2)$ since $6n^2$ is $O(n^2)$ and $\log n$ is $O(n)$ is $O(n^2)$. Here we are also using transitivity, rule 4, and the rule for sums. We can conclude that $T(n)$ is $\Theta(n^2)$ as we did previously. If $T(n)$ were merely $O(6n^2 + \log n)$, we still could conclude that $T(n)$ is $O(n^2)$, but not that $T(n)$ is $\Theta(n^2)$. For example, $T(n)$ might equal $2n$.

As another example, let $T(n)$ be $O(9n^5 + 2^n)$. Both terms are $O(2^n)$, so the sum $T(n)$ is $O(2^n)$. By rule 2, note $9n^5 + 2^n$ is $O(2^n)$, so $T(n)$ is $\Theta(2^n)$.

In practice, we usually simplify the calculation as we go. If in the previous example we discovered that the first part of an algorithm took $9n^5$ steps, we would immediately simplify to $O(n^5)$, which would eventually be compared with $O(2^n)$.

1.7 Analyzing Algorithms

In the final section of this chapter we consider the number of steps required by common segments of algorithms. We assume that a formal expression of the algorithm to be analyzed is available. In practice, the analysis of algorithms can often be done before formalizing the algorithm. For example, we can see the most natural algorithm for finding the maximum of n elements has $\Theta(n)$ time complexity even before it is expressed in a particular language. We will often take this approach in later chapters.

There is a subtle distinction between this sort of analysis of algorithms and analyzing a problem in the sense of concluding that any algorithm solving the problem must have a certain time complexity. For example, any algorithm for finding the maximum of n elements must have time complexity $\Omega(n)$. But for a problem like multiplication of two polynomials of degree n, even though the most natural algorithm would involve $O(n^2)$ multiplications of coefficients, it is possible to do better (see Section 2.14).

We proceed with the analysis of program segments.

Any segment for which each statement is executed only once will have a total number of statements executed that is independent of n. In other words, such a segment has $O(1)$ time complexity.

A segment such as that in Figure 1.4a, where the dash represents a segment

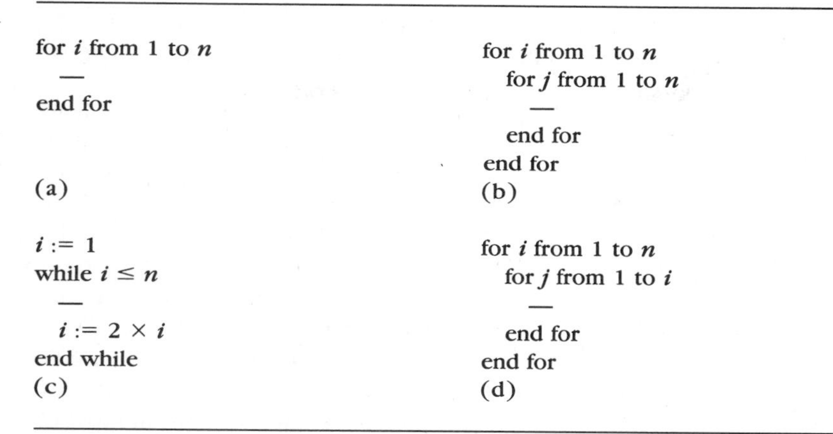

Figure 1.4 Sample Algorithm Segments for Analysis

with no loops,[8] will require n repetitions of the $O(1)$ segment represented by the dash. So the entire segment takes time $n \times O(1)$ or $O(n)$.

Similarly the inner **for** loop of the segment of Figure 1.4b has $O(n)$ time complexity. Since it is iterated n times, the double loop has $n \times O(n)$ or $O(n^2)$ time complexity.

Logarithmic behavior is exemplified by the segment of Figure 1.4c. The value of i will change from 1 to 2, 4, 8, and so on, until it reaches or exceeds n. The number of repetitions will be the number of times it takes to double 1 until n is reached, which is just $\log_2 n$. More precisely, the loop will be executed $1 + \lfloor \log_2 n \rfloor$ times. So the loop has time complexity $O(\log_2 n)$ or $O(\log n)$.

Figure 1.4d illustrates a slightly more complicated case. The inner **for** loop will be executed once when $i = 1$, twice when $i = 2$, and so on. Altogether it will be executed $1 + 2 + 3 + \cdots + n$ times. By rule 10 of the previous section, this sum is $\Theta(n^2)$. So although the statements inside the loops of example d are executed fewer times than in example b, asymptotically there is no difference in running time.

Fortunately these examples illustrate most situations that arise in at least the worst-case analysis of the algorithms we will see. Average-case analysis is usually more difficult. In analyzing worst-case behavior, most complications result from combining certain of the preceding cases. For example, if example b had a triple loop instead of a double loop, it would have time complexity $O(n^3)$.

Occasionally some confusion arises about when Θ can be used to describe an algorithm and when only O or Ω is appropriate. In general, the number of

[8] Note that loops should not be present indirectly, for example, in the form of an assignment statement whose right-hand side is a function calculated using a loop whose number of iterations depends on n. An example might be a vector sum or character string comparison.

steps of an algorithm actually executed depends in a complicated way on the input, so we can only specify an upper bound on the time required, using the O-notation, or a lower bound, using the Ω-notation.

Sometimes, as in the examples of this section, the number of steps is at least asymptotically independent of the input, so we can use the Θ-notation, which is more precise. In addition, we can use the Θ-notation under the assumption that the input has a particular form. A common example is a particular form known to give the best- or worst-case. Descriptions of these cases can often be expressed in terms of Θ. Also, many algorithms have best- and worst-case time complexities that are both $\Theta(f(n))$ for the same function f. Then we can conclude that the algorithm has time complexity $\Theta(f(n))$ in any case.

Probably the best way to clear up confusion in this regard is to remember that O is a generalization of $\leq$ and corresponds to having an upper bound, that Ω is a generalization of $\geq$ and corresponds to having a lower bound, and that Θ is a generalization of $=$ and can be used if the upper and lower bounds coincide.

1.8 Exercises

Solutions of exercises marked with an asterisk appear in Appendix C; references are given in that appendix for those marked with a double asterisk.

*1. What is 25 mod 7? 100 mod 11? 12345 mod 100? 69 mod 23? 54321 mod 2?

*2. Find the following values: $\lceil 3.5 \rceil$, $\lfloor 3.5 \rfloor$, $\lceil 3 \rceil$, $\lfloor 3 \rfloor$, $\lceil -0.5 \rceil$, $\lfloor -0.5 \rfloor$, $\lceil -5 \rceil$, $\lfloor -5 \rfloor$.

3. Find 110110 $\wedge$ 100101. Find 010101 $\vee$ 111000.

4. Construct a table like that of Figure 1.3 for other values of n. Then modify your table to use actual units of time (e.g., microseconds) instead of ratios.

5. Show that for the seven functions of Figure 1.3, each function stands in the O-relation to the succeeding function.

*6. Find the simplest function f such that the given functions is $\Theta(f)$.

a) $n^3 + n^2 + n$
b) $6n + 2n^4 + 3n^5$
c) $n^2 + n \log n$
d) $(\log n) + n$
e) $n + \sqrt{n}$
f) $(\log n) + \sqrt{n}$
g) $1/n$
h) $\log n^2$
i) $n + (\log n)^2$
j) $(n + 2)(n^2 + n)$
k) $n + n \log n$

*7. How does the function $\log_2 \log_2 n$ compare asymptotically with each of the functions in Figure 1.3? What about $n!$? Add these two functions to the table of Figure 1.3.

*8. What is the time complexity of each of the following program segments?

a)
```
for i from 1 to n
   for j from 1 to 5
      —
   end for
end for
```

b)
```
for i from 1 to n
   for j from 1 to i + 1
      —
   end for
end for
```

c)
```
for i from 1 to n
   for j from 1 to 6
      for k from 1 to n
         —
      end for
   end for
end for
```

d)
```
for i from 1 to n
   for j from 1 to i
      for k from 1 to n
         —
      end for
   end for
end for
```

*9. Construct a program segment with time complexity $\Theta(n \log n)$.

*10. What is the worst-case time complexity for the most natural algorithm to

a) find the average (arithmetic mean) of n integers?

b) multiply two polynomials of degree n?

c) print an n by n multiplication table?

11. Show that $\log n$ is $O(\sqrt{n})$ directly from the definition of the O-relation.

12. Is 3^n $O(2^n)$?

13. What is the difference between saying that an algorithm has worst-case time complexity $\Theta(f)$ and saying that it has time complexity $O(f)$?

*14. Suppose that f is $O(g)$ and $f \neq \Theta(g)$. Why is a program with time complexity $O(f)$ not necessarily better than one with time complexity $O(g)$? Give several reasons.

15. Without using rule 5, verify the example in the next-to-last paragraph of Section 1.5., page 13.

16. Suppose that an algorithm is composed of $f(n)$ steps, that the fastest possible step takes time m, and that the slowest takes time M. Show that the time required by the algorithm is $\Theta(f(n))$.

*17. Show that the second definition of f is $O(g)$ in Section 1.5 follows from the first under the assumption that f and g are positive-valued functions defined on the positive integers.

18. Show that we can define the Ω-relation by replacing the $\leq$ sign by a $\geq$ sign in the definition of the O-relation.

19. Show that, of the seven functions of Figure 1.3, none is in the Θ-relation to any other.

**20. Without looking at Appendix A, prove each of the theorems of this chapter about O-notation.

21. Show that rules 1, 2, 3, and 6 for manipulating O-notation remain true if "Θ" replaces "O."
22. If $x \bmod y = m$, show that $(x + cm) \bmod y = m$ for any c.
*23. If $x \bmod y = m$, $x \bmod a = 0$, and $y \bmod a = 0$, show that $m \bmod a = 0$.
*24. If $x \bmod y = m$, $m \bmod z = n$, and $y \bmod z = 0$, show that $x \bmod z = n$.
25. a) If $x < 0$ and $x \bmod m = 0$, show that $-x \bmod m = 0$

    b) If $x < 0$ and $x \bmod m = a > 0$, show that $-x \bmod m = m - a$.
*26. Find functions f and g such that 1) f is $O(g)$ 2) f is not $\Omega(g)$ and 3) $f(n) > g(n)$ for infinitely many n.
27. Can you extend the definition of *mod* to polynomials? How many properties of *mod* remain true?

Additional Reading—Chapter 1

For information on data structures and data types beyond that in Appendix D, there is a large number of useful texts mentioned in the Bibliography, such as Horowitz and Sahni (1976), Reingold and Hansen (1983), Smith (1987), Standish (1980), and Tenenbaum and Augenstein (1981). Stubbs and Webre (1985) is typical of more recent texts in its emphasis on abstract data types. The three-volume set of Knuth (1973, 1981, and 1973) is both a virtual encyclopedia and the foundation on which most of the later texts rest.

Aho, Hopcroft, and Ullman (1983) and Wirth (1976) cover both data structures and algorithms. Texts on algorithms include Aho, Hopcroft, and Ullman (1974), Baase (1988), Brassard and Bratley (1988), and Horowitz and Sahni (1978).

Induction and Recursion

2.1 Induction Proofs and Recursive Definition

One useful way of solving a large problem is to divide the problem into pieces, handle each piece, and combine the solved pieces into a solution for the entire problem. This is called the *divide-and-conquer* approach.

Divide-and-conquer is a very natural technique for data structures, since they are by definition composed of pieces. When a data structure of finite size is divided, ultimately the pieces become so small that they can no longer be divided. Thus a divide-and-conquer algorithm for an operation defined on a structured data type typically has the following sort of definition:

Def 2.1 Process(S):

1. If S is small, process S directly;
2. Otherwise,
   divide S into pieces,
   process each piece recursively,
   and combine the processed pieces.

Of course, the definitions of "small," "directly," "divide," and "combine" depend on the particular data type. Quite often the algorithms for checking smallness, for dividing, and for processing directly are trivial, and the only real work is in figuring how to combine the processed pieces.

Since the processing procedure or function in step 2 is the same as that being defined, divide-and-conquer algorithms are recursive in the sense of Chapter 1. The great advantage of recursion in programming is that the natural divide-and-conquer solutions can very easily be specified and converted into provably correct programs. If your experience with recursion was a couple of weeks crammed into the end of a course in Pascal, you may not yet find recursion very appealing. One goal of this text is to convince you that recursion is as useful and as pleasant as iteration.

All our recursive definitions will have the form of the preceding informal definition above: 1) a part that handles small pieces and 2) a recursive part that gives the definition on large data objects in terms of that on smaller data objects. Often we leave implicit the "else" or "otherwise" before part 2.

Although we usually don't think of integers as data structures, if we are willing to think of the pieces of a positive integer $n > 1$ as being 1 and $n - 1$, then we may make recursive definitions for functions defined on integers. For example, the *exponential* function has the recursive definition,

Def 2.2 $a^n =$

1. 1, if $n = 0$
2. $a \times a^{n-1}$, if $n > 0$.

For example, $2^1 = 2 \times 2^0 = 2 \times 1 = 2$

$2^2 = 2 \times 2^1 = 2 \times 2 = 4$ and

$2^3 = 2 \times 2^2 = 2 \times 4 = 8$

Notice we can easily turn this definition into an algorithm in any high-level language supporting recursion. In our earlier terminology, the recursive definition serves as a specification, and it is easy to go from a recursive definition to a recursive algorithm. For this reason we generally will not give explicit procedures for recursive definitions of operations and will use the same numbering system for definitions as for algorithms, which are given procedurally. Proofs of correctness for the corresponding algorithms will be omitted.

We may define an analogous concept for function composition:

Def 2.3 $f^n(x) =$

1. $f(x)$ if $n = 1$
2. $f(f^{n-1}(x))$ if $n > 1$.

Another definition we have already used informally is the Σ definition for sums. A recursive definition would be

Def 2.4 $\sum_{k=a}^{n} t_k$ is

1. t_a, if $a = n$
2. $t_n + \sum_{k=a}^{n-1}$ if $a < n$.

For example, $\sum_{k=1}^{1} k = 1$

$\sum_{k=1}^{2} k = 2 + \sum_{k=1}^{1} k = 2 + 1 = 3$. Also,

$\sum_{k=1}^{3} k = 3 + \sum_{k=1}^{2} k = 3 + 3 = 6$

Even with recursion, algorithm design remains largely an art, although it has at least some elements of a science.

The concept of mathematical induction is very closely related to that of recursion. "Recursion" is typically used in the context of definition or specification, while "induction" is typically used in discussing proofs. In fact, inductive proofs are typically used for correctness even for algorithms specified iteratively. In the remainder of this section, the primary topic is induction; the rest of the chapter deals mostly with recursive definition of particular data structures and their operations.

The main problem in proving correctness of an algorithm is that there are generally infinitely many data structures of the input type(s) expected by the algorithm, and we need to show the algorithm handles each one correctly. Induction provides a way of dealing with an infinite number of objects simultaneously.[1]

Strictly speaking, inductive proofs deal with properties of nonnegative integers. For us, these integers will most commonly represent sizes of data structures. (Recall our convention for using n to stand for the size of a data structure.) We first need to define what we mean by a "property."

Def 2.5 A *property* is a Boolean-valued function defined on the set of nonnegative integers.

For example, "x is odd," "r is at least 20," and "n is a perfect square" are properties. We will use notation $P(n)$ to mean that property P is true for integer n. Then the principle of *mathematical induction* may be stated as follows:

If 1) $P(0)$

and 2) $P(n)$ whenever $n > 0$ and $P(n - 1)$

then $P(n)$ is true for all nonnegative integers n.

To use the induction principle, we first establish $P(0)$, and then prove that P is true of n under the assumption that P is true of $n - 1$. Once this is done, we may conclude that P holds for all nonnegative integers. In induction proofs, the part corresponding to part 1 of the principle is called the *basis,* or the *base case,* and the part corresponding to part 2 is called the *induction step.* The assumption that P is true for values smaller than n is generally called the *induction hypothesis.*

We may start with a base case of $P(a)$ if we are willing to settle for a proof that P holds for all integers greater than or equal to a. (When you have finished this section, you might try to prove this claim. A proof is given in Appendix C as the answer to Exercise 41.)

---

[1] You may have seen infinity handled by the use of ellipsis, symbolized by "..." and meaning "and so on." This can help give an intuitive understanding, but it can't work in general. For example, you may have seen the sequence 1, 2, 4, ... and expected the next terms to be 8, 16, 32, ... In Section 2.17 we discuss a sequence that begins 1, 2, 4, ... whose next terms are 7, 12, 20, ... Another example showing the danger of generalizing from a limited number of examples is given as Exercise 8.

Often the result to be proved is expressed explicitly as a property of n, and the induction hypothesis is just the theorem itself. However, sometimes this is not true, or a slightly more general property is better suited for an induction proof (e.g., the parsing algorithm of Section 2.18). Thus when doing an induction proof it is a good idea to state the induction hypothesis explicitly.

Our first example of an induction proof, a simple result that has important consequences for analyzing algorithms, is shown in Example 1. (If you want a more amusing first example, you might try Exercises 29–31, preferably before looking at their solutions in Appendix C.) We first need the following definition:

Def 2.6 The *power set* 2^S of a set S is the set of all subsets of S.

Our example will justify the notation 2^S.

Example 1: If S is a set and $|S| = n$, then $|2^S| = 2^n$.

Before proceeding with the proof, we look at an example. If $n = 3$, say $S = \{a, b, c\}$, then the induction hypothesis tells us that any set of size 2, say $\{a, b\}$, has $2^2 = 4$ subsets. For $\{a, b\}$, these subsets are just $\emptyset$, $\{a\}$, $\{b\}$, and $\{a, b\}$. Notice all of these sets are subsets of the original set S, and the remaining subsets of S are those containing the additional element c of S. In fact, they are just those subsets resulting from adding c to the four subsets of $\{a, b\}$. The inductive proof merely generalizes these observations.

Proof by induction on n. The induction hypothesis is just the statement of the theorem.

Basis: If $n = 0$, then $S = \emptyset$ and $2^S = \{\emptyset\}$. Thus $|2^S| = 1 = 2^0$.

Induction step: If $n > 0$, then we may choose c in S. Let $T = S - \{c\}$. Then by induction, $|2^T| = 2^{n-1}$. The subsets of S that do not contain c are precisely the elements of 2^T. Since each subset of S containing c may be obtained from a different subset of T by adding c, there must also be 2^{n-1} such subsets. If P is the set of subsets of S that contain x, and Q is the set of subsets of S that do not, then P and Q are disjoint and their union is 2^S. So we have

$$|2^S| = |P| + |Q| = 2^{n-1} + 2^{n-1} = 2 \times 2^{n-1} = 2^n$$ ■

Another form of the induction principle is the *strong form of induction.* It is particularly useful for data structures and may be stated as follows:

If 1) $P(0)$

and 2) $P(n)$ whenever $n > 0$ and $P(k)$ is true for all integers k with $0 \leq k \leq n - 1$,

then $P(n)$ is true for every nonnegative integer n.

The advantage of the strong form is that in proving the induction step we may assume P holds for any k less than n, not just for $n - 1$. The proof that this form is equivalent to the original form of induction is left as Exercise 40, which is solved in Appendix C.

We now consider a few examples that will be useful later.

First we show that we can get a more precise special case of rule 10, Section 1.6. At this point the right-hand side of the formula must be taken as given. We describe how to find it in Section 2.17.

Example 2: $\sum_{k=1}^{n} k = n(n + 1)/2$

Proof by induction on n. The induction hypothesis is just the statement of the theorem.

Basis: For $n = 1$, the sum consists only of the term 1, while the right-hand side is $1(1 + 1)/2 = 1$.

Induction step: By definition, the sum on the left is

$$n + \sum_{k=1}^{n-1} k$$

By induction, this is $n + (n - 1)(n - 1 + 1)/2 = n + (n - 1)n/2$. With a common denominator, the sum becomes $2n/2 + (n - 1)n/2$. Factoring out $n/2$, we get $= (2 + n - 1)(n/2) = n(n + 1)/2$ as desired. ■

Notice the sum is $\Theta(n^2)$, as promised in Section 1.6. An even simpler example occurs when all the terms to be summed are equal to 1.

Example 3: $\sum_{k=1}^{n} 1 = n$

Proof by induction on n. The induction hypothesis is just the statement of the theorem.

Basis: For $n = 1$, the sum consists only of the term 1, which is equal to n.

Induction step: By definition, the sum on the left is

$$1 + \sum_{k=1}^{n-1} 1$$

By induction, this is $1 + (n - 1)$, or n. ■

Another important sum, not quite the same as that of rule 10, is the sum of the first n powers of a given number b. This sum is useful in analyzing exponential growth. Again a method for determining the right-hand side will be given in Section 2.17.

Example 4: For $b \neq 1$ and $n \geq 0$, $\sum_{k=0}^{n} b^k = \frac{(b^{n+1} - 1)}{(b - 1)}$

Proof by induction on n. The induction hypothesis is just the statement of the theorem.

Basis: If $n = 0$, then the sum is $b^0 = 1$. Since this equals $\frac{(b^{0+1} - 1)}{(b - 1)}$, we have shown the basis.

Induction step: for $n > 0$, the sum is $b^n + \sum_{k=0}^{n-1} b^k$. By induction, this equals $b^n + \frac{(b^n - 1)}{(b - 1)}$. With a common denominator, the sum becomes

$$\frac{[(b^{n+1} - b^n) + (b^n - 1)]}{(b - 1)} = \frac{(b^{n+1} - 1)}{(b - 1)}, \text{ as desired.}$$

Thus we have shown by induction that the sum of the first n powers of b is $\frac{(b^{n+1} - 1)}{(b - 1)}$ ■

Two useful properties of sums are easy to prove using induction. They have been left as (solved) exercises. They are

1. $\sum_{k=a}^{n} (f_k + g_k) = \sum_{k=a}^{n} f_k + \sum_{k=a}^{n} g_k$
2. $\sum_{k=a}^{n} (c \times f_k) = c \times \sum_{k=a}^{n} f_k$

Now we may verify rule 10 of Section 1.6.

Example 5: For each integer $r \geq 1$, there exists $c > 0$ such that $\sum_{k=1}^{n} k^r \leq cn^{r+1}$

Proof by induction on n. The induction hypothesis is the theorem itself. We will wait to choose c until we see what values of c are necessary to make the basis and induction step work.

Basis: If $n = 1$, the left-hand side is just $1^r = 1$. The right-hand side is just $c \times 1^{r+1}, = c$, so any c as large as 1 will work.

Induction step: For $n > 1$, we may assume by induction that

$$\sum_{k=1}^{n-1} k^r \leq c(n - 1)^{r+1}$$

This left-hand side differs from the left-hand side of the inequality we want by the additional term n^r of the sum, so we add this term to both sides to get

$$\sum_{k=1}^{n} k^r \leq c(n - 1)^{r+1} + n^r$$

It is now enough to show that

$$c(n - 1)^{r+1} + n^r \leq cn^{r+1}$$

We may simplify by subtracting the first term on the left from both sides, and it is now enough to show

$$n^r \leq c[n^{r+1} - (n - 1)^{r+1}]$$ ■

This inequality warrants a separate induction proof.

Example 6: For each integer $r \geq 1$, there is a $c > 0$ such that for all $n \geq 1$, $n^r \leq c[n^{r+1} - (n-1)^{r+1}]$

Proof by induction on r. The induction hypothesis is the statement of the theorem.

Basis: If $r = 1$, the left-hand side is n. The right-hand side is $c[n^2 - (n-1)^2] = c[2n - 1]$, so any c as large as 1 will work.

Induction step: We may assume that $n^{r-1} \leq c[n^r - (n-1)^r]$ by induction. To get the left-hand side to be what we want, we multiply by n to get

$$n^r \leq c[n^{r+1} - n(n-1)^r]$$

So it's enough to show that

$$c[n^{r+1} - n(n-1)^r] \leq c[n^{r+1} - (n-1)^{r+1}] \text{ or}$$

$$n^{r+1} - n(n-1)^r \leq n^{r+1} - (n-1)^{r+1} \text{ or}$$

$$-n(n-1)^r \leq -(n-1)^{r+1} \text{ or}$$

$$n(n-1)^r \geq (n-1)^{r+1} \text{ or}$$

$$n \geq n - 1$$

To get the last inequality we divided by $(n-1)^r$. If this value is 0, then $n = 1$ and the original inequality was trivially true for any $c \geq 1$. Since $n \geq n - 1$ for any n and c (greater than 1, to satisfy the basis), we are done with Example 6. ■

To complete the proof of Example 5, we choose $c \geq 1$ so that both the basis and the induction step are true; with such a choice we are done. ■

One useful form of mathematical induction is the *minimal counterexample* principle. This principle states that if P is a property of the natural numbers that is not true for all natural numbers, then there must be a smallest natural number for which the property is false. This principle is equivalent to ordinary induction; the proof is left as a solved exercise. The principle is most commonly used in proofs by contradiction: the property P is assumed not to be true, a minimal counterexample is considered, and a contradiction is found, usually to minimality. In considering properties of data structures, it is particularly useful to consider a smallest "bad" data structure and then to show there is a smaller "bad" structure of the same type.

For a numerical example, consider the property of factorizability into primes:

Example 7: Every positive integer is the product of one or more primes.

Proof Assume that the property is not always true. Then there is a smallest number n for which it is not true. The number n itself cannot be prime, since otherwise the property would be true of n. Thus $n = pq$, where $p < n$ and $q < n$. It must be the case that one of $\{p, q\}$, say p, does not satisfy the property, since otherwise n would. But now p is a counterexample to the minimality of n, and the proof is complete. ■

We may now apply some of the preceding results to the analysis of algorithms. For example, we have the following corollary of Example 1:

Corollary Any algorithm that in the worst case must inspect all subsets of a set of size n has worst-case time complexity $\Omega(2^n)$.

The next two examples show ways in which this unfortunate situation might occur.

Example: The ***knapsack*** problem gets its name from the problem of a backpacker who has a knapsack of fixed capacity M; a choice of n items to include, each of which has a size w_i (usually called the weight); and a value p_i (usually called the profit), which can be realized only if the item is included in the knapsack. There are two forms of the problem, depending on whether pieces of items can be included or not. If so, the problem is called the *continuous knapsack problem*. Otherwise, we call the problem the *zero-one knapsack problem*, or ZOK for short.

Since ZOK is a more interesting problem in some sense than the continuous knapsack problem, some authors use the phrase "the knapsack problem" to refer only to the zero-one version. The abbreviation ZOK is not universal.

The problem is to maximize the total profits of the included items subject to remaining within the capacity of the knapsack. Let x_i be the amount of item i to be included. In the continuous version, x_i must lie in the interval $[0, 1]$. In the zero-one version, x_i must lie in the **set** $\{0, 1\}$, hence the name. If amount x_i of item i is included, we are allowed to count $x_i p_i$ as the profit of the ith item.

More precisely, we want to find the set $\{x_i\}$ that maximizes $\sum_{i=1}^{n} x_i p_i$ subject to $\sum_{i=1}^{n} x_i w_i \leq M$. Note in the zero-one case these sums are simply the sums of the profits included and the sums of the weights included respectively.

For example in the zero-one case: if $M = 7$ and there are three items with respective weights 2, 3, and 4 and respective profits 6, 5, and 9, the maximum profit of 15 would come from including the first and third item. A suboptimal profit of 14 would come from including the last two items, even though that would completely fill the knapsack. In the continuous case, 1 unit of the second item could be added to the optimal zero-one solution to get an optimal profit of $15 + (1/3)5 = 16\,2/3$.

One obvious algorithm for the zero-one version uses a *generate-and-test* strategy, that is, would generate every possible subset of items and test each one. The preceding corollary implies this algorithm would have time complexity $\Omega(2^n)$.

Of course, some subsets might not have to be tested. For example, if the subset T exceeds the capacity M, then so will any subset containing T. But it is not obvious that this sort of simplification will give a better time complexity than $\Theta(2^n)$. In fact, this problem appears to be intractable, an issue that will be explored further in Chapter 5.

An efficient algorithm for the continuous version of the problem will be presented in Section 2.19.

A second example arises from a problem in logic.

Def 2.7 A *Boolean expression* of n variables $\{x_i\}_{i=1}^n$ is either

1. a variable x_i or
2. an expression of one of the following forms: $(E), E \wedge F, E \vee F$, or $\sim E$, where E and F are Boolean expressions.[2]

The expression $E \wedge F$ is called a *conjunction;* here E and F are called *conjuncts.* The expression $E \vee F$ is called a *disjunction;* here E and F are called *disjuncts.*

For example, each of the following expressions is a Boolean expression in three variables:

$$x_1 \wedge (x_2 \vee \sim x_3) \wedge (\sim x_1 \vee x_3)$$

$$(x_1 \vee \sim x_2 \vee (x_3 \wedge \sim x_1)) \wedge (\sim x_1 \vee \sim x_3)$$

Example: The *Boolean satisfiability problem,* given a set $\{x_k\}_{k=1}^n$ of n Boolean variables and a Boolean expression in those variables as an instance, asks whether there is an assignment of truth values to $\{x_k\}$ (i.e., a function from $\{x_k\}$ to $\{T, F\}$) such that the entire expression has the value T.

For example, the first expression has the satisfying assignment ($x_1 = x_2 = x_3 =$ T), while the second has satisfying assignment ($x_1 =$ T, $x_2 = x_3 =$ F).

A generate-and-test algorithm for this problem would consider all possible truth assignments to the n variables until all possible assignments are considered or a satisfying assignment is found. Since each assignment may be identified with the set of all variables, which get the value T in that assignment, there are 2^n possible assignments. In the worst case, when there is no satisfying assignment, this algorithm will test all 2^n assignments. So it has time complexity $\Omega(2^n)$.

For an example of worst-case behavior, the expression

$$(x_1 \vee x_2) \wedge (x_1 \vee \sim x_2) \wedge (\sim x_1 \vee x_2) \wedge (\sim x_1 \vee \sim x_2)$$

has no satisfying assignment.

This problem, like ZOK, appears to be intractable. The reason that no such algorithm is likely is discussed in Chapter 5. The Boolean satisfiability problem plays a special role in the discussion of that chapter on the existence of efficient algorithms for a large class of problems.

Finally, we make one more recursive definition of a function that will be useful in analyzing algorithms, the factorial function. Just as the exponential function arises naturally in considering sets and their subsets, the factorial

---

[2] Compare this with the analogous definition of an *algebraic expression* in n variables as either 1) a variable x_i or 2) an expression of one of the forms $(E), -E, E + F, E - F, E \times F$, or E/F, where E and F are algebraic expressions. We leave as an exercise the definitions of the value of algebraic or Boolean expressions given values for each of the variables.

```
function factorial(n: integer): integer;
begin
  if n = 0 then factorial:= 1 else factorial:= n × factorial(n − 1);
end;
```

(a) The factorial function in Pascal

```
(define (factorial x)
  (cond ((zero? x)     1        )
    (else   (× x (factorial(− x 1))))))
```

(b) The factorial function in Lisp

Figure 2.1 The Factorial Function in Two High-Level Languages

function arises naturally in considering the concept of permutations of a set. Permutations themselves are discussed in the next section.

Def 2.8 If n is a nonnegative integer, then *Factorial*(n) =

1. Factorial(0) = 1
2. Factorial(n) = $n \times$ Factorial($n - 1$), for $n > 0$.

A more common notation for Factorial(n) is $n!$, although many programming languages would not allow this syntax. In the new notation,

$$0! = 1,$$

$$1! = 1 \times 0! = 1 \times 1 = 1$$

$$2! = 2 \times 1! = 2 \times 1 = 2$$

$$3! = 3 \times 2! = 3 \times 2 = 6 \text{ and}$$

$$4! = 4 \times 3! = 4 \times 6 = 24$$

Intuitively, $n!$ is just the product of the first n integers.

To illustrate the simplicity of going from a recursive definition to a program in a given language—assuming a knowledge of the language—procedures for computing the factorial function in Pascal and in the Scheme dialect of Lisp are given in Figure 2.1.

2.2 Recursive Definition of Data Structures

We suggested in the previous section that data structures of a given type can be defined recursively in terms of smaller data structures of the same type. Beginning with this section, we discuss the case of sequences, lists, trees, and other more specialized types.

Recall that a typical recursive definition for a data structure operation has the following form, which we may call a *schema* for this special sort of recursive definitions:

Def 2.9 Process(S):

1. If S is small, process S directly;
2. Otherwise,
   divide S into pieces
   process each piece recursively
   combine the processed pieces.

The division into parts 1 and 2 reflects the schema of a general recursive definition.

Typically, one of two cases occurs when dividing data structures. In one common situation, there is one small piece and one large piece. In this case it is relatively easy to find a correct iterative algorithm. The divide-and-conquer approach seldom improves efficiency, at least asymptotically, in this case. (It may make the algorithm easier to design, though, as opposed to more efficient to execute when converted to a program.)

In the other common situation, there are two or more large pieces. Here divide-and-conquer often helps in both finding an algorithm quickly and finding a quick algorithm.

Analysis of divide-and-conquer algorithms, except for some intuitive observations, can be more difficult than for iterative algorithms. We defer general analysis of divide-and-conquer algorithms to Sections 2.16 and 2.17.

Beginning with Chapter 3 we present more procedural algorithms. Although they are often easier to analyze, they are harder to prove correct.

Particular examples, of course, require definitions of particular types of data structures. We therefore proceed directly to consideration of individual data types.

2.3 Sequences

Intuitively, sequences are simply lists with a first element, a second element, and so on. Examples include text (a sequence of characters) and files (a sequence of records). Formally, we make the following definition:

Def 2.10 A *sequence* of elements either

1. is empty or
2. consists of an element called the *head,* corresponding to the first element of the sequence, and a sequence called the *tail,* corresponding to the rest of the sequence.

Note that this definition may also be considered to specify an operation on a data type—the operation corresponding to the Boolean-valued function

Is-a-sequence. Nevertheless, we will not number definitions of data types as we do other operations.

Also notice that if A is a sequence with head a and tail b, we write A as $(a \,.\, B)$. The empty sequence is written as $()$. The head of S is denoted *head*(S) and the tail of S as *tail*(S). Short sequences may be defined by listing all their elements in parentheses; for example, the sequence of the first five squares is $(1\ 4\ 9\ 16\ 25)$. Occasionally the use of parentheses would be confusing. In these cases we will use angle brackets just as ordinary brackets are used in ordinary text. Then the preceding sequence would be written as $\langle 1\ 4\ 9\ 16\ 25\rangle$.

All elements of a sequence can be required to be of the same type. This can be enforced as follows:

Def 2.11 A *sequence of elements of type T* either

1. is empty or
2. consists of a head of type T and a tail that is a sequence of elements of type T.

For example, $(3\ 4)$ is a sequence of integers, since 3 is an integer and (4) is a sequence of integers. This latter fact is true since 4 is an integer and $()$, being empty, is a sequence of integers by the basis of the recursive definition.

We may now define some of the common functions and operations on sequences. Their definitions will follow the divide-and-conquer schema given above.

If we make the reasonable assumption that the head and tail operations can be implemented in time $O(1)$, these new operations will have time complexity $O(n)$ unless specified otherwise. This is true since each element of the sequence can then be processed in bounded time. Since each function and operation is applied to the tail, which is finite and strictly smaller, all will terminate.

We begin with equality.

Def 2.12 Two sequences S and T are *equal* iff

1. they are both empty or
2. the head of S equals the head of T and the tail of S equals the tail of T.

Here of course we assume that equality has been defined for elements of a sequence.

As an example, $(3\ 4)$ and $(3\ 5)$ are equal iff $(4) = (5)$, since $3 = 3$. But (4) and (5) are not equal, since their heads are different. Thus the original sequences are unequal.

The best-case time complexity of a corresponding equality algorithm would be $O(1)$, for example, if the two heads were not equal.

We next define membership in a sequence. Here, as in many of the following examples, we write the function name before its arguments instead of between them. The arguments are in the same order as usual so that, for example, ElementOf(x, S) means "x is an *element* of S," or "x is a *member* of S" or simply "x is in S."

Def 2.13 *ElementOf*(x, S) is

1. FALSE if S is empty
2. TRUE if S is nonempty and x = head(S)

2′. ElementOf(x, tail(S)) otherwise.

For example, 4 is an element of (3 4) because 4 is an element of (4), because 4 is the head of (4).

The corresponding algorithm has best-case time complexity O(1), since x may appear as the head of S.

The concept of the *length* of a sequence is a natural one to define. It is useful, for example, in text processing when words or other strings are represented as sequences of characters.

Def 2.14 If S is a sequence, then *Length*(S) is

1. 0, if S is empty
2. 1 + Length(tail(S)), if not.

For example, Length(⟨1 2 3⟩) =

$$1 + \text{Length}(\langle 2\ 3\rangle) =$$

$$1 + (1 + \text{Length}(\langle 3\rangle)) =$$

$$1 + (1 + (1 + \text{Length}(\langle\rangle)) =$$

$$1 + (1 + (1 + 0)) =$$

$$1 + (1 + 1) =$$

$$1 + 2 =$$

$$3$$

To define the kth element of a sequence S, we use the following function:

Def 2.15 *Kth*(k, S) is

1. undefined, if S is empty

1′. head(S) if k = 1

2. Kth(k − 1, tail(S)) otherwise.

For example, the third element of (1 2 3) is the second element of (2 3) is the first element of (3) is 3. The third element of (1 2) is the second element of (1) is the first element of (), which is undefined.

The corresponding algorithm has time complexity O(k). If k is equally likely to have any value between 1 and the length n of S, then the average number of recursive calls is about $n/2$, so the average-case time complexity is $\Theta(n)$.

We often use the notation a_k for the kth element of the sequence (a). If the sequence (a) has length n, we sometimes write it as $(a_k)_{k=1}^{n}$. If the permissible values of the indices are clear, they may be omitted, for example, (a_k). Occasionally a set of n elements will be written as $\{x_k\}_{k=1}^{n}$. We only use this notation if we are considering an arbitrary ordering of the set elements since, strictly speaking, sets are unordered.

Some divide-and-conquer algorithms for sequences are most naturally defined only on nonempty sequences. In this case it is useful to have a definition of a nonempty sequence.

Def 2.16 A *nonempty sequence* consists of either

1. a single element {the head} or
2. a head that is a single element and a tail that is a nonempty sequence.

Showing that this definition is equivalent to the original definition of a sequence when the sequence is nonempty is left as an exercise.

Def 2.17 *Traversal* of a nonempty sequence may be accomplished by

1. visiting the head and
2. traversing the tail recursively.

Recall that the precise meaning of "visit" depends on the application. For example, a traversal algorithm to print all elements of a sequence could be written

Def 2.18 Traverse(S) {print the elements of a sequence S}
if S is nonempty then

1. print head(S)
2. traverse(tail(S)).

One of the most common operations on two sequences involves "gluing them together." This operation is another common text operation. For example, (1 2 3) and (4 5) may be combined to get (1 2 3 4 5). Formally, this operation is called *concatenation,* or *appending.* Its result is another sequence, and it is defined as follows:

Def 2.19 *Concatenation*(S, T) is

1. T, if S is empty
2. (head(S) . Concatenation(tail(S), T)) otherwise.

So Concatenation ($\langle 1\ 2\rangle\langle 3\ 4\ 5\rangle$) =

$$\langle 1\ .\ \text{Concatenation}\ (\langle 2\rangle, \langle 3\ 4\ 5\rangle)\rangle =$$

$$\langle 1\ .\ \langle 2\ .\ \text{Concatenation}(\langle\rangle, \langle 3\ 4\ 5\rangle)\rangle\rangle =$$

$$\langle 1\ .\ \langle 2\ .\ \langle 3\ 4\ 5\rangle\rangle\rangle =$$

$$\langle 1\ 2\ 3\ 4\ 5\rangle$$

The corresponding algorithm would have time complexity proportional to the size of S. It is possible to get O(1) time complexity depending on the representation of the sequence; the details are left as an exercise.

Concatenation is an example of an operation that could take an arbitrary number of arguments. For example, (1 2), (3), and (4 5 6) could be concatenated to get (1 2 3 4 5 6). Many data types have such operations. In these cases the operands may be represented as a sequence. It is simple to describe recursively how the list of operands is processed; a general schema is

1. for two operands, use an appropriate binary operator;
2. otherwise apply the binary operator to a) the head of the list and b) the result of applying the general operator to the tail of the list.

For example, we may define a function to find the sum of the elements in a nonempty sequence S of numbers as follows:

Def 2.20 *Sum*(S) is {S is a nonempty sequence to be summed}

1. a if S is the sequence (a) {i.e., if S has length 1}
2. head(S) + Sum(tail(S)) otherwise.

For example, Sum($\langle 1\ 2\ 3\rangle$) =

$$1 + \text{Sum}(\langle 2\ 3\rangle) =$$

$$1 + (2 + \text{Sum}(\langle 3\rangle)) =$$

$$1 + (2 + (3 + \text{Sum}(\langle\rangle)) =$$

$$1 + (2 + (3 + 0)) =$$

$$1 + (2 + 3) =$$

$$1 + 5 =$$

$$6$$

We leave as an exercise the proof that this defines the same function as does the Σ notation.

The maximum of a nonempty sequence of integers is easy to define using the suggested approach.

Def 2.21 If S is nonempty, then *Max*(S) is

1. $\text{Max}(\langle a\rangle) = a$
2. $\text{Max}(\langle a\ .\ S\rangle) = a \quad \text{if } a > \text{Max}(S)$

2′. $\text{Max}(\langle a\ .\ S\rangle) = \text{Max}(S) \quad \text{if } a \le \text{Max}(S)$.

The minimum may be defined in a similar manner. Both functions may be defined for any type (e.g., real, character) for which a total order may be defined.

As an example, finding Max($\langle 1\ 9\ 4\rangle$) involves comparing 1 with Max($\langle 9\ 4\rangle$), which involves comparing 9 with Max($\langle 4\rangle$) = 4. Since $9 > 4$, Max($\langle 9\ 4\rangle$) = 9. Since $1 \le 9$, Max($\langle 1\ 9\ 4\rangle$) = Max($\langle 9\ 4\rangle$) = 9.

We next consider the example that motivated considering a sequence of arguments.

Def 2.22 If the elements of S are all sequences, then a function *Concatlist* may be defined by:

1. $\text{Concatlist}(\langle\rangle) = \langle\rangle$
2. $\text{Concatlist}(\langle s \,.\, S\rangle) = \text{Concatenation}(s, \text{Concatlist}(S))$.

This is the natural generalization of the binary Concat operator. Its time complexity is subject to the same caveats as Concat. Abbreviating Concat(S, T) by $S \,.\, T$, we have the example

$$\text{Concatlist}(\langle 1\ 2\rangle\ \langle 3\ 4\ 5\rangle\ \langle 6\rangle) =$$
$$\langle 1\ 2\rangle \,.\, \text{Concatlist}(\langle 3\ 4\ 5\rangle\ \langle 6\rangle) =$$
$$\langle 1\ 2\rangle \,.\, (\langle 3\ 4\ 5\rangle \,.\, \text{Concatlist}(\langle 6\rangle)) =$$
$$\langle 1\ 2\rangle \,.\, (\langle 3\ 4\ 5\rangle \,.\, (\langle 6\rangle, \text{Concatlist}(\langle\rangle))) =$$
$$\langle 1\ 2\rangle \,.\, (\langle 3\ 4\ 5\rangle \,.\, (\langle 6\rangle \,.\, \langle\rangle)) =$$
$$\langle 1\ 2\rangle \,.\, (\langle 3\ 4\ 5\rangle \,.\, \langle 6\rangle) =$$
$$\langle 1\ 2\rangle \,.\, \langle 3\ 4\ 5\ 6\rangle =$$
$$\langle 1\ 2\ 3\ 4\ 5\ 6\rangle$$

A nonempty sequence is *sorted* in nondecreasing order iff every element (except for the last) is no larger than its immediate successor. We may formulate this intuitive idea as a recursive definition. We assume a $>$ operator for comparing elements.

Def 2.23 *Sorted*(S) is {S is a nonempty sequence}

1. TRUE if tail(S) is empty
2. FALSE if head(S) $>$ head(tail(S))
3. Sorted(tail(S)) otherwise.

For example, Sorted($\langle 1\ 3\ 2\rangle$) has the same value as Sorted($\langle 3\ 2\rangle$), which is FALSE, since $3 > 2$. Sorted($\langle 1\ 2\ 3\rangle$) has the same truth value as Sorted($\langle 2\ 3\rangle$), which has the same truth value as Sorted($\langle 3\rangle$), which is TRUE.

Note that Sorted has best-case time complexity $O(1)$.

Both sorting and insertion into a sorted sequence are common database operations. A single element may be inserted into an empty or sorted list as follows:

Def 2.24 *InsertIntoSorted*(x, S) is

1. $\langle x\rangle$ if S is empty
2. $\langle x \,.\, S\rangle$ if $x \leq \text{head}(S)$

2′. $\langle \text{head}(S) \,.\, \text{InsertIntoSorted}(x, \text{tail}(S))\rangle$ otherwise.

For example, InsertIntoSorted(2, $\langle 1\ 3\ 4\rangle$) =

$$\langle 1 \,.\, \text{InsertIntoSorted}(2, \langle 3\ 4\rangle) =$$
$$\langle 1 \,.\, \langle 2\ 3\ 4\rangle\rangle =$$
$$\langle 1\ 2\ 3\ 4\rangle$$

This operation also has best-case time complexity O(1). It may be made the basis of a sorting algorithm called *insertion sort.*

Def 2.25 *InsertionSort*(S) is

1. ⟨⟩ if S is empty
2. InsertIntoSorted(head(S), InsertionSort(tail(S))) otherwise.

As an example, with InsertionSort abbreviated as Sort and InsertIntoSorted as Insert,

Sort(⟨3 1 2⟩) =

Insert(3, Sort(⟨1 2⟩)) =

Insert(3, Insert(1, Sort(⟨2⟩))) =

Insert(3, Insert(1, Insert(2, Sort(⟨⟩)))) =

Insert(3, Insert(1, Insert(2, ⟨⟩))) =

Insert(3, Insert(1, ⟨2⟩)) =

Insert(3, ⟨1 2⟩) = ⟨1 2 3⟩

For this algorithm, processing the head does not have O(1) time complexity, since it involves a call to InsertIntoSorted. This call has time complexity $O(k)$ for a list with k elements, so the overall time complexity is $O\left(\sum_{k=1}^{n} k\right)$ or $O(n^2)$, by rule 10 of Section 1.6. Note that if the input sequence is already sorted, each application of InsertIntoSorted requires only time O(1). This is the best case for InsertionSort, so its time complexity is $\Omega(n)$.

Merging two sorted lists is another common database operation. We may merge two sorted lists into a larger sorted list, containing exactly the elements of the concatenation of the two lists.

Def 2.26 *Merge*(S, T) is {will be sorted if S and T are}

1. S if T is empty

1′. T if S is empty

2. ⟨head(S) . Merge(tail(S), T)⟩ if head(S) ≤ head(T)

2′. ⟨head(T) . Merge(S, tail(T))⟩ otherwise.

For example, Merge(⟨2 3⟩, ⟨1 4 5⟩) =

⟨1 . Merge(⟨2 3⟩, ⟨4 5⟩)⟩ =

⟨1 . 2 . Merge(⟨3⟩, ⟨4 5⟩)⟩ =

⟨1 . 2 . 3 . Merge(⟨⟩, ⟨4 5⟩)⟩ =

⟨1 . 2 . 3 . ⟨4 5⟩⟩ =

⟨1 2 3 4 5⟩

Here the sum of the lengths of the two arguments decreases by 1 at each recursive call, so the worst-case time complexity of Merge is proportional to the sum of the two input lengths. We can see that the best case requires time proportional to the length of the shorter input sequence.

The merge operation may also be made the foundation of a sorting algorithm. Intuitively we begin with a sequence of sorted sequences, or *runs.* Then we merge pairs of these runs until there are half as many runs as before. This operation—denoted *Mergelist* in the following—is repeated until there is a single run, which is the merged list.

The initial runs may be created trivially as sequences of size 1 (or as the following *CreateRuns* algorithm does, as sequences of size 2). Perhaps a more common application is in sorting files or other sequences stored externally. Mergesort is useful in this context because it minimizes references to external storage. In this case the runs may be created by a sorting algorithm that sorts as many elements as can be handled internally.

The function *Mergesort* simply creates the runs and then passes them to a recursive function *RecursiveMergesort* that calls Mergelist until there is a single run remaining.

Def 2.27 *Mergesort*(S) is RecursiveMergesort(CreateRuns(S)).

Def 2.28 *RecursiveMergesort*(S) is {sorts a sequence of runs}

1. head(S) if Length(S) = 1
2. RecursiveMergesort(Mergelist(S)) otherwise.

Def 2.29 *Mergelist*(S) is {merges elements of a list 2 at a time}

1. S if Length(S) $<$ 2
2. ⟨Merge(head(S), head(tail(S))) . Mergelist(tail(tail(S)))⟩ otherwise.

Def 2.30 *CreateRuns*(S) is

1. ⟨⟩ if S is empty

1′. ⟨⟨head(S)⟩⟩ if tail(S) is empty

2. ⟨⟨head(S), head(tail(S))⟩ . CreateRuns(tail(tail(S)))⟩ if head(S) $<$ head(tail(S))

2′. ⟨⟨head(tail(S), head(S)⟩ . CreateRuns(tail(tail(S)))⟩ otherwise.

We can see that Mergelist has time complexity $\Theta(n)$, where n is the total number of elements in all the runs. We can also see that RecursiveMergesort will make $\Theta(\log r)$ calls to Mergelist, where r is the number of runs, since at each call the number of runs is cut in half. Since $r = \Theta(n)$, the time complexity of RecursiveMergesort is $\Theta(n \log n)$. Since Mergesort requires only a $\Theta(n)$ call to CreateRuns before calling RecursiveMergesort, we may conclude that Mergesort has $\Theta(n \log n)$ time complexity. Note this is asymptotically better than InsertionSort.

As an example, consider the sequence ⟨8 5 4 9 1 7 6 3 2⟩. When CreateRuns (here abbreviated CR) is applied to the sequence, we get

$$\text{CR}(\langle 8\ 5\ 4\ 9\ 1\ 7\ 6\ 3\ 2\rangle) =$$

$$\langle\langle 5\ 8\rangle\ .\ \text{CR}(\langle 4\ 9\ 1\ 7\ 6\ 3\ 2\rangle)\rangle =$$

$$(\langle 5\ 8\rangle \,.\, (\langle 4\ 9\rangle \,.\, \mathrm{CR}(\langle 1\ 7\ 6\ 3\ 2\rangle))) =$$

$$(\langle 5\ 8\rangle \,.\, (\langle 4\ 9\rangle \,.\, (\langle 1\ 7\rangle \,.\, \mathrm{CR}(\langle 6\ 3\ 2\rangle)))) =$$

$$(\langle 5\ 8\rangle \,.\, (\langle 4\ 9\rangle \,.\, (\langle 1\ 7\rangle \,.\, (\langle 3\ 6\rangle \,.\, \mathrm{CR}(\langle 2\rangle))))) =$$

$$(\langle 5\ 8\rangle \,.\, (\langle 4\ 9\rangle \,.\, (\langle 1\ 7\rangle \,.\, (\langle 3\ 6\rangle \,.\, (\langle 2\rangle))))) =$$

$$(\langle 5\ 8\rangle\ \langle 4\ 9\rangle\ \langle 1\ 7\rangle\ \langle 3\ 6\rangle\ \langle 2\rangle)$$

Then when this sequence is passed to RecursiveMergesort (here abbreviated RM), we get the following:

$$\mathrm{RM}((\langle 5\ 8\rangle\ \langle 4\ 9\rangle\ \langle 1\ 7\rangle\ \langle 3\ 6\rangle\ \langle 2\rangle)) =$$

$$\mathrm{RM}((\langle 4\ 5\ 8\ 9\rangle\ \langle 1\ 3\ 6\ 7\rangle\ \langle 2\rangle)) =$$

$$\mathrm{RM}((\langle 1\ 3\ 4\ 5\ 6\ 7\ 8\ 9\rangle\ \langle 2\rangle)) =$$

$$\mathrm{RM}((\langle 1\ 2\ 3\ 4\ 5\ 6\ 7\ 8\ 9\rangle)) =$$

$$\langle 1\ 2\ 3\ 4\ 5\ 6\ 7\ 8\ 9\rangle$$

One application of sequences is to text processing. We begin with the following definition:

Def 2.31 A *character string* is a sequence of characters.

Many of the preceding operations—such as insertion, deletion, length, kth, and concatenation—are useful in text processing. It is also useful to have an algorithm for comparing two strings to see which string precedes the other. This allows, for example, sorting of strings.

The usual order for strings, alphabetical order, can be generalized without difficulty to what is called *lexicographical order.* Both methods assume an ordering for individual characters, but in alphabetical order the character ordering is restricted to letters.

The character ordering for general lexicographical ordering is machine dependent. Usually strings of bits, which may be interpreted as binary numbers, are associated with characters, and the order for characters is induced by that on the binary numbers. This association of characters with bit strings is called a *character code.* The most common character code is ASCII, for American Standard Code for Information Interchange, and is presented in Appendix B.

Assuming a character ordering, lexicographical ordering can be defined as follows on two character strings S and T (i.e., on sequences of characters):

Def 2.32 *LessOrEqual*(S, T) is {compares in lexicographical order}

1. TRUE if S is empty
1′. FALSE if T is empty and S is not
2. TRUE if neither is empty and head(S) < head(T)
2′. FALSE if neither is empty and head(S) > head(T)
2″. LessOrEqual(tail(S), tail(T)) if neither is empty and head(S) = head(T).

For example, writing character strings in quotes instead of as sequences,

LessOrEqual("LOOP," "LIST") =

(by 2″) LessOrEqual("OOP," "IST") =

(by 2′) FALSE

This definition will return TRUE for two equal strings. Two character strings are equal iff they are equal as sequences. Its best-case time complexity is $O(1)$ (if $\text{head}(S) \neq \text{head}(T)$).

Records in a database may be considered sequences of attributes. For example, a student record may be represented as the pair (BLACK, MARY) and another as (WHITE, JOHN). Using our definition of lexicographical order, the first record would precede the second. If the order of the attributes were switched, then (JOHN, WHITE) would precede (MARY, BLACK).

Many different orderings can be imposed on a set of n elements. Some problems ask for the optimal ordering of such a set. One example is the *Traveling Salesperson Problem* (formerly the *Traveling Salesman Problem;* in any case commonly abbreviated *TSP*). In the original version of this problem, a salesman must visit each of n cities, ending in the city from which he began. A cost (e.g., a mileage) is given for getting from each city to the next. The cost of the tour is simply the sum of the costs of the n city-to-city trips. The problem is to minimize the cost of the tour.

This problem can be formulated as a directed graph problem. Cities are represented by vertices, and the cost of an edge represents the cost of a trip from one city to another.

The TSP problem also illustrates a situation common to many problems. A solution is desired that is optimal, subject to certain constraints. Solutions that satisfy the constraints but are not necessarily optimal are called *feasible.* Here the feasible solutions are those consisting of n cities, or vertices, each of which is directly accessible from its predecessor in the tour—where the predecessor of the first city is the last city. An *optimization* problem is one that asks for the best solution (in some sense) among all feasible solutions.

A related problem is the *Hamilton circuit* or *Hamiltonian cycle* problem. This problem takes as instance a directed graph and asks whether any cycle visits each of the n vertices exactly once and returns to the starting vertex. This problem may be more likely to occur in electrical or pipeline networks, when there are no direct connections between many pairs of stations.

For each problem, a generate-and-test algorithm would investigate all possible orderings of the n cities. An ordering of a finite set is called a *permutation.* We now define permutations more formally and relate them to the analysis of our proposed algorithm:

Def 2.33 A sequence T is a *permutation* of a set S iff

1. $S = \varnothing$ and $T = ()$ or
2. $T = (a \cdot U)$, where a is an element of S and U is a permutation of $S - \{a\}$.

There are six permutations of the set {1, 2, 3}. They are (1 2 3), (1 3 2), (2 1 3), (2 3 1), (3 1 2), and (3 2 1). Using the definition, we see that (3 1 2) is a permutation, since 3 is an element of {3 1 2} and (1 2) is a permutation of {1 2}. This latter statement is true since 1 is a member of {1 2} and (2) is a permutation of {2}—since 2 is a member of {2} and ( ) is a permutation of { } by part 1 of the definition.

One may also talk about a permutation of a sequence, using the membership and deletion operators appropriate for sequences. The set of permutations of a set is the same as the set of permutations of a permutation of the set; the proof of this claim has been left as an exercise.

Analysis of the generate-and-test algorithm depends on counting the number of permutations of a given set. We can relate this count to the definition of the factorial function at the end of the last section.

Theorem There are exactly $n!$ different permutations of a set S of n elements.

Proof by induction on n. The induction hypothesis is just the statement of the theorem.

Basis: If $n = 0$, then ( ) is the only permutation.

Induction step: For $n > 0$, and for each choice of a first element x of the permutation, there are by induction $(n - 1)!$ different choices for the tail of the permutation (corresponding to the permutations of $S - \{x\}$, a set of size $n - 1$). Since two permutations with different first elements are different, each of the n choices for the first element gives $(n - 1)!$ permutations, for a total of $n(n - 1)!$ permutations, all of which must be different. But $n! = n(n - 1)!$ ■

Corollary Any algorithm that in the worst case looks at all permutations of a set of n elements has time complexity $\Omega(n!)$.

It is not immediately clear how fast the factorial function grows, say, compared to exponential functions; this question is addressed in Section 2.16. In particular, we will see its growth is not bounded by that of any polynomial function. As is true for ZOK and the Boolean satisfiability problem, both TSP and the Hamiltonian circuit problem appear intractable. These problems will also be discussed in Chapter 5.

2.4 Binary Trees and Ordered Trees

In addition to the family trees and corporate hierarchies described in Section 1.3, tree structures have applications to algebra and algebraic expressions, languages, translation, compiling, and files and file organization. Tree structures are also useful in searching, sorting, and analyzing algorithms in a wide range of fields. A simple glance at the figures throughout this text, especially in the first part of Chapter 4, will show the importance of trees to algorithms and to computer science in general.

We begin by considering a particularly simple, yet general, type of tree. *Binary trees* are useful in representing certain types of algebraic expressions and algorithms, and in searching and sorting.

Def 2.34 A *binary tree* either

1. is empty or
2. consists of a node called the *root R*, a left subtree (abbreviated *LST*) that is a binary tree, and a right subtree (abbreviated *RST*) that is a binary tree. The roots (if any) of the subtrees are called *children* of the node. The node is called the *parent* of its children. A *leaf* is a node with no children. An *internal node* is a node that is not a leaf.

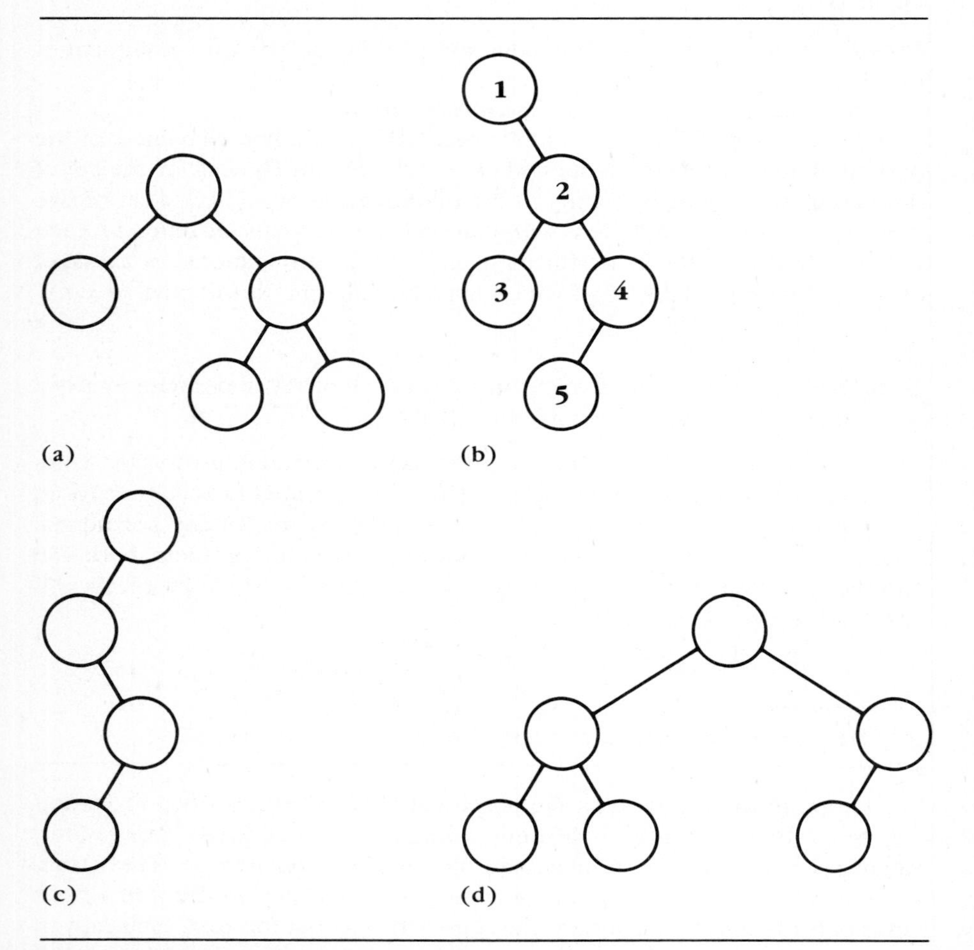

Figure 2.2 Examples of Binary Trees

Def 2.35 A node is a *member* of a binary tree iff

1. it is the root of the tree or
2. it is a member of the left subtree or a member of the right subtree.

In a typical application of binary trees, each member node would be implemented as a node or record to hold data. Each node would also have a left pointer field to point to its left child and a right pointer field to point to its right child. These links represent the structure of the data, as opposed to the data itself, which would be stored in a data field in each node (see Appendix D).

In diagrams of binary trees, we draw left children below and to the left of their parents and right children below and to the right of their parents. Figure 2.2 shows examples of binary trees.

Other terms, such as *ancestors, descendants,* and *siblings,* may be defined for trees following the natural analogy with parents and children. Their definitions are left as exercises. Sometimes we identify subtrees with the children at their roots, speaking, for example, of an empty child or a subtree of a node.

In the tree of Figure 2.2b, identifying nodes with the data they contain, we may say that 2 is a parent of 4 and an ancestor of 5. Thus 5 is a descendant of 2. There are two leaves, 3 and 5. The root is 1; it and 2 and 4 are the internal nodes. The nodes containing 3 and 4 are siblings. The left child of 4 is 5; the right subtree is empty. (See Figure 2.3 for an illustration of levels.) A binary tree, in which each internal node has exactly two children, has exactly one more leaf than internal node.

The divide-and-conquer schema suggests that functions defined on trees will most naturally be defined recursively in terms of their values on the subtrees. Since binary trees, as we have defined them, are finite and the subtrees are smaller, these algorithms will terminate.

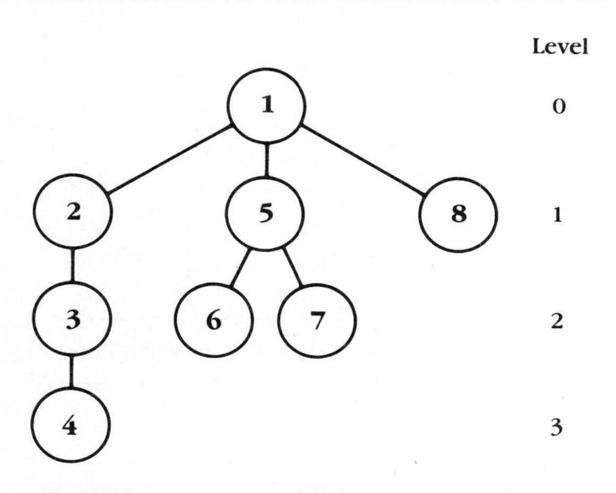

Figure 2.3 Levels in an Oriented Tree

In some cases it is convenient to allow trees to have more than two children. Thus we make the following definition:

Def 2.36 An *ordered tree* either

1. is empty or
2. consists of a node called the *root* and a sequence of zero or more subtrees, each of which is an ordered tree.

Children, parents, and so on, are defined for ordered trees as for binary trees.

In Figure 2.4 the same hierarchical structure is represented by four ordered trees differing only in the order of the children of certain nodes.

Notice a binary tree is not precisely a special case of an ordered tree in which left subtrees correspond to first subtrees and right subtrees to second subtrees. This is true since a binary tree may have a right subtree without having a left subtree, while an ordered tree may not have a second subtree without having a first subtree.

We may use instead the following definition for ordered trees, equivalent except for empty trees:

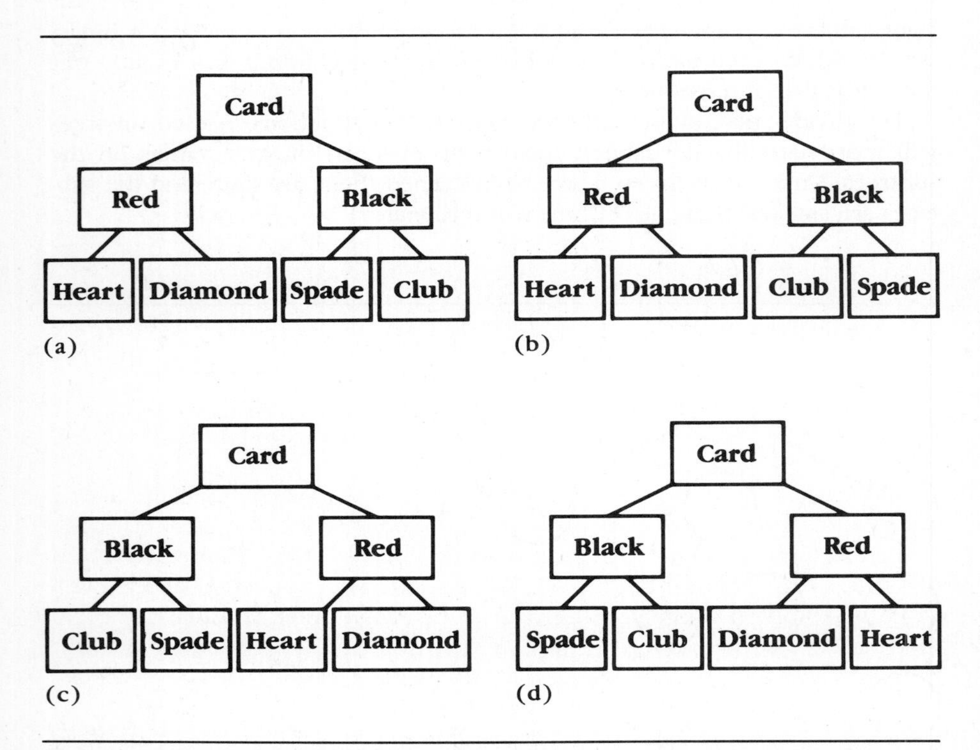

Figure 2.4 Four Different Ordered Trees Representing the Same Hierarchy

Def 2.37 A nonempty *ordered tree* either

1. consists of only a root or
2. consists of a root and a sequence of 1 or more subtrees, each of which is a nonempty ordered tree.

We can now make three simple definitions. We state them for binary trees. Defining membership in binary trees and extending the definitions to ordered trees are left as exercises. In each case the function being defined is recursively applied to both subtrees; only for the **Size** function is this likely to be useful in practice. The other two definitions are mere terminology.

Def 2.38 If T is a binary tree T, then *Size*(T) is

1. 0, if T is empty
2. 1 + Size(Left(T)) + Size(Right(T)), if T is not empty.

The Size function merely counts the number of nodes in the tree. Another way of measuring the size of a tree corresponds to what is most naturally called the height of the tree. More precisely,

Def 2.39 If T is a binary tree, *Height*(T) is

1. -1 if T is empty
2. 1 + Max{Height(Left(T)), Height(Right(T))}, if T is not empty.

The value of -1 for the height of an empty tree is merely a convenience so that the height of a tree with exactly one node is 0. Other writers consider the height of a tree with exactly one node to be 1.

Note that the height of a tree is a property of a tree as a whole. The corresponding definition for a single node is worth making. In this case we talk about the *level* of a node N in a tree T. The level of a node is also known as its *depth*.

Def 2.40 *Level*(N, T) is

1. 0, if N is the root of T
2. 1 + Level(N,Left(T)), if N is a member of Left(T)
3. 1 + Level(N,Right(T)), if N is a member of Right(T).

If Level(N, T) = k, we say that node N is on level k in the tree T. The level of a node is just the number of edges between it and the root. Also, the height of a tree is just the maximum level of any node in the tree. Note that lower-numbered levels appear higher in diagrams. Although important as terminology, the definition is often not particularly useful in practice, since determining whether a node is a member of the left subtree or the right subtree may not be easy.

The tree of Figure 2.3 has height 3 and size 8.

It is not difficult to show a child node is at a level numbered one higher than its parent.

Theorem If X is the parent of Y in a binary tree T, then Level(Y, T) $= 1 +$ Level(X, T).

Proof by induction on the level of X.

Basis: If X is at level 0, then Y is the root of a subtree of X and is at level 0 in the subtree. Hence it is at level 1 in the original tree.

Induction step: If X is at level $k > 0$, then X and Y must be in the same subtree S of the root. By induction Level(Y, S) $= 1 +$ Level(X, S), so Level(Y, T) $= 1 +$ Level(Y, S) $= 2 +$ Level(X, S) $= 1 +$ Level(Y, T). ∎

Tree traversal, even for binary trees, is not so simple as sequence traversal; tree traversal has three pieces instead of two and no natural order among the pieces. Even if the usual custom of traversing the left subtree before the right one is observed, three possible traversal orders remain, corresponding to the three possible times at which one could visit the root. More precisely, we have the following definitions:

Def 2.41 To perform a *preorder traversal* of a nonempty binary tree T, {assuming a **visit** procedure}

1. visit the root of T
2. traverse Left(T) recursively
3. traverse Right(T) recursively.

Def 2.42 To perform an *inorder traversal* of a nonempty binary tree T, {assuming a **visit** procedure}

1. traverse Left(T) recursively
2. visit the root of T
3. traverse Right(T) recursively.

Def 2.43 To perform a *postorder traversal* of a nonempty binary tree T, {assuming a **visit** procedure}

1. traverse Left(T) recursively
2. traverse Right(T) recursively
3. visit the root of T.

The tree of Figure 2.5 has preorder traversal (X Y P C F R S), inorder traversal (P Y C F X S R), and postorder traversal (P F C Y S R X).

Note that the numbering of the steps of each traversal algorithm no longer corresponds to the numbering of our divide-and-conquer schema. This is partly because the processing of the root, which corresponds to the basis step in the schema, may no longer be done at the beginning.

Inorder traversal may also be called *symmetric order traversal.* The names of the traversal methods are derived from an application to algebraic expressions, which will be discussed in Section 4.1.

Each traversal method has its particular applications. Perhaps the most amusing application of preorder traversal is to succession to the British throne. This rather special purpose application is relegated to the exercises. Inorder

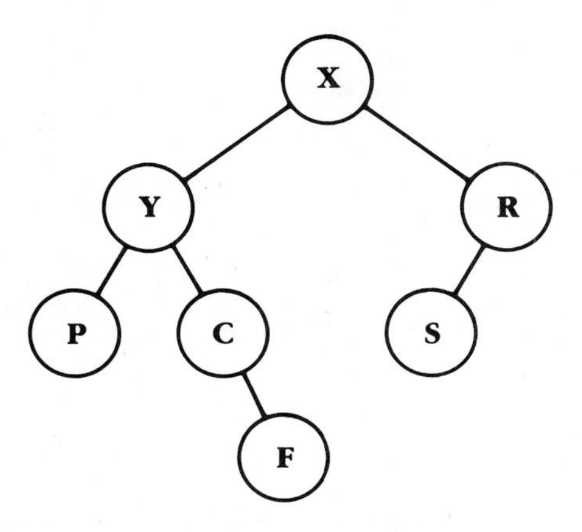

Figure 2.5 A Binary Tree to Be Traversed

traversal has application to sorting. Postorder traversal is useful in evaluating algebraic expressions.

Several properties of these traversals have been left as exercises. It is easy to generalize preorder and postorder to arbitrary ordered trees, but not so easy for inorder. The time complexity for each type of traversal is $\Theta(n)$. All three traversal methods visit leaves in the same order. It is possible to recover a binary tree given the results of any two types of traversals, but not the result of just one traversal.

2.5 Binary Search Trees

So far we have no natural way of letting the structure of a tree correspond to the ordering of the data in the way that a sorted sequence does. This correspondence can be achieved for trees, and the resulting techniques and their generalizations give powerful methods for searching, sorting, and file organization in general.

We start with the following definition:

Def 2.44 A *binary search tree* T is a binary tree that either

1. is empty or
2. consists of a *root* containing a data element x such that
   a. each data element y of Left(T) satisfies $y \le x$
   b. each data element z of Right(T) satisfies $z \ge x$, and
   c. Left(T) and Right(T) are binary search trees.

Intuitively, smaller keys are in the left subtree and larger keys are in the right subtree. In particular, if we identify nodes with the data they contain, a

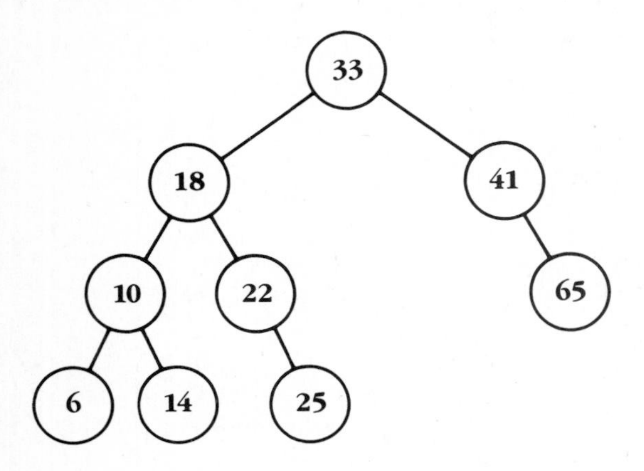

Figure 2.6 A Binary Search Tree

left child is smaller than and a right child larger than its parent. A binary search tree is pictured in Figure 2.6.

Binary search trees, and other types of search trees, are useful in implementing files. We need some terminology.

Def 2.45 A *file* is a set of records. A *record* is a sequence (a_i) of *attributes,* each drawn from a particular *domain* (D_i). The D_i are the same for a fixed i, for all records in the file. If a domain D_i has the property that no two records have the same attribute from that domain (i.e., that each attribute uniquely identifies the record), then each attribute is called a *key.*[3]

When searching for a record, enough information must be provided to the search algorithm to uniquely identify a record. This is the function of a key. For example, an employee record may consist of the attributes (John, Doe, 000-00-0000, Rural Route 1, Anytown, IL, 61000). The domains of the third and seventh attributes could be the set of integers; the other domains would be the set of character strings. The Social Security number could serve as a key; it is likely to be a unique identifier.[4] Searches for employee records could then be done by Social Security number. In a small company the family name could serve as a key.

---

[3] While it is possible that no domain defines a key, some set of domains always does. Certainly the entire set of domains does define a key if there are no duplicate records. To simplify the discussion, we will assume that a single domain is sufficient for unique identification.

[4] Although Social Security numbers are designed to be unique identifiers, this is not always the case. The story is that in the early days of the program, a wallet manufacturer decided to provide sample Social Security cards, all with the same number, in his products instead of the usual blank identification card. Since the program was new, many people thought this was their official card. Attrition is slowly taking care of this problem.

Def 2.46 An *index* for a file is another file composed of records; each record of the index file consists of a key for and a pointer to the physical location of a record in the file being indexed.

The index is smaller than the underlying file, which is an advantage because search in the smaller index may be more efficient than search in the larger file. This advantage is especially important whenever searching requires reference to external memory. Search trees can be used to organize the files themselves, but are more commonly used for indexing.

One advantage of the binary search tree is that it has a natural search algorithm. If x is the key in the root of a binary search tree and we are looking for x, then we may quit.[5] If we are looking for a key smaller than x, then it must be in the left subtree. Similarly, any key larger than x must be in the right subtree. We may then proceed recursively as in the following more formal algorithm, where we assume the node containing the key is to be returned. In programming languages with explicit pointer variables, a pointer to the node would be a more natural choice.

Def 2.47 *Search*(k, T) {**data**(T) returns the data in the root of T}

1. if Empty(T), return unsuccessfully
2. if k = data(T), then return the root of T

2′. if $k <$ data(T), then Search(k, Left(T))

2″. if $k >$ data(T), then Search(k, Right(T)).

For example, a search for the key 25 in the binary search tree of Figure 2.6 would descend left after comparing 25 with 33, right after comparing 25 with 18, and right again after comparing 25 with 22. Then the key would be found. A search for the key 21 would also compare with 33, 18, and 22, but the left subtree of the node containing 22 is empty, so the search would be unsuccessful.

Insertion into a binary search tree is similar to search: There is no choice where to insert—except for the case of duplicate keys, in which we arbitrarily insert into the right subtree.

Def 2.48 *Insert*(k, T) {**data**(T) returns the data in the root of T}

1. if Empty(T), let data(T) equal k
2. if $k <$ data(T), then Insert(k, Left(T))

2′. if $k \geq$ data(T), then Insert(k, Right(T)).

Figure 2.7 shows the tree resulting from inserting the key 50 into the tree of the previous figure. A search is made for 50, comparing 50 with 33, 41, and 65. The left subtree of the node containing 65 is found to be empty, so the new key is inserted as the left child of this node.

---

[5] We are assuming here that only one occurrence of the key need be found. This assumption will be true if the keys are actually unique file identifiers. The threaded tree representation described in Section 4.6 will allow easy location of all nodes containing a nonunique key once one such node is found.

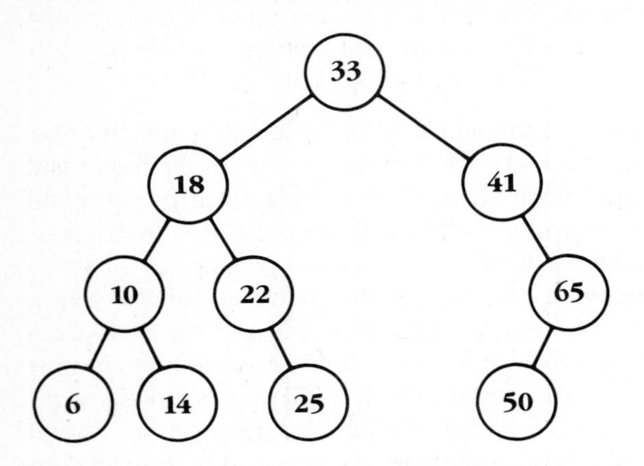

Figure 2.7 The Binary Search Tree of Figure 2.6 After Insertion of the Key 50

According to the preceding definition, this last step requires inserting the new key k into an empty subtree. If this insertion is not done carefully, there will be no way to connect the new node with the rest of the tree. If the algorithm is implemented recursively, one natural approach is to have the insertion procedure return a tree containing the new element, which is then assigned to the left subtree in (2) and the right subtree in (2′). In (1), a node containing k is constructed and returned. If the algorithm is implemented iteratively, there is the danger of trying to follow an empty pointer. In this case, call by value-result may be used; see Appendix D.

Binary search trees can also be used for sorting, as implied by the following theorem:

Theorem If a binary search tree is traversed in inorder, the keys are visited in sorted order.

Proof by induction on the number n of nodes in the tree. The induction hypothesis is that if x is visited before y in the traversal, then $x \le y$. Establishing this hypothesis is sufficient to prove the theorem.

Basis: There is nothing to prove if $n = 0$ or $n = 1$.

Induction step: Either one of $\{x, y\}$ is in the root or neither key is in the root. If x is in the root, then since x is visited before y, y must be in the right subtree. By definition of a binary search tree, $x \le y$. Similarly if y is in the root, then x must be in the left subtree, and $x \le y$.

If neither key is in the root, then they may be in different subtrees or the same subtree. If they are in the same subtree, then since the subtrees are also traversed in inorder, we are done by induction.

So assume that x and y are in different subtrees, and that z is the key in the root. Since x is visited before y, x must be in the left subtree and y in the right subtree. But then by the definition of a binary search tree, $x \le z \le y$ as desired. ∎

Clearly, the preceding algorithms for binary search tree search and insertion require time proportional to the level of the nodes being sought. Their worst-case time complexities are therefore given by the height of the tree. The analysis of these algorithms, then, depends on knowing how large the height h can be for a tree with n nodes. Alternatively, we may ask how large or small n can be for a given height h.

The worst case is easy to determine. A tree with n nodes may have height as large as $n - 1$. For example, this case occurs if each node in the tree has a right child but no left child. A more general example is shown in Figure 2.8. Binary search in this case reduces to sequential search. An easy induction shows that a tree of height h must have at least $h + 1$ nodes, so that $n - 1$ is the largest possible height for a tree with n nodes.

Finding the best case is not so trivial but not difficult either. Attempts to pack the largest number of nodes into trees of various heights lead to the trees of Figure 2.9. We see that for $h = 1$, we can have $n = 3$; for $h = 2$, $n = 7$; for $h = 3$, $n = 15$; for $h = 4$, $n = 31$. This pattern suggests we try to establish the following:

Theorem A nonempty binary tree of height h has at most $2^{h+1} - 1$ nodes.

Proof It's enough to show that level j has at most 2^j nodes. Then on all levels, there are at most $\sum_{j=0}^{h} 2^j = 2^{h+1} - 1$ nodes. This lemma may be proved by induction on j. ■

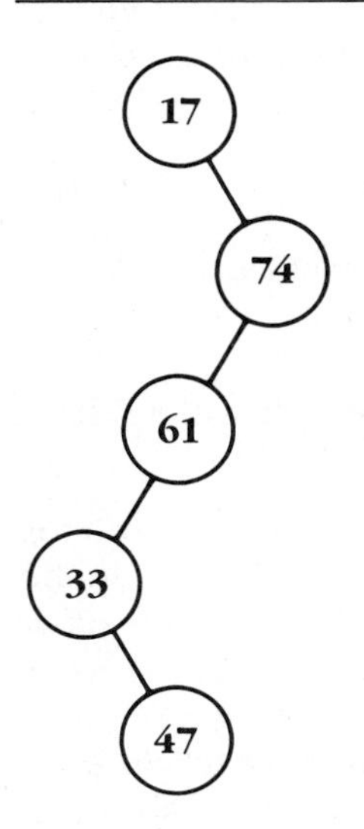

Figure 2.8 Example of a Worst-Case Binary Search Tree

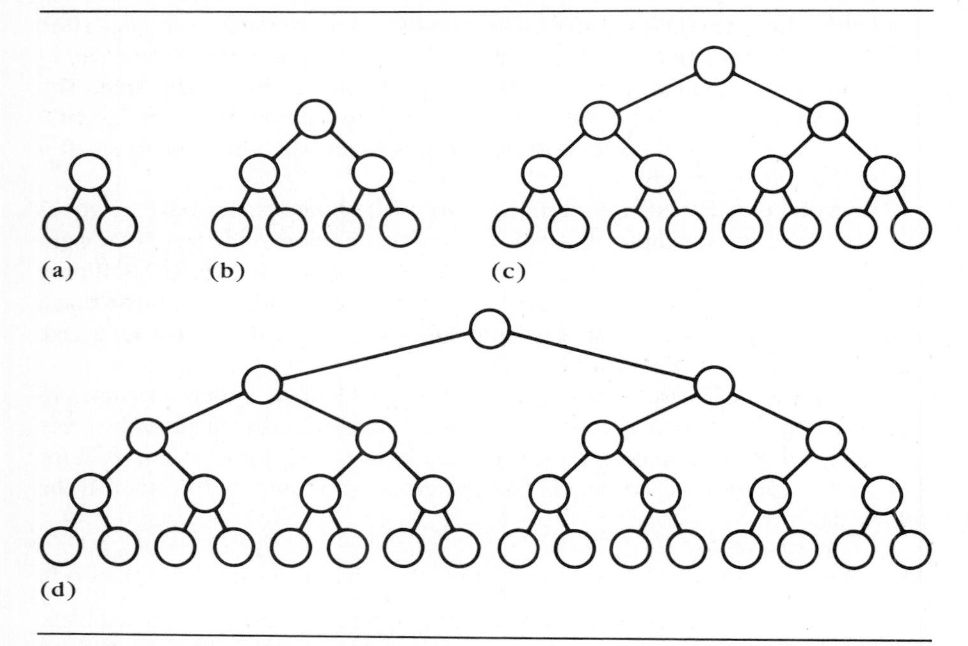

Figure 2.9 Completely Full Binary Trees

Lemma Level j of a binary tree has at most 2^j nodes.

Proof of lemma by induction on the level. At level 0, it is clear there is just $1 = 2^0$ node. For level j, we have by induction that level $j - 1$ contains at most 2^{j-1} nodes. So the nodes on level $j - 1$ can have at most $2 \times 2^{j-1}$ children. But these children are precisely the nodes on level j, so there can be at most 2^j such nodes. ∎

Corollary The height of a tree of n nodes is $\Omega(\log n)$.

Proof If h is the height, then $n \leq 2^{h+1} - 1$ by the previous theorem. So $h \geq \log_2(n + 1) - 1$. ∎

Corollary The preceding recursive search and insertion algorithms have worst-case time complexity $\Omega(\log n)$ and $O(n)$.

To summarize: The worst-case time complexity for search and insertion is $\Theta(n)$. We next show that the best possible worst-case time is $\Theta(\log n)$ in the sense that the height of a binary search tree with n nodes is $\Omega(\log n)$. We show later in this section that if each of the $n!$ insertion orders of the n keys is equally likely, the average-case search and insertion time is $O(\log n)$.

The trees of Figure 2.9 appear to be the best possible trees of height h, so they deserve a name and a formal definition. The name we use is not completely standard.

Def 2.49 A binary tree of height h is *completely full* iff

1. it is empty (i.e., $h = -1$) or
2. its left and right subtrees are each completely full trees of height $h - 1$.

It appears from the figure that all leaves of a completely full tree lie on the same level. This fact is not hard to verify inductively.

Theorem In a completely full tree of height h, all leaves lie on level h.

Proof by induction on h.

Basis: If $h = 0$, there is only one node in the tree. This node is on level 0 and is a leaf.

Induction step: For $h > 0$, the root is not a leaf. All other leaves must be leaves of one of two subtrees that are completely full trees of height $h - 1$. By induction their leaves all lie on level $h - 1$ of the subtree, which is level h of the original tree. ∎

We are still trying to verify the best-case behavior of complete trees of height h. The desired result is a corollary of the following theorem:

Theorem In a completely full tree with height h, if $j \leq h$, then there are exactly 2^j nodes on level j.

Proof by induction on j. We include as part of the induction hypothesis that a node on level j is the root of a complete tree of height $h - j$.

Basis: If $j = 0$, only the root is on level j, so that there is $1 = 2^0 = 2^j$ node on level $j = 0$. This node is certainly the root of a complete tree of height $h - j = h - 0 = h$.

Induction step: If $0 < j \leq h$, by induction each node on level $j - 1$ is the root of a complete tree of height $h - (j - 1) > 0$. Since the subtrees of each node on level $j - 1$ have height $h - (j - 1) - 1 = h - j \geq 0$, these trees cannot be empty. Thus each of the 2^{j-1} nodes on level $j - 1$ has two children. Since each node on level j must have a parent on level $j - 1$, there are exactly $2 \times 2^{j-1} = 2^j$ nodes on level j. ∎

And finally the desired corollary:

Corollary A completely full tree of height h has exactly $2^{h+1} - 1$ nodes.

Proof Since level j contains 2^j nodes, there are $\sum_{k=0}^{h} 2^k = 2^{h+1} - 1$ nodes in each of the levels from 0 to h. The sum is evaluated by Example 4 of Section 2.1. ∎

It follows from this corollary that many values of n exist for which there is no completely full tree with n nodes. We may generalize the definition of completely full trees to include trees with other sizes, such as those of Figure 2.10. The following term is more standard than the previous one:

Def 2.50 A binary tree of height h is *complete* iff

1. it is empty (i.e., $h = -1$) or
2. its left subtree is complete of height $h - 1$ and its right subtree is completely full of height $h - 2$ or

2′. its left subtree is completely full of height $h - 1$ and its right subtree is complete of height $h - 1$.

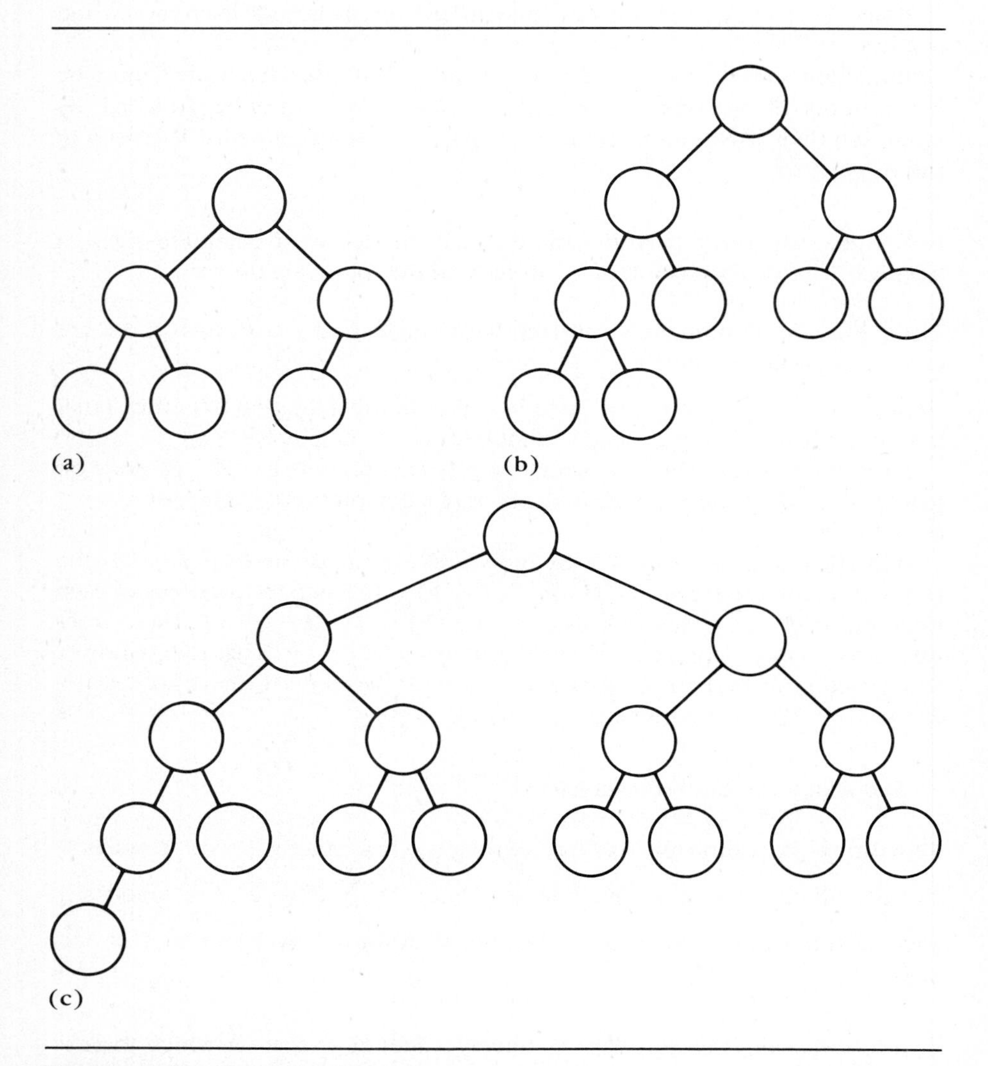

Figure 2.10 Complete Trees

Intuitively, all leaves in a complete tree are on at most two adjacent levels, and all nodes at the lowest level are as far to the left as possible. The proof of the first claim is left as a solved exercise.

Note in both (2) and (2′) the original tree has height h, since the maximum subtree height is $h - 1$. By (2′) and a trivial induction, completely full trees are complete.

You may convince yourself that a unique complete tree of n nodes exists for any n by drawing complete trees for the first few integers n. Each new node can go into only one place if the tree is to remain complete. In fact, you may be able to determine under what situations (2) of the definition applies and under what circumstances (2′) applies. Formally, we have the following theorem:

Theorem If $n \geq 1$ and $2^h \leq n < 2^{h+1}$, then there is a complete tree of height h on n nodes.

Proof by induction on h. The induction hypothesis is just the statement of the theorem.

Basis: If $h = 0$, then $1 = 2^0 \leq n < 2^{0+1} = 2$, so $n = 1$. Clearly there is a unique complete tree of height 0 when $n = 1$.

Induction step: Let $h > 0$ and let $M = 2^h + 2^{h-1}$. First suppose that $n \geq M$. Then there is a tree whose left subtree is a completely full tree of height $h - 1$ and thus has $2^h - 1$ nodes. This leaves $R = n - (1 + 2^h - 1) = n - 2^h$ nodes for the right subtree. Since $n \geq M, R = n - 2^h \geq M - 2^h = 2^{h-1}$. On the other hand, since $n < 2^{h+1}, R = n - 2^h < 2^{h+1} - 2^h = 2^h$. Thus $2^{h-1} \leq R < 2^h$, so by induction there is a complete subtree of height $h - 1$ on R nodes. We may let this subtree be the right subtree of the desired complete tree by (2′) of the definition of a complete tree.

If $n < M$, the argument is similar. In this case, we are aiming at part 2 of the definition of a complete tree. The desired tree will have a completely full right subtree of height $h - 2$ and $2^{h-1} - 1$ nodes, leaving $L = n - (1 + 2^{h-1} - 1) = n - 2^{h-1}$ nodes for the left subtree. Since $n < M, L = n - 2^{h-1} < M - 2^{h-1} = 2^h$. But since $n \geq 2^h, L = n - 2^{h-1} \geq 2^h - 2^{h-1} = 2^{h-1}$. So $2^{h-1} \leq L < 2^h$, and there is by induction a complete tree of height $h - 2$ with L nodes, which may serve as the left subtree of our desired tree. ■

The uniqueness of the complete tree on n nodes may be extracted from the fact that there is only one h with $2^h \leq n < 2^{h+1}$ and from the fact that n may not be both greater than or equal to M and less than M. The details are left as an exercise. In fact, we also have the following

Corollary The complete tree on n nodes has height $\lfloor \log_2 n \rfloor$.

Proof For any n, there is exactly one h such that $2^h \leq n < 2^{h+1}$. By the previous theorem, this h is the height of the complete tree on n nodes. The inequality implies that $h \leq \log_2 n < h + 1$. We are done by definition of the floor function. ■

Complete trees are of use partly because under reasonable assumptions about what the average case is, complete trees give best possible average-case behavior for search and insertion. (However, compare the binary search trees of Section 2.18.) To show this, we need one other simple property of complete trees.

Theorem If T is a nonempty complete tree of height h, then T has exactly 2^j nodes on level j for $j < h$.

Proof by induction on h. The induction hypothesis is just the statement of the theorem. We have seen that completely full trees are complete and have exactly 2^j nodes on level j.

Basis: Suppose $h = 0$. Here there is no level j with $j < h$, so there is nothing to prove. Next suppose $h = 1$. The only level j with $j < h$ is level 0, and only the root can be on level 0. Level 0 has $1 = 2^0$ nodes.

Induction step: For $h > 0$, first assume we are in case 2′ of the definition of a complete tree. By induction, for all $j < h - 1$, there are 2^j nodes on level j of the left subtree and 2^j nodes on level j of the right subtree. Since for $j > 0$ the number of nodes on level j of T is equal to the number on level $j - 1$ of Left(T) plus the number of level $j - 1$ of Right(T), there must be $2^{j-1} + 2^{j-1} = 2^j$ as desired. For $j = 0$ we have $1 = 2^0$ nodes on level j.

Now assume that case 2 of the definition applies. Here for $j < h - 1$ there are 2^j nodes on level j of Left(T). Also for $j \leq h - 2$ (note that equality is allowed here), there are 2^j nodes on level j of Right(T). As before, the number of nodes on level $j > 1$ of T is the $2^{j-1} + 2^{j-1}$ for $j \leq h - 1$. Once again level 0 has exactly 1 node, so the proof is complete. ■

We have seen that the time required to search for a given key is proportional to the level on which the key is found. Thus if all keys are equally likely to be sought, the average-case successful search time is proportional to

$$1 + (1/n) \sum_{x \text{ in } T} \text{level}(x) = (1/n) \sum_{k=0}^{h} k\text{n}(k)$$

where h is the height of T and n(k) is the number of nodes on level k. This sum defined by the Σ expression is called the *internal path length* and often denoted I.

It is not difficult to see that for any n, the complete tree of n nodes gives the best possible average-case successful search time.

Theorem The complete tree of n nodes has the minimum internal path length of any binary tree of n nodes.

Proof by minimal counterexample. Suppose that T is the complete tree on n nodes. If T' has shorter internal path length, it must have some $n(k)$ different than that of T, and therefore there is some smallest value k_0 of k for which the n-functions disagree. For all trees of minimal internal path length, let T' be the one for which the value of k_0 is maximal. In other words, T' is that minimal tree whose n-function agrees with T as long as possible.

Suppose that $k_0 = h$. Let m be the number of nodes on level h in T. If T' has more than m nodes on level h, then T' has more than n nodes, since T has only n levels and the two trees have the same number of nodes on all other levels. If T' has exactly m nodes on level h, then T' has no other nodes on higher-numbered levels and must therefore have the same internal path length as T. These contradictions show that T' must have fewer nodes on level h than T does. But now the internal path length of T' must be greater than that of T, since the two trees have the same number of nodes on all levels with lower numbers than h, since all m remaining T nodes are at level h, and since some of the m remaining T' nodes are at levels with higher numbers than h.

Thus $k_0 < h$. Since T has the maximum number $2\binom{k}{0}$ of nodes at level k_0, T' must have strictly fewer nodes on that level. We can get a contradiction to the minimality of T' as follows: Take a leaf from the highest-numbered level of T'. Move it to level k_0. [There is a place for it, since $k_0 > 0$, T' (like T) has the maximum number of nodes at level $k_0 - 1$ and T' does not have the maximum number of nodes at level k_0.] The resulting tree has smaller internal path length than does T'. This contradiction proves the theorem. ■

It is now easy to find an expression for the best possible internal path length for a tree of n nodes. For $j < h$, $\mathrm{n}(j) = 2^j$. All remaining nodes are on level h, so $\mathrm{n}(h) = n - \sum_{j=0}^{h-1} 2^j = n - (2^h - 1)$, by Example 4 of Section 2.1. So in the best case, $I = h(n - 2^h + 1) + \sum_{j=0}^{h-1} j2^j$. In Section 2.17 this sum is shown to be $\Theta(n \log n)$, so that the average over n nodes is $\Theta(\log n)$.

Fortunately we needn't go through the same computation in the unsuccessful case. If there are n nodes in a binary tree, then there are $n + 1$ intervals into which an unsuccessful search could be directed. Each corresponds to an empty child—here we are identifying children with subtrees—of a node. Sometimes it is useful to indicate these $n + 1$ positions in tree diagrams by dummy nodes called *external nodes;* we have done so in Figure 2.11. The nodes actually in the tree are drawn as circles; the external nodes are drawn as squares.

If an unsuccessful search is assumed to be equally likely to reach any external node, then the average-case number of key comparisons in an unsuccessful search in a tree of height h is $(1/(n + 1))E$, where $E = \sum_{k=0}^{h} k\mathrm{e}(k)$, and $\mathrm{e}(k)$ is the number of external nodes on level k. E is called the *external path length.* Fortunately there is a close relationship between E and I, which you might want to explore by looking at particular examples before reading any further.

Theorem For a binary tree with n nodes, $E = I + 2n$.

Proof by induction on n. The induction hypothesis is just the statement of the theorem.

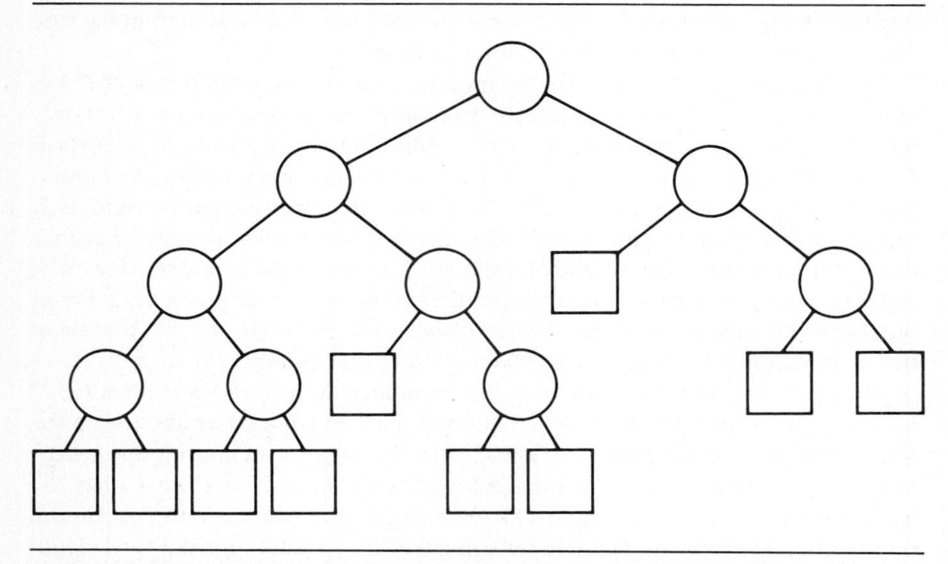

Figure 2.11 The Binary Search Tree of Figure 2.6 with External Nodes

Basis: If $n = 1$, then there is a node on level 0 and two external nodes on level 1. Thus $I = 0$ and $E = 2$, and $E = I + 2 \times 1$.

Induction step: For $n > 1$, the tree must have a leaf on level h, where h is the height of the tree. If the leaf is removed, then the resulting tree has internal path length $I - h$. Two external nodes at level $h + 1$ disappear and are replaced by an external node at level h. Thus the new external path length is $E - 2(h + 1) + h = E - h - 2$. The resulting tree has $n - 1$ nodes, so by induction we can apply the relation between its external and internal path lengths to get $E - h - 2 = I - h + 2(n - 1)$. Thus $E - 2 = I + 2n - 2$, so $E = I + 2n$. ∎

If for a given n, a tree is chosen that will minimize the average-case successful search time $(1/n)(1 + I)$, then I and hence E will be minimized. So the average-case unsuccessful search time $E/(n + 1)$ will also be minimized. Thus complete trees minimize both types of search.

Finding the trees with n nodes that maximize E and I is left as an exercise.

Finally we show that the natural search algorithm in binary search trees has logarithmic time complexity. Strictly speaking, we show this only for successful search. The corresponding result for the closely related insertion and unsuccessful search algorithms are left as exercises.

Theorem The search algorithm of this section for binary search trees has time complexity $O(\log n)$ in the successful case.

Proof [based on an idea of Knuth (1973, vol. 3)]: Let $A(n)$ be the average-case successful search time in a binary search tree with n nodes. If I is the average internal path length over all sequences of n insertions, then since the sequences are assumed equally likely, we see $A(n) = 1 + I/n$, or $I = nA(n) - n$.

$A(n + 1)$ may be computed in terms of $A(n)$ as follows: For a binary search tree of n nodes, there is probability $1/(n + 1)$ that the last node inserted will be sought. This node corresponds to an external node in a binary search tree of n nodes. As in the preceding analysis, the expected level of this node is $E/(n + 1) = \dfrac{I + 2n}{n + 1}$. However, since we are doing a search instead of insertion, one additional comparison will be necessary to identify this node as the one sought.

Thus with probability $1/(n + 1)$ the search will have average cost $1 + \dfrac{I + 2n}{n + 1}$, and with probability $n/(n + 1)$ the search will be among the first n nodes inserted and will thus have average cost $A(n)$. So our expected search cost $A(n + 1)$ is

$$A(n)n/(n + 1) + [1 + (I + 2n)/(n + 1)]/(n + 1) =$$

$$nA(n)/(n + 1) + [1 + (nA(n) - n + 2n)/(n + 1)]/(n + 1) =$$

$$[nA(n) + (n + 1)/(n + 1) + nA(n)/(n + 1) + n/(n + 1)]/(n + 1) =$$

$$[nA(n)(n + 2)/(n + 1) + (2n + 1)/(n + 1)]/(n + 1) =$$

$$A(n)(n^2 + 2n)/(n^2 + 2n + 1) + (2n + 1)/(n + 1)^2 \leq$$

$$A(n) + 2/(n + 1) \text{ and}$$

$$A(n + 1) - A(n) \leq 2/(n + 1)$$

Since $A(1) = 2/(1 + 1)$, $A(n + 1) \leq \sum_{k=1}^{n+1} 2/k$. By the integral test from calculus this is $O(\log n)$, and we are done. ■

2.6 Balanced Search Trees

Complete trees are not as useful in practice as they may appear, since generally we do not have control over how the tree grows. For example, in a binary search tree the place where a new element is to be inserted is already determined. However, we do have sufficient control in one application of binary trees. This application of complete binary trees to the representation of priority queues is presented in Section 2.8.

For binary search trees, the major issue is that the height need not be logarithmic. Informally, we may define a *balanced* binary tree to be one that has nearly the minimum possible height. This informal definition may be made precise in different ways by defining "size" differently, but the formal definitions all tend to have the following recursive structure:

Def 2.51 A binary tree is *balanced* iff

1. it is empty or
2. its left subtree and its right subtree are nearly the same size in some sense and both subtrees are balanced.

Balanced trees represent a compromise between complete binary trees and arbitrary binary trees. The insertion and deletion operations become more complicated and may require extra overhead, but the search and insertion algorithms avoid the $\Theta(n)$ worst-case behavior of general search trees.

Many insertions and deletions will leave balanced trees still balanced. In those cases where these trees become unbalanced, a small amount of work (e.g., $O(\log n)$) will suffice to rebalance the tree. Thus the height may be kept logarithmic with relatively little effort. Even this effort, or overhead, is not painful for operations on externally stored files, where moving to a new level in the tree typically requires a separate external access. Minimizing these accesses, which are much slower than accesses to internal memory, by minimizing the tree height is much more important than avoiding a small number of internal operations.

One simple example of a balanced binary tree, illustrated in Figure 2.12, is now defined.

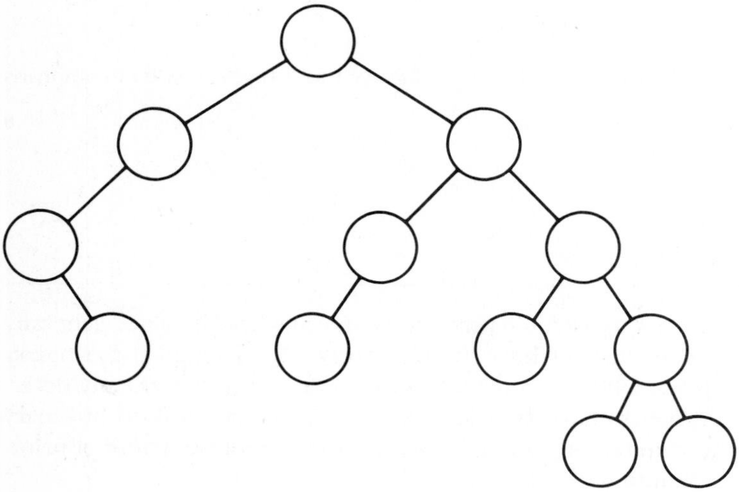

Figure 2.12 An AVL Tree

Def 2.52 An *AVL tree* is a binary search tree that

1. is empty or
2. has left and right subtrees, whose heights differ by at most 1, and are both AVL trees.

The name AVL tree comes from their appearance in a paper by Adelson-Velsky and Landis (1962).

We show in Section 2.17 that AVL trees have logarithmic height. The insertion and deletion algorithms, together with their rebalancing components, are described in Section 4.7.

One way of reducing the height of a search tree with n keys is to relax the restriction requiring only one key per node. If m keys are present in a node, they divide the interval of possible keys into $m + 1$ subintervals. Each of these intervals may then correspond to a subtree.

We are led to the following definition:

Def 2.53 An *m-way search tree*

1. is empty or
2. consists of a root containing keys $(k_i)_{i=1}^{j}$ for some $1 \leq j < m$ and a sequence of subtrees $(T_i)_{i=0}^{j}$ such that
   a. if k is a key in T_0 then $k \leq k_1$
   b. if k is a key in T_i and $0 < i < j$, then $k_i \leq k \leq k_{i+1}$
   c. if k is a key in T_j then $k > k_j$ and
   d. all T_i are nonempty m-way search trees or all T_i are empty.

Notice the parameter m represents the maximum number of children possible for any node of an m-way search tree. Also, each node that is not a leaf has one more child than key.

A 4-way search tree is shown in Figure 2.13.

We may represent a node with j keys and $j + 1$ children with the notation $[(k_i)_{i=1}^{j}, (T_i)_{i=0}^{j}]$. If we include the dummy keys $k_0 = -\infty$ and $k_{j+1} = \infty$ in each node, then the preceding conditions (a) through (c) collapse for all i to

$$\text{if } k \text{ is a key in } T_i, \text{ then } k_i \leq k \leq k_{i+1}$$

It is not difficult to show that for $m > 2$, most keys in an m-way search tree are in leaves. The proof is left as an exercise.

There are balanced m-way search trees called *B-trees,* first described by Bayer and McCreight (1972). More precisely, we have the following definition:

Def 2.54 A *B-tree of order m* is an m-way search tree such that

1. all leaves are on the same level
2. all nodes except for the root and the leaves have at least $m/2$ children and at most m children. The root has at least 2 children and at most m children.

Figure 2.14a shows a B-tree of order 3. B-trees of order 3 are sometimes called 2-3 trees.

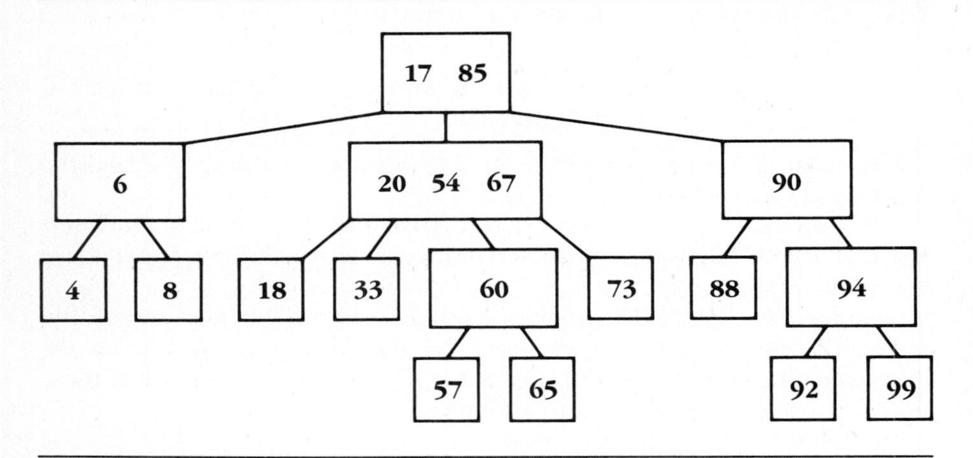

Figure 2.13 A 4-Way Search Tree

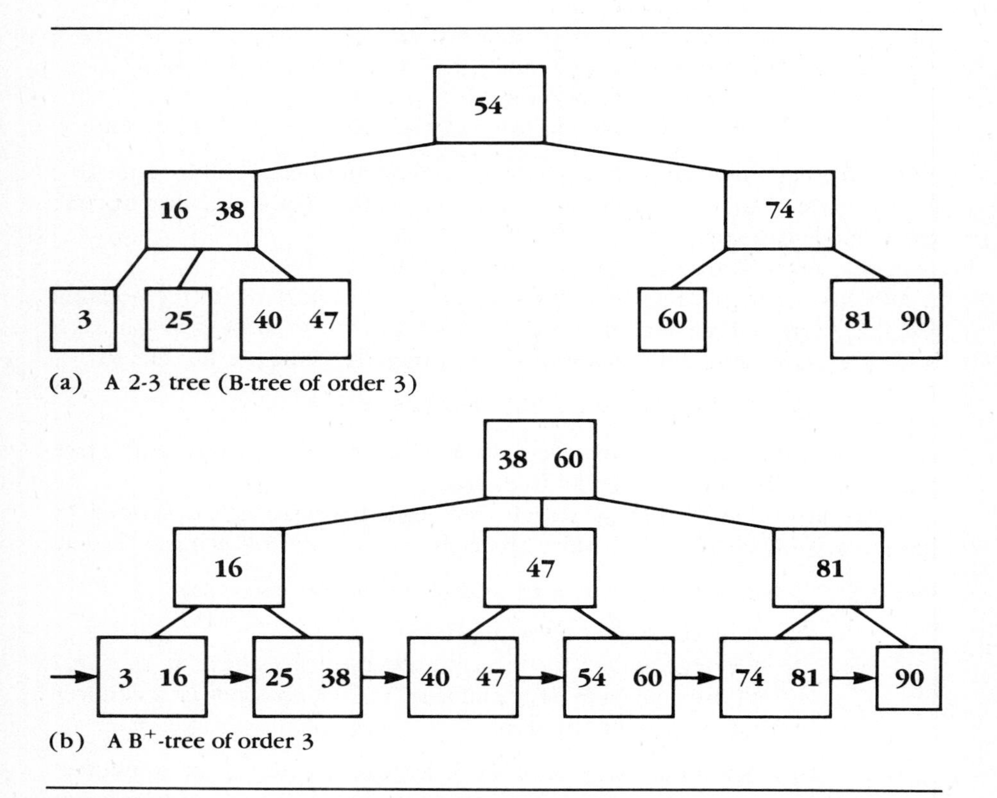

Figure 2.14 A B-Tree and a B^+-Tree with the Same Keys

It is sometimes useful to take the view that all keys in internal nodes are dummies, as in Figure 2.14b. Notice here the keys must be duplicated in the leaves. If the leaves are linked together in increasing order of the keys, then once the first key is found, traversal may be accomplished without visiting the lower-numbered levels at all. These modified B-trees are called *B^+-trees.*

Both B-trees and B^+-trees are often used to index files. In this case a pointer to the external location of a file record would be stored together with that record's key.

The search algorithm for an m-way search tree generalizes in a natural way the search algorithm for a binary search tree. We have written it as a Boolean function; it is a routine exercise to modify it to produce, say, a pointer to the desired key.

Def 2.55 *Search*(k, T) is

1. FALSE, if T is empty
2. TRUE, if k appears as a key in the root of T

2′. Search(k, T_i), if k_i is the key in the root of T for which $k_i \leq k < k_{i+1}$.

Note we are assuming that duplicate keys, if any, are inserted to the right. If keys in nodes are treated as dummies, then condition 2 holds only if the root of T is a leaf.

As an example, imagine searching for the key 25 in the B-tree of Figure 2.14. The first comparison is with 54 in the root. Since $25 < 54$, we must descend left to the node containing 16 and 38. Since $16 < 25 < 38$, we descend to the node between 16 and 38. This contains the key 25.

The time complexity of the search algorithm implied by the preceding definition would, as for ordinary binary search, be proportional to the height of the tree. The worst-case height of a B-tree or a B^+-tree may be determined as for a binary search tree: by first finding the minimum number of nodes in a tree of height h.

A tree of height h will have 1 node on level 0 and may have as few as 2 nodes on level 1. After that, the number of nodes goes up by a factor of at least $(m/2)$ per level, since each node has at least $m/2$ children. So the minimum number of nodes outside the root is

$$2 + \sum_{j=1}^{h-1} 2(m/2)^j =$$

$$2 \times \sum_{j=0}^{h-1} (m/2)^j =$$

$$2[(m/2)^h - 1]/[(m/2) - 1]$$

Each node may have as few as $[(m/2) - 1]$ children, so the number n of keys, including keys in the root, satisfies

$$n \geq 1 + 2[(m/2)^h - 1] = 2(m/2)^h - 1$$

Thus $(m/2)^h \leq (n + 1)/2$, and

$$h \leq \log_{m/2} (n + 1)/2$$

Consequently the worst-case height is about $\log_{m/2} n$. This is an improvement over balanced search trees of a factor of

$$(\log_2 n)/(\log_{m/2} n) = \log_2 m/2$$

However, we have ignored the cost of searching for the appropriate subtree T_i. This cost will be approximately $\log_2 m/2$ for each level of the tree, which essentially cancels the height advantage.

Nevertheless all is not lost. Even for externally stored files, searching for the correct T_i will be an internal search. Reducing the height of the tree, moreover, will reduce the number of external accesses. The extra internal comparisons and assignments are a small price to pay for a reduction in the dramatically slower external file operations.

The best-case height of a B-tree may be shown to be roughly $\log_m n$ by an argument similar to the preceding argument.

Note that B-tree heights may be quite small. For example, if $n = 4$ and $m = 100$, there will be about $2 \times 50^{4-1} = 2 \times 50^3 = 250{,}000$ keys in even the worst-case B-tree. To reduce the height below this, n must decrease quite substantially or m must increase quite substantially.

We postpone the question of insertion into B-trees until Section 4.8. One topic to be resolved there is what condition(s) will allow more than one key to be put in a node without allowing all keys to end up in the same node.

2.7 Complete Trees and Level-Order Traversal

The unique complete tree of n nodes for a positive integer n can be represented by a diagram like Figure 2.15. If there are n nodes in the tree, then the nodes labeled 1 to n in the figure will be those present in the tree.

The diagram has other interesting properties. First, it suggests a way of assigning sequential storage to a complete tree. A complete tree of, say, 18 nodes can be stored in 18 consecutive locations of a 1-dimensional array. No space for pointers is required. This representation may be called a *sequential storage representation,* or *array representation,* as well as a *consecutive storage representation.*

In the absence of pointers, there must be a way of finding both children of a node. The figure suggests a simple way of finding the children given the index of a node: The left child of node k has index $2k$ and the right child of node k has index $2k + 1$. In both cases the child exists iff its index is at most n. In addition, we get for free an algorithm for finding the parent of a node. The parent of node k has index $\lfloor k/2 \rfloor$. This index will correspond to a node unless

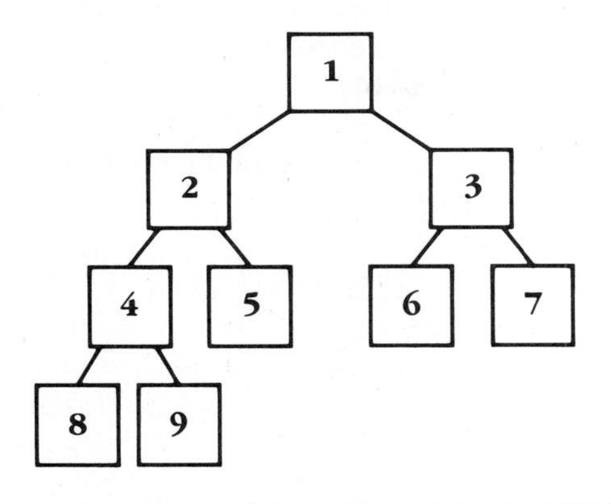

Figure 2.15 Array Indices in the Consecutive Storage Representation of a Complete Binary Tree

$k = 1$, that is, unless k is the root. This behavior is exactly what we would expect for a function that finds indices of parents.

Notice the size of the tree must be stored and maintained so that the calculated child index may be compared with it.

The figure also suggests a new traversal method for binary trees and, in fact, for ordered trees. It is clear from the figure that all nodes on a level have lower indices than any node on any lower level. The nodes on a level are traversed from left to right. The idea of "left to right" ordering is an intuitive notion that still needs to be formalized, but otherwise we have a good specification for the sequence that should result from the new method of traversal.

If we identify a level with a sequence of nodes, then calling the following algorithm with parameter (T) will serve to traverse T in level order.

Def 2.56 If S is a sequence of nodes, then *LevelOrder*(S) is

1. () if $S = ()$
2. Concat(S, LevelOrder(ConcatChildren(S))) otherwise

where

ConcatChildren(S) =

1. () if $S = ()$
2. Concat(Children(H), ConcatChildren(T)), if $S = (H\ .\ T)$

and

Children(H) for a node H is just a sequence of the nonempty children of H, in order.

ConcatChildren concatenates the lists of children for all nodes in the given level. It corresponds to the ConcatList function discussed in Section 2.3. Clearly this definition can be applied to noncomplete binary trees as well as to general ordered trees.

For example, if we identify trees, the nodes at their roots, and the indices of the nodes, we have for the tree of Figure 2.15

ConcatChildren(⟨2 3⟩) =

Concatenate(⟨4 5⟩, ConcatChildren(⟨3⟩)) =

Concatenate(⟨4 5⟩, Concatenate(⟨6 7⟩, ConcatChildren(⟨⟩))) =

Concatenate(⟨4 5⟩, Concatenate(⟨6 7⟩, ⟨⟩)) =

Concatenate(⟨4 5⟩, ⟨6 7⟩) =

⟨4 5 6 7⟩

and

LevelOrder(⟨1⟩) =

⟨1 . LevelOrder(⟨2 3⟩)⟩ =

⟨1 . 2 . 3 . LevelOrder(⟨4 5 6 7⟩)⟩ =

⟨1 . 2 . 3 . 4 . 5 . 6 . 7 . LevelOrder(⟨8 . 9⟩)⟩ =

⟨1 . 2 . 3 . 4 . 5 . 6 . 7 . 8 . 9 . LevelOrder(⟨⟩)⟩ =

⟨1 . 2 . 3 . 4 . 5 . 6 . 7 . 8 . 9⟩

Traditionally level-order traversal is implemented by means of a queue. The queue is initialized to contain the root, and then a node is deleted from the queue and its children added until the queue is empty. The design of such an algorithm is left as Exercise 111, which is solved in Appendix C.

To prove the desired relationship of level-ordering numbering and level-order traversal, we first define a notion of "to the left of," at least for nodes on the same level.

Def 2.57 In a binary tree, node M is *to the left of* node N iff

1. there is a node X with left child M and right child N or
2. the parent of M is to the left of the parent of N.

Theorem In a complete binary tree, if node X with level-order index k has a left child, then the index of the left child is $2k$. If X has a right child, then the index of the right child is $2k + 1$.

Proof Suppose that X is on level r and X has a left child. Then by the properties of complete trees, level r and all lower-numbered levels have the maximum number of nodes. In particular, there are $2^r - 1$ nodes on lower-numbered levels. Thus by the definition of k, there are $k - (2^r - 1) - 1$ nodes on level r visited before X. So $2^r - [k - (2^r - 1) - 1] - 1 = 2^{r+1} - k - 1$ nodes remain on level r, to be visited after X.

Suppose we can show that all nodes on level r visited before X have exactly two children. Then the nodes visited before the left child of X are

k	up to node X
$2^{r+1} - k - 1$ and	from level r after node X
$2(k - (2^r - 1) - 1) = 2k - 2^{r+1}$	from level $r + 1$

This gives a total of $2k - 1$ nodes. Thus the left child of X must have index $2k$, and the right child (if any) must have the next index $2k + 1$. So it is enough to show the following claim: If Y is to the left of X in a complete binary tree and X has a descendant on level $h + 1$, then Y has the maximum number of descendants on level $h + 1$.

Proof of claim: by induction on the level s of X. The induction hypothesis is the statement of the theorem.

Basis: X cannot be on level 0, since Y is to the left of X.

Induction step: Suppose the theorem is true for nodes at lower-numbered levels than s. If X and Y are children of the same node, then by the definition of complete, either 1) the height of the subtree rooted at X is less than the subtree rooted at Y, and Y is complete, in which case the conclusion follows; or 2) the subtrees rooted at X and Y have the same height but the subtree rooted at Y is completely full. In this case all levels of this latter subtree are full, and the conclusion follows.

So we may assume that X and Y are not children of the same node. Then by definition of "to the left of," the parent Q of Y is to the left of the parent P of X. P has a descendant on level $h + 1$, since X does. By induction, Q has the maximum number of descendants; so Y does also.

This completes the proof of the claim and hence of the theorem. ■

The consecutive storage representation for binary trees may be used for noncomplete trees, but it is no longer space efficient. A glance at Figure 2.16 suggests that the worst case happens when each node in the tree is the right child of its parent. In this case we easily verify there is only one node on each level and the index of the node on level s is $2^{s+1} - 1$. If there are n nodes in such a tree, then there are $n - 1$ levels, so the index of the node on the lowest level is $2^n - 1$. Thus the space requirement may grow exponentially in the worst case.

Level order is closely related to a search technique called *breadth-first search,* to be discussed in Section 4.11.

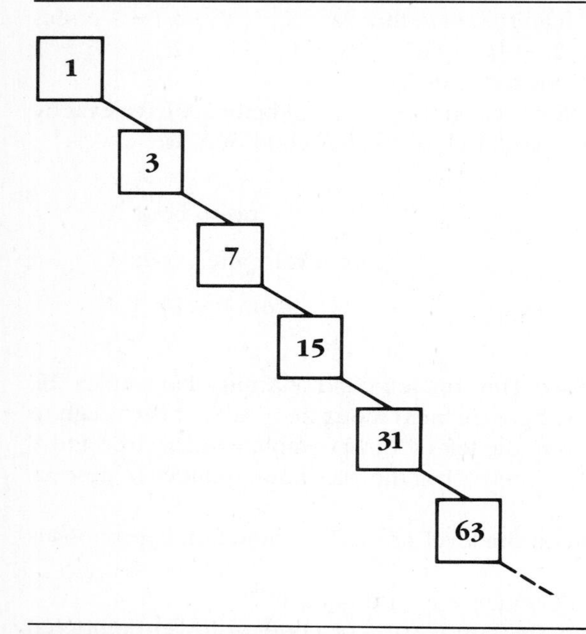

Figure 2.16 Worst-Case Space Requirements for the Consecutive Storage Representation

2.8 Heaps

In some applications of binary search trees, we have enough flexibility to put nodes wherever we want. One example occurs in the representation of priority queues. We begin with a definition similar to that of a binary search tree.

Def 2.58 A binary tree has the *heap property,* or is a *heap,* iff

1. it is empty or
2. the key in the root is larger than that in either child and both subtrees have the heap property.

If a node in part 2 of the definition does not have one or both subtrees, then part 2 holds automatically for that child. For example, in the heap of Figure 2.17 the subtrees rooted at 24, 29, and 76 still have the heap property. If desired, the word "larger" in the definition can be replaced by the word "smaller"; properties analogous to those that follow will hold.

It is not difficult to show that the largest key of a heap is contained in the root.

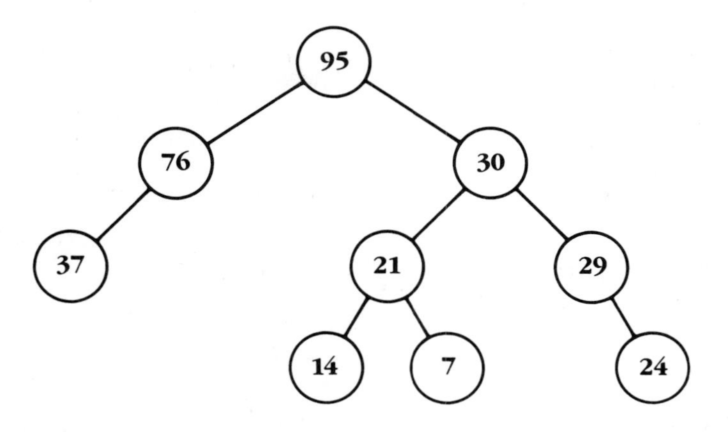

Figure 2.17 A Binary Tree with the Heap Property

Theorem The largest key of a binary tree with the heap property is stored in the root.

Proof by induction on the height h of the tree.

Basis: If the height of the tree is 0, then there is only one node, which is necessarily the largest.

Induction step: For larger heights, we know by induction that the largest key in each subtree is stored in the root of that subtree. By the definition of a heap, the key in the root is larger than that in each of its children (i.e., the roots of each of its subtrees), so it must be the largest key overall. ■

In general when the key in the root is deleted, the root itself cannot be deleted. Suppose any convenient leaf is deleted and its key transferred to the root. Clearly, the heap property may now be violated, as in Figure 2.18b, but we can construct an algorithm that restores the heap property, as in Figure 2.18c.

The only possible violation of the heap property before restoration is that the key K in the root may be too small. If so, and if we exchange K with the larger child key, then the new root key will be larger than both of its child keys. The only possible violation of the heap property will again be that K, now moved down one level, is too small. This may be handled by a recursive call. Eventually the heap property will obtain in the tree rooted at K, perhaps only because K has no children.

Since in the deletion algorithm it doesn't matter where the new node is deleted, the sequential storage representation may be used. If the original heap has size n in this representation, then the deletion algorithm will delete the node at position n. The example of Figure 2.18 behaves this way.

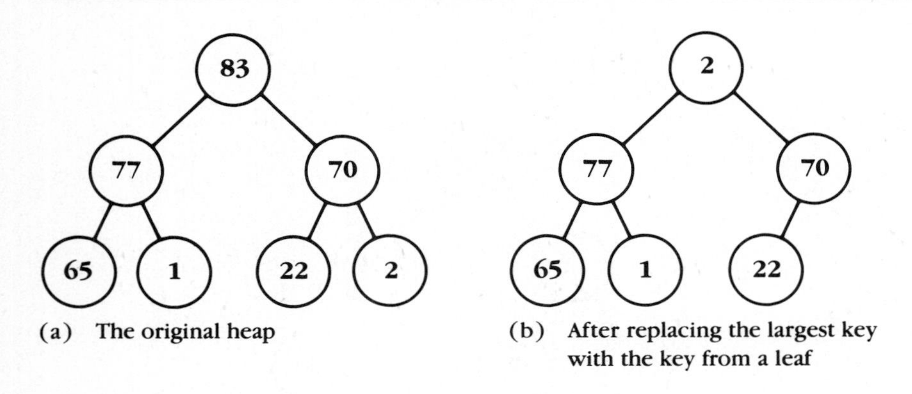

(a) The original heap

(b) After replacing the largest key with the key from a leaf

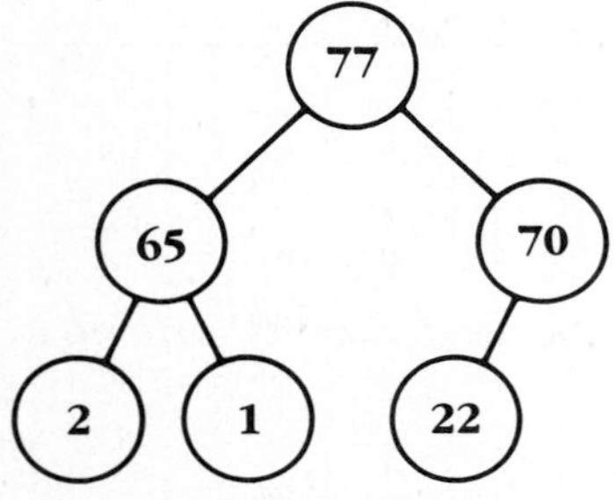

(c) After moving down the key moved to the root

Figure 2.18 Deletion from a Heap

In the consecutive storage representation it is easy to define functions to return the left child, right child, parent, and to determine whether a subtree is empty. Each function takes as argument an array index. To check emptiness we need to assume that the size of the heap is available, so the heap needs to be passed as well. We will freely use the functions below in the algorithms of this section.

left(k) is $2k$

right(k) is $2k + 1$

parent(k) is $\lfloor k/2 \rfloor$

empty(H, k) is ($k > 0$ and $k \leq$ size(H))

Notice in the consecutive storage representation a node is a leaf iff it has no left child.

The Delete and MoveDown algorithms use the preceding approach.

```
 1. Delete(H)                      {returns and removes the largest key
 2.                                 from the heap H, restoring the heap
 3.                                 property. Assumes the size of the
 4.                                 heap is available as size(H)}
 5. save := data(H, 1)                            {save largest key}
 6. data(H, 1) := data(H, size(H))         {move key from leaf to root}
 7. size(H) := size(H) - 1                      {update size counter}
 8. MoveDown(H, 1)                            {restore heap property}
 9. return save                              {and return deleted key}
```

Algorithm 2.59(a) Delete from a Heap

```
 1. MoveDown(H, k)               {restores the heap property to the heap
 2.                                    after a key from a leaf replaces
 3.                                              the key in position k
 4.                                  It assumes functions left and right
 5.                                that find children, a function empty,
 6.                                 and a function data(H, k) that gets
 7.                               the data in the root of the subtree at
 8.                                                position k of heap H}
 9. if not empty(H, k) then               {if the subtree at k is a leaf}
10.    largechild := left(k)                          {then do nothing}
11.    if not empty(H, right(k)) then                      {else find}
12.       if data(H, right(k)) > data(H, left(k)) then
13.          largechild := right(k)                     {larger child}
14.       end if
15.    end if
16.    if data(H, k) < data(H, largechild) then        {if this child is}
17.       swap(data(H, k), data(H, largechild))      {larger than root,}
18.       MoveDown(largechild)            {swap and apply MoveDown}
19.    end if                                             {to subtree}
20. end if                                           {if not, we're done}
```

Algorithm 2.59(b) MoveDown

Since the parameter to MoveDown is growing by at least a factor of 2 at every call, and since it cannot be less than n, MoveDown and hence Delete will terminate in time $O(\log n)$. In the worst case, the key from the root will be moved down to a leaf, but the algorithm may terminate earlier if the heap property is restored before reaching a leaf.

The example of Figure 2.18 illustrates this algorithm. We next show correctness.

Theorem The Delete algorithm returns a heap.

Proof We have already seen that it returns and removes the correct key.

In the heap H resulting from the algorithm, let T be a subtree of smallest height without the heap property. In particular, the root of T must have at least one child. Since deletion of a leaf from a tree cannot result in a violation of the heap property, it must be that T fails to be a heap owing to the new key in the root or to the operation of MoveDown. In either case, the heap property cannot fail at the root of T, since the operation of MoveDown guarantees that the children of T's root have smaller keys than does T's root itself. So the heap property for T must fail because one of T's children is not a heap. This violates minimality of T and proves correctness of the algorithm. ■

Insertion into a heap works in the opposite way. In this case, as in the priority queue operation we are trying to implement, the argument is a key k that is to be inserted into the heap so that the heap property remains true. It can be inserted into any available location of the tree as long as the heap property is eventually restored. Note that the only violation of the heap property involves the new key, which needs to be moved up until it is larger than either child.

In the consecutive storage representation, we simply increase the size n of the heap by 1 and include the new key into the new nth position before restoring the heap property.

The counterpart of the MoveDown operation may be called *MoveUp*. It works by comparing the key to be moved up to its parent and changing places with the parent as long as the key is larger than its parent.

The insertion algorithm, with auxiliary procedure MoveUp, are given as Algorithms 2.60a and b.

An example consistent with the consecutive storage representation is shown in Figure 2.19.

---

1. Insert(X, H) {inserts key X into heap H represented in
2. consecutive storage and restores heap property}
3. size(H) := size(H) + 1 {update size counter}
4. data(H, size(H)) := X {insert data in next available spot}
5. MoveUp(H, size(H)) {and restore heap property}

Algorithm 2.60(a) Insert

```
 1. MoveUp(H, k)                        {restores heap property after insertion of
 2.                                     a key in position k of a heap represented
 3.                                            using consecutive storage. Assumes
 4.                                     functions data and empty as in MoveDown}
 5. if not empty(H, parent(k)) then                     {if heap property fails}
 6.    if data(H, k) > data(H, parent(k)) then          {and key larger than}
 7.       swap(data(H, k), data(H, parent(k))              {parent, then swap}
 8.       MoveUp(H, parent(k))              {with parent and keep moving up}
 9.    end if
10. end if
```

Algorithm 2.60(b) MoveUp

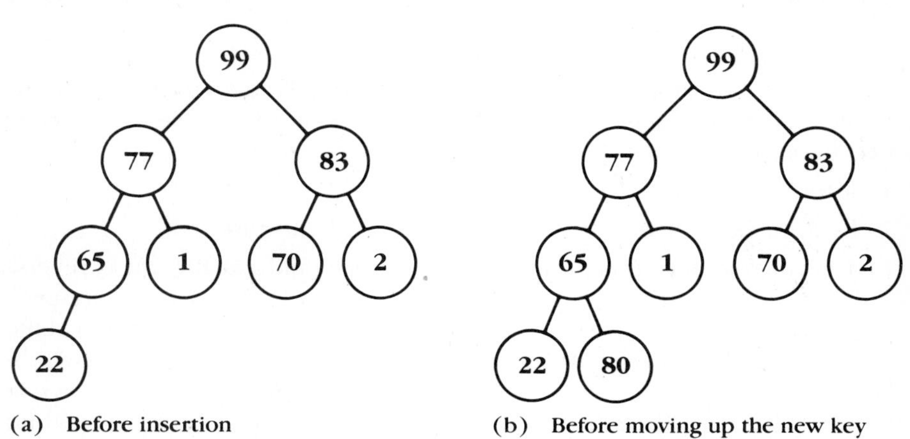

(a) Before insertion

(b) Before moving up the new key

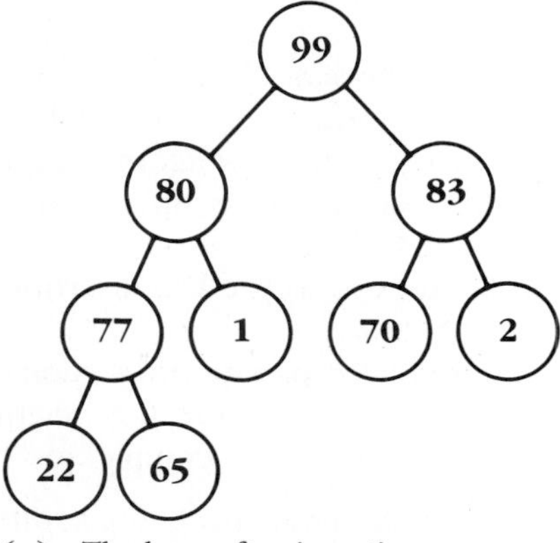

(c) The heap after insertion

Figure 2.19 Insertion into a Heap

The algorithm terminates in time $O(\log n)$, since the value of k decreases by at least a factor of 2 at every call. We need the following theorem:

Theorem The result of the insertion algorithm 2.60 is a heap.

Proof Let T be a counterexample of minimal height. By symmetry, we may assume that the insertion of a key k was made into the left subtree. By minimality, the left and right subtrees of T have the heap property. Thus T must fail to be a heap because the key in its root is less than one of its child keys. The offending child key can't be in the right subtree, since the right subtree is unchanged and the MoveUp algorithm can only increase the key in the root. So the left child key x is larger than the root key r. Since this was not the case before insertion, x must have been moved into the left child position by the insertion algorithm. Since the algorithm did not terminate before reaching the left child, it must have gone on to compare x to r. In this case it would not have left the smaller key r in the root. This contradiction proves the theorem. ■

A useful observation is that a priority queue may be used to sort according to the following *Heapsort* algorithm. We do not express it in the most precise way because it will not be the final version of Heapsort.

1. Insert each of the items to be sorted.
2. Perform as many deletions as the number of items inserted.

According to our assumptions, this algorithm would actually sort in reverse order, since the first item deleted would be the largest. The heap representation avoids this problem in a very simple way. Recall that in the first step of the deletion algorithm the key in the root is saved. If this key is saved in position n of the representing array (i.e., if the deletion algorithm swaps the keys in positions 1 and n), then the largest key, which is the nth smallest, will go into position n. Since this position is no longer part of the heap, it will remain there throughout all subsequent deletions.

We postpone a complete example until a more efficient strategy for the n insertions can be discussed. Figure 2.20b does illustrate a sequence of deletions from a heap in consecutive storage representation. Note that even though the largest element is deleted first, the sequence ends up sorted in increasing order.

If complete trees are used, it is a simple matter to analyze the insertion, deletion, and Heapsort algorithms.

Both insertion and deletion have worst-case time complexity proportional to the height of the tree. In the consecutive storage representation, this height is just $\Theta(\log n)$. So insertion and deletion have $O(\log n)$ time complexity.

The insertion stage of Heapsort requires time proportional to $\sum_{k=1}^{n} \log k$, due to the insertions. Deletion requires time proportional to the same sum. Clearly this sum is at most $\sum_{k=1}^{n} \log n$, so is $O(n \log n)$. Thus our current version of Heapsort requires time $O(n \log n)$.

After heapifying \ Location	1	2	3	4	5	6	7	8
Nothing (original input)	65	22	2	99	1	70	83	77
Subtree rooted at 4	65	22	2	99	1	70	83	77
Subtree rooted at 2	65	99	2	77	1	70	83	22
Subtree rooted at 3	65	99	83	77	1	70	2	22
Subtree rooted at 1	99	77	83	65	1	70	2	22

(a) Heapifying

n \ Location	1	2	3	4	5	6	7	8
8	99	77	83	65	1	70	2	22
7	83	77	70	65	1	22	2	<u>99</u>
6	77	65	70	2	1	22	<u>83</u>	<u>99</u>
5	70	65	22	2	1	<u>77</u>	<u>83</u>	<u>99</u>
4	65	2	22	1	<u>70</u>	<u>77</u>	<u>83</u>	<u>99</u>
3	22	2	1	<u>65</u>	<u>70</u>	<u>77</u>	<u>83</u>	<u>99</u>
2	2	1	<u>22</u>	<u>65</u>	<u>70</u>	<u>77</u>	<u>83</u>	<u>99</u>
1	<u>1</u>	<u>2</u>	<u>22</u>	<u>65</u>	<u>70</u>	<u>77</u>	<u>83</u>	<u>99</u>

(b) Successive deletions from a heap

Figure 2.20 Array Representation of Stages of Heapsort

We cannot improve on this asymptotic result, but we can be more efficient during the insertion stage. Note that if the input is given in an array, the array may be interpreted as a tree in the consecutive storage representation. This tree need not have the heap property. The preceding sequence of n insertions will give this tree the heap property by moving each new key up from the bottom of the heap.

However, it is more efficient to create a heap by moving each key down using MoveDown rather than MoveUp because the average key in a complete binary tree is closer to the bottom than to the top. In a binary tree of height h, there are at most 2^k nodes on level k, so that the sum of the distances to the lowest level is at most $\sum_{k=0}^{h} 2^k(h-k)$.

We may rewrite this sum as $\sum_{k=0}^{h} 2^k h - \sum_{k=0}^{h} k2^k$. The first sum is just $h\sum_{k=0}^{h} 2^k = h(2^{h+1}-1)$ (see Example 4 of Section 2.1). For the second sum, we make the following claim:

Theorem $\sum_{k=0}^{h} k2^k = 2 + (h-1)2^{h+1}$.[6]

[6] It will also be relatively simple to derive this formula with the techniques of Section 2.17. We do so in that section.

```
1. Heapify(H, k)          {gives heap property to the tree H rooted at k
2.                          in the consecutive storage representation
3.                          Assumes functions left, right, and empty}
4. if not empty(H) then
5.    Heapify(left(H))
6.    Heapify(right(H))
7.    MoveDown(H)
8. end if
```

Algorithm 2.61 Heapify

Proof by induction on h.

Basis: $\sum_{k=0}^{0} k \times 2^k = 0 = 2 + (0 - 1)2^{0+1}$.

Induction step: By induction, the sum from 0 to $h - 1$ is $2 + (h - 2)2^h$. Then the sum from 0 to h is

$$2 + (h - 2)2^h + h2^h =$$

$$2 + (2h - 2)2^h =$$

$$2 + (h - 1)2^{h+1}$$

■

The difference between the two sums is therefore $2^{h+1} - h - 2$, which is $\Theta(2^h)$ or $\Theta(n)$. Thus the algorithm that successively moves keys down to create a heap has time complexity $O(n)$. This algorithm, given as Algorithm 2.61, is called ***heapify***. In the algorithm, the original call would be with second parameter 1.

An iterative Heapify algorithm would just apply MoveDown to the heaps rooted at locations $\lfloor n/2 \rfloor$ through 1.

An example using the consecutive storage representation is shown in Figure 2.20.

Finally, the Heapsort algorithm is given as Algorithm 2.62.

```
1. Heapsort(H)            {sorts the sequence of elements between 1 and
2.                                              size(H) in the array H}
3. Heapify(H, 1)
4. for k from size(H) to 1 by −1
5.    H[k] := Delete(H)
6. end for
```

Algorithm 2.62 Heapsort

2.9 Arrays and Sequences

There are a number of useful divide-and-conquer algorithms not so closely associated with recursively defined data structures as those earlier in this chapter. Many of these algorithms are associated with arrays. The assumption of this text is that you have experience with arrays but are not necessarily familiar with their formal properties. From our point of view, the fundamental array operations are *Put* and *Get,* where Get(*index, array*) returns the value associated with the given index in the given array and Put(x, *index, array*) associates the value x with the given index in the given array.

In our algorithmic notation, assignment of Get(*index, array*) to a variable x is written $x := array[index]$ and Put(x, *index, array*) is written $array[index] := x$.

Arrays may be represented in terms of linked lists and the two preceding operations defined recursively. However in this representation, these operations do not have O(1) time complexity, as do the corresponding primitive operations available in most language implementations. In this text we assume the Put and Get operations have time complexity O(1).

The easiest way to represent sequences as arrays is to store the kth element of the sequence in location k of the array. Often we will assume this representation. However, we will have occasion to consider a sequence and one or more subsequences at the same time, so generally we will need to specify the respective indices *low* and *upp* of the first and last element of the sequence. Then the kth element of the sequence will be naturally stored at location $low + k - 1$.

The index x may be an n-tuple of integers instead of a single integer. The number n is called the *dimension* of the array. An array of 1 dimension is sometimes called a *vector.* An array of 2 dimensions is sometimes called a *matrix.* Methods of implementing arrays of dimension greater than 1 are discussed in the exercises.

In diagrams of arrays, lower-numbered array locations are shown at the left. Sometimes we use the words “left” and “right” when discussing arrays with this convention in mind.

In array-based algorithms we assume that the size n of a sequence represented by an array A can be obtained as the value of size(A). Also, these algorithms are not generally written as functions, but typically operate instead by changing values of array locations. In many cases these algorithms are quite distinct from their specifications.

2.10 Binary Search

If a sequence is known to be sorted, then there is a straightforward divide-and-conquer method for searching for a given key. Suppose the first comparison is between the given key and mth element. The index m divides the sequence into two pieces: those elements with smaller index, which must be less than or equal to the mth element, and those with larger index, which must be greater than or equal to the mth element. If the given key is not equal to the mth element, then only one of these pieces needs to be searched. If the given key is less than the mth element, only the subsequence of elements from the 1st to the $(m - 1)$st need be searched. If the given key is greater than the mth element, only the subsequence of elements from the $(m + 1)$st to the nth need be searched. If the desired key is the mth key, then m may be returned.

This approach is inefficient in a linked representation, since finding the mth element does not have $O(1)$ time complexity. In the array representation, it is most efficient to choose as the index m a value halfway between the lowest and highest indices. In recursive calls to the function defined by this ***binary search*** algorithm, these indices will not necessarily be 1 or n.

The binary search Algorithm 2.63 is given as a procedure rather than as a specification. It uses the local variable *mid* for m above. In the usual array representation of a sequence, the original call to it would be BinarySearch (*key, In,* 1, n).

The algorithm will terminate because the value of $(high - low)$ decreases by at least 1 at each recursive call, and the algorithm terminates when this difference becomes negative.

For example, consider an array *In* whose entries from the 1st to the 8th form the sequence (2 4 7 12 18 25 28 31). When searching for 18, initially $low = 1$ and $high = 8$, so *mid* is 4.

$18 > In[4] = 12$, so *low* becomes 5, *high* 8, and *mid* 6.

$18 < In[6] = 25$, so *low* becomes 5, *high* 5, and *mid* 5.

$18 = In[5]$, so the algorithm returns 5.

```
 1. BinarySearch(key, In, low, high)        {returns the index in the
 2.                                    array In at which the element key
 3.                                    is found, if there is such an index
 4.                                                between low and high}
 5. if low > high then return unsuccessfully
 6. {else}
 7. mid := ⌊(low + high)/2⌋                           {find the midpoint}
 8. if key = In[mid] then return mid
 9. {else} if key < In[mid] then return BinarySearch(key, In, low,
    mid − 1)
10. {else if key > In[mid]} return BinarySearch(key, In, mid + 1, high)
```

Algorithm 2.63 Recursive Binary Search

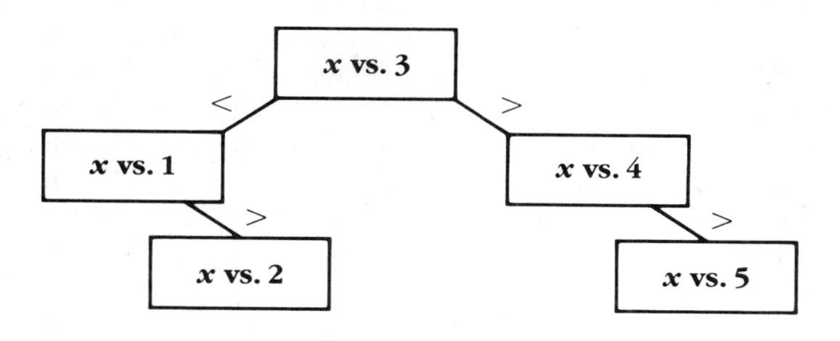

Figure 2.21 A Decision Tree for the Binary Search Algorithm (When $n = 5$)

If the search had been for 19, the comparisons with 12 and 25 would have yielded the same results. Then with $low = high = mid = 5$, $19 > In[5]$, so *low* would be set to 6 and *high* would remain 5. At the next recursive call, the algorithm would terminate unsuccessfully.

The relation of binary search to the binary search trees of Section 2.5 is suggested in Figure 2.21. The tree of this figure shows the sequence of comparisons performed for various positions of the sought element. If the result of a comparison is "less than," then the comparison in the left subchild is made. If "greater than," then the comparison in the right subchild is made. If "equal to," then no further comparison is necessary.

For example, if the key in position 2 is sought, comparisons will first be made with the elements in positions 3 and 1. Finding the key in position 4 requires only one previous comparison, which involves position 3.

The tree of the figure is called a *decision tree.*

2.11 Quicksort

An algorithm related to Mergesort, which has been shown empirically to be the fastest-known algorithm in many cases, is called *Quicksort.* Quicksort is a divide-and-conquer algorithm in which the work is done in the "divide" phase, and the "conquer" phase is trivial. In the terminology of Merritt (1985), Quicksort is "hardsplit/easyjoin."

For the "conquer" phase of Quicksort to be trivial, the input sequence has to be preprocessed in the "divide" phase so that no element in the first piece is larger than any element of the second piece. Then the recursive sorts may be done independently; when the recursive sorts are completed, so is the main sort. This preprocessing is difficult to do efficiently if the two pieces are to have the same size, so generally the pieces have different sizes. Thus the *partition* algorithm, which does the preprocessing, must determine the sizes

```
1. Quicksort(A, low, high)        {assumes a sequence stored between
2.                                 locations low and high in an array A.
3.                    A partition algorithm is assumed; see Section 3.5}
4. if high − low > 0 then
5.    mid := Partition(A, low, high)
6.    Quicksort(A, low, mid − 1)
7.    Quicksort(A, mid + 1, high)
8. end if
```

Algorithm 2.64 Quicksort

of the two pieces as well as move small elements to the left and large elements to the right.

Since there is nothing inherently recursive about the partition problem, we defer discussion of partition algorithms to Section 3.5.[7] To show that the Quicksort algorithm is correct, we need only a formal specification of the partition algorithm.

Assuming that the input to the partition algorithm is an array A and indices *low* and *high*, after partitioning the following should be true:

There is an index *mid* returned by partition such that

1. for $low \leq j < mid$, $A[j] \leq A[mid]$ and
2. for $mid < j \leq high$, $A[j] \geq A[mid]$.

Then if the subsequence to left of position j and that to the right of position *mid* are sorted, the entire sequence will be sorted.

With these assumptions about partition, we now have Algorithm 2.64 for Quicksort.

In the best case, the two pieces of the sequence are of equal size and the analysis of Mergesort (see Section 2.3 or 2.17) also applies to Quicksort, giving a best-case time of $\Theta(n \log n)$. In the worst case, the $\Theta(k)$ partition algorithm reduces from a problem of size k to a problem of size $k - 1$. Thus the overall time complexity in this case is $\sum_{k=1}^{n} k$, which is $\Theta(n^2)$.

[7] There is no reason why interested students or instructors can't turn to the presentation of our partition algorithm immediately after this section. To understand the correctness proof of the algorithm, however, it might be useful to have gone over the general discussion of invariants and correctness proofs for iterative algorithms earlier in Chapter 3.

2.12 Bucket Sort

There is a natural divide-and-conquer algorithm for sorting a sequence in lexicographical order. In this section we assume that each element in the sequence is a character string, but the discussion of this section generalizes to any element type that can be compared lexicographically. We write $a_i(j)$ for the jth character of the ith string.

The "divide" phase assigns each string to one of several *buckets* based on their first character. These buckets are just sequences, so they may be implemented as linked lists. For example, if $a_i(1) = k$, a_i would be assigned to bucket k. The buckets may then be sorted recursively, with recursion terminating, say, when there is only one item to sort. Finally, the sorted buckets are concatenated in character order. We can easily see this approach will correctly sort the sequence.

Unfortunately, there is a serious flaw in this algorithm. An entire set of buckets—or set of header nodes if the buckets are implemented as linked lists—must be allocated for each recursive call. In general the same set of buckets cannot be used for two different calls, so the space requirements for pure recursive bucket sort turn out to be impractical.

There is, however, a bottom-up version of recursive bucket sort that avoids these flaws. We discuss it in Section 3.8 under the name of *radix sort.*

2.13 Hashing

The divide-and-conquer approach using buckets is also the basis of an efficient search strategy. The classical example deals with the set of identifiers used in a program, since repeated reference will need to be made to their values (at run time in the interpreted case) or to their associated memory locations (at compile time in the compiled case). The closest analogy with bucket sort would be to allocate a bucket for each possible initial character and to insert each string into the bucket associated with its initial character. Unfortunately, in practice programmers often used many variables with the same initial characters, so that searches in large buckets were both frequent and slow.

The solution has been to define a function from keys to bucket indices that destroys, one hopes, any nonrandomness in the set of keys. Such a function is called a *hash function,* presumably because it is supposed to make hash of the set of keys. A table of buckets for use with a hash function is called a *hash table.* A hash table used to store symbolic identifiers and their locations or values is called a *symbol table.*

Assigning keys to buckets by using a hash function reduces the value of buckets for sorting, since the set of keys may no longer be sorted merely by sorting each bucket and concatenating. Thus hash tables tend to be used when searching is more common than sorting, as in symbol tables. If sorting is necessary—as is often the case when symbol tables are printed for the benefit of programmers—a general purpose sorting algorithm can be used.

In what follows we will assume that b buckets are indexed by integers from 0 through $b - 1$. We will also assume that keys are integers. Character strings represented in an n-bit character code can easily be converted to integers by interpreting the sequence of character codes as an integer in radix 2^n. For example, the string "THE" in an 8-bit ASCII code (see Appendix B) corresponds to the sequence (84, 72, 69) and thus to the integer $84 \times 256^2 + 72 \times 256 + 69$.

Since for a particular value of b the hash function must have integer values between 0 and $b - 1$, and since it is pointless to have in this range values that the hash function does not attain, the choice of b may be considered part of the definition of the hash function. In other words, the choice of a hash function and the choice of the number of buckets are not independent. In fact, certain hash functions work well for some values of b and poorly or not at all for others.

At least from a mathematician's point of view, the most natural way of assigning to some integer K a number between 0 and $b - 1$ is to reduce the number mod b. Ordinarily this would require a time-consuming division, whose time cost might be hidden from a high-level language programmer. However, one case where reduction mod b is simple, even in assembly language, is the case where b is a power of 2. Reduction of a binary integer mod 2^k yields the last k bits of the number in exactly the same way that the last k digits of a decimal number will be the remainder when it is divided by 10^k.

Although this ease of computation may seem to be an argument for having a table size b equal to a power of 2, in fact such a choice for b is not advantageous. For example, if $b = 2^k$ and a k-bit character code is in use, reduction mod b corresponds to extraction of the last k bits, which is the same as the extraction of the last character. Using the last character as the hash function value can have the same disadvantages as the use of the first character in a string.

If a k-bit code is in use and $b = 2^j$ where $j < k$, then all keys with the same last character still go to the same bucket, along with all keys that end with certain other characters. If $j > k$, then keys with a given last character may not go to the same bucket, but they can go to only $1/2^k$ of the possible buckets.

So this *division* method works best if b is not a power of 2. Later we shall see that in some cases taking b to be prime is advantageous. In fact a value of b near a power of 2 may be undesirable.

For an example of the division method, assume that $b = 197$ and that our character string is "THE" with the preceding encoding. If the hash function is called h, then $\mathrm{h}(K) = 5505024 + 18432 + 69 \bmod 197 = 5523525 \bmod 197 = 39$. Naturally this calculation can run more slowly than others based on simple bit operations.

In *folding,* a result between 0 and $2^k - 1$ is obtained by dividing the key into segments and operating on those segments to produce a k-bit number. The name *folding* apparently comes from thinking of storing the key on a strip and of folding the strip until certain of the bits are aligned as in Figure 2.22a. Note that folding in this sense requires certain substrings of bits to be reversed; this reversal is not done in practice because of the time required and because it does not improve the behavior of the function.

An example of folding is illustrated in the example of Figure 2.22b. Here we divide a 32-bit key into substrings of length 5 starting from the right. Of course 2 bits will be left over. If these seven 5-bit numbers are added, the result will not generally be a 5-bit number, but any carry beyond the rightmost 5 bits may be ignored. Thus if the hash function is called h, our example shows that h(10111100100100101000101010101100) = 10001.

In folding, the bit substrings generally do not have different lengths because this can introduce some bias into the results. For example given K = 10110111 and the table size as 2^4, then if we take substrings of sizes 1, 4, 2, and 1, beginning at the right, we get the sum of Figure 2.22c. In this sum the longest substring is more important than the others in the sense that its left-most bit is likely to be the leftmost bit of the sum.

Sometimes the "exclusive-or" operation is used in folding instead of addition. The "and" and "or" operations turn out to be unacceptable; the reason is left as an exercise.

Folding gives results from 0 through $2^k - 1$, so it is naturally used with $b = 2^k$. The result of folding may be reduced modulo some value other than a power of 2, in effect combining hashing with folding, but the resulting function will be somewhat biased toward lower-numbered buckets. Again the details are left as an exercise.

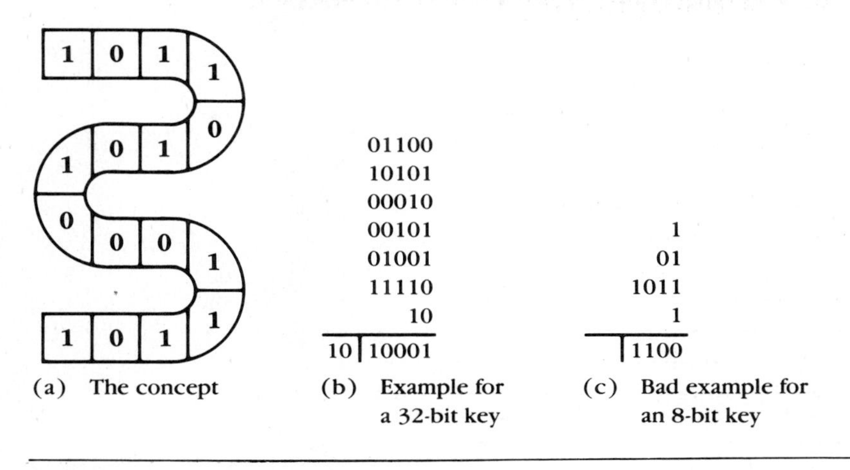

Figure 2.22 Folding

Another type of hash function is constructed by multiplying the key by some other number. The simplest choice for the second number is the key K itself. Of course K^2 is likely to be much larger than b, so it must be reduced to a value from 0 to $b - 1$.

The easiest way to make K^2 smaller is to extract a fixed number j of bits. (Recall that K may be interpreted as a binary number.) This use of j bits corresponds to a table size $b = 2^j$. Unfortunately, the first j bits depend too heavily on the first few bits of K, and the last j bits depend too heavily on the last few bits of K (see Figure 2.23). These simplest choices are therefore not appropriate. The best choice is usually the middle j bits, since a change of a single bit anywhere in K is likely to result in change somewhere in the middle j bits of K^2. This method is called the ***midsquare*** method.

A second possibility is to let the second factor be fixed. This second factor must be chosen with care, however, since some factors—such as a large power of 2—will not result in a desirable distribution of the keys. One method that has been suggested is to let $h(x) = \lfloor b(cx \bmod 1) \rfloor$ where reduction mod 1 means taking the fractional part of a number and c is between 0 and 1. A favorite choice for c is the "golden ratio"

$$\frac{\sqrt{5} - 1}{2}$$

but others are possible. [See Reingold and Hansen (1983), 337–39.]

It may also be tempting to combine the multiplication and division meth-

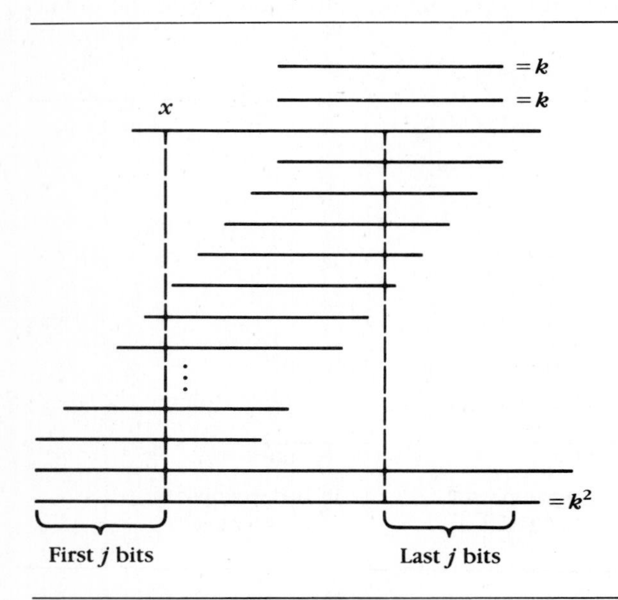

Figure 2.23 Finding the First and Last j Bits of a Square

ods, that is, to reduce K^2 mod b. This will result, however, in half or fewer of the buckets being used. (See the exercises for an example.)

As in bucket sort, buckets (i.e., hash table locations) may be implemented as linked lists. This implementation is called *chaining*. Another possibility is to allow only one entry per table location. If the hash function returns the index of an occupied location, the insertion algorithm needs to search for another location. For each key, the insertion algorithm will consider in order a sequence of table locations, called a *probe sequence*. (Compare the everyday experience of looking for a parking place.)

A good probe sequence will check all table locations. In general there is less conflict if probe sequences of different keys are different. A second hash function h_2 can be used to generate a probe sequence. If the original hash function is called h, then one probe sequence would have as its $(k + 1)$st term $h + kh_2$ mod b; that is, each index in the sequence would be obtained from the previous term by adding h_2 (mod b). The use of two hash functions is called *double* hashing.

It is a good exercise to show that for a given key K, if $h_2(K)$ and b have no common divisors, then this probe sequence will visit the entire table. In particular, if b is a power of 2 and h_2 is always odd, or if b is prime and h_2 ranges from 1 to $b - 1$, then all table locations will be visited. A third possibility is for h_2 to have the constant value 1; this is called *linear probing*.

2.14 Multiplying Polynomials and Large Integers

We can easily see informally that the standard algorithm for multiplying large integers has time complexity $\Theta(mn)$ if the two factors have respectively m and n digits, since each digit has to be multiplied by each other digit and each such multiplication requires a bounded number of additions. The same principle applies to general polynomials of one variable of respective degrees m and n.

We assume the natural representation in which the coefficient of x^n is stored as the $(n + 1)$st element of the representing sequence. In some applications most coefficients are zero, and it is more efficient to omit these terms from the sequence. In this case, the exponent as well as the coefficient has to be stored.

Here we are assuming all multiplications and additions have the same cost, which may not be a good assumption. For example, coefficients of polynomials may be extremely large or may be high-precision decimals. Very large integers may be represented in a radix different from 10, say in radix 2^w, where w is the word size of a particular machine. In any of these cases, multiplication will be more costly than addition, so that an algorithm using fewer multiplications may be preferred even if it uses more additions.

We illustrate a divide-and-conquer approach for integers. Assume that two integers with at most $2k$ digits in radix r are to be multiplied. Then these

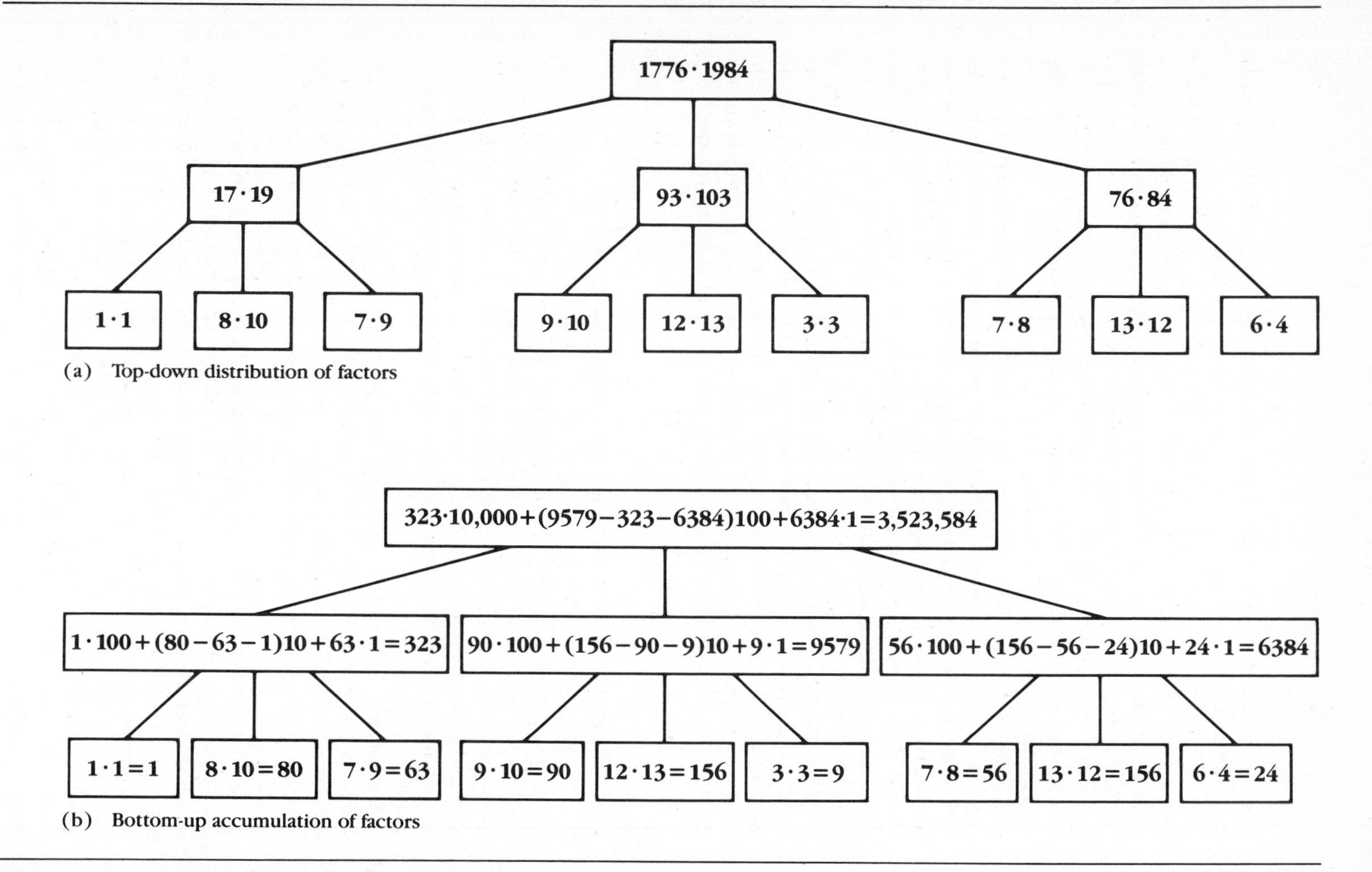

Figure 2.24 Multiplication of Large Integers

integers may be written $a2^k + b$ and $c2^k + d$, and their product may be written

$$(ac)r^{2k} + (ad + bc)r^k + (bd)$$

where a, b, c, and d have at most k digits. Thus we have reduced the original problem to 4 subproblems of half the original size. A similar argument applies to polynomials of degree at most $2k - 1$, and a, b, c, and d polynomials of degree at most $k - 1$ (i.e., with at most k coefficients).

Unfortunately, as we will see in Section 2.17, this does not give an asymptotic improvement in the number of multiplications. (If you're curious, try counting the number of multiplications when k is a small power of 2.) But it does suggest there is room for improvement. There are not 4 parenthesized terms in the product, but only 3. The middle term $ad + bc$, which equals $(a + b)(c + d) - ac - bd$, may be obtained with only a single multiplication. Thus if we calculate the three products ac, bd, and $(a + b)(c + d)$, at the cost of a few extra additions and subtractions we can find the original product. We have therefore reduced the original problem to 3 problems of half the size; in Section 2.17 we show that this gives an asymptotic improvement. (Again, for now you might want to check the situation when k is a small power of 2.)

The tree of Figure 2.24a shows for the multiplication of 1776 by 1984 the three smaller multiplications necessary for each larger multiplication. Part b of the figure shows how the smaller products are combined to form the larger product.

The same principle can be applied to matrix multiplication to reduce 8 multiplications to 7. The details are left as Exercise 136, which you can try after you've read Section 2.17.

2.15 Mergesort

If Quicksort is the prototypical hardsplit/easyjoin algorithm, then Mergesort is the prototypical easysplit/hardjoin algorithm. Given a merge algorithm, a divide-and-conquer approach is natural for Mergesort. We divide the sequence in half, sort each piece recursively, then merge the sorted pieces together.

The reason this approach was not taken for the version of Mergesort in Section 2.3 is that finding the midpoint of a sequence need not have $O(1)$ time complexity if an array representation is not used. If an array representation is used, then we get the Mergesort Algorithm 2.64.

An iterative merge algorithm is given in Section 3.4. Interested readers and instructors may read that section now, although as for Quicksort the correctness proof may be difficult to follow without having read Section 3.1. That merge algorithm is similar in specification and spirit to that of Section 2.3. Notice the merge algorithm does all the work.

The intuitive analysis of Section 2.3 still applies to suggest that the time complexity of Mergesort is $\Theta(n \log n)$. A more precise argument will be given in Section 2.17.

```
1. Mergesort(S, low, high)            {sorts a sequence stored between
2.                                     locations low and high of array S}
3. if low < high then
4.    mid := ⌊(low + high)/2⌋                        {find the midpoint}
5.    Mergesort(S, low, mid)                         {sort the first half}
6.    Mergesort(S, mid + 1, high)                   {sort the second half}
7.    Merge(S, low, mid, mid + 1, high)       {merge the sorted pieces}
8. end if
```

Algorithm 2.65 Mergesort

2.16 Guessing and Verifying the Time Complexity of an Algorithm

Simple guessing is useful in dealing with algorithms in two ways. We may design an algorithm that guesses a solution; or given an algorithm, we may guess, at least asymptotically, its time complexity. This guess should then be verified.

In this section we consider time complexity and show how guesses may be verified recursively. In Sections 2.18 through 2.20 and in much of Chapters 4 and 5, we consider designing algorithms as sequences of guesses.

Our first examples will deal solely with functions that might represent the time complexity of an algorithm rather than with an algorithm itself.

For example, consider $\sum_{k=1}^{n} k^2$. We saw in Section 2.1 that this sum is $O(n^3)$. A reasonable guess would be the sum is also $\Omega(n^3)$. This example shows how easy it can be to construct a hypothesis. We may guess that the best known algorithm represents the best possible time complexity, or that the worst case is also the best case, or that the average case is the same asymptotically as the worst case, or that the time complexity is as suggested by empirical observation.

For the current problem, formulation of a hypothesis more suitable for an inductive proof leads to the following theorem:

Theorem There is a constant $c > 0$ such that

$$\sum_{k=1}^{n} k^2 \geq cn^3$$

Proof Certainly there is a value for c that makes the inequality true for $n = 1$. We need to show the same c can work for all values of n. Thus we consider the induction step first and then see how well the choice of c given there works for the basis.

Consider the inequality of the theorem to be an induction hypothesis. We may assume its truth for a sum of $n - 1$ terms. We begin with the sum for n terms, express it in terms of the sum for $n - 1$ terms, and then apply the induction hypothesis. This gives:

$$\begin{aligned}\sum_{k=1}^{n} k^2 &= n^2 + \sum_{k=1}^{n-1} k^2 \\ &\geq n^2 + c(n - 1)^3 \\ &= cn^3 + (1 - 3c)n^2 + 3cn - 1\end{aligned}$$

We would like to make this sum greater than cn^3. This is only possible as long as the coefficient $(1 - 3c)$ of the n^2 term is nonnegative. This will hold if $c \leq 1/3$. To make sure that the remaining contribution $3cn - 1$ is also nonnegative, we need $c \geq 1/(3n)$, which is certainly true for $c \geq 1/3$. Our two constrains on c suggest a choice of $c = 1/3$.

We can easily verify this choice of c will work for the basis step as well as for the induction step in an induction proof, so we have the desired result. Note that the value of 1/3 for c follows naturally from the integral test (compare the proof of theorem 10 in Appendix A). ■

We next consider another sum that occurs naturally in algorithms, $\sum_{i=1}^{n} \log i$. In this example we assume all logarithms are to base 2. By theorem 9 of Appendix A, this assumption is irrelevant to the asymptotic result.

One reason that this sum occurs naturally in analyzing algorithms is that it is just the logarithm of $n!$. We have seen that $n!$ occurs when considering permutations of a set of size n.

As seen in Section 2.17, an upper bound for this sum is easy to find. Since each term $\log i$ is at most $\log n$, and since there are n terms, the sum must be bounded by $n \log n$. A natural guess for an asymptotic lower bound is this same function.

The given sum is somewhat awkward to analyze because the terms are seldom integers. But by writing out a few terms, we see that a natural and simple lower bound appears if we round each logarithm down to the nearest integer. In this case the term j appears 2^j times, assuming n is at least 2^{j+1}. To simplify the following calculations, we assume $n = 2^k - 1$ for some k. Then each integer j less than k appears 2^j times, and no other integer appears in the sum. More precisely, we have (for $n = 2^k - 1$) $\sum_{i=1}^{n} \log i \geq \sum_{j=0}^{k-1} j\, 2^j$. We would like this sum to be $\Omega(n \log n)$, or $\Omega(k2^k)$. We are led to the following theorem:

Theorem There is a constant $c > 0$ such that for k sufficiently large, $\sum_{j=0}^{k-1} j2^j \geq ck2^k$.

Proof by induction on k. Again it is easier to consider the induction step before the basis, so that we may find a suitable value for c. With an inductive hypothesis given by the inequality of the theorem, we may assume (replacing k by $k - 1$)

$$\sum_{j=0}^{k-2} j2^j \geq c(k-1)2^{k-1} \text{ so}$$

$$\sum_{j=0}^{k-1} j2^j = (k-1)2^{k-1} + \sum_{j=0}^{k-2} j2^j$$

$$\geq (k-1)2^{k-1} + c(k-1)2^{k-1}$$

$$= (1 + c) \times (k-1)2^{k-1}$$

We would like this value to be greater than $ck2^k$. This will happen if

$$(1 + c) \times (k-1)2^{k-1} \geq ck2^k \text{ or}$$

$$(1 + c)(k-1) \geq 2ck \text{ or}$$

$$k + kc - 1 - c \geq 2ck \text{ or}$$

$$k - 1 \geq kc + c \text{ or}$$

$$(k-1)/(k+1) \geq c$$

For $k \geq 2$ the left-hand side is at least 1/3, so that we may choose $c = 1/3$. For $k = 3$ we have the desired inequality, since $\sum_{j=0}^{3-1} j2^j = 0 + 2 + 8 = 10 \geq \frac{24}{3} = \frac{1}{3} \times 3 \times 2^3$. Thus we may use $k = 3$ as the basis and get the desired result for any larger k. ■

The theorem is now proved, but we still need to handle the case when n is not of the form $2^k - 1$. In this case it is true that for some k, $2^k - 1 < n < 2^{k+1} - 1$. So if we let $K = 2^k - 1$, since $k + 1 \geq \log_2 (n + 1)$, we have

$$\sum_{i=1}^{n} \log i \geq$$

$$\sum_{i=1}^{K} \log i \geq$$

$$ck2^k \geq$$

$$\frac{c(-1 + \log_2 (n+1))(n+1)}{2}, \text{ which is } \Omega(n \log n)$$

We may therefore conclude that $\log n! = \sum_{i=1}^{n} \log i$ is $\Theta(n \log n)$.

Next we consider an actual problem for which one might want an algorithm, namely, finding both the maximum and minimum of a set of n numbers.

A straightforward algorithm for finding the maximum (see the algorithm for finding the minimum in Section 2.3) would begin by comparing the first two elements and then comparing each successive element with the largest so far. This would require $n - 1$ comparisons. If the analogous algorithm is applied to find the minimum, then $2(n - 1)$ comparisons will be required to solve the original problem.

Clearly we cannot do any better asymptotically than this $O(n)$ algorithm, since any algorithm must at least look at all n items, which requires at least $\lceil n/2 \rceil$ comparisons. However, we may ask whether an algorithm exists that gives a lesser time complexity in the nonasymptotic sense.

We already have a general strategy we can apply to this problem, the divide-and-conquer strategy. Suppose n is even and we may therefore partition our set into two equal halves and find the maximum and minimum of each half. It will take two extra comparisons, one for the two maxima and one for the two minima, to solve the original problem.

This suggests the following theorem:

Theorem If n is a power of 2, the preceding divide-and-conquer algorithm requires at most $cn + d$ comparisons.

Proof by induction on n. Again we consider the induction step first. Assuming the induction hypothesis with $n/2$ replacing n, then $n/2$ is a power of 2 and an upper bound on the number of comparisons required for a set of size n is

$$c\left(\frac{n}{2}\right) + d \qquad \text{(for the first piece)}$$

$$+\, c\left(\frac{n}{2}\right) + d \qquad \text{(for the second piece)}$$

$$+\, 2 \qquad \text{(for the ``combine'' stage)}$$

This total of $cn + 2d + 2$ will be less than or equal to $cn + d$ iff $2d + 2 \leq d$, or $d \leq -2$.

We still need to consider the basis step. If we begin with $n = 2$, for which 1 comparison is required, then we need a value of c such that $1 \leq cn + d = 2c + d \leq 2c - 2$. We see that $2c \geq 3$, so the best we can do is to take $c = 3/2$, in which case $d \geq -2$. Combining this inequality with our previous constraint on d, we get a time complexity of $3n/2 - 2$ comparisons, an improvement over our original algorithm. ■

Unfortunately, to justify this analysis, we need to assume not only that our original value of n is even, but that it remains even whenever it is divided by 2, since each piece created by divide-and-conquer must also have even size for our analysis to apply. This assumption requires that n be a power of 2. For example, if $n = 3$, the problem requires more than $3 \times 3/2 - 2 = 2.5$ comparisons.

We may recover our time complexity by using a slightly different divide-and-conquer algorithm. The crucial observation is that $n = 2$ is an especially favored case, since we need only make one comparison. This observation suggests in a divide-and-conquer algorithm we might make one of the pieces have size 2. A further advantage is that if n is even, the other piece will also have even size.

Mimicking the proof of the previous theorem, we see that whether n is even or odd, the new divide-and-conquer algorithm is linear if

$$2 + 1 + [c(n - 2) + d] \leq cn + d \text{ or}$$

$$3 + cn - 2c + d \leq cn + d \text{ or}$$

$$3 - 2c \leq 0$$

The smallest c that can make this true is once again 3/2. With n even, the base case is the same as before.

If n is not even, then the base case is $n = 1$, which of course requires 0 comparisons. For $c = 3/2$ and $0 \leq c \times 1 + d$, the best we can do for d is to choose $d = -3/2$. Verification of the induction step is a matter of showing that

$$2 + 1 + [3/2(n - 2) - 3/2] \leq (3/2)n - 3/2$$

which is also trivial to do.

In summary, our final divide-and-conquer algorithm requires

$$(3/2)n - 2 \quad \text{comparisons if } n \text{ is even}$$

$$(3/2)n - 3/2 \quad \text{comparisons if } n \text{ is odd.}$$

Note that the constant terms differ by very little; in fact we may combine the formulas to get a time complexity of $\lceil (3/2)n - 2 \rceil$ comparisons.

Finally, we consider the average-case behavior of quicksort. As its name suggests, empirical evidence for Quicksort has shown that the average-case time complexity is closer to the best case than to the worst case. This evidence suggests that the average-case time complexity of Quicksort is $\Theta(n \log n)$ rather than $\Theta(n^2)$. This argument is much simpler than the typical average-case analysis, partly because it is not so difficult to define average-case input as for many problems, but largely because we are able to guess what the average-case time complexity might be. In some cases average-case analysis is so difficult that it is simply not done, and empirical evidence is sought instead.

We assume that all permutations of n keys are equally likely as input to the quicksort algorithm.[8] We further assume there exists an $O(n)$ partition algorithm. With these assumptions, we now establish the following theorem:

Theorem There is a $c > 0$ such that the average-case number of comparisons required by quicksort is at most $cn \log_2 n$ for sufficiently large n.

Proof by induction on n. Again we consider the induction step first.

---

[8] You may decide for yourself how realistic this assumption is. For example, it implies the probability that an input sequence of 10 elements is already sorted is $1/10! = 1/3628800$.

Let $A(n)$ be the average-case number of comparisons required by Quicksort for a sequence of size n. Our assumption is that partition requires at most cn comparisons, so

$$A(n) \leq cn + L(n) + R(n)$$

where $L(n)$ and $R(n)$ are respectively the average-case time for the left subsequence and the average-case time for the right subsequence. Our assumption that all permutations are equally likely means the element returned by partition is equally likely to end up in any position from 1 to n, so that each of the two subsequences is equally likely to have any length from 0 to $n - 1$.

Thus

$$A(n) \leq cn + (1/n) \sum_{k=0}^{n} A(k) + (1/n) \sum_{k=0}^{n} A(k)$$

$$\leq cn + (2/n) \sum_{k=0}^{n} A(k)$$

Our proof will be by induction on n, with induction hypothesis $A(n) \leq dn \log n$. (In this section we assume all logs are natural logs; recall that the asymptotic result is independent of the base of the logarithm.) So we may assume $A(k) \leq dk \log k$, [except for $k = 0$ we note $A(0) = 0$] and the equation for $A(n)$ becomes

$$A(n) \leq cn + (2/n) \sum_{k=1}^{n-1} dk \log k$$

$$\leq cn + (2d/n) \int_{k=1}^{n} k \log k \, dk$$

This last inequality is true since $k \log k \leq n \log n$ on the interval $[n - 1, n]$. The integral may be evaluated using integration by parts to get

$$\int k \log k \, dk = (k^2/2)(\log k) - \int (k^2/2)(1/k)\, dk$$

$$= (k^2/2)(\log k) - \int (k/2)\, dk$$

$$= (k^2/2)(\log k) - k^2/4$$

Evaluating from 1 to n, we get

$$(n^2/2)(\log n) - n^2/4 + 1/4$$

Substituting into the inequality for $A(n)$, we get

$$A(n) \leq cn + (2d/n)[(n^2/2)(\log n) - n^2/4 + 1/4]$$

$$= cn + dn \log n - dn/2 + d/2n$$

We are done if we can show that this last expression is bounded above for sufficiently large n by $dn \log n$. But this is equivalent to showing that for some choice of d, $(c - d/2)n + d/2n \leq 0$.

This inequality can be achieved by choosing d larger than $2c$, say $d = 2c + 1$. Then $(c - d/2)n < -n/2 < -1/2$, and for sufficiently large n, $d/2n < 1/2$. ■

2.17 Recurrence Relations

The divide-and-conquer strategy for designing algorithms has a natural counterpart in the analysis of algorithms. Suppose we have a particular algorithm in mind. Let t_n be a measure of the time required by the algorithm for input of size n. Usually we think of t_n as measuring the worst-case time, in which case t_n considered as a function of n is just the time complexity of the algorithm. If the algorithm is a divide-and-conquer algorithm, then a natural equation relating t_n to values t_i earlier in the sequence is likely to exist. Such an equation is called a *recurrence relation.*

For example in sequential search, a problem of size n may be solved by a single O(1) test of equality followed by solving a sequential search problem of time n

$$t_n = t_{n-1} + c$$

Here c is O(1).[9]

For another example, in binary search within an array, one O(1) comparison reduces a problem of size n to a problem of size $\lceil n/2 \rceil$. So

$$t_n = t_{\lceil n/2 \rceil} + c$$

with c as above.

A third example is InsertionSort, where a single insertion may require $O(n)$ time. So

$$t_n = t_{n-1} + cn$$

with c as above.

A fourth example is Mergesort, where an $O(n)$ merge reduces a problem of size n to two problems of size $\lceil n/2 \rceil$. Here

$$t_n = 2t_{\lceil n/2 \rceil} + cn$$

with c as above.

A recurrence relation may be thought of as a recursive definition of the function $t_n = t(n)$. Like other recursive definitions, it needs a basis part if the function is to be properly defined. In this context, the basis part, which is a

[9] If testing equality has time complexity O(1), then it is more nearly correct to say that c, and therefore the right-hand side of the recurrence relation, is an upper bound for the time complexity instead of an exact measure of the time complexity. In this case, however, the solution of the recurrence relation will give an upper bound for the desired time complexity, which is all we expect. (See Exercise 130.)

definition of t_n for small n, is called an *initial condition,* or a *boundary condition.* If you are familiar with differential equations, this much—and in fact our entire discussion of recurrence relations—should sound familiar.

To determine initial conditions for the preceding algorithms, we determine how much time is required for a problem of size 1. So we have initial conditions of:

Sequential search:	$t_1 = d$ for some constant d
Binary search:	$t_1 = d$ for some constant d
InsertionSort:	$t_1 = 0$
Mergesort:	$t_1 = 0$

Recurrence relations have other uses as well. For example, to find the fewest nodes in an AVL tree of height h, we can set up and solve the recurrence relation

$$t_h = 1 + t_{h-1} + t_{h-2} \text{ where}$$

$$t_0 = 1 \text{ and } t_1 = 2$$

Notice we need two initial conditions here. Just knowing t_0 is not enough to determine t_1 or t_2.

The general equation says that a minimal AVL tree of height h has 1 root node, a minimal left or right subtree of height $h - 1$, and a minimal second subtree whose height is the minimum possible, namely $h - 2$.

Here, for example, $t_3 = 1 + t_2 + t_1 = 1 + 1 + 2 = 4$, $t_4 = 1 + t_3 + t_2 = 1 + 4 + 2 = 7$. Thus the sequence beginning $(1, 2, 4, \ldots)$ is not composed of powers of 2. In fact, this sequence is more closely related to another sequence you might have seen, the Fibonacci sequence. If you have not seen it before, don't worry; the Fibonacci sequence will be discussed in the next section.

Another example is the sum $\sum_{j=0}^{n} j2^j$ discussed in regard to the heapify algorithm in Section 2.8, which may be defined as a function t_n of n by

$$t_n = t_{n-1} + n2^n, \text{ and } t_0 = 0$$

A solution to a recurrence relation is an equation $t_n = \mathrm{f}(n)$, where f is some algebraic function. We describe below a method for solving many common recurrence relations. Our discussion follows Lueker (1980).

Let E be the function defined on sequences that throws away the first term.

For example, $\mathrm{E}(\langle 1\ 2\ 4\ 8\ 16 \ldots\rangle) = \langle 2\ 4\ 8\ 16\rangle$, that is,

$$\mathrm{E}(\langle 2^k\rangle_{k=0}^{n}) = \langle 2^k\rangle_{k=1}^{n}$$

Formally, $\mathrm{E}(\langle t_n\rangle_{n=a}^{\infty}) = \langle t_n\rangle_{n=a+1}^{\infty}$. Another useful operator on sequences is the *scalar multiplication* operator that multiplies the entire sequence by a constant. It works by multiplying each term of the sequence by a constant, that is,

$$c\langle t_n\rangle_{n=a}^{\infty} = \langle ct_n\rangle_{n=a}^{\infty}$$

Finally, one may add two sequences term by term, for example,

$$\langle t_n\rangle_{n=a}^{\infty} + \langle u_n\rangle_{n=a}^{\infty} = \langle t_n + u_n\rangle_{n=a}^{\infty}$$

Recurrence relations may be expressed as equations on sequences; for example,

$$t_n = t_{n-1} + c$$

may be expressed as

$$\begin{aligned}\langle t_n\rangle &= \langle t_{n-1}\rangle + \langle c\rangle \text{ or}\\ \mathrm{E}\langle t_{n-1}\rangle &= \langle t_{n-1}\rangle + \langle c\rangle \text{ or}\\ \mathrm{E}\langle t_n\rangle - \langle t_n\rangle &= \langle c\rangle\end{aligned}$$

where we make a simple variable substitution. This last equation is often written

$$(\mathrm{E} - 1)\langle t_n\rangle = \langle c\rangle$$

(note that $1\langle t_n\rangle = \langle t_n\rangle$). $(\mathrm{E} - 1)$ looks like a polynomial in E; with the natural interpretation of E^n as the composition of n applications of E, expressions like $(\mathrm{E} - 1)$ can be treated like ordinary polynomials.[10]

The recurrence relation for insertion sort thus becomes

$$(\mathrm{E} - 1)\langle t_n\rangle = \langle cn\rangle$$

and that for the AVL trees becomes

$$(\mathrm{E}^2 - \mathrm{E} - 1)\langle t_n\rangle = \langle 1\rangle$$

The solution to a recurrence relation of the type we are describing is simplest if the right-hand side of the equation is zero. You may see such an equation called a *homogeneous* equation. For example, let us consider the simple recurrence

$$t_n = 2t_{n-1},\ t_0 = 1$$

or in our new notation,

$$(\mathrm{E} - 2)\langle t_n\rangle = \langle 0\rangle$$

Calculating the first few terms should suggest that a solution to this recurrence is just $t_n = 2^n$. This may be easily verified by induction. The general solution turns out to be $t_n = c2^n$; the proof has been left as an exercise. The operator $(\mathrm{E} - 2)$ is said to *annihilate* any such sequence. In general an operator F *annihilates* $\langle t_n\rangle$ iff $F(\langle t_n\rangle) = \langle 0\rangle$.

The proof of the exercise just mentioned may be generalized to show that $(\mathrm{E} - b)$ annihilates precisely those sequences of the form $\langle cb^n\rangle$.

---

[10] Formally, E and the scalar multiplication operators generate a commutative ring whose operations are addition and function composition.

Annihilators may be found for more complicated sequences. To see what $(E - 1)^2$ annihilates, we first note that $(E - 1)^2 = (E - 1)(E - 1)$ annihilates $\langle t_n \rangle$ iff $(E - 1)$ is an annihilator of $(E - 1)\langle t_n \rangle$. Since we know precisely what $(E - 1)$ annihilates, we get $(E - 1)\langle t_n \rangle = \langle b \times 1^n \rangle = \langle b \rangle$ for some constant b. By definition of $(E - 1)$, we observe that $\langle t_{n+1} - t_n \rangle = \langle b \rangle$. Writing out the first few terms suggests that $\langle t_n \rangle = \langle bn + c \rangle$ for arbitrary constants b and c. Verification of this claim is an easy induction exercise (solved as Exercise 133) and another illustration of the feasibility of first guessing and then verifying a solution.

We have seen that $(E - 1)$ annihilates any polynomial of degree 0, and $(E - 1)^2$ any polynomial of degree 0 or 1. In general, $(E - 1)^k$ annihilates any polynomial of degree $k - 1$ (compare differentiation), and $(E - c)^k$ annihilates precisely sequences $\langle c^n p(n) \rangle$, where p is a polynomial of degree $k - 1$ or less. This last result actually includes all of our previous results.

The sum of two sequences is annihilated by the product (i.e., the composition) of their annihilators. For example, $(E - 1)(E - 2)$ annihilates precisely the sequences $\langle b + c2^n \rangle$.

Arbitrary constants in the general solution to a recurrence relation may be evaluated by considering the initial conditions. For example, we have seen that the recurrence relation

$$(E - 2)\langle t_n \rangle = \langle 0 \rangle$$

has the general solution $t_n = c2^n$. If in addition we know that $t_0 = 1$, we can substitute this in the general solution to get $1 = t_0 = c2^0 = c$, which gives us the value 1 for c. Thus $t_n = 2^n$ for the given recurrence relation and initial condition.

Recurrence relations whose right-hand sides are not zero may be solved by the preceding method simply by finding an annihilator for the right-hand side and applying it to both sides. Since the right-hand side is now zero, we know that our sequence is among those annihilated by the new left-hand side.

We may use this technique to analyze two of the divide-and-conquer algorithms mentioned at the beginning of the section, as well as to handle the recurrences associated with AVL trees and the heapify algorithm.

Example: sequential search

The recurrence relation is $(E - 1)\langle t_n \rangle = \langle c \rangle$

The annihilator of the right-hand side is $(E - 1)$, which we may apply to both sides to get $(E - 1)^2\langle t_n \rangle = \langle 0 \rangle$. This gives a solution of $t_n = 1^n(p + qn) = (p + qn)$. We see immediately that t_n is $O(n)$, giving a time complexity of $O(n)$. The unknowns p and q may be evaluated knowing that $t_0 = d$ and (using the recurrence) that $t_1 = d + c$. This gives an exact solution in terms of c and d of $t_n = cn + d$.

Example: InsertionSort

The recurrence relation is $(E - 1)\langle t_n \rangle = \langle cn \rangle$

The annihilator of the right-hand side is $(E - 1)^2$, which we may apply to both sides to get $(E - 1)^3\langle t_n\rangle = \langle 0\rangle$. This gives a solution of $1^n(p + qn + rn^2)$ and a consequent time complexity of $O(n^2)$. It is easy to check that the exact solution in terms of c and n is $(c/2)(n^2 + n - 2)$.

Example: height of AVL trees

The recurrence relation for the minimum size of an AVL tree of height h is $(E^2 - E - 1)\langle t_h\rangle = \langle 1\rangle$.

The annihilator of the right-hand side is $(E - 1)$, which we may apply to both sides to get $(E - 1)(E^2 - E - 1)\langle t_h\rangle = \langle 0\rangle$. To get a solution we may use the quadratic formula to factor $(E^2 - E - 1)$ into $(E - a)(E - b)$, where $a = (1 + 5^{1/2})/2 \approx 1.6$ and $b = (-1 + 5^{1/2})/2 \approx -0.6$. The solution is then $1^h p + a^h q + b^h r$. Since $|a| > |b|$ and $|a| > |1|$, this function is $\Theta(a^h)$, giving an approximate height bound of $\Omega(1.6^h)$ for n. Since for sufficiently large n, $n \geq c \times 1.6^h$, we have that $h \leq \log_{1.6}(n/c)$. Thus h is $O(\log n)$, as promised.

Example: heapify

Recall that in the analysis of heapify, the variable h represented the height of a complete tree with n nodes. The $\Theta(n)$ time complexity of heapify depended on the fact that $\sum_{k=0}^{h} k2^k = 2^{h+1} - h - 2$. We may obtain this result by using recurrence relations, identifying t_n with $\sum_{k=0}^{n} k2^k$.

The recurrence then becomes

$$t_n = t_{n-1} + n \times 2^n$$

which becomes $(E - 1)\langle t_n\rangle = \langle n2^n\rangle$. The right-hand side is annihilated by $(E - 2)^2$. Applying this operator to both sides, we get

$$(E - 1)(E - 2)^2\langle t_n\rangle = \langle 0\rangle$$

Thus $t_n = p + 2^n(q + rn)$. With $t_0 = 0$, it is easy to see that $t_1 = 2$ and $t_2 = 10$ and to evaluate the constants p, q, and r to get $t_n = 2 + 2^n(2n - 2)$.

We still do not have techniques sufficient to solve the recurrences for binary search and Mergesort. Note that these divide-and-conquer algorithms, since they divide the underlying data structure into nearly equal parts, correspond to the second class of divide-and-conquer algorithms discussed in Section 2.2.

The approach we use is called *domain transformation.* We do not need the general technique discussed in Lueker (1980), but instead we use a special case based on the following observations.

Suppose first that n is a power of 2. Then successive divisions by 2 leave no remainder. In particular, $\lceil n/2 \rceil = n/2$. It is easy to see that when evaluating t_n in terms of $t_{n/2}$, the only subscripts we ever consider are those that are a power of 2. In other words, we are only considering a subsequence $\langle s_u\rangle$ of $\langle t_n\rangle$, where u is a power of n. More precisely, $t_n = s_{2^u}$. Note that $s_{u-1} = t_{n/2}$, so that we may rewrite our recurrence in terms of s_u. Of course, any occurrences of n in the recurrence also need to be restated in terms of u. Once the recurrence is solved for u, the relation between n and u will allow solution for n.

The general technique for domain transformation works in the same way, except that it is not always obvious how u and n are related. In general, another recurrence relation relating u and n may need to be established and solved.

If only an asymptotic solution is sought, then usually the solution for n a power of two will generalize to other n. We now consider examples.

Example: binary search

Here (assuming n is a power of 2), the recurrence relation is $\langle t_n \rangle - \langle t_{n/2} \rangle = \langle c \rangle$. Setting $n = 2^u$ and $s_u = t_n$, we get $\langle s_u \rangle - \langle s_{u-1} \rangle = \langle c \rangle$. This becomes $(\mathrm{E} - 1)\langle s_u \rangle = \langle c \rangle$. As seen in the sequential search example, the general solution to this recurrence is $s_u = p + qu$. In terms of n, this becomes $t_n = p + q(\log_2 n)$. So t_n is $O(\log n)$. In other words, for sufficiently large n, there is a constant c such that an input of size n requires time at most $c \times \log_2 n$.

If n is not a power of 2, then there is some k such that $2^k \leq n < 2^{k+1}$. So $k \leq \log_2 n < k + 1$. By the previous paragraph we know that an input of size 2^{k+1} requires time at most $c(k + 1)$. Input of size n cannot require any more time, if only because we may add dummy elements. So $t_n \leq c(k + 1) \leq c(1 + \log_2 n)$. Thus t_n is $O(\log n)$ for arbitrary n.

Example: Mergesort

Here if n is a power of 2 the recurrence relation becomes $\langle t_n \rangle = 2\langle t_{n/2} \rangle + \langle cn \rangle$. Setting $n = 2^u$ and $s_u = t_n$, we get $\langle s_u \rangle = 2\langle s_{u-1} \rangle + \langle c2^u \rangle$. This becomes $(\mathrm{E} - 2)\langle s_u \rangle = \langle c2^u \rangle$. The right-hand side is annihilated by $(\mathrm{E} - 2)$, so that we get $(\mathrm{E} - 2)^2\langle s_u \rangle = \langle 0 \rangle$. The general solution is

$$s_u = 2^u(qu + r)$$

In terms of n, we get

$$t_n = n(q \times \log_2 n + r), \text{ which is } O(n \log n)$$

If n is not a power of 2, then as in the previous example there is some k with $2^k \leq n < 2^{k+1}$ and input of size 2^{k+1} requires time at most $c(2^{k+1}(k + 1))$. Since $k \leq \log_2 n$ and $2^{k+1} < 2n$, the time requirement for input of size n is at most $c(2n(1 + \log_2 n))$. Thus we get $O(n \log n)$ behavior in the general case.

Example: polynomial multiplication (see Section 2.14)

For the original divide-and-conquer algorithm, the recurrence relation is $\langle t_n \rangle = 4\langle t_{n/2} \rangle + \langle cn \rangle$, where cn represents the $O(n)$ time complexity of the additions. Setting $n = 2^u$ and $s_u = t_n$, we get

$$\langle s_u \rangle = 4\langle s_{u-1} \rangle + \langle c2^u \rangle \text{ or } (\mathrm{E} - 4)\langle s_u \rangle = \langle c2^u \rangle$$

Again the right-hand side is annihilated by $(\mathrm{E} - 2)$, so we get

$$(\mathrm{E} - 2)(\mathrm{E} - 4)\langle s_u \rangle = \langle 0 \rangle$$

We conclude that $s_u = p2^u + q4^u = p2^u + q(2^u)^2 = pn + qn^2 = \Theta(n^2)$, since it is easy to check that $q \neq 0$. This does not give an improvement over the straightforward polynomial multiplication algorithm. But recall a second algorithm in Section 2.14 that required only 3 multiplications. In the corresponding recurrence, the factor of 4 is replaced by 3. An argument otherwise identical to the preceding analysis applies to show that

$$s_u = p2^u + q3^u$$

Now with logs taken to the base 2,

$$3^u = (2^{\log 3})^u = 2^{u \log 3} = (2^u)^{\log 3} = n^{\log 3}$$

Since $\log_2 3$ is approximately 1.6, we have a time complexity of about $\Theta(n^{1.6})$.

2.18 Saving Solutions

The Fibonacci sequence is closely related to the sequence used in the previous section to describe the height of AVL trees. It has many interesting properties, some suggested in Exercise 21.

Def 2.66 The nth *Fibonacci number* f_n is
$f_1 = 1, f_2 = 1,$
$f_n = f_{n-1} + f_{n-2}$ for $n > 2$.

We may use the methods of the previous section to find the time complexity of calculating the Fibonacci numbers directly from their definition.

If t_n is the time required to calculate f_n, then $t_1 = c$, $t_2 = c$, and $t_n = t_{n-1} + t_{n-2}$. Since the unit of time is arbitrary, we may choose $c = 1$. In this case the recurrence for t_n is the same as that for f_n. A recurrence relation similar to that used in the AVL tree example shows that t_n grows exponentially with n.

On the other hand, if you were to calculate the values of the first few f_n, you would probably choose to write down successive values for f_n, beginning with f_1, and then refer to earlier values when computing later values in the sequence. This algorithm would not seem to require exponential time. We can easily formalize this algorithm by recording values of f_n in a 1-dimensional array, like that of Figure 2.25, as they are calculated.

Comparing these two algorithms, we see that the enormous time complexity of the first results from the very large number of repeated steps. For exam-

n	1	2	3	4	5	6	7	8	...
f(n)	1	1	2	3	5	8	13	21	...

Figure 2.25 The Fibonacci Sequence

ple, all the steps in the calculation of f_{n-2} are superfluous, since f_{n-2} must have been calculated in order to calculate the previous summand f_{n-1}.

This example suggests a general principle: In divide-and-conquer problems, it may be profitable to record solutions to all smaller problems, so that they can be referred to as needed.

One more numerical example may be useful. For $0 \leq m \leq n$, the ***binomial coefficient*** $\binom{n}{m}$ is the coefficient of $x^m y^{n-m}$ in the expansion of $(x + y)^n$.

Binomial coefficients are of importance in probability and statistics in part because $\binom{n}{m}$ gives the number of subsets of size m possessed by a set of size n. It has a recursive definition:

Def 2.67

1. $\binom{n}{0} = 1, \binom{n}{n} = 1$
2. $\binom{n}{m} = \binom{n-1}{m-1} + \binom{n-1}{m}$ if $m, n > 0$ and $m < n$.

That $\binom{n}{m}$ gives the number of subsets of size m of a set of size n is clear from part (1) of the definition if $m = 0$ or $m = n$. In general, the two terms of part (2) correspond to the number of subsets containing a fixed element x and those not containing x.

The values of $\binom{n}{m}$ for small m and n are given in Figure 2.26a. This figure is more commonly presented in the form shown in Figure 2.26b, which is merely a 45-degree rotation of the previous figure. Note that each element not equal to 1 is the sum of the two elements immediately above it. This figure is known as *Pascal's triangle.*

As in the Fibonacci example, the time complexity of calculating $\binom{n}{m}$ directly from the definition is given by $\binom{n}{m}$ itself; the proof has been left as an exercise. It can be shown that $\binom{n}{m} = \frac{n!}{(n-m)!m!}$ and that this function grows rapidly with n and m; for example, $\binom{n}{n/2} \geq 2^n/n$ if n is even. (Again the proof is left as an exercise.)

A more efficient algorithm would simply construct that portion of Pascal's triangle lying at or above row n. There are $\Theta(n^2)$ such entries, and each requires $O(1)$ time to calculate and record. Thus the new algorithm has time complexity $\Theta(n^2)$.

Next we consider how saving partial solutions can improve more general algorithms.

n \ m	0	1	2	3	4	5	6	7
0	1							
1	1	1						
2	1	2	1					
3	1	3	3	1				
4	1	4	6	4	1			
5	1	5	10	10	5	1		
6	1	6	15	20	15	6	1	
7	1	7	21	35	35	21	7	1

(a) Binomial coefficients $\binom{n}{m}$ for small values of n and m

```
              1
            1   1
          1   2   1
        1   3   3   1
      1   4   6   4   1
    1   5  10  10   5   1
  1   6  15  20  15   6   1
1   7  21  35  35  21   7   1
```

(b) Pascal's triangle

Figure 2.26 Binomial Coefficients and Pascal's Triangle

2.19 Dynamic Programming

The idea suggested in the previous section of saving the solutions to subproblems is the basis of a powerful technique called *dynamic programming.* In a dynamic programming algorithm, like a divide-and-conquer algorithm, a larger problem can be divided into subproblems. Unlike the divide-and-conquer case, it is not clear in advance how the larger problem should be divided. In many cases, if we were to consider all possible ways of dividing the problem, the problem would become intractable.

If we imagine all possible ways of dividing a large problem, it is apparent that small related problems could occur as subproblems in many different ways. If each subproblem has the same solution no matter how it occurs in a larger problem, then the solution can be saved the first time it is computed and merely looked up afterward. This is illustrated by the examples of the previous section.

Dynamic programming is usually applied to optimization problems. Dynamic programming algorithms typically take the form of a sequence of choices about how to divide up the problem. Although there will not be enough information available to make the optimal choice, it will be true that if the optimal choice could be made, an optimal solution to the subproblems would yield an optimal solution to the original problem. Horowitz and Sahni (1978) call this the *principle of optimality.*

A typical dynamic programming algorithm, like any recursive method, will actually construct solutions to the smallest problems first, then use those to construct solutions to successively larger problems. Eventually, and often relatively quickly, the solution to the original problem will be constructed. With this in mind, often it is easier to present dynamic programming algorithms non-recursively.

Optimal Binary Search Trees

In Section 2.5 we showed that if all keys were equally likely to be sought, a complete tree would minimize the expected successful search time. In this section we consider the case when key k_i is sought with probability p_i $(1 \leq i \leq n)$ and a key between k_i and k_{i+1} is sought with probability q_i $(1 \leq i \leq n-1)$. For completeness we let q_0 be the probability of searching for a key less than k_1 and q_n be the probability of searching for a key greater than k_n. We may assume that $k_i < k_{i+1}$. There is no reason why the search tree in this case should be balanced.

In general, we want quicker access to frequently sought nodes than to other nodes, which means that the more probable keys should appear higher in the tree. More precisely, we want to minimize the expected search time averaged over both internal and external nodes. Since internal node i, say on level s_i, has probability p_i and external node i, say on level t_i, has probability q_i, the sum to be minimized is

$$\left(\sum_{k=1}^{n} p_i(1 + s_i)\right) + \left(\sum_{k=0}^{n} q_i t_i\right)$$

Figure 2.27 shows two binary search trees for the same set of weights. The first has cost $[6 + 2(5 + 12)] + [2(1 + 2 + 3 + 4)] = 60$, while the second has cost $[12 + 2(6) + 3(5)] + [1(4) + 2(3) + 3(1 + 2)] = 58$. Thus the second has a lower cost even though the first is better balanced.

A divide-and-conquer approach would first choose which key to include in the root. By the binary search property, each remaining internal and external node is assigned either to the left subtree or to the right subtree. We have therefore divided the original problem into two problems of the same type (except that the sum of the p's and q's need not be 1, but we will have no need of this assumption).

Based on the information available at the time of the choice, there is no way to know which choice is correct. However, let $C(i, j)$ be the optimal cost

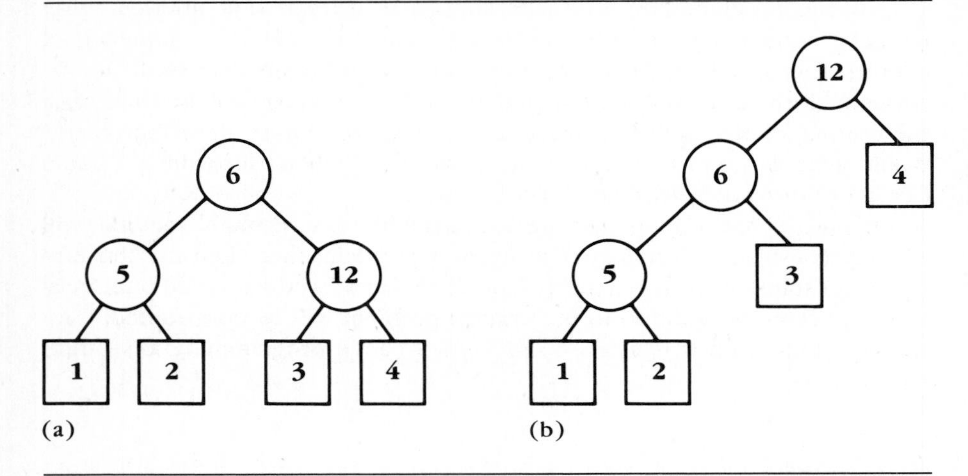

Figure 2.27 Two Possible Binary Search Trees for the Same Set of Weights

$\sum_{k=i+1}^{j} p_i(1 + s_i) + \sum_{k=i}^{j} q_i t_i$ of the subtree containing internal nodes $i + 1$ through j and external nodes i through j. If internal node r is chosen to be the root of this subtree, then the left subtree will have internal nodes $i + 1$ through $r - 1$ and external nodes i through $r - 1$. The right subtree will have internal nodes $r + 1$ through j and external nodes r through j (see Figure 2.28).

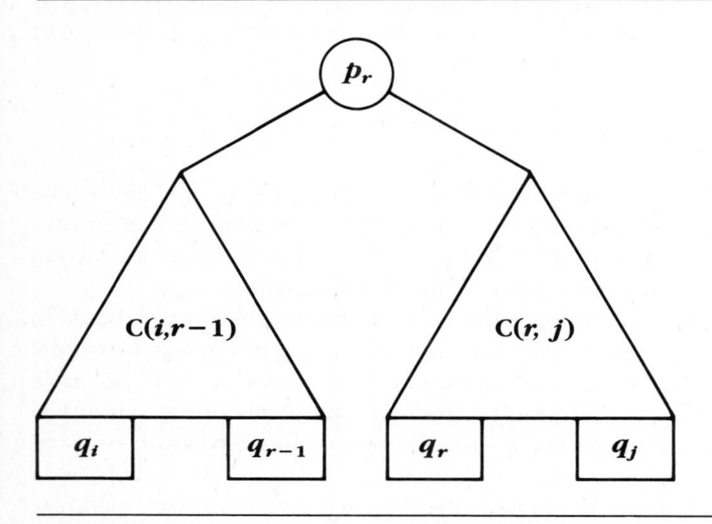

Figure 2.28 A Subtree of an Optimal Binary Search Tree

Also, $C(i, i) = 0$, since there is only a single external node on level 0. The minimum expected cost we seek is given by $C(0, n)$.

We also see that the overall tree can only be optimal if the two subtrees are optimal, so the principle of optimality applies. We would like to find $C(i, j)$ in terms of the cost of the two subtrees.

Given the cost $C(i, r - 1)$ and $C(r, j)$ of the optimal subtrees, if k_r is chosen as the root, we see that except for the root itself, each internal node and each external node is one level lower in the original tree than in the subtree containing it. Since the values of s_i and t_i increase by 1, the cost contributed by the left subtree is

$$\sum_{k=i+1}^{r-1} p_i(1 + 1 + s_i) + \sum_{k=i}^{r-1} q_i(1 + t_i)$$

an increase of

$$\sum_{k=i+1}^{r-1} p_i + \sum_{k=i}^{r-1} q_i \text{ over } C(i, r - 1)$$

Similarly, the contribution of the right subtree is

$$C(r, j) + \sum_{k=r+1}^{j} p_i + \sum_{r+1}^{j} q_i$$

The contribution from the root itself is just

$$p_r$$

Note that the sums involving p_i and the sums involving q_i can be collapsed into a single sum each. Thus the overall cost of the tree, given the choice of k_r as the root, is just

$$C(i, r - 1) + C(r, j) + \sum_{k=i+1}^{j} p_i + \sum_{k=i}^{j} q_i$$

Note these last two terms depend only on i and j and not on r, so we may call their sum $S(i, j)$.

We may conclude that $C(i, j) = S(i, j) + \min_r \{C(i, r - 1) + C(r, j)\}$, where $i < r \leq j$. Note that $S(i, i) = q_i$ and if $j > i$, $S(i, j) = S(i, j - 1) + p_j + q_j$.

Although our discussion has assumed a top-down divide-and-conquer strategy, the principle of optimality suggests dynamic programming and a bottom-up approach. Smaller subtrees correspond to values of i and j that are close together. Recall the initial conditions give $C(i, i)$, and $C(0, n)$, gives the minimum cost. This leads to an iterative algorithm, given as Algorithm 2.68, where at iteration s, $C(i, i + s)$ is found for $i \geq 0$ and $i + s \leq n$.

Note that so far we have only considered how to find $C(0, n)$, not how to find the tree giving the minimum cost. The tree can be reconstructed after $C(0, n)$ is found if for each i and j the value of $\text{Root}(i, j)$ is assigned to be the value of r giving the minimum for $C(i, j)$.

```
1. OptimumBST(p, q)      {Takes arrays p and q, indexed from 0 to n
2.                        and 1 to n respectively, giving the
3.                        probability of a search's terminating at
4.                        respectively the ith internal and the ith
5.                        external node. Returns 2-dimensional
6.                        arrays C, Root, and S as discussed in the text}
7. S[0, 0] := q[0]                                    {initialize S}
8.   for i := 1 to n
9.     for j := 0 to i − 1 do
10.      S[j, i] := S[j, i − 1] + p[i] + q[i]         {for all i and j}
11.    end for
12.    S[i, i] := q[i]
13.    C[i, i] := 0                                   {and C[i, i] for all i}
14.  end for
15.  for d from 1 to n                        {for subtree sizes from 1 to n}
16.    for i from 0 to n − d                  {for all possible initial nodes i}
17.      j := i + d                           {find the final node j}
18.      min := ∞                             {and find C[i, j] as discussed}
19.      for k from i + 1 to j                {in the text}
20.        if C[i, k − 1] + C[k, j] < min then
21.          min := C[i, k − 1] + C[k, j]
22.          Root[i, j] := k                  {also keeping track of the root}
23.        end if
24.      end for
25.      C[i, j] := min + S[i, j]
26.    end for
27. end for
28. return C, S, Root
```

Algorithm 2.68 Optimum Binary Search Tree

Figure 2.29 treats the instance with values 5 for n; values 20, 15, 10, 5, and 10 for p_1 through p_5; and values 5, 10, 10, 5, 5, and 5 for q_1 through q_5. It gives a minimum-cost binary search tree and the values of C, S, and Root for this instance. In some cases there is more than one possible choice for Root$[i, j]$; in each of these cases the figure shows the smallest.

As an example of the construction of the table entries, we consider C[0, 5]. It is S[0, 5] + min{C[0, 0] + C[1, 5], C[0, 1] + C[2, 5], C[0, 2] + C[3, 5], C[0, 3] + C[4, 5], C[0, 4] + C[5, 5]} = 100 + min{0 + 155, 35 + 95, 95 + 45, 135 + 20, 170 + 0]} = 100 + 130 = 230. The value of the root giving the minimum 35 + 95 is 2, so this becomes the value of Root[0, 5].

To construct the tree from the table, we begin with Root$[0, n]$. As we have just seen, this value is 2. So p_2 corresponds to the root of the tree, the left subtree has cost C[0, 2 − 1], and the right subtree has cost C$[2, n]$.

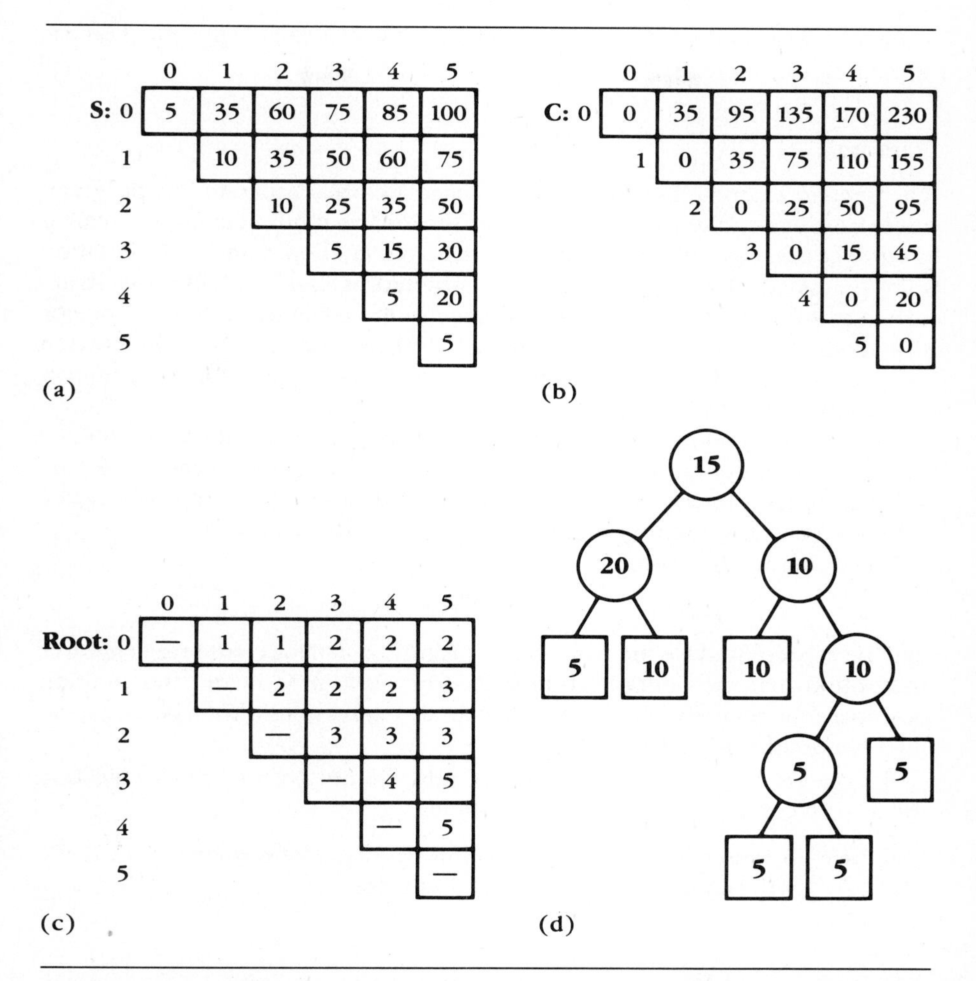

(a)

S:	0	1	2	3	4	5
0	5	35	60	75	85	100
1		10	35	50	60	75
2			10	25	35	50
3				5	15	30
4					5	20
5						5

(b)

C:	0	1	2	3	4	5
0	0	35	95	135	170	230
1		0	35	75	110	155
2			0	25	50	95
3				0	15	45
4					0	20
5						0

(c)

Root:	0	1	2	3	4	5
0	—	1	1	2	2	2
1		—	2	2	2	3
2			—	3	3	3
3				—	4	5
4					—	5
5						—

(d)

Figure 2.29 Example of the Dynamic Programming Algorithm for Optimal Binary Search Trees

Root[0, 2 − 1] and Root[2, n] give the respective roots 1 and 3 of these subtrees, and so on.

The time complexity of Algorithm 2.68 is dominated by that of the triple **for** loop of lines 15–27. This triple loop requires time $\sum_{d=1}^{n} \sum_{i=0}^{n-d} d$, which is $O(n^3)$. An exact solution may be obtained by evaluating the inner sum to get $d(n - d + 1)$ and evaluating the resulting sum $\sum_{d=1}^{n} d(n - d + 1) = \sum_{d=1}^{n} dn - \sum_{d=1}^{n} d^2 + \sum_{d=1}^{n} d$ by the techniques of Section 2.17. It is left as an exercise to show this sum is $n(n + 1)(n + 2)/6$, which is $\Theta(n^3)$. Thus Algorithm 2.68 has $\Theta(n^3)$ time complexity.

Parsing

Our second example hs to do with language. Before a program in a programming language or a sentence or other expression in a human language — called a *natural language* in this context — can be understood, some representation of its grammatical structure must be constructed, at least implicitly. This structure may be expected to be closely related to the meaning of the sentence or the action to be taken by the program. Here we are assuming the written language is defined as the set of character strings consistent with the grammar, a common practice for programming languages.

Informally, a *parsing* algorithm for a particular grammar will take a sentence or string as input and decide whether that input is legal with respect to the given grammar. It may also attempt to return a data structure (typically, a tree called a *parse tree*) representing the structure of the sentence.

For example, the sentence

$$x + y \times z$$

may be represented by the parse tree of Figure 2.30. Notice the tree captures the notion that the multiplication should be performed before the addition (compare the recursive definition of algebraic expressions). We have not specified the grammar used.

To analyze the parsing problem precisely, we need a precise specification for it. So we make the following definitions:

Def 2.69 A *sentence* over a set T is a finite sequence of elements from T. We will assume throughout that T is finite.

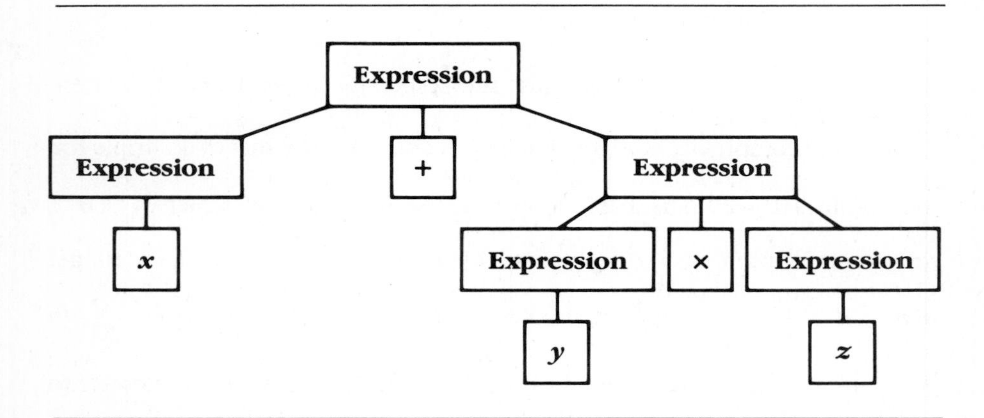

Figure 2.30 Parse Tree for the Expression $x + y \times z$

A *context-free grammar* over T consists of a finite set V of *variables,*[11] a distinguished element S (called the *start symbol*) of V, and a set P of *productions*. Each production has a *left-hand side* that is a variable and a *right-hand side* that is either 1) an element of T or 2) an ordered pair of elements of V.

The productions correspond to rewrite rules used to generate sentences from the start symbol. Using arrows to point from left-hand to right-hand sides of productions and angle brackets to enclose variable names, we may define the following productions for sample grammar for a small fragment of English:

$$
\begin{aligned}
&\langle\text{sentence}\rangle &&\rightarrow \langle\text{noun phrase}\rangle\ \langle\text{verb phrase}\rangle\\
&\langle\text{noun phrase}\rangle &&\rightarrow \langle\text{determiner}\rangle\ \langle\text{noun}\rangle\\
&\langle\text{verb phrase}\rangle &&\rightarrow \langle\text{verb}\rangle\ \langle\text{noun phrase}\rangle\\
&\langle\text{noun}\rangle &&\rightarrow \langle\text{boy}\rangle\\
&\langle\text{noun}\rangle &&\rightarrow \langle\text{girl}\rangle\\
&\langle\text{verb}\rangle &&\rightarrow \langle\text{likes}\rangle\\
&\langle\text{determiner}\rangle &&\rightarrow \langle\text{the}\rangle
\end{aligned}
$$

Here ⟨sentence⟩ is the start symbol.

Informally, a grammar G generates a string x over T iff x can be obtained from the start symbol by applying a succession of rewrite rules. For example, the sentence "the boy likes the girl" can be obtained as follows:

⟨sentence⟩
⟨noun phrase⟩ ⟨verb phrase⟩
⟨determiner⟩ ⟨noun⟩ ⟨verb phrase⟩
the ⟨noun⟩ ⟨verb phrase⟩
the boy ⟨verb phrase⟩
the boy ⟨verb⟩ ⟨noun phrase⟩
the boy likes ⟨noun phrase⟩
the boy likes ⟨determiner⟩ ⟨noun⟩
the boy likes the ⟨noun⟩
the boy likes the girl

The parse tree corresponding to this derivation is given in Figure 2.31.

In general, we make the following definitions:

Def 2.70 If α and β are strings over the set $V \cup T$, and string concatenation is indicated by juxtaposition, then

$\alpha\, A\, \beta \Rightarrow \alpha\, a\, \beta$ iff there is a production $A \rightarrow a$ in G

$\alpha\, A\, \beta \Rightarrow \alpha\, B\, C\, \beta$ iff there is a production $A \rightarrow BC$ in G

($\Rightarrow$ may be read "derives in one step")

$\alpha \overset{*}{\Rightarrow} \beta$ iff

1. $\alpha \Rightarrow \beta$ or
2. For some string γ over $V \cup T$, $\alpha \overset{*}{\Rightarrow} \gamma \Rightarrow \beta$

[11] More precisely, what we are defining is a context-free grammar in Chomsky normal form.

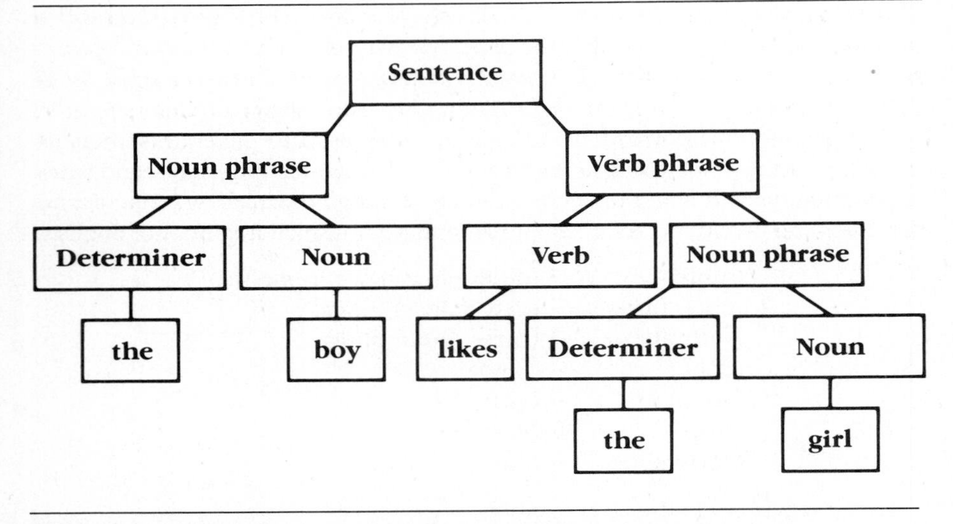

Figure 2.31 Parse Tree for the Sentence "The Boy Likes the Girl"

The relation $\stackrel{*}{\Rightarrow}$ may be read "derives in one or more steps."[12] For example,

$$\text{the } \langle\text{noun}\rangle\ \langle\text{verb phrase}\rangle \Rightarrow \text{the boy } \langle\text{verb phrase}\rangle$$

$$\langle\text{sentence}\rangle \stackrel{*}{\Rightarrow} \text{the boy likes } \langle\text{noun phrase}\rangle$$

$$\text{the } \langle\text{noun}\rangle\ \langle\text{verb phrase}\rangle \stackrel{*}{\Rightarrow} \text{the boy likes the } \langle\text{noun}\rangle$$

Finally,

Def 2.71 A string x over T is in L(G), the *language generated by G* iff S $\stackrel{*}{\Rightarrow} x$, where S is the start symbol for G.

According to this definition, the sentence "the boy likes the girl" is in the language generated by G because

$$\langle\text{sentence}\rangle \stackrel{*}{\Rightarrow} \text{the boy likes the girl.}$$

It is membership in the language generated by a grammar that we may test by a dynamic programming algorithm.

Note that if the length of x is greater than 1, the first production applied to S in the derivation of x must be of the form $S \rightarrow AB$. In this case x is in L(G) iff $x = yz, A \stackrel{*}{\Rightarrow} y$, and $B \stackrel{*}{\Rightarrow} z$. Thus we have reduced the parsing of x to the smaller problems of parsing y and z. However, these smaller problems are not strictly speaking of the same type, since neither A nor B is required to be the start

[12] The relation denoted by $\stackrel{*}{\Rightarrow}$ is called the *transitive closure* of that denoted by $\Rightarrow$. More generally, if R is a relation, then the transitive closure R* is defined by R*(a, b) iff (1) R(a, b) or (2) R*(a, c) for some c with R(c, b); or equivalently R(a, c) for some c with R*(c, b).

symbol. So our algorithm will answer a more general question, namely the question, "for what variables A does $A \stackrel{*}{\Rightarrow} x$?"

For a fixed string x, let $S(i,j)$ be the set of variables deriving the substring of x beginning at position i and ending at position j. If c is the ith character of x, then $S(i, i) = \{A \mid A \rightarrow c$ is a production of the grammar$\}$. If $i < j$, then $S(i,j) = \{C \mid$ for some k, $C \rightarrow AB$ is a production of the grammar, A is in $S(i, k)$ and B is in $S(k + 1, j)\}$. This captures formally the insight of the previous paragraph that A should derive the first half of the string, B should derive the second half, and C should derive A followed by B.

If the start symbol S is a member of $S(0, n)$, then the string can be derived. To recover a parse, each element of $S(i, j)$ should be accompanied by the values of k, A, and B that justified its inclusion in the set.

The iterative version, given as Algorithm 2.72, of the dynamic programming algorithm previously described calculates the entries in a table for S in increasing order of $j - i$. For a given i and j, all permissible values of k must be considered, and for all k, all pairs of variables (A, B) from $S(i, k) \times S(k + 1, j)$ must be considered. Given a pair (A, B), the final step is to determine which variables derive AB.

Note that the size of $S(i, k)$ and $S(k + 1, j)$ are independent of the length n of the string x. Thus the time required to find $S(i, j)$ is proportional to the number of choices for k, which is $\Theta(j - i)$. If d is defined to be $j - i$, the number of leaves in the subtree, then there are $d - 1$ possible choices for k, and i may range from 1 to $n - d$. The overall time complexity is measured by the sum over all d of the time to find $S(i, j)$ for all i and j differing by d. For a fixed d, this time is proportional to $\sum_{i=1}^{n-d} d - 1 = (d - 1)(n - d)$.

The overall time complexity is then

$$\sum_{d=1}^{n-1} (d - 1)(n - d) = \sum_{d=1}^{n-1} n(d - 1) - \sum_{d=1}^{n-1} d(d - 1)$$

Each sum is $\Theta(n^3)$; the recurrence relation techniques of Section 2.17 may be used (the details are left as an exercise) to show that the leading coefficients are different. Thus the difference is $\Theta(n^3)$, and this is the time complexity of the CYK parsing Algorithm 2.72.[13]

The table of Figure 2.32a gives the table for S for the input string *abbaa* in a grammar whose productions are $S \rightarrow BS$, $S \rightarrow AS$, $S \rightarrow BA$, $A \rightarrow SA$, $A \rightarrow AA$, $B \rightarrow SS$, $B \rightarrow BB$, $B \rightarrow AB$, $A \rightarrow a$, and $B \rightarrow b$. Since S is a member of $S(1, 5)$, the string is generated by the grammar; one parse tree for the string is given in part b of the figure.

It is worth explaining some sample entries in the table. $S(1, 1)$ contains only A, because only A derives a, the first symbol of the string, in one step. $S(1, 2)$

[13] The name CYK comes from Cocke, Younger, and Kasami. You may see other permutations of these initials.

```
1. CYK(string, p)               {p represents the grammar; see below
2.             Productions are stored in an array p indexed from 0 to np.
3.                 Each production has integer-valued fields lhs, first, and
4.               second, containing respectively the variables on the left
5.              hand side, at the beginning of the right-hand side, and at
6.                   the end of the right-hand side of the production. A
7.         function terminalRHS is assumed which will determine whether
8.                 the right-hand side of a production consists of a single
9.                          terminal. If so, the terminal is given by first
10.            The input string is stored in an array indexed from 1 to n.
11.                      The size of the string is assumed to be available via
12.              size(string). S[i, j] contains variable A iff A derives the
13.            substring beginning at position i and ending at position j.
14.          Functions are assumed which will add a variable to a set and
15.               determine whether a variable is a member of a set. The
16.                                      algorithm returns the array S.
17.                      It also returns arrays where, left, and right that,
18.              whenever A is in S[i, j, k], contain (1) in where[i, j, k] a
19.         number m such that B derives the substring from character i
20.         to character m, C derives the string from character m + 1 to
21.           character j, and A → BC is a production (2) B in left[i, j, k]
22.                    and (3) C in right[i, j, k]. These values will allow
23.             reconstruction of the parse tree. There is a parse from the
24.                                      given input string iff S is in S[0, n]}
25. n := size(string)
26. for i from 1 to n
27.    for j from 1 to n
28.       S[i, j] := {}                                     {initialize S}
29.    end for
30. end for
31. for i from 1 to n                      {for each character in the string}
32.    for j from 1 to np do                       {check each production}
33.       if terminalRHS(p[j]) and first(p[j]) = string[i] then
34.          Add(lhs(p[j], S[i, i])                    {and add its LHS if it}
35.       end if                                    {derives the character}
36.    end for
37. end for
38. for d from 1 to n − 1                          {for each substring length}
39.    for i from 1 to n − d              {for each substring of that length}
40.       j := i + d                                      {find the endpoint}
41.       for k from i to j − 1                         {and for each possible}
42.          for m from 1 to np                          {midpoint, try each}
43.             if member(first(p[m]), S[i, k]) and
                   member(second(p[m]), S[k + 1, j]) then
```

```
44.                 Add(lhs(p[m]), S[i,j])                {production for that}
45.                 where[i, j, lhs(p[m])] := k                {midpoint. If it}
46.                 left[i, j, lhs(p[m])] := first(p[m])
47.                 right[i, j, lhs(p[m])] := second(p[m])
48.             end if                          {works, update S, where,}
49.         end for                                    {left, and right}
50.     end for
51.   end for
52. end for
53. return S,where,left,right
```

Algorithm 2.72 CYK Algorithm for Parsing

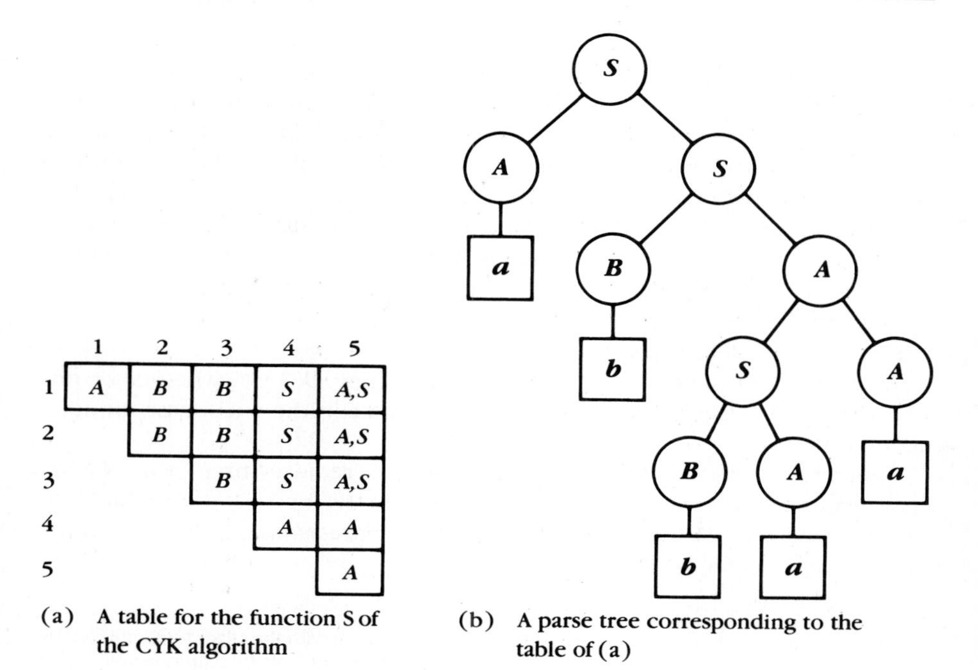

	1	2	3	4	5
1	A	B	B	S	A,S
2		B	B	S	A,S
3			B	S	A,S
4				A	A
5					A

(a) A table for the function S of the CYK algorithm

(b) A parse tree corresponding to the table of (a)

Figure 2.32 Example of the CYK Algorithm for Parsing

contains all variables deriving AB, since $S(1, 1) = \{A\}$ and $S(2, 2) = \{B\}$. Only B qualifies.

An entry may be included in $S(1, 5)$ by pairing a variable from $S(1, 1)$ with one from $S(2, 5)$; or variables from $S(1, 2)$ and $S(3, 5)$; or variables from $S(1, 3)$ and $S(4, 5)$; or variables from $S(1, 4)$ and $S(5, 5)$. These pairs are AA, AS, BA, BS, BA again, and SA. The only productions with any of these pairs on the right-hand side have A or S on the left-hand side, so $S(1, 5) = \{A, S\}$.

The design of an algorithm to construct a parse tree from the sets $S(i, j)$ is left as an exercise.

All-Pairs Shortest Paths

Our final example deals with directed graphs. Given a directed graph G containing vertices x and y, a natural question would be what is the shortest or cheapest path between x and y? To be more precise, we make the following definitions:

Def 2.73 A *path* in a directed graph is just a sequence of edges $\langle(v_i, w_i)\rangle_{i=1}^{m}$ such that $w_i = v_{i+1}$ for $1 \le i < n$.[14] If $v_1 = w_n$ then the path is a *cycle.* If there are no other repetitions of vertices, then the path or cycle is said to be *simple;* in the following we assume all paths and cycles are simple unless otherwise specified. A graph is *connected* iff there is a path from every vertex to every other vertex.

To analyze graph algorithms we need to specify the representation for the graph. The *adjacency list* representation provides for each vertex v a list of all vertices w in the graph such that (v, w) is an edge.[15] The *adjacency matrix* representation imposes an ordering on the vertices and provides a 2-dimensional array whose (i, j) entry is nonzero iff (v_i, v_j) is an edge. If each edge has a cost, this cost can be the (i, j) entry. Otherwise, 1 is usually used as the entry.

Figure 2.33 shows a directed graph with edge costs and two adjacency matrices representing the graph. In the first matrix, no cost information is stored; all entries are 0 or 1. In the second, the cost information is stored. An entry of ∞ is used to represent a missing edge.

Adjacency matrices allow determination in time $O(1)$ of whether or not two vertices are adjacent. This requires $\Theta(n)$ time in the worst case for adjacency lists.

Finding all vertices adjacent to a given vertex requires time $\Theta(n)$ if adjacency matrices are used. Adjacency lists have better best-case behavior. It is easier to add new vertices in the adjacency list representation, since the vertices themselves may be stored in a list.

---

[14] Any undirected graph may be considered to be a directed graph with the edge $\{v, w\}$ replaced by (v, w) and (w, v). Thus paths may be defined for undirected graphs simply by requiring undirected edges $\{v_i, w_i\}$.

[15] If desired, a second list for v can contain all w such that (w, v) is an edge.

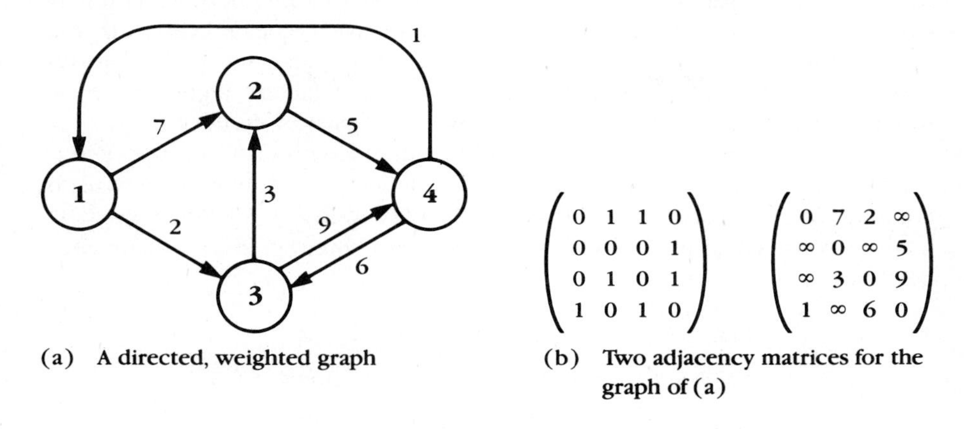

Figure 2.33 A Directed, Weighted Graph and Two Adjacency Matrices

In the cheapest path problem, we assume a nonnegative cost for each edge (a missing edge is considered to have infinite cost) and define the cost of a path from x to y as the sum of the cost of the edges making up the path. The shortest path is simply the special case of the cheapest path when all edges have cost 1. Unfortunately, many authors do not make this distinction between shorter paths and cheaper paths.

In the graph of Figure 2.33, the shortest path from vertex 1 to vertex 4 has length 2, but the cheapest path has cost $2 + 3 + 5 = 10$.

In general, let $C(x, y)$ be the cost of the edge from x to y and $P(x, y)$ be the cost of the cheapest path from x to y. Assume there are n vertices in the graph. The number of paths from x to y is too large to enumerate efficiently, so sequential search through all possible path costs is not practical. However, each path from x to y of length greater than 1 is composed of shorter paths. If z is an intermediate vertex on the cheapest path from x to y, the paths from x to z and from z to y must be minimal as well. In other words, $P(x, z) + P(z, y) = P(x, y)$, so the principle of optimality holds.

There is no way to choose in advance a vertex z that is guaranteed to be on the cheapest path. However, the principle of optimality suggests a dynamic programming approach. The subproblem of finding the cheapest path from x to z and from z to y is more complicated than the original problem since we now must consider all vertices in the graph and not just x and y. We will do so for the rest of this section. The problem of finding the shortest path—by which we actually mean the cheapest path, the name of the problem being too well established to modify—between every x and every y is called the *all-pairs shortest paths* problem.

However, the subproblem is simpler than the original problem in the sense that the path lengths are shorter, so we may construct a recurrence relation based on the length of the path. We can easily see that no cheapest path may have length greater than n; any such path would have a repeated vertex w, and the cycle from w to itself could be omitted from the cheapest path. Unfortunately, this approach does not give us the most efficient algorithm, even though it will find the cheapest path between all pairs of vertices.

If we define $R(x, y, t)$ to be the cheapest path of length at most t from x to y, then knowing the vertex z occurring before y on this path would allow us to compute $R(x, y, t)$ as the sum of $R(x, z, t-1)$ and $R(z, y, 1)$. Since we don't know z, we need to minimize this sum over all choices of z. The recurrence becomes

$$R(x, y, 1) = C(x, y) \text{ if } x \neq y \text{ and } 0 \text{ if } x = y$$

$$R(x, y, t) = \min_z\{R(x, z, t-1) + R(z, y, 1)\}$$

$P(x, y)$ is just $R(x, y, n)$, so we may evaluate $R(x, y, t)$ in increasing order of t. Such a bottom-up calculation of P would require an outer loop for t, an inner loop for z, and intermediate loops for x and y. Since each loop would need to be iterated n times, we would get an algorithm with time complexity $\Theta(n^4)$ for the problem of finding the cheapest paths between all pairs of points.

This algorithm can be improved; see Exercise 4.83.

A more efficient algorithm can be constructed by combining the loops for z and t. In the preceding algorithm, there was no particular reason for requiring z to be the next-to-last vertex on the path. We may instead number the vertices, which is necessary for the adjacency matrix representation anyway, and consider that intermediate vertex z with the highest number. By defining $R(x, y, t)$ to be the cost of the cheapest path with no intermediate vertex numbered higher than t, we don't have to worry about path lengths. The paths from x to vertex t and from vertex t to y have no intermediate vertex numbered higher than $t - 1$; therefore, if we identify vertices x and y with their indices, our recurrence is

1. $R(x, y, 0) = C(x, y)$ if $x \neq y$ and 0 if $x = y$
2. $R(x, y, t) = \min\{R(x, y, t-1), R(x, t, t-1) + R(t, y, t-1)\}$

In other words, if allowing the intermediate vertex t gives us a cheaper path, we use this new path cost. Otherwise, we use the old value.

Notice the old value is best if $x = t$ or $y = t$. This means that in the iteration corresponding to t, the terms $R(x, t, t-1)$ and $R(t, y, t-1)$ do not change. The term $R(x, y, t-1)$ may change, but it is irrelevant for calculating $R(u, v, t)$ if (u, v) is different from (x, y). This means that the same array for R may be used for any value of t, so that the array need be only 2-dimensional. Consequently, the third subscript is dropped in the all-pairs Algorithm 2.74. In the algorithm, we have not attempted to skip the cases $x = y$ or $x = t$ or $y = t$.

The algorithm is based on that of Floyd (1962).

The time complexity of this algorithm is dominated by the three nested loops and is therefore $\Theta(n^3)$.

```
1. ShortestPaths(C)             {C is an n by n array. C(i, j) represents
2.                               the cost of the edge from vertex i to vertex j
3.                               in a directed graph. The algorithm returns an
4.                               n by n array R where R(i, j) represents the cost
5.                               of the cheapest path from vertex i to vertex j}
6. for i := 1 to n
7.    for j := 1 to n
8.       R(i, j) := C(i, j)                 {with no intermediate vertices,}
9.    end for                               {path costs are edge costs}
10. end for
11. for t := 1 to n                         {consider t as the highest
12.                                          numbered intermediate vertex}
13.    for i := 1 to n
14.       for j := 1 to n                   {for all pairs of vertices}
15.          if R(i, t) + R(t, j) < R(i, j) then
16.             R(i, j) := R(i, t) + R(t, j)        {if allowing t gives a}
17.          end if          {cheaper path, then include t in the path}
18.       end for
19.    end for
20. end for
21. return R
```

Algorithm 2.74 All-Pairs Shortest Paths

The entries of the matrix R at the end of each iteration are given in Figure 2.34 for the graph of Figure 2.33. The final matrix gives the cheapest path from every vertex to every other vertex.

2.20 Recursive Verification of Greedy Algorithms

We have seen that many useful algorithms can be formulated as finite sequences of guesses. Boolean satisfiability, ZOK, the TSP, the Hamiltonian circuit problem, and the problems of the previous section all have this property.

Unfortunately, it is seldom possible to make a correct sequence of guesses without ever guessing wrong. Generally, optimality of the sequence of guesses can't be checked until most or all of the guesses have been made. For example, in ZOK, guessing to include a weighty item might preclude the inclusion of two smaller items with a larger combined profit.

Sometimes it is useful to imagine a mythical being with perfect foresight able to make the correct guess in every case. This being would require time $\Theta(n)$ to find a solution to ZOK. The decision procedure for this mythical being

$$\begin{pmatrix} 0 & 7 & 2 & \infty \\ \infty & 0 & \infty & 5 \\ \infty & 3 & 0 & 9 \\ 1 & \infty & 6 & 0 \end{pmatrix}$$

(a) Original adjacency matrix

$$\begin{pmatrix} 0 & 7 & 2 & \infty \\ \infty & 0 & \infty & 5 \\ \infty & 3 & 0 & 9 \\ 1 & 8 & 3 & 0 \end{pmatrix}$$

(b) After allowing intermediate vertex 1

$$\begin{pmatrix} 0 & 7 & 2 & 12 \\ \infty & 0 & \infty & 5 \\ \infty & 3 & 0 & 8 \\ 1 & 8 & 3 & 0 \end{pmatrix}$$

(c) After allowing intermediate vertices 1 and 2

$$\begin{pmatrix} 0 & 5 & 2 & 10 \\ \infty & 0 & \infty & 5 \\ \infty & 3 & 0 & 8 \\ 1 & 6 & 3 & 0 \end{pmatrix}$$

(d) After allowing intermediate vertices 1, 2, and 3

$$\begin{pmatrix} 0 & 5 & 2 & 10 \\ 1 & 0 & 8 & 5 \\ 9 & 3 & 0 & 8 \\ 1 & 6 & 3 & 0 \end{pmatrix}$$

(e) After allowing intermediate vertices 1, 2, 3, and 4

Figure 2.34 Example of the All-Pairs Shortest Paths Algorithm

could not be determined in advance; it corresponds to a *nondeterministic* algorithm. The concept of nondeterminism can be formalized; we do so at the end of Chapter 4.

Since we do not have nondeterministic machines, we must think of nondeterminism as an ideal to be approached. Some of the techniques of Chapters 4 and 5 can be thought of as ways of simulating nondeterminism.

In this section, we consider those problems for which correct guesses can be made solely on the basis of the information available at the time of the guess. Since these guesses are made on the basis of short-term calculations rather than on the basis of a concern about future choices, algorithms based on this approach are often called *greedy* algorithms.

The design of a greedy algorithm is often straightforward. The hard part about using a greedy algorithm, as you might have guessed, is proving the algorithm does indeed solve the problem it is supposed to solve. Fortunately, there is a proof technique that often works.

The method of proof used in the examples is a variation of the minimal counterexample argument. In general the greedy solution is feasible but not obviously optimal. We assume the greedy solution is not optimal and choose an optimal solution as close as possible to the greedy solution according to some measure of closeness. We then show, by using the greedy nature of the solution, that the optimal solution can be modified so that it remains optimal but is closer to the greedy solution. This contradiction disproves our assumption that the greedy solution is not optimal.

As you might expect, the hard part about proving correctness of a greedy algorithm lies in defining a measure of closeness between two feasible solutions. In constructing a proof, usually we start with an optimal solution, modify the optimal solution so that it intuitively seems closer to a greedy solution, and only then determine a measure of closeness corresponding to the intuitive insight.

We consider four examples below. Several examples of problems for which the natural greedy approach does not work are considered in the exercises.

Minimum Cost Spanning Trees

Our first example deals with networks. Imagine a group of islands that are not linked by any bridges. Someone proposes to connect them in such a way that any island is reachable from any other and with the minimum construction cost consistent with doing so. Assume the cost of connecting any pair of islands is known or can be estimated. Is there an algorithm for finding the least expensive method of connection? Analogous problems might arise for pipelines, communications, and other networks.

A more abstract and general statement of the problem would run as follows: Given an undirected graph with a positive cost assigned to each edge, find a subgraph on the same set of n vertices that is connected and has the minimal cost over all such subgraphs. Here the cost of a subgraph is the sum over all edges in the graph of the edge costs. We introduce one term:

Def 2.75 A connected subgraph with no cycles containing all vertices of the original graph is called a *spanning tree.*[16]

In this new terminology we are looking for the minimum cost spanning tree. Note that the solution to our problem cannot have a cycle and need only have $n - 1$ edges if the graph has n vertices.

Figure 2.35 shows a graph and two of its spanning trees.

One approach to such an algorithm would be to sort the edges by cost and to include the $n - 1$ cheapest. Unfortunately this approach will not work in general, because the $n - 1$ cheapest edges may not form a connected subgraph.

---

[16] We haven't yet explained how a tree can be a graph. A *free tree* is a connected graph with no cycles. An *oriented tree* is a free tree with a root. Note that this choice of a root defines a direction "away from the root" for all edges, giving the usual concept of a tree. More precisely, the parent/child relation may be defined by saying 1) the root has no parent and 2) the children of a node are all the adjacent nodes except for its parent. Of course, we also say x is the parent of y iff y is the child of x. In addition, if an order is imposed on the set of children of each node, the result is an ordered tree. A free tree can correspond to many different ordered trees.

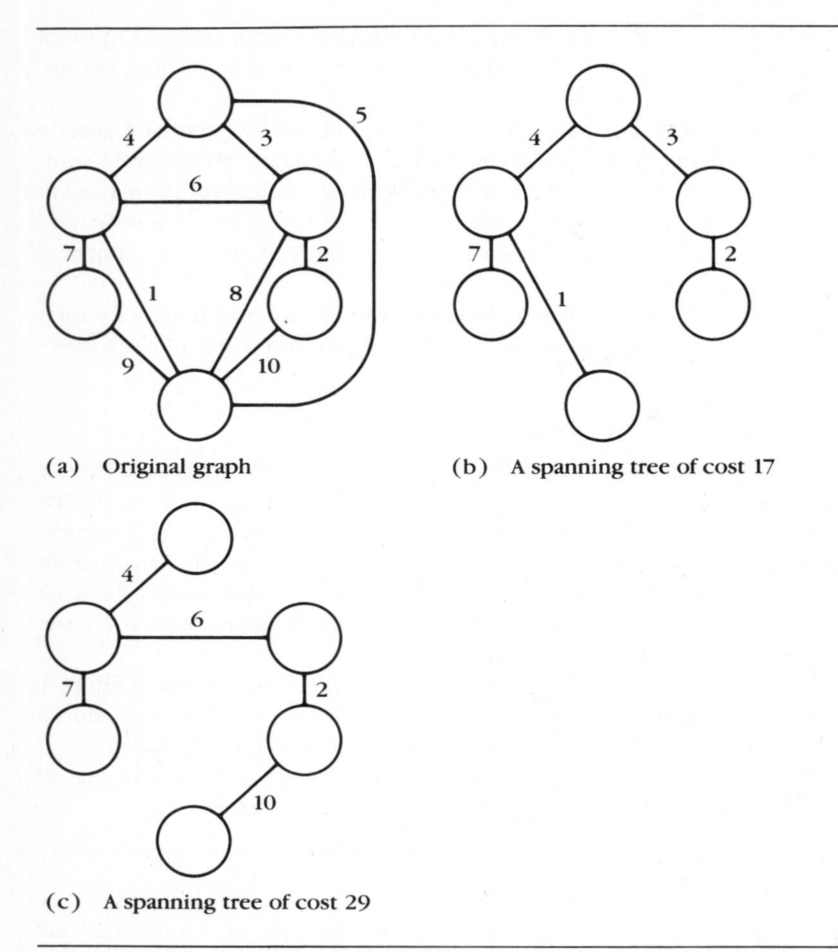

Figure 2.35 An Undirected Graph and Two Spanning Trees

The next simplest approach would be to consider edges in order of cost starting with the cheapest and to include those that do not complete cycles. Such an algorithm would terminate when $n - 1$ edges have been included.

This approach will work, although we defer until Section 4.9 the question of how to efficiently determine whether or not inclusion of an edge will complete a cycle. In Algorithm 2.76, attributed to Kruskal (1956), we have not specified how to determine cheapest remaining edge or how to determine whether an edge completes a cycle.

Note that a heap can be used to store the edges. If there are e edges in the graph, the heap can be constructed in $O(e)$ time and the next cheapest edge can be found in $O(\log e)$ time. In the worst case, $\Theta(e)$ edges must be considered. In Section 4.9 we learn that the test for completion of a cycle takes time $O(\log e)$, so that the $O(e \log e)$ cost for finding and considering $\Theta(e)$ edges is

```
 1. MinimumSpanningTree(G, n, c)  {Given a graph G with n vertices
 2.                  and cost matrix c, finds a minimum cost spanning tree.
 3.               Assumes algorithm for AddEdge for adding an edge to the
 4.      tree, GetNextEdge for getting the cheapest remaining edge, and
 5.              Cycle for determining whether an edge completes a cycle}
 6. T := {}
 7. for i := 1 to n − 1
 8.    repeat
 9.       (x, y) := GetNextEdge (G)
10.    end repeat if not Cycle (x, y, G)
11.    AddEdge(x, y, T)
12. end for
```

Algorithm 2.76 Minimum Spanning Tree

the dominant cost of the algorithm. Here e is $O(n^2)$, so $\log e$ is $O(\log n^2)$ or $O(2 \log n)$ or $O(\log n)$, and $e \log e$ is $O(n^2 \log n)$.

For example, Figure 2.36a shows a graph with 11 edges. The integer weights range from 1 to 11 with no duplication. The edges with weights 1 and 2 may be included, but the edge with weight 3 completes a cycle. The edges with weights 4 and 5 may be included, but the edge with weight 6 completes a cycle. The edge with weight 7 may be included; this is the fifth and last edge of the spanning tree. The spanning tree is shown in (b) of the figure.

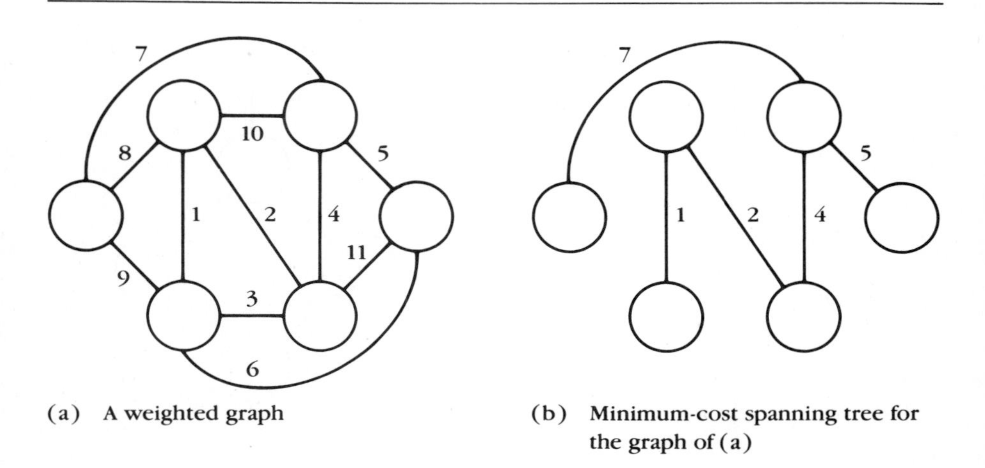

(a) A weighted graph

(b) Minimum-cost spanning tree for the graph of (a)

Figure 2.36 Example of Kruskal's Algorithm for Finding Minimum-Cost Spanning Trees

Theorem Kruskal's algorithm always gives an optimal solution.

Proof Clearly the algorithm will return a spanning tree if one exists. Let T be the spanning tree returned by the greedy algorithm. Assume T is not the minimum-cost spanning tree. In this case, for every optimal tree there will be a cheapest edge from T that is not in the optimal tree. Let O be an optimal solution for which this edge is as costly as possible. In other words, if we imagine the edges of an optimal solution to be generated in order of cost, let O be the optimal solution that differs from T at the lastest possible position.

Consider the cheapest edge x from T that is not in O. Since O is a tree, addition of the edge x to O will create a cycle. Since T contains no cycles, this cycle must contain an edge y not in T. Consider the subgraph P obtained by adding x to O and deleting y (see Figure 2.37). Since every vertex in the aforementioned cycle remains connected to every other such vertex, as well as to every other vertex in the graph, P is connected. Since P has $n - 1$ edges, P must be a spanning tree.

Now the edge y must be at least as expensive as x, since otherwise the greedy algorithm would have considered y before x and included it in T (y could not have been rejected for completing a cycle, since without x it does not complete a cycle in O). Consequently, P has cost at least as small as O. Since O is optimal, its cost is at least as small as P; hence, P and O have the same cost, and P is optimal. In addition P, which includes x, is closer to T than O is, violating our choice of O and leading to a contradiction. We therefore conclude that T is optimal. ■

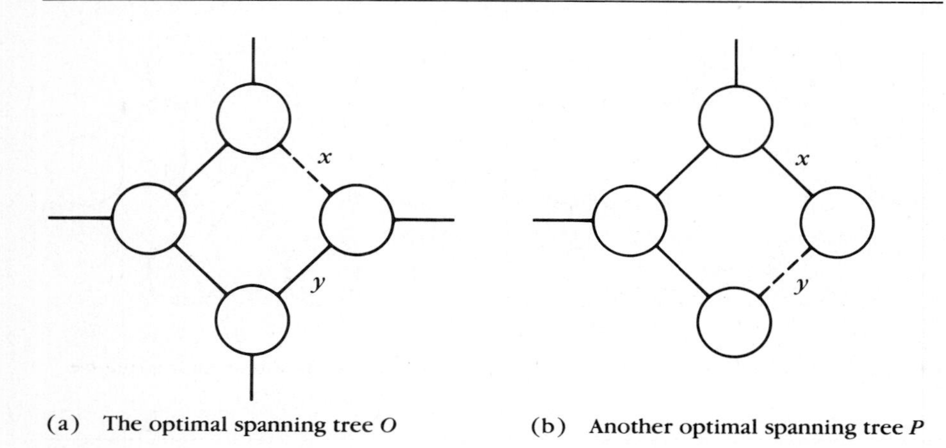

(a) The optimal spanning tree O (b) Another optimal spanning tree P

Figure 2.37 Modifying Optimal Spanning Trees

Minimum Length Encoding

Our next example deals with text processing. For most computational applications, text is usually stored as a sequence of characters, each represented by a fixed-length string of bits, such as the ASCII code mentioned in Section 2.3 and defined in Appendix B. In this section we assume each character is encoded as a string of bits.

The use of the words "code" or "encode" does not imply secrecy. In computer science the word *encryption* is used instead of "encoding" to mean that secrecy is desired.

ASCII and other fixed-length character codes have the great advantage of ease in determining where the code for one character ends and where the next begins. A potential disadvantage is that the more common codes require just as much space as the less common codes. If the code for the more common characters were shorter, then length of the average encoded text would presumably decrease. If you are familiar with international Morse code, compare the encoding of the letters "E" and "T," requiring one bit each, with that of such uncommon letters as "J" and "Q," which require four bits each. Note that even the international Morse code does not encode text as a string of bits, since an encoded message consists of dots, dashes, and spaces.

To minimize the expected length of encoded text, we would need a large amount of information on what characters are likely to occur together. We may formulate a simpler minimization problem by asuming the probability of occurrence of any character is independent of its location in the text. With this assumption, minimizing the expected length of an encoded text is the same as minimizing the expected length of a single character code.

In our version of the minimal length encoding problem, we assume a fixed, finite set of characters, each with a given probability of occurrence. We then ask how to construct a code so that encoding and decoding can be done unambiguously and how to find such a code with the minimum expected length.

The following example shows how ambiguity might be a problem. Suppose the character "A" had code 00 and the character "B" had code 000. The encoded string 000000 could then stand either for "AAA" or "BB." In this section we indicate the concatenation of two strings x and y as simply xy.

The problem with the example is that the code for "A" is a prefix of the code for "B."[17] Upon reading the first two 0's, it is not clear whether we have reached the end of the code for "A" or are still in the middle of the code for "B." We say a character code has the *prefix property* iff it is not the case that the code for one character is the prefix of the code for another. The character code in the example does not have the prefix property. A character code with the prefix property is subject to ambiguity in decoding.

[17] Recall that if x and y are strings of bits or characters, then x is a *prefix* of y iff there is some string z such that $xz = y$, that is, if x occurs at the beginning of y. Additionally, x is a *suffix* of y iff there is some string z such that $y = zx$.

A character code with the prefix property has a natural representation as a binary tree. Each node in the tree is associated with a string of bits. The root is associated with the empty string. The left child of the node for string x is associated with the string $x0$ (i.e., x concatenated with 0), and the right child with the string $x1$. The prefix property insures that characters and their codes are associated only with leaves.

In the tree of Figure 2.38, the character "A" is encoded as 010, "E" as 00, "N" as 110, "O" as 111, "S" as 011, and "T" as 10. We easily see from the codes themselves, or because of the tree, that this character code has the prefix property.

The decoding algorithm for such a tree is relatively simple. Begin at the root of the tree and read the encoded string one bit at a time. Descend left after reading a 0 bit and right after reading a 1 bit. Upon reaching a leaf, output the associated character and return to the root of the tree. Turning this description into a formal algorithm is left as an exercise.

For example, the string 1000110 could be decoded as "TEN" without ambiguity.

If there are n characters $\{c_i\}_{i=1}^{n}$, character c_i has probability p_i of occurrence, and the code for c_i has length s_i, then the expected code length is $\sum_{i=1}^{n} p_i s_i$. It is this sum that we want to minimize. Note that this sum, like the sum minimized by the optimal binary search tree Algorithm 2.68, is merely a weighted external path length.

In the binary tree interpretation, the more frequently occurring characters should occur higher up in the tree, since characters higher in the tree have shorter code lengths (i.e., appear at lower-numbered levels).

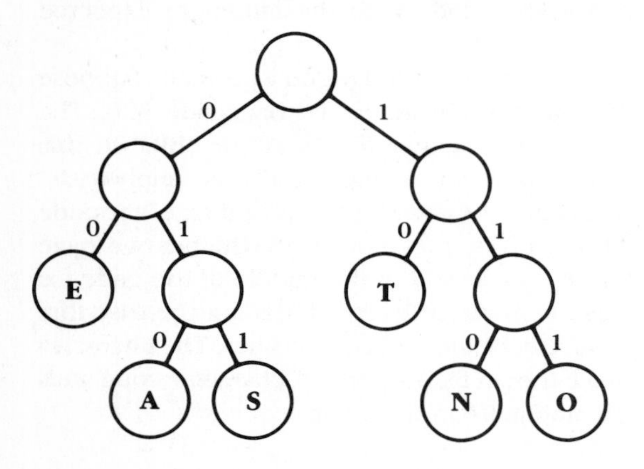

Figure 2.38 Binary Tree Representation of a Character Code with the Prefix Property

The natural divide-and-conquer approach would ask which characters are to appear in the left subtree and which in the right, corresponding to construction of the binary tree from the top down. This approach is awkward, since there is no particular reason to assign any character to either of the subtrees.

Instead we consider construction of the tree from the bottom up. We begin with a set, or *forest,* of n trees of one node each; each node contains a single character and each node is assigned a weight proportional to the probability of the character. Since we expect and will eventually prove that the two least frequent characters should be at the higher-numbered levels of the tree, we combine the corresponding nodes by creating a new internal node and making the two least-weighty nodes its children. It doesn't matter which is the left child. The resulting tree is assigned a weight equal to the sum of the weights of the two nodes. We have thus reduced the problem from one with n trees to one with $n - 1$ trees.

In general given k trees, we find the two with the smallest weights and combine them by making them the two children of a new root. To the resulting tree we assign a new weight equal to the sum of the two original weights. This leaves a set of $k - 1$ trees and their weights. When there is only one tree left, we are done.

Algorithm 2.77, attributed to D. Huffman (1952), embodies this greedy approach.

Figure 2.39 shows the initial stages and the final stage of the workings of the algorithm for an instance with 6 characters. The character "A" has weight 28, "B" has weight 5, "C" 15, "D" 16, "E" 29, and "F" 7. The respective codes turn out to be 10, 0100, 011, 00, 11, and 0101.

Finally, we need to show correctness.

Theorem The Huffman algorithm is correct.

Proof by minimal counterexample. Suppose it is not correct. Let n be the smallest number of characters for which the greedy algorithm fails. Let T be the nonoptimal tree constructed by the greedy algorithm, and let O be an optimal tree for the same set of characters and weights.

Let c_1 with weight p_1 be the least-frequently-occurring character in the character set. We claim that c_1 must appear on the lowest level of O. If not, let c be a character with weight p appearing on the lowest level h (nodes on the lowest level must be leaves and thus must correspond to characters). Assume that c_1 appears on level j, so that $h > j$. If c and c_1 swap locations in O, then in the sum representing the cost measure, the terms $p_1 j$ and ph are replaced by $p_1 h$ and pj. This changes the total cost by $pj + p_1 h - (p_1 j + ph) = p(j - h) + p_1(h - j) = (p - p_1)(j - h)$. Since the first factor of this increment is nonnegative and the second is negative, the product is nonpositive. Thus the swap has not increased the total cost over the optimal cost of O, and the resulting tree is also optimal.

We may apply the same argument to the second-cheapest node in O and assume it also lies on the lowest level. Note the lowest level must contain at

```
 1. Huffman(n, forest)              {Takes a number n of characters and
 2.                          an array forest of size n, each element being a
 3.                          pointer to a binary tree of size 1. Each node of
 4.                          each tree contains one of the characters stored in
 5.                          a char field and its probability or weight stored
 6.                          in a weight field. Assumes a function Findmin
 7.                          which given a number i returns the two indices in
 8.                          forest of the tree whose root has minimum weight
 9.                          over all indices less than or equal to i. Returns
10.                          a binary tree with minimal weighted internal path
11.                                                                   length.}
12. i := n                                {until all characters have been}
13. while i > 1                           {combined into one binary tree}
14.    (min1, min2) := Findmin(i)         {find the two smallest weights}
15.    temp := Get                        {and combine the corresponding}
16.    left(temp) := forest[min1]         {trees into a new tree whose}
17.    right(temp) := forest[min2]        {weight is the sum of the old}
18.    weight(temp) := weight(forest[min1]) + weight(forest[min2])
19.    char(temp) := dummy                {weights and which has a dummy
                                                               character}
20.    forest[min1] := temp               {replace the two trees by}
21.    forest[min2] := forest[i]          {the new tree, which decreases}
22.    i := i - 1                         {the number of trees by 1}
23. end while
24. return forest[1]
```

Algorithm 2.77 Minimal Length Encoding

least two leaves that are siblings. In fact, since the cost measure depends only on the level of a character, not on its precise position in the tree, we may assume that the cheapest and second-cheapest nodes are siblings on the lowest level.

Let $\{p_i\}_{i=1}^{n}$ be the set of probabilities in our counterexample and assume that p_1 and p_2 are the cheapest probabilities. If $n < 3$, we are done. Otherwise since we may assume p_1 and p_2 belong to the lowest level, by minimality, the greedy solution to the instance with probabilities $\{p_1 + p_2\} \cup \{p_i\}_{i=3}^{n}$ is optimal. (We may assume $n \geq 3$; the contrary case is trivial.) Because of the way the greedy algorithm works, the corresponding tree T' differs from T only by having the subtree whose children are leaves with weights p_1 and p_2 replaced by a single node with the weight $p_1 + p_2$. The corresponding change in the cost from T to this optimal tree (see Figure 2.40) is $(p_1 + p_2)(h - 1) - (p_1h + p_2h) = (p_1 + p_2)(h - 1 - h) = -(p_1 + p_2)$. Call this value K.

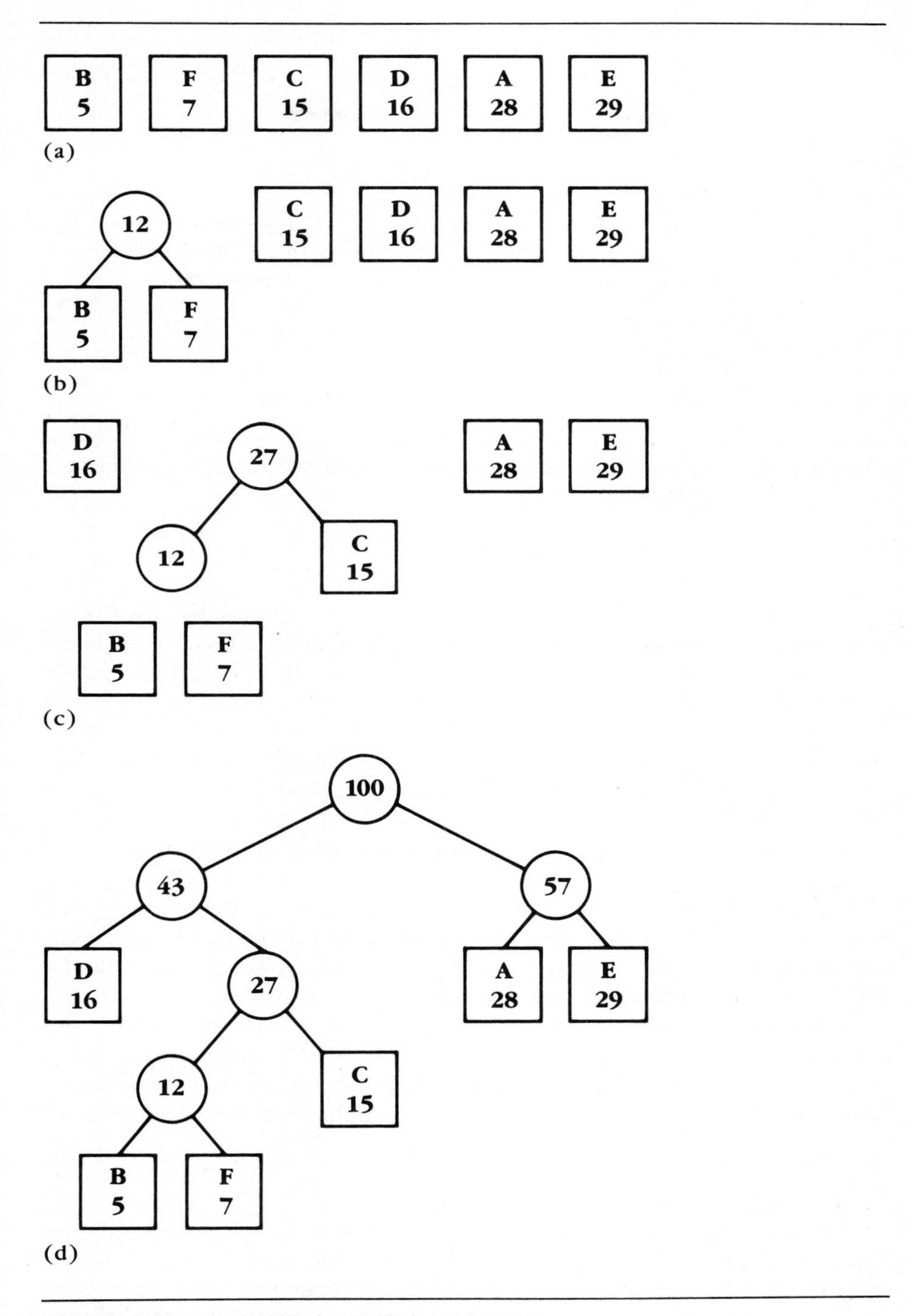

Figure 2.39 Stages in the Huffman Algorithm

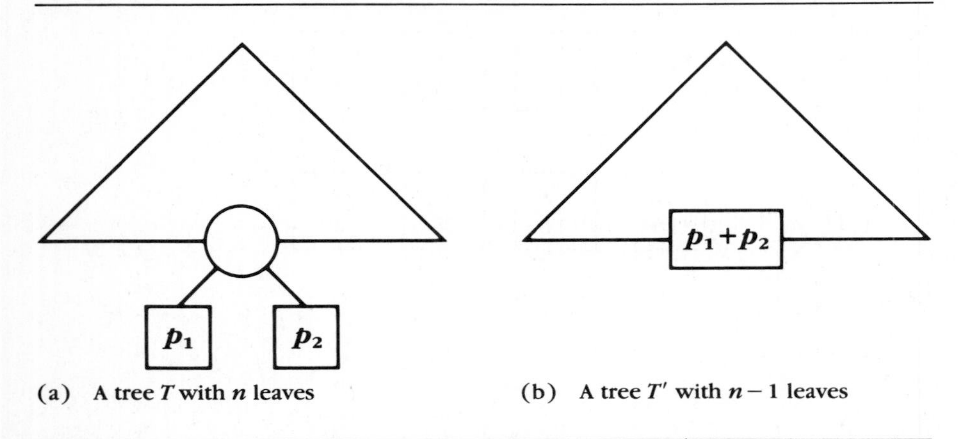

(a) A tree T with n leaves (b) A tree T' with $n-1$ leaves

Figure 2.40 Applying the Induction Hypothesis in the Correctness Proof for the Huffman Algorithm

The same argument applied to the modified tree O shows that if it has the leaves with weights p_1 and p_2 replaced by a node with these leaves as children, the cost of the resulting tree O' differs from that of O by K; note K is negative.

By optimality of T', the cost of T' must be at most that of O'. Thus $K +$ (the cost of T) $=$ (the cost of T') $\leq$ (the cost of O') $= K +$ (the cost of O). Subtracting K from both sides and using the optimality of O, we see that T must also be optimal. This contradiction completes the proof. ■

The Huffman algorithm also has another application, this time to file merging.

Suppose there is a set of n runs of sizes $\{p_i\}_{i=1}^{n}$ that are to be merged into one large run. Suppose only two runs can be merged at once, and suppose the merge algorithm used has cost $r + s$ when merging two runs of sizes r and s. (This cost is consistent with our analysis of merge algorithms.)

In this case we may represent the merge pattern as a tree where two runs are siblings in the tree if they are merged (see Figure 2.41). If an original run of size p_i is on level s_i of the tree, its elements will participate in s_i merges for a total cost attributable to that run of $p_i s_i$. When this cost is summed over all original input runs, we get the exact weighted path length $\sum_{i=1}^{n} p_i s_i$ that the Huffman algorithm was developed to minimize.

The analysis of the Huffman algorithm is relatively straightforward. The main loop is executed $\Theta(n)$ times. If a heap is used to store the weight of each tree, then each iteration requires $O(\log n)$ time to find the cheapest weights and to insert the new weights. Therefore the time complexity is $O(n \log n)$. Note this result is not affected by the $O(n)$ time required to initialize the heap.

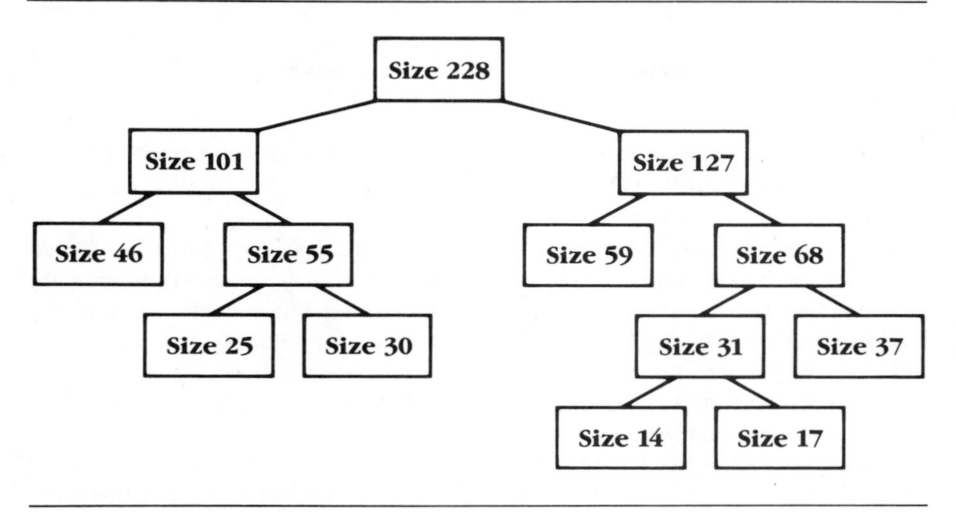

Figure 2.41 Optimal Merge Pattern for Runs of Sizes 14, 17, 25, 30, 37, 46, and 59

Continuous Knapsack

For our next example of a problem solvable by a greedy algorithm, we recall that an instance of the continuous knapsack problem consists of a set of n items, a weight w_i for each item, a profit p_i for each item, and a capacity M. We want to know how much of item i to include. Algebraically, we want a set $\{x_i\}_{i=1}^{n}$ with $0 \le x_i \le 1$ such that $\sum_{i=1}^{n} x_i p_i$ is maximal subject to $\sum_{i=1}^{n} x_i w_i \le M$.

Intuitively, we prefer to include items with high profits or low weights. Try to convince yourself that a greedy algorithm that attempts to include only the most profitable items or only the least weighty items will not be correct. This exercise suggests that the *profit density* p_i/w_i ought to be the controlling variable.

In the zero-one case, including several items of smaller density may be preferable to including one weighty item of large density and having space left over in the knapsack. This choice isn't a problem in the continuous case, since the weighty item of large density may be included together with a fraction of some less dense item.

Thus a greedy algorithm for the continuous knapsack problem could be constructed as follows: Sort the items in nonincreasing order of profit density. Include each item in order until some item doesn't fit (include as much of this item as will fit). If all items fit, there is no problem.

For example, consider the instance with 5 items, capacity $M = 10$, respective weights 3, 1, 4, 5, and 2, and respective profits 15, 4, 12, 10, and 2. Note these items are sorted by profit density. We can include all of the first three items, at which time 2 units of capacity are left. Since the next item has weight 5, only 2/5 of it can be included, giving a profit of $15 + 4 + 12 + (2/5)10 = 35$.

Theorem The preceding greedy algorithm, formalized as Algorithm 2.78, gives an optimal solution.

Proof Assume not. Then there is a nonoptimal greedy solution G for an instance of the continuous knapsack problem. We may assume the items are numbered in order of nonincreasing profit density. We may also assume the optimal solution fills the knapsack, since it could only fail to do so if the sum of all the weights was less than the capacity. In this case, however, the greedy algorithm would find the optimal solution that includes every item.

For each optimal solution, consider the lowest-numbered item entirely included in G that is not entirely included in the optimal solution. Let O be the optimal solution for which this number has its maximum value i. If there are several such optimal solutions, choose the one containing the fewest items with numbers larger than i. Notice such a choice can be made even if there are an infinite number of different optimal solutions.

This is one of those nonintuitive measures of closeness described at the beginning of this section that results from first trying to make an optimal solution closer to the greedy solution and then figuring out what measure will give this definition of "closer."

Now assume that O contains an amount $x_i w_i$ of item i. By assumption, $x_i < 1$. There must be some item j such that O contains more of j than G does — otherwise O does not fill the knapsack and can't be optimal. The profit density of item j can't be larger than that of item i, because then the greedy algorithm would have considered j before i. In this case, however, the greedy algorithm could not have included all of i but only part of j.

Let $x_j w_j$ be the amount of item j contained in O.

First assume that $(1 - x_i)w_i \leq x_j w_j$. We may then modify O by replacing $(1 - x_i)w_i$ units of item j with an equal amount of item i. The solution contains all of item i and will still satisfy the weight constraint, but the profit density of i is at least as large as that of j, so the profit cannot decrease. On the other hand, the profit cannot increase because of the optimality of O. We have constructed an optimal solution with a later first difference from G than O has, contradicting the choice of O.

Finally, if $(1 - x_i)w_i > x_j w_j$, then consider the feasible solution obtained from O by replacing all $x_j w_j$ units of item j with the same number of units of item i. As in the previous case, the profit remains unchanged, so this new solution is optimal. The new solution has $x_i w_i + x_j w_j$ units of item i; by assumption this amount is less than $x_i w_i + (1 - x_i)w_i = w_i$, so that the first difference between this new solution and G is still at item i. This new optimal solution has no units of item j, which violates the second condition in the choice of O.

```
 1. ContinuousKnapsack(n, w, p, M)
 2.           {Given arrays w and p containing respectively the weights
 3.                and profits of a given number n of items, and given a
 4.           capacity M, returns an array x determining how much of
 5.                     each item should be included. It is assumed that
 6.                        p[i]/w[i] ≥ p[i + 1]/w[i + 1] for 1 ≤ i < n.}
 7. filled := 0                     {filled keeps track of how much capacity}
 8. i := 1                          {has been used, profit of the profit so far,}
 9. profit := 0;                                  {and i of the current item}
10. while filled < m and i ≤ n          {if there is room in the knapsack}
11.    if filled + w[i] ≤ m them            {and items remaining, then if}
12.       x[i] := 1                  {there is room for an entire new item,}
13.       profit := profit + p[i]         {then include it and update the}
14.       filled := filled + w[i]         {profit and capacity filled so far}
15.    end then
16.    else
17.       x[i] := (m − filled)/w[i]          {else include as much of the}
18.       profit := profit + x[i] × p[i]       {item as will fit, update the}
19.       filled := m;                    {profit. The knapsack is now full}
20.    end if
21.    i := i + 1;                      {Go back and consider the next item}
22. end while
23. return x
```

Algorithm 2.78 Continuous Knapsack

In either case we get a contradiction to the assumption that there is a greedy nonoptimal solution; therefore, the greedy solution is always optimal. ■

Algorithm 2.78 has time complexity $O(n)$ and assumes that the n input items are sorted by profit density. If an $O(n \log n)$ algorithm is needed to sort, the time complexity of the overall algorithm will also be $O(n \log n)$.

Single Source Shortest Paths

In our final example we consider a less ambitious version of the shortest paths problem of the previous section. Here we consider a single vertex v in a directed graph and ask what are the shortest paths and their costs from v to every other vertex in the graph. We assume all edge costs $c(x, y)$ are positive.

For the directed graph of Figure 2.42a, vertex 1 has been chosen as the starting vertex. The cheapest path to vertex 2 consists of a single edge and has cost 6. The cheapest path to vertex 3 has length 1 and cost 2. The cheapest path to vertex 4 goes through vertex 3 and has cost 3. The cheapest path to

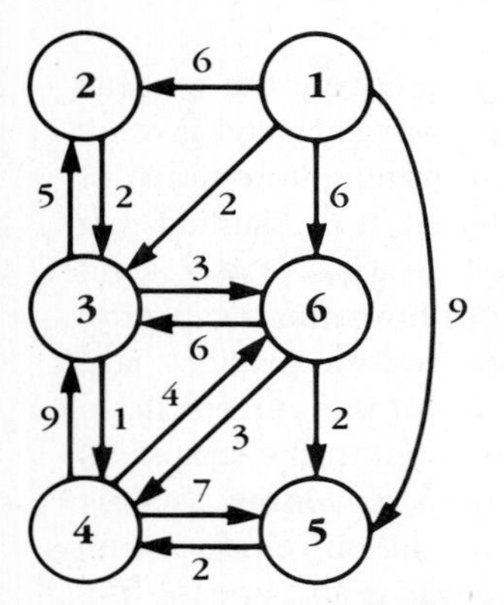

(a) A directed, weighted graph

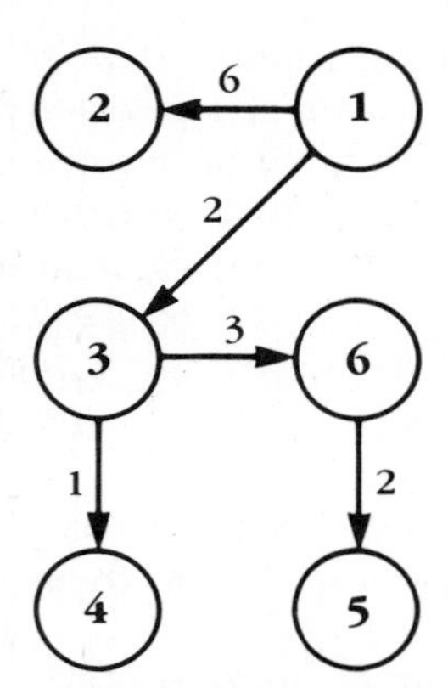

(b) Single source spanning tree for the digraph of (a)

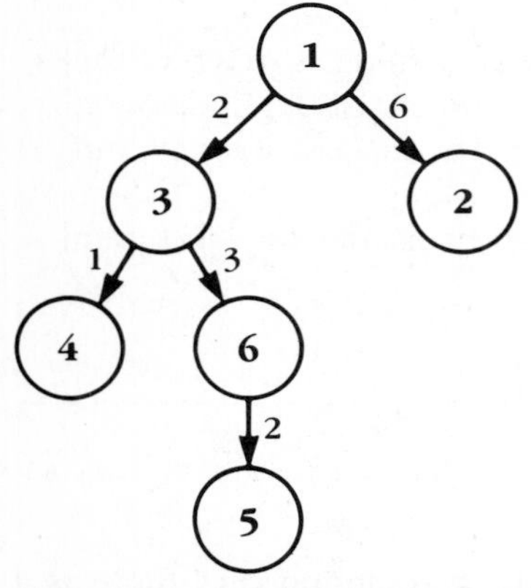

(c) The oriented tree of (b) with levels clearly indicated

Figure 2.42 A Single-Source Spanning Tree

vertex 5 goes through vertices 3 and 6 and has cost 7. The cheapest path to vertex 6 goes through vertex 3 and has cost 5.

Note the subgraph consisting of all of the vertices and those edges occurring in some shortest path form a spanning tree, as pictured in part b of the figure.[18] Here the direction of each edge is away from the source vertex 1. In

[18] It is not quite true that every collection of shortest paths will make up a spanning tree. A counterexample may be constructed if there are two or more shortest paths to the same node. Even so, some collection of shortest paths will still make up a spanning tree. The details are left as an exercise.

part c of the figure, the spanning tree is exhibited as an oriented tree in the more usual way.

If the value of each node is defined to be the cost of the shortest path to the node, then the spanning tree forms a heap in which smaller elements are parents of larger elements. This suggests that a greedy algorithm for the single source problem could generate a spanning tree one vertex at a time in order of these costs. At any point in such a greedy algorithm, a subtree T rooted at the source vertex would have been generated already, and the next vertex to be added would be a child of one of the nodes already in T.

For any particular T, consider the cheapest cost of adding a new vertex to T. For the cheapest new vertex x, this cost would be the sum of the cost of reaching some parent node w plus the cost $c(w, x)$ of the edge from w to x. The first term has already been calculated; we assume it has been saved as the value of COST(w).

As a new vertex y is added to T, it is possible y would be a better parent than the old best-parent w. This may be ascertained for all x not in T in time $O(n)$ simply by comparing COST(w) + $c(w, x)$ to COST(y) + $c(y, x)$. In fact, the first of these sums could be maintained as the value of COST(x). Then COST(z) is consistently equal to the cheapest cost of reaching z by a path whose intermediate vertices are all in T, whether or not z is in T.

In general COST(z) does not give the cost of the cheapest path to z. We show in the following that COST(z) does give the cost of the cheapest path to z_0 if COST(z_0) $\leq$ COST(z) for all remaining vertices z. In this case z_0 is the proper vertex to add to T.

Initially T contains the source vertex v and no edges, COST(v) = 0, and COST(x) = $c(v, x)$ for all $x \neq v$. The greedy Algorithm 2.79, attributed the Dijkstra (1959), repeatedly chooses that vertex x not in T with the cheapest value of COST and updates COST as described in the previous paragraph. This greedy algorithm then terminates when all vertices have been added to T. Its time complexity is $\Theta(n^2)$.

Figure 2.43 shows the contents of COST for vertices not in T. The value of COST(x) does not change once x is included in T. Note that the vertex included is always the one not in T with the minimum value of COST. The construction of the spanning tree is shown in Figure 2.44.

For example, after the first vertex, vertex 3, is included with cost 2, COST(4) becomes COST(3) + $c(3, 4) = 2 + 1 = 3$. COST(6) becomes COST(3) + $c(3, 6) = 2 + 3 = 5$. COST(2) < COST(3) + $c(3, 2) = 2 + 5$, so COST(2) remains unchanged. Similarly COST(5) < COST(3) + $c(5, 3) = 2 + \infty$, so COST(5) remains unchanged.

Finally, we show correctness of Algorithm 2.79.

Theorem Dijkstra's Algorithm 2.79 finds the shortest path from vertex v to all other vertices.

Proof We divide the proof into two parts. First, we show if for all x, COST(x) does indeed give the cost of the cheapest path to x all of whose

```
1. SingleSource(n, v, c)             {given a set of n vertices, a source
2.                       vertex v and an n x n cost array c, this algorithm
3.            returns the cost of the cheapest path from v to every other
4.                       vertex. The vertices are indexed from 1 to n; T[i]
5.                                 is true iff the vertex i is included in T}
6. for i from 1 to n
7.    cost[i] := c[v, i]                        {initialize the cost matrix}
8.    T[v] := false;                                   {T is initially empty}
9. end for
10. T[v] := true                          {now include the source vertex in T}
11. newest := v                      {the newest node is used to update cost}
12. for i from 2 to n do                    {choose the ith vertex of T and see}
13.    leastcost := ∞                               {if inclusion of the latest}
14.    for j from 1 to n                     {node in T allows cheaper access}
15.       if not T[j] then                             {to any node not in T}
16.          if (cost[newest] + c[newest, j] < cost[j]) then
17.             cost[j] := cost[newest] + c[newest, j]
18.          end if                        {note lines 14–17 are useless if i = 2}
19.          if cost[j] < leastcost then                      {and keep track of}
20.             cheapest := j                               {the vertex not in T}
21.             leastcost := cost[j]                   {with the lowest access}
22.          end if                                          {cost, and that cost}
23.       end if
24.       T[cheapest] := true                           {include this vertex in T}
25.       newest := cheapest                               {and record it as the}
26.    end for                                                  {newest vertex}
27. end for
28. return cost
```

Algorithm 2.79 Single-Source Shortest Paths

intermediate vertices are in T, then the algorithm will select the next vertex correctly. Second, we show that COST always does have the correct value.

Suppose that for some T, COST(x) has the correct value for all x. In addition, suppose that the cheapest path to a vertex not in T is to vertex z, while COST(y) < COST(z). By definition of COST(z), this cheapest path to z must have intermediate vertices not in T. Let u be such an intermediate vertex. But then the cost of the path to z is greater than the cost of the path to u. Since u is also not in T, this contradicts the choice of z; thus we complete the first part of the proof.

Now we need to show that COST always has the correct value for each vertex. Suppose not. Since initially no intermediate vertices exist in T (v_0 can't

$$\begin{pmatrix} 0 & 6 & 2 & \infty & 9 & 6 \\ \infty & 0 & 2 & \infty & \infty & \infty \\ \infty & 5 & 0 & 1 & \infty & 3 \\ \infty & \infty & 9 & 0 & 7 & 4 \\ \infty & \infty & \infty & 2 & 0 & \infty \\ \infty & \infty & 6 & 3 & 2 & 0 \end{pmatrix}$$

(a) Adjacency matrix for the directed graph of Figure 2.42a

	1	2	3	4	5	6	
Before iteration 1		6	2	∞	9	6	(include vertex 3)
Before iteration 2		6		3	9	5	(include vertex 4)
Before iteration 3		6			9	5	(include vertex 6)
Before iteration 4		6			7		(include vertex 2)
Before iteration 5					7		

(b) Successive values of the COST function

Figure 2.43 Trace of the Greedy Algorithm for the Single-Source Shortest-Paths Problem

be an intermediate vertex on a path from v_0), COST is initially correct. Consequently, there must be some vertex w such that COST is correct just before inclusion of w in T but incorrect just afterward. In particular, COST(w) is the cost of the cheapest path to w, since its value was correct just before the inclusion of w and does not change afterward. Suppose x is the vertex such that COST(x) is incorrect after inclusion of w.

By the choice of w, COST(x) was correct before the inclusion of w. First suppose inclusion of w does not change the cost of the cheapest path to x with all intermediate vertices in T. Then this cost is at most COST(w) + $c(w, x)$, since the sum is just the cost of one such path. So COST(x) is unchanged by the algorithm after the inclusion of w and is still correct.

Now suppose COST(x) should change; that is, the inclusion of w in T has allowed a cheaper path to x with all intermediate vertices in T. Then w must be on this path. Suppose there is some vertex z between w and x on this path. Then z, like all intermediate vertices, is in T. The cheapest path to z does not go through w, since z was included in T before w, so the cheapest path through z to x cannot go through w either. Therefore, there is no vertex between w and x.

The new cheapest path to x goes through T to w and then directly to x. This path has cost COST(w) + $c(w, x)$, which by assumption is less than COST(x). In this case, however, the algorithm will replace COST(x) by COST(w) + $c(w, x)$, and therefore works correctly. ∎

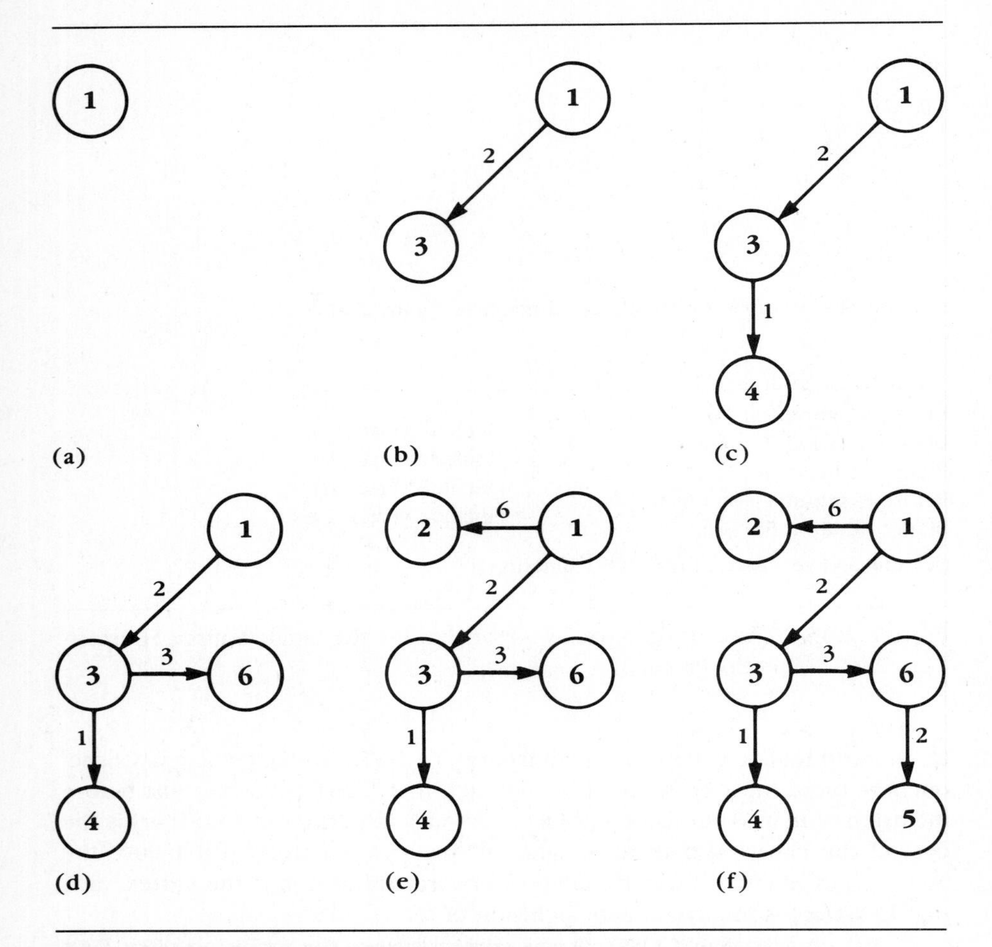

Figure 2.44 Construction of Single-Source Spanning Tree

2.21 Exercises

Solutions of exercises marked with an asterisk appear in Appendix C; references are given in that appendix for those marked with a double asterisk.

Induction and Recursion — Traces

*1. Evaluate the following sums:

a) $\sum_{k=1}^{5} k$

b) $\sum_{k=1}^{6} k^2$

c) $\sum_{k=2}^{5} 1/k$

d) $\sum_{k=1}^{5} 2$

*2. Find $n!$ for $1 \leq n \leq 10$.

*3. Find the first few terms of the sequence defined by $a_n = a_{n-1} + 2^n$, $a_0 = 1$. What if 2^n is replaced by 3^n?

*4. List all elements of the power set of $\{1, 2, 3, 4\}$.

*5. List all permutations of $\{1, 2, 3, 4\}$.

*6. Find the first 15 Fibonacci numbers.

*7. Find all values of $\binom{n}{m}$ for $n \leq 10$.

8. Find a value for n such that $n^2 - n - 41$ is not prime. Find the smallest value for n such that $n^2 - n - 41$ is not prime.

*9. *Ackermann's function* $A(r, s)$ may be defined recursively as follows:

a) $A(0, s) = 2s$, $A(r, 0) = 0$, $A(0, 1) = 2$
b) $A(r, s) = A(r - 1, A(r, s - 1))$ for $s > 1$

Find the values of $A(r, s)$ for $0 \leq r \leq 3$, $0 \leq s \leq 4$. You may write a program to do this if you want. Be prepared for an unacceptable time or space requirement for calculating $A(3, 4)$.

10. Give an example to show that the lexicographical ordering of integers differs from the usual numerical ordering.

11. Give an example to show that buckets can't necessarily be reused in recursive bucket sort.

12. Draw a decision tree for the binary search algorithm for an input sequence of 9 keys. First assume that all your searches are successful, then consider the case in which searches may be unsuccessful.

13. Use the recursive definitions of Section 2.3 to evaluate

a) Length(⟨4 6 5 2⟩)
b) ElementOf(5, ⟨3 6 7⟩)
c) ElementOf(6, ⟨2 4 6 8⟩)
d) Equal(⟨2 3 4⟩, ⟨2 3 4⟩)
e) Equal(⟨2 3 4 5⟩, ⟨2 3 5 4⟩)
f) kth(3, ⟨4 5 7 1⟩)
g) kth(4, ⟨2 3 9⟩)
h) Sum(⟨5 1 9 7⟩)
i) Max(⟨5 1 9 7⟩)
j) Concat(⟨3 5 7⟩, ⟨9 1 6⟩)
k) Sorted(⟨3 5 8 9⟩)
l) Sorted(⟨3 5 1 8⟩)
m) LessOrEqual(⟨3 6 7⟩, ⟨4 8⟩)
n) LessOrEqual(⟨3 8⟩, ⟨3 8 9⟩)
o) LessOrEqual(⟨3 6 2⟩, ⟨3 5 1⟩)
p) LessOrEqual(⟨3 8 9⟩, ⟨3 8⟩)

14. Trace the merge algorithm on the two sorted sequences (1 4 9 16 25) and (6 12 18 24 30).

15. Trace the InsertionSort and Mergesort algorithms on the sequence (8 5 4 9 1 7 6 2 3).
16. Trace the recursive binary search algorithm for the sorted sequence (1 4 7 11 13 15 16 17 22 25) for input keys 4, 1, 2, 25, 28, and 22.

Induction Proofs

17. Show that $0^n = 0$ for all n. Show that $1^n = 1$ for all n.

*18. Show that if $p < q$, then $p^n < q^n$ for all $n > 1$.

*19. Show that $n! > 2^n$ for $n \geq 4$.

20. For a fixed b, find an integer N depending only on b such that $n! > b^n$ whenever $n > N$.

*21. For the Fibonacci sequence (f_k), show that

a) $\sum_{k=1}^{n} f_k = f_{n+2} - 1$

b) $\sum_{k=1}^{n} f_k^2 = f_n f_{n+1}$

c) for $n \geq 1, f_{n+1}f_{n+2} = f_n f_{n+3} + (-1)^n$

d) for $n \geq 2, f_{n-1}f_{n+1} = f_n^2 + (-1)^n$

*22. Prove formally the following properties of binomial coefficients:

a) $\sum_{m=0}^{n} \binom{n}{m} = 2^n$

b) $\binom{n}{m} = n!/[m!(n-m)!]$

c) $\binom{n}{m}$ is the coefficient of $a^m b^{n-m}$ in the expansion of $(a+b)^n$

d) $\binom{n}{n/2} > \binom{n}{k}$ for n even, $k \neq n/2$

e) $\binom{n}{n/2} = \Omega(2^n/n)$ for even n

*23. Show by induction that every nonnegative integer n has a decimal representation. Show that this representation is unique. Can your proof be generalized to bases other than 10?

24. Show by induction that n and the sum of the digits in the decimal representation of n have the same residue modulo 9.

25. Show by induction that $|P \times Q| = |P||Q|$.

*26. Show by induction that

a) $\sum_{k=a}^{n} (f_k + g_k) = \sum_{k=a}^{n} f_k + \sum_{k=a}^{n} g_k$

b) $\sum_{k=a}^{n} cf_k = c \sum_{k=a}^{n} f_k$

27. What's wrong with the following induction proof that $6n = 0$ for all $n \geq 0$?
Basis: If $n = 0$ then $6n = 0$.
Induction step: If $n > 0$ then let $n = a + b$. By induction, $6a = 0$ and $6b = 0$. So $6n = 6(a + b) = 6a + 6b = 0$.

28. Show that
   a) the exclusive-or function is associative, and
   b) the exclusive-or function applied to n arguments has value TRUE if an odd number of the arguments have value TRUE and FALSE otherwise.

*29. Show by induction that if the population of the world is greater than the number of hairs on anybody's head, and everybody's head has at least one hair, then two people have exactly the same number of hairs.

*30. Show that the result of the previous problem is not true if there are completely bald people in the world. What goes wrong with the induction proof?

*31. Show by induction that every positive integer is interesting.

32. Show by induction that the length of the concatenation of two lists is the sum of the lengths of the individual lists.

33. Show by induction that a spanning tree for a graph with n nodes has $n - 1$ edges.

34. Why does the base case of the max-min argument in Section 2.16 consider $n = 2$ and not $n = 1$?

35. Prove the following facts about Ackermann's function (see Exercise 9):
   a) $A(0, s) = 2 + A(0, s - 1)$
   b) $A(1, s) = 2 \times A(1, s - 1)$
   c) $A(2, s) = 2^{A(2,s-1)}$
   d) $A(3, s)$ is as defined in Figure 2.45. Is it clear now why calculating $A(3, 4)$ was so difficult?

36. The following function will be useful in Chapter 4. $\alpha(m, n) = \min \{r \geq 1 \mid A(r, 4\lceil m/n \rceil) > \log_2 n\}$, where A is Ackermann's function defined in Exercise 9. Show that if $n < 2 \times A(3, 4)$ (a pretty safe assumption!) and $m \neq 0$ then $\alpha(m, n) \leq 3$. In other words, α is an unbounded function that is bounded for all practical purposes.

37. Show that the time complexity of finding binomial coefficient $\binom{n}{m}$ directly from the definition is $\Theta\binom{n}{m}$.

*38. Can the Σ notation for sums be defined by removing the first element instead of the last from the sum? If so, are the definitions equivalent?

$$A(2, s) = 2^{2^{\cdot^{\cdot^{\cdot^{2}}}}} \Big\} s \text{ times, for } s \geq 1$$

(a)

$$A(3, s) = 2^{2^{\cdot^{\cdot^{\cdot^{2}}}}} \Big\} A(3, s-1) \text{ times, for } s \geq 2$$

(b)

$$A(3, 4) = 2^{2^{\cdot^{\cdot^{\cdot^{2}}}}} \Big\} 65{,}536 \text{ times}$$

(c)

Figure 2.45 Evaluating Ackermann's Function

*39. Show that the minimal counterexample principle is equivalent to ordinary induction.

*40. Show that the strong form of mathematical induction is equivalent to ordinary induction.

*41. Show that if establishing $P(a)$ is done in the basis step instead of establishing $P(0)$, mathematical induction will prove P true for all integers greater than or equal to a.

Recursive Definitions and Algorithms

42. Give a recursive definition of a set of size n. Give recursive definitions of the common set operations using this definition of a set.

43. Give formal definitions for "ancestor," "descendant," and "sibling" in binary trees. Are the definitions different in ordered trees?

44. Give formal definitions for "member," "height," "level," and "size" for ordered trees.

45. Design and implement an algorithm for finding the minimum element of a sequence.

46. Design and implement a recursive function that searches for an element in a sequence. What would be a reasonable value to return?

47. Give a recursive definition of "greater than" for two strings of "less than (and not equal to)."

48. Give a recursive definition of "greater than or equal to" for two strings.

49. Define formally a function "and" on two strings of bits so that $1100 \wedge 1010 = 1000$. Do the same for "or" and "exclusive or." Do the same for "not" for a single string of bits.

50. If sets are represented as lists without duplicate elements, give a recursive algorithm for finding
    a) the union of two sets
    b) the intersection of two sets
    c) the Cartesian product of two sets represented as lists, and
    d) the difference of two sets.

    What is the time complexity of each of your algorithms?
51. Design algorithms for
    a) determining whether a list has duplicate elements and
    b) removing duplicate elements from a list.
52. Design and implement algorithms for the following operations on sequences:
    a) insertion into position k
    b) deletion from position k
    c) deleting the first occurrence of a given element, and
    d) replacement of the element in position k.
53. Design and implement algorithms to find the immediate predecessor and immediate successor of a given element in a given sequence. Design and implement algorithms to determine whether a given element appears anywhere before another given element in a given sequence. Design and implement an algorithm to return the subsequence of a given sequence beginning with a given element. What if the input element appears more than once in the sequence in any of these operations?
54. Are any of the algorithms of the previous problem simplified if you assume
    a) doubly linked lists?
    b) header nodes?
    c) call by reference or by value-result?
55. Design and implement a recursive bucket sort algorithm.
56. Implement the algorithm of Section 2.14 for multiplying large integers. How must your algorithm be changed if you want to use it to multiply polynomials?
57. What is the time complexity for the Concatenation algorithm (implied by Definition 2.19) of Section 2.3?
58. Is there a representation for sequences in which Concatenation has O(1) time complexity? If so, design such an algorithm for that representation.
59. Show that the head and tail operations do not have O(1) time complexity if sequences are implemented as lists of pairs, for example, if (6 7 8) may be implemented as ((2 7) (1 6) (3 8)) or ((3 8) (1 6) (2 7)).

60. Show that
    a) "less than" or "equal to" is equivalent to "less than or equal to" and
    b) "greater than" is equivalent to "not less than or equal to."
61. Show the definition for a nonempty sequence given in the text is equivalent to the definition of a sequence, except when the sequence according to this second definition is empty. Do the same for the definition of nonempty ordered trees.
62. Write a recursive algorithm that prints out all of the permutations of a given list of elements. Can you print them out in lexicographical order?
63. Show that the set of permutations of a set is the same as the set of permutations of any permutation of the set.
64. Choose a value for k and a fixed Boolean function of k variables and write a recursive program to print the truth table of the function. Can you modify your program to accept different values of k or different functions?

*65. Does InsertionSort have time complexity $\Theta(n^2)$?

Binary Trees

*66. List all the different binary trees on 4 nodes. What is their average height?

*67. Show that the leaves of a binary tree are visited in the same order in each of the three recursive traversal algorithms.

68. Show formally that the height of a tree is equal to $\max\{\text{level}(N)\}$ where N ranges over all nodes in the tree.
69. Design and implement algorithms
    a) to copy a binary tree
    b) to determine whether one binary tree is equal to another
    c) to print a diagram corresponding to a binary tree, and
    d) to traverse a binary tree and return the list of keys as a sequence.
70. If the definition in the text for the height of a binary tree is converted to an algorithm in the most natural way, what is the time complexity?

*71. What's wrong with the following definition of a binary search tree?

A binary search tree T is a binary tree which either

a) is empty or
b) consists of a root R containing a data element x such that
   (1) the data element y of Left(T) satisfies $y \leq x$
   (2) the data element z of Right(T) satisfies $z \geq x$ and
   (3) Left(T) and Right(T) are binary search trees.

*72. The binary search algorithm given as Algorithm 2.80 has one fewer comparison than Algorithm 2.63. What's wrong with it?

```
1. BinarySearch(key, In, low, high)        {returns the index in the
2.                                     array In at which the element key
3.                                      is found, if there is such an index
4.                                               between low and high}
5. if low = high then                     {if the sequence has size 1}
6.    if key = In[low] then               {then check whether the}
7.       return low                                  {single element}
8.    else                                   {is the desired element}
9.       return unsuccessfully
10.   end if
11. end if
12.                                            {else} {otherwise}
13. mid := ⌊(low + high)/2⌋                   {find the midpoint}
14. if key ≤ In[mid] then return BinarySearch(key, In, low, mid)
                                              {else if key > In[mid]}
15. return BinarySearch(key, In, mid, high)
```

Algorithm 2.80 Bad Recursive Binary Search

73. Give a recursive definition of the *reflection* T^{-1} of a binary tree. Make sure that the two trees of Figure 2.46 are reflections of one another according to your definition. How do the inorder, preorder, postorder, and level-order traversals of the reflection compare to those of the original tree?

74. Find a binary tree of n nodes
    a) with the maximum external path length and
    b) with the maximum internal path length.

    Prove that your answer is correct.

75. Show that the stack size during a recursive traversal is at worst proportional to the height of the tree.

76. Show that a tree of height h must have at least $h + 1$ nodes.

77. Show that a binary tree with k nodes has $k + 1$ external nodes.

78. Can a binary tree be constructed given just its preorder traversal? just its inorder traversal? just its postorder traversal? What if you are given two of these traversals? Does it matter which two? What if you are given all three? What if you are given only the level-order traversal? What if you are given the level-order traversal and one of the recursive traversals?

79. Define the weight of a tree to be the number of nodes in the tree excluding the root. Can you reconstruct a binary tree given only the preorder traversal and the weight of the subtree rooted at each node?

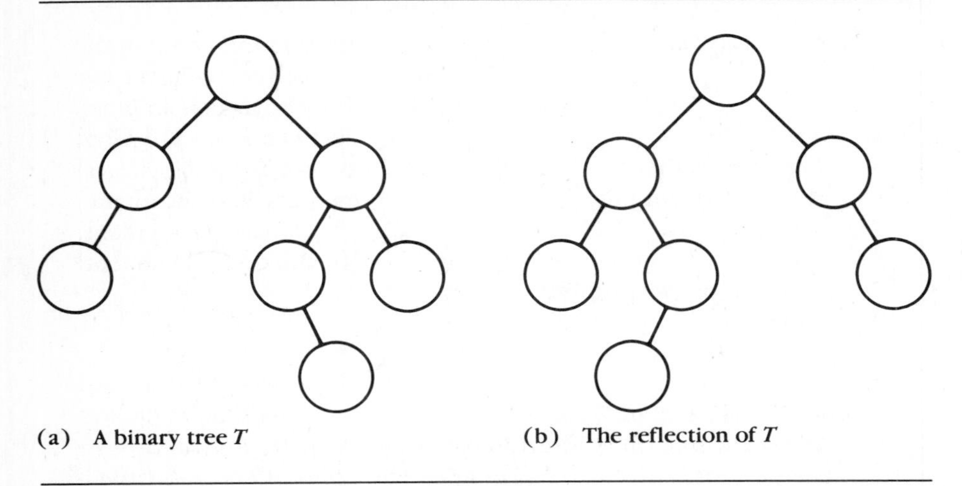

(a) A binary tree T (b) The reflection of T

Figure 2.46 A Binary Tree and Its Reflection

What if, instead of the weight, you were given the number of subtrees of each node? How can you use the results of this exercise to find a space-saving representation for a binary tree?

Binary Search Trees

*80. Which of the trees of Figure 2.47 are binary search trees? Which are AVL trees?

*81. Replace the keys in the tree of Figure 2.5 by the integers from 1 to 7 so that they will be visited in numerical order during

a) a preorder traversal
b) an inorder traversal
c) a postorder traversal and
d) a level-order traversal.

82. Construct all external nodes for the binary search tree of Figure 2.8. To which interval does each correspond?

83. Trace insertions of the keys 20, 10, 1, 2, 80, and 40 into the binary search tree of Figure 2.8.

*84. Trace the operation of binary tree sort on the input sequence (7 21 16 18 3 20 14 10 4 2).

85. Give three possible sequences of insertions (beginning with an empty tree) that would give the binary search tree of Figure 2.6.

86. Sketch all possible AVL trees with 4 nodes.

*87. List all of the 4! binary search trees obtained by inserting the elements of $\{1, 2, 3, 4\}$ in all possible orders. What is the average height of these

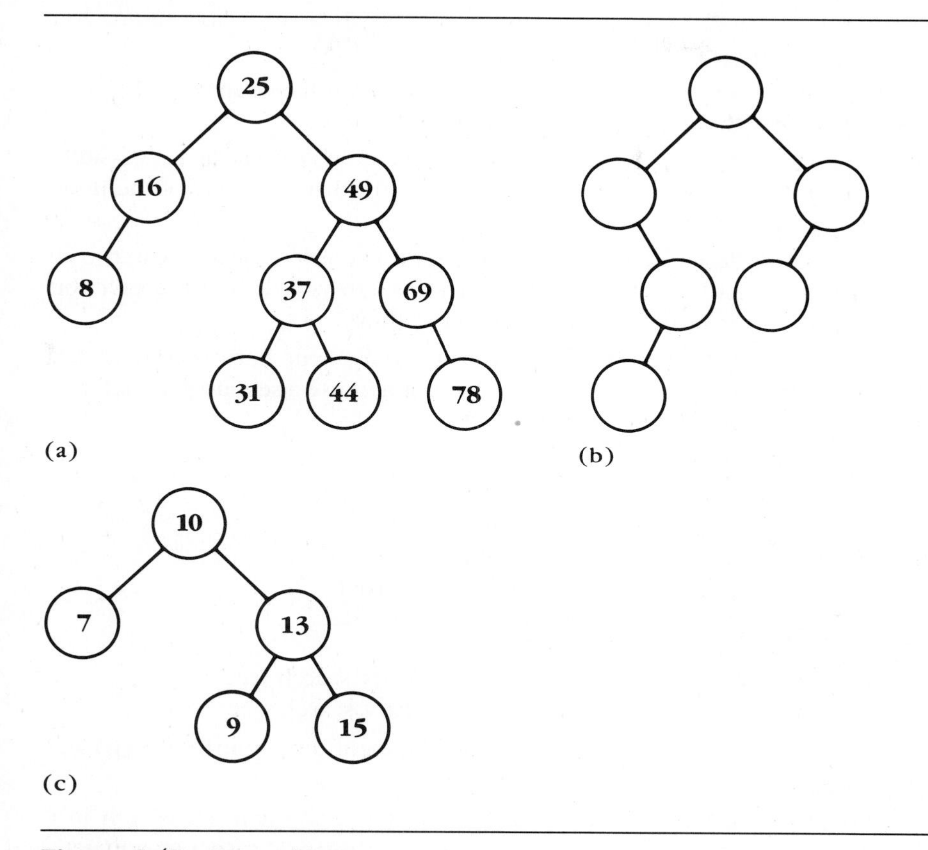

Figure 2.47 Binary Trees

trees? What is the average internal path length? Are some binary trees represented more than once? Are the trees represented the most more likely to be balanced or to be unbalanced? What does this suggest about the average-case behavior of search in a binary search tree?

*88. Is a binary search tree a 2-way search tree?

89. If keys are inserted in a different order into a binary search tree, will the tree necessarily be the same?

90. Design an algorithm
    a) to construct the reflection of a binary tree and
    b) to determine whether or not one binary tree is the reflection of another.

91. Use induction to show that binary tree traversal has $O(n)$ time complexity.

92. A sorting algorithm is *stable* iff equal keys appear in the same order in

the output as they do in the input. Is binary tree sort stable? If not, can it be modified to be made stable?

93. Give an example of a binary search tree with the heap property. Can your tree have an arbitrary number of nodes?

94. Show formally that the number of key comparisons in successfully searching a binary search tree is one more than the level of the node sought.

95. Can you design a procedure to insert a key into a binary search tree that explicitly calls a search algorithm? Can you do so for a deletion procedure? If so, implement the algorithms.

96. Can you write a single algorithm to implement both searching and insertion in a binary search tree? You might use a second parameter to distinguish between the two cases.

97. Design and implement algorithms to

    a) insert a given key into a binary search tree
    b) traverse a given binary tree in inorder
    c) traverse a given binary tree in preorder
    d) traverse a given binary tree in postorder
    e) traverse a given binary tree in level order
    f) implement binary tree sort
    g) find the maximum element in a binary search tree
    h) find the minimum element in a binary search tree.

    Determine the time complexity of the algorithm you use for (g) and (h).

**98. We have seen that the average-case time complexity of search in a binary search tree is $\Theta(\log n)$, the same as the best case. Can you get a more precise estimate of how much worse the average case is than the best case?

Ordered Trees

99. Show several oriented trees equivalent to the same free tree. Show several ordered trees equivalent to the same oriented tree.

100. The ordered tree of Figure 2.48 represents the British royal family in 1984. Give the succession to the throne (recall that this succession is determined by a preorder traversal) among those represented by nodes. What is the earliest that any newborn could fall in the succession, assuming everyone else survives? If the blanks represent the deceased and edges without children represent unnamed royals, which unnamed royal is first in the order of succession?

101. For ordered trees, define

    a) membership
    b) size

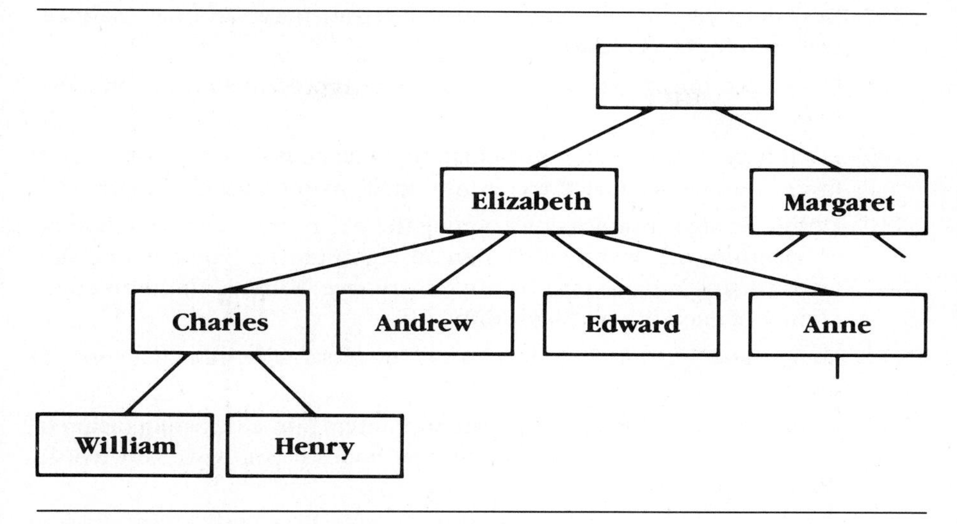

Figure 2.48 British Royal Family, 1984

c) level
d) height
e) preorder and postorder traversal (why can't you define inorder traversal?)
f) level-order traversal.

102. Define "left of" in an ordered tree (see the definition of "left of" in Section 2.7).

103. Design a representation for ordered trees and design algorithms Empty(T) and KthChild(k, T) in this representation. Using these algorithms, design algorithms to traverse an ordered tree, find the size of an ordered tree, and determine whether a node of an ordered tree is a leaf.

104. Suppose that for an ordered tree you were given operations to find the first child of a node and the next sibling of a node [which given the kth child of a node returned the $(k + 1)$st child of the node). Design an algorithm to find the kth child of a node in terms of these operations.

*105. Given an ordered tree T, consider a binary tree T' on the same nodes such that for any node N, left(N) is the first child of N and right(N) is the next sibling of N. T' is said to be the *binary tree representation* of the ordered tree T. Sketch the binary tree representation of the ordered tree of Figure 2.3.

*106. Find the ordered trees represented by the binary trees of Figure 2.6 if the keys 41 and 65 are removed.

107. Why does the root of a binary tree that represents an arbitrary ordered tree have only a single child?

108. Does each binary tree correspond to the representation of some ordered tree?

109. Is a binary tree its own representation when considered as an ordered tree? You may assume there are no nodes with only right children.

110. Design an algorithm for constructing the binary tree representation of an arbitrary ordered tree. Design an algorithm for constructing the ordered tree represented by any binary tree. What is the time complexity of each of your algorithms?

*111. Design and implement a level-order traversal algorithm that uses a queue.

112. Design an algorithm to insert parent pointers into an existing ordered tree. You may copy the keys into new nodes if you wish. Is it worth doing so?

Complete Trees

113. Sketch the complete tree on 19 nodes.

114. Sketch the complete binary tree whose level-order traversal gives the sequence (4 7 8 2 9 1 0 5 6).

115. Sketch the binary tree whose sequential representation consists of the sequence of keys (3 5 7 1 2 8 4 9 0).

*116. In the consecutive storage representation of a binary tree, where would you find
   a) the parent
   b) the left child and
   c) the right child of the node stored at location 13? (You may assume that the children exist.)

117. Suppose that a tree in which up to k children per node are allowed is traversed in level order. What is the worst-case queue size as a function of the number of nodes?

118. Show that the worst-case space requirement for a sequentially represented binary tree is the case discussed in regard to Figure 2.16.

119. Generalize the sequential representation of binary trees to the representation of k-ary trees.

120. Generalize the definition of complete binary trees to ordered trees in which each node may have at most m children.

121. Design an algorithm that given a sorted sequence will construct a complete binary search tree for the elements of the sequence.

122. Show that all leaves in a complete tree are on at most two adjacent levels.

*123. Show that completely full trees are complete.

124. Show that level-order traversal of a complete binary tree in the consecutive storage representation visits the nodes in increasing order of their index. Generalize to ordered trees.

Recurrence Relations

125. Find exact solutions to the following recurrence relations:

a) $t_n = 3t_{n-1} + 3n + 1,\ t_0 = 0$
b) $t_n + t_{n/2} = 2^n,\ t_2 = 3$
c) $t_n - t_{n/2} = 2n,\ t_1 = 1$

126. Find a formula for the nth Fibonacci number.

127. Use recurrence relations to find a formula in terms of n for the sum

a) of the first n integers
b) of the first n squares and
c) of the first n cubes.

128. Verify the exact solution given to the recurrence relation for insertion sort in Section 2.17 in terms of the constants c and d.

129. Use recurrence relations

a) to check correctness of the last example of Section 2.16
b) to check correctness of the max-min example of that section and
c) to find the exact value of the sum at the end of the analysis of the CYK algorithm in Section 2.19.

*130. Suppose that $(t_n) \le \mathrm{P}(t_{n-1}) + \mathrm{f}(n),\ t_0 = c$, where P is a polynomial in E with nonnegative coefficients. Suppose $(s_n) = \mathrm{P}(s_{n-1}) + \mathrm{f}(n),\ s_0 = c$. Show that $t_n \le s_n$ for all n. This result allows us to solve certain recurrences involving inequalities instead of just equalities.

*131. Find the general solution to $(\mathrm{E} - b)\langle t_n\rangle = \langle 0\rangle$ without using the techniques of Section 2.17.

*132. Show that the general solution to $(\mathrm{E} - 1)\langle t_n\rangle = \langle b\rangle$ is $t_n = bn + c$.

*133. Show that $t_n = c2^n$ is the general solution to the recurrence $(\mathrm{E} - 2)\langle t_n\rangle = \langle 0\rangle$. [Hint: Let a general solution be $\mathrm{f}(n)2^n$ and determine the possibilities for f.]

*134. Suppose that the cost of multiplying a b-bit number by a c-bit number is bc and that x^n has $n|x|$ bits. What is the asymptotic cost of finding x^n assuming $x^{n/2}$ is known? How does this compare to the cost of finding x^n directly from the definition? You may assume x is a power of 2.

*135. Refigure the previous problem, replacing x^n by $x^n \bmod M$, where M is a fixed positive integer.

136. Show that multiplication of the matrices $\begin{pmatrix} A & B \\ C & D \end{pmatrix}$ and $\begin{pmatrix} E & F \\ G & H \end{pmatrix}$ requires only 7 multiplications of the entries. [Hint: Consider the products

$(A + D)(E + H)$, $A(F + H)$, $D(G - E)$, $(C + D)E$, $(A + B)H$, $(C - A)(E + F)$, and $(B - D)(G + H)$. How many additions and subtractions are necessary to find the product matrix?]

*137. If n is a power of 2, use the previous problem to design a divide-and-conquer algorithm for multiplying two matrices. Set up and solve a recurrence relation for the time complexity of this algorithm. Is your answer better than $O(n^3)$? [See Strassen (1969).]

138. Show that the E and $\circ$ operators generate a commutative ring.

Design Techniques

139. Trace the single-source shortest-path algorithm for the graph of Figure 2.42a for all possible starting vertices.

140. Solve the continuous knapsack instance with $M = 15$, $w_i = i$ for i from 1 to 7, and profits 3, 12, 12, 10, 8, 20, 35 respectively. What is the optimal zero-one solution for this instance?

*141. Is the following Boolean expression satisfiable?
$(x_1 \vee x_2') \wedge (x_1 \vee x_3) \wedge (x_1' \vee x_2) \wedge (x_2' \vee x_3') \wedge (x_1' \vee x_2' \vee x_3)$

*142. Solve the instance of the TSP consisting of a graph with vertices 1 through 5 and the cost of (v, w) equal to vw.

143. Does the graph of Figure 1.1(b) have a Hamiltonian circuit?

*144. Find the optimal binary search tree for $(p_i)_{i=1}^{4} = (0.2\ 0.1\ 0.2\ 0.1)$ and $(q_i)_{i=0}^{4} = (0.05\ 0.1\ 0.1\ 0.05\ 0.1)$.

145. If the production $A \rightarrow BS$ is added to the grammar of Section 2.19, apply the CYK algorithm to the new grammar and the string *bbaba*.

146. Apply the all-pairs shortest-paths algorithm to the graph of Figure 2.42a. Is your solution consistent with that of n applications of the single-source algorithm?

*147. Apply the Huffman algorithm to the 26 letters of the English alphabet, where the probabilities are as given below:

A	.078	I	.075	Q	.005	Y	.020
B	.013	J	.006	R	.057	Z	.002
C	.030	K	.009	S	.073		
D	.042	L	.038	T	.082		
E	.107	M	.029	U	.032		
F	.025	N	.072	V	.016		
G	.018	O	.072	W	.020		
H	.058	P	.018	X	.003		

*148. Find the optimal merge tree for a set of runs of sizes 7, 9, 11, 20, 21, 40, and 47.

149. Find the minimum-cost spanning tree for the graph of Figure 2.35a.

150. Do fixed-length character codes have the prefix property?

151. Describe precisely the $\Theta(n^4)$ algorithm mentioned in Section 2.19 for shortest paths. How can you find the actual paths, as opposed to their costs?

*152. Can the single-source shortest-paths algorithm be used to solve the all-pairs shortest-paths question? If so, what is the time complexity of the resulting algorithm?

153. Show that in an undirected graph the all-pairs shortest-paths algorithm chooses a minimum-cost edge $\{v, w\}$ for each vertex v. Can an all-pairs shortest-paths algorithm be designed simply by choosing such a cheapest edge for each vertex? What if all edge costs are different?

154. Show that if codes may be assigned to pairs of letters, then the Huffman algorithm is no longer optimal. Can you assume that the probability of each pair of characters is the product of the probabilities of the individual characters?

155. Design and implement an algorithm for decoding a string encoded by the Huffman algorithm.

156. Give an algorithm for recovering the parse tree after application of Algorithm 2.72, given the values for the arrays *where*, *left*, and *right*.

157. Formulate a greedy algorithm for ZOK and show it is not always correct. Do the same for the TSP and Hamiltonian circuit problems.

**158. Construct a dynamic programming algorithm for the TSP.

159. Show that an arbitrary collection of single-source shortest paths doesn't necessarily give a spanning tree. Show that the single-source shortest-paths algorithm of the text does give a spanning tree.

Heaps

*160. Does the tree in Figure 2.49 have the heap property?

*161. Trace the insertion of the sequence of keys (10 50 85 80 100) into the heap of Figure 2.19a.

*162. Trace 3 deletions in the heap whose sequential representation is (45 36 18 19 10 4 17 8).

*163. Trace the following operations on the heap given in the previous problem:
insert 40, delete, insert 50, insert 1, delete, delete.

*164. Trace the operation of the heapify algorithm on the tree in Figure 2.50.

*165. Trace the heapsort algorithm on the sequence (7 1 9 2 6 3 4 5 8).

*166. What heap results if MoveUp is used instead of MoveDown for the example of Figure 2.20?

*167. How many key comparisons and exchanges are required to create a heap from the unsorted data in the example of Figure 2.20 if the keys are inserted one at a time into a priority queue instead of using the

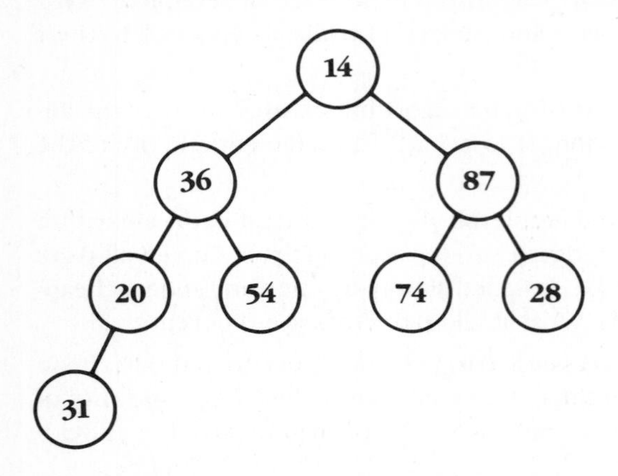

Figure 2.49 A Binary Tree

heapify algorithm? Compare this number with the number used by the heapify algorithm. How may comparisons and exchanges are made if Moveup instead of Movedown is used in the heapify algorithm?

*168. Why must the MoveDown procedure exchange with the larger of the child keys?

169. Does insertion followed by deletion always give the same heap? What if the key inserted is larger than any then in the tree?

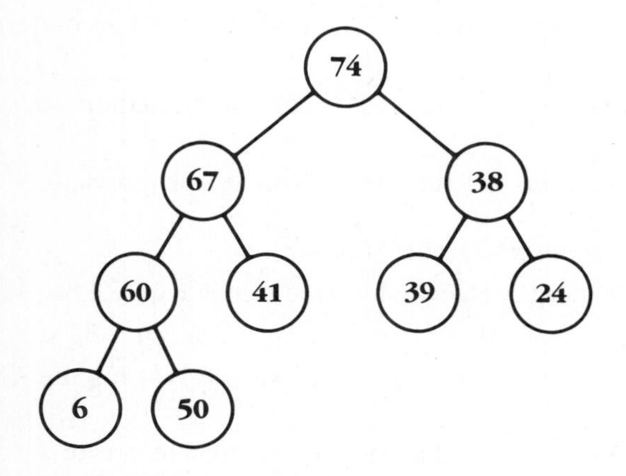

Figure 2.50 A Binary Tree

170. Implement a priority queue with operations Insert and Delete. Test with your own test program.

171. Implement the Heapsort algorithm. Test with your own test program. Include these in your test program:

    a) a case where the input sequence is already sorted
    b) a case where the input sequence is sorted in reverse order
    c) a case with duplicated keys and
    d) the case of Exercise 8.

    Include in your procedure a count of the number of comparisons and exchanges made between keys. How does the number of comparisons compare with the theoretical results of this chapter?

*172. In deletion from a heap, you might think in the average case half the possible key comparisons will require an exchange. In fact, more than half will. Why?

173. Show that Heapsort has time complexity $\Theta(n \log n)$.

Hashing

*174. What is the average-case successful search time for the hash table of Figure 2.51?

*175. What is the average-case unsuccessful search time for the table of Figure 2.51 if

    a) all buckets are equally likely to be requested?
    b) buckets 7, 8, and 9 are twice as likely to be requested as any of the others?

*176. Given a hash table of size 16, calculate the first 5 terms of the probe sequence for double hashing if

    a) $h_1(K_1) = 1$, $h_2(K_1) = 3$
    b) $h_1(K_2) = 5$, $h_2(K_2) = 9$
    c) $h_1(K_3) = 12$, $h_2(K_3) = 1$
    d) $h_1(K_4) = 0$, $h_2(K_4) = 7$

*177. Suppose in a hash table of size 16, locations 0, 1, 4, 5, 6, and 14 are occupied. Trace the insertions of the keys K_1, K_2, K_3, and K_4 with hash function values as in the previous problem for

    a) linear probing
    b) double hashing.

*178. Suppose a hash table of size 8 is used with linear probing. Find the expected number of comparisons required for the next insertion if all hash function values are equally likely and there are already keys in

    a) positions 1, 2, 4, and 6
    b) positions 1, 3, 5, and 7
    c) positions 1, 2, 3, and 6
    d) positions 1, 2, 3, and 4.

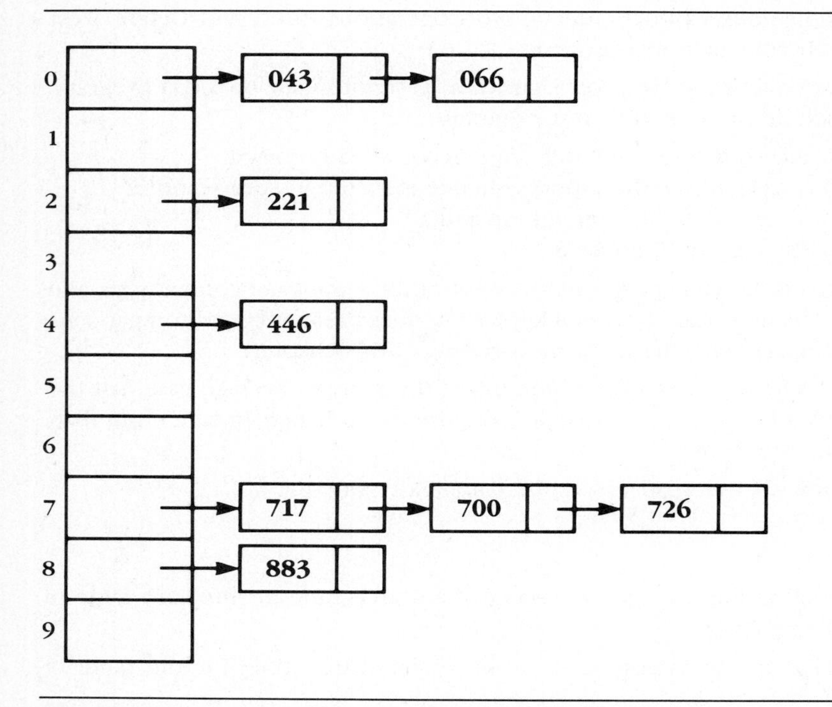

Figure 2.51 A Hash Table (with a Bad Hash Function)

*179. In a hypothetical programming language where all symbols consist of a letter followed by 0 to 5 digits, how many table locations would be required to give each possible symbol a separate bucket?

*180. Why are the "and" and "or" functions worse choices than "exclusive or" for folding?

181. What are the advantages and disadvantages of using a large hash table size?

182. Why can't we delete from a hash table just by zeroing out the entry to be deleted?

183. Why is $x^2 \bmod b$ a bad choice for a hash function?

184. If $x \bmod b$ is chosen for a hash function, why is it a bad idea to choose b near a power of 2?

185. In those hash functions calculated by multiplying the key by a constant, reducing mod 1, and multiplying by the table size, which of the constants $\{0.5, 0.04, 1/3, \pi, \sqrt{2} - 1, (1 + \sqrt{5})/2\}$ work well?

**186. Is it possible to keep the keys on each probe sequence in sorted order using chaining? Using double hashing?

*187. In the case where the distribution of keys is known or can be approximated in advance, the method of *digit analysis* is feasible. It works as follows for a table size of 2^k: Consider all the keys as binary numbers and determine the probability of finding a 1 bit in a given position. Determine those k bit positions for which this probability is closest to 1/2. Define $h(K)$ to be the value obtained by choosing the k bits in the specified positions of K. For example, if keys 1010, 0011, 1001, and 0101 each were expected to occur with probability 1/4 each, then the probability of a 1 in each bit position would be from left to right 1/2, 1/4, 1/2, and 3/4. We would choose the first and third bits from the left to get the respective hash function values 11, 01, 10, and 00. (We don't always get such a neat distribution of hash function values.) If a hash table with 4 buckets is used, which bit positions would be best for digit analysis for

a) the squares of the binary numbers from 0 to 1111?
b) the keys 100101, 010101, 111101, 110001, 100010, 011001, 010000, 110101?
c) the ASCII codes of the 26 letters (see Appendix B) if they are accessed with the frequencies of Exercise 147.

188. Implement a hash table using chaining. Construct your own hash function. Use character strings for keys. Use

a) a table size of 197 and
b) a table size of 256.

189. Do the previous problem using double hashing.

*190. Show that the probe sequence $h, h + c, h + 2c, \ldots, h + (b - 1)c$ visits all table locations iff $(c, b) = 1$.

*191. What is wrong with the following argument? If b buckets are implemented as a linked list in a hash table with n keys, then the average number of comparisons in successfully searching a bucket of size k is $(k + 1)/2$. Since the average bucket has size n/b, the average number of comparisons in a successful search is $(1 + n/b)/2$. Is a similar argument correct for unsuccessful search?

*192. Find an equation relating the average-case number of comparisons in successful and unsuccessful search for hashing with chaining similar to the second relation between S and U in the analysis of binary search trees of Section 2.16. Use this equation to show that the average-case number of comparisons in successful search is $1 + (n - 1)/2b \approx 1 + n/2b$, where n is the number of keys and b the number of buckets.

*193. Suppose that for every term of every probe sequence, there is probability $1/b$ that the term will refer to any given bucket. What is the average number of key comparisons to be expected

a) for unsuccessful search and
b) for successful search?

194. How could hashing be used to an array indexed by character strings?

195. How could hashing be used to implement a dynamic *sparse array* (i.e., an array in which most elements have a default value such as zero)?

*196. The ***hash signature*** may be defined for a set S in terms of h(s) for each s in S, where h is a hash function. For example, h(S) may be defined to be $\sum_{s \text{ in } S} \mathrm{h}(s) \bmod b$. Can you think of any applications of hash signatures?

New Topics

*197. Suppose that a given key K is to be sought between locations *low* and *upp* of an array A. If $A[low]$ and $A[upp]$ are known, can you think of a better choice for a place to look for K than at location $\lfloor (low + upp)/2 \rfloor$? Under what conditions would your choice be an improvement? What would be a disadvantage of using your choice.

198. Modify the binary search algorithm according to your answer to the previous problem. Implement your new algorithm.

199. Implement a data type *polynomial* (of one variable) with operations addition, subtraction, and multiplication. First represent each polynomial as a sequence, with the kth element corresponding to the coefficient of x^{k-1}. Can you think of any other operations to include?

200. Suppose you expected most of the coefficients of the preceding polynomials to be zero. Polynomials in which most coefficients are zero are called *sparse.* Compare the sparse arrays of Exercise 195. Could you design a more efficient implementation?

201. A *Gray code* is a sequence of length 2^n such that

   a) every element is a string of n bits
   b) any two elements of the sequence are different and
   c) any two consecutive elements differ in exactly one bit. Here the first and last elements are considered to be consecutive. For example, the sequence (00, 01, 11, 10) is a Gray code for $n = 2$. Construct a Gray code for 3-bit numbers. Can you think of any applications for a Gray code?

*202. Use the divide-and-conquer approach to design an algorithm to construct a Gray code for any given n.

*203. One sorting method with a recursive flavor involves dividing the n keys into r groups of s (thus $rs = n$ if the last group has the full s elements), finding the smallest key in each group, storing these keys in an array A of size r, and then finding the smallest key in A. If this smallest of the n keys comes from group k, we can replace it in A by the smallest remaining element in group k. The smallest element of A is now the smallest remaining key of the input. By repeating this process we can sort the entire input sequence.

a) Assuming that the minima are found by a simple sequential search, what is the time complexity of this algorithm in terms of r and s?
b) What would be the best choices for r and s in this algorithm?

204. Design an algorithm to implement the sorting method of the previous problem.

*205. Can quicksort be modified to give $O(n)$ best-case time complexity?

206. The ***bubblesort*** sorting algorithm works as follows: At iteration k, compare pairs of elements from left to right beginning with the pair in positions k and $k + 1$. Swap any pair found out of order. Convince yourself that after iteration $n - 1$, the input sequence is sorted. Can you think of any improvements to bubblesort?

207. What is the worst-case time complexity of bubblesort? What is the best-case time complexity?

*208. Show that if each of the $n!$ input permutations is equally likely, the average key in any sorting algorithm moves a distance of about $n/3$ positions. What does this say about the average-case time complexity of such an algorithm? What does this say about bubblesort?

**209. One algorithm that can move keys more than one position in a sequence per exchange was first described in Shell (1959) and is usually known as *shellsort.* In one simple version, the $n/2$ subsequences consisting of every $(n/2)$th element are first sorted. Then the $n/4$ subsequences consisting of every $(n/4)$th element are sorted, and so on, until just one subsequence is sorted. For example, if $n = 8$, the first $8/2 = 4$ sublists correspond to positions (0 4), (1 5), (2 6), and (3 7); the next $8/4 = 2$ sublists to (0 2 4 6) and (1 3 5 7). The set $\{n/2, n/4, \ldots, 1\}$ of distances between compared keys is called a set of *increments;* other values may be used for the increments as long as they decrease to a final value of 1. Write and test a program which implements shellsort

a) where the increments are $\{2^k, 2^{k-1}, \ldots, 1\}$ for appropriate k and
b) with an arbitrary but fixed set of increments.

*210. Give an intuitive argument why on the average shellsort might run faster than $\Theta(n^2)$.

211. Why are decreasing powers of 2 a bad choice of increments for shellsort?

212. Design and implement a *text* data type with operations suitable for a text editor. Represent your text as a string of characters. Is it worthwhile instead to make your text a string of paragraphs, which are themselves strings of sentences, which are themselves strings of words? Can you think of any way of improving your implementation?

Additional Reading — Chapter 2

Texts mentioned in the Bibliography on discrete mathematics or combinatorics such as Ross and Wright (1988), Sahni (1985), or Tucker (1980) typically cover mathematical induction, permutations, recurrence relations and the related technique of generating functions, and other discrete topics (i.e., topics that do not depend on the properties of the real numbers). Lueker (1980) covers recurrence relations in a readable way and more deeply than in this text.

Abelson and Sussman (1985) apply a recursive approach to a large part of computer science. Horowitz and Sahni (1978) and Stubbs and Webre (1985) in their Appendix B, treat data structures and data types in a more formal, axiomatic way.

The algorithms text of Horowitz and Sahni (1978) stresses design techniques such as divide-and-conquer and dynamic programming.

Good references for files and databases are Date (1981), Ullman (1980), and Wiederhold (1983).

The texts of Knuth (1973), Reingold and Hansen (1983), and Standish (1980) cover hashing more thoroughly than in this text.

Parsing is covered in a formal way in Hopcroft and Ullman (1979) and Lewis and Papadimitriou (1981). Parsing of natural languages, the topic of Pereira and Shieber (1987), is also covered in many books and articles on computational linguistics such as Grishman (1986).

Coding theory as exemplified by the Huffman algorithm is covered in Peterson and Weldon (1972) and MacWilliams and Sloane (1978).

Iteration

3.1 Introduction and Invariants

We have seen that the representation of a sequence in terms of an array allows O(1) access to the kth element of the sequence, allowing for more efficient algorithms in some cases than other representations (e.g., linked representations). For instance, search for a given element in a sorted sequence requires only O(log n) time. The natural way of processing an array or subarray of known size is iteratively. In many cases, however, a divide-and-conquer algorithm is still appropriate. One topic of this chapter is the relation between divide-and-conquer, or recursive thinking, and iterative algorithms.

This chapter is shorter than the previous chapter partly because iterative algorithms should be relatively familiar to readers and partly because analysis of iterative algorithms is typically easier. The simple techniques of Section 1.7, rather than the use of recurrence relations, are often sufficient.

The difficulties with iterative algorithms arise when moving from a specification to an algorithm or program and when proving that the algorithm or program is correct. Since iterative algorithms as well as recursive algorithms generally deal with data structures of arbitrary size, induction proofs of correctness are often needed. The arbitrary size means loops may be executed an unpredictable number of times, suggesting there is no more obvious variable on which to perform induction than the number of iterations itself.

To use this approach, we need a formal specification of the desired result and the input assumptions of the algorithm. In addition, we need a property P, where $P(k)$ is to represent the state of the computation after the kth iteration. In an inductive proof, since P is to be true for all k, we call P an *invariant* (or a proposed invariant) of the loop.

We can then prove correctness as follows:

1. show that $P(0)$ follows from the initial assumptions
2. show that $P(k - 1)$ implies $P(k)$
3. show that the desired result follows from $P(n)$, where n is the last iteration.

The value of n need not be known; it is enough to show that when the condition arises that terminates the loop, the desired result follows from the invariant. This argument assumes that a terminating condition does eventually occur. Proof of termination may require a separate argument.

In the case where termination of the loop does not terminate the algorithm, the execution of statements following the loop needs to be taken into consideration.

How to find the right P is not obvious; on the other hand, the process of finding P is not totally mysterious. Generally P follows relatively simply from even an informal description of how the algorithm is supposed to work. For example, if the algorithm is supposed to construct a data structure with a given property, $P(k)$ may say that the data structure existing after the kth iteration has the property.

One notational convention will be important. Often we need to consider the value of a variable both at the beginning and at the end of an iteration. Since the value of the variable may change during the loop, use the notation x_{old} for the value of x at the end of the previous iteration and the notation x_{new} for the value of x at the end of the current iteration.

You may find some of the algorithms in this chapter simple, familiar, and obviously correct. They are included mainly to give you practice in correctness proofs and the use of invariants.

3.2 Finding the Minimum in an Array

As a first example, consider the problem of finding the minimum element in an array of n elements. Intuitively, a simple approach is to keep track of the smallest element seen so far and the corresponding array index.

This approach suggests a loop to be iterated n times. At each iteration, the current value of the array is compared with the smallest value seen so far. If the current value is smaller, then the smallest value so far and the corresponding array index are updated. An algorithm embodying this approach is given as Algorithm 3.1.

If the array is called *In* and is indexed from 1 to n, then a more formal specification of the approach above would be to find an element *min* and an index *minindex* such that

$$1 \leq minindex \leq n,\ min = In[minindex], \text{ and } min \leq In[j] \text{ for } 1 \leq j \leq n$$

We now show correctness.

```
Minarray(In)              {returns the smallest element min of an array}
                          {In and the first location minindex of min}
1. min = ∞                {initialize min so that the first}
2. for k from 1 to n      {element seen will be minimum}
3.    if In[k] < min then {for each element, if it is}
4.       min := In[k]     {smaller than the current minimum}
5.       minindex := k    {update the minimum and its index}
6.    end if
7. end for
8. return min and minindex
```

Algorithm 3.1 Finding the Minimum in an Array

Theorem Algorithm 3.1 correctly finds the minimum value in an array.

Proof The idea that *min* should keep track of the smallest value seen so far means *min* will have a value no greater than $In[j]$ for j between 1 and k. Its location minindex will also be between 1 and k, suggesting the following invariant: at the end of the kth iteration, $1 \le minindex \le k$, $min = In[minindex]$, and $min \le In[j]$ for $1 \le j \le k$. At the end of the first iteration, these statements are all necessarily true. If they hold after the nth and last iteration, then the algorithm specification will be met. (Notice the specification is the same as the invariant with n replacing k.) Since the loop is executed a fixed number of times, the algorithm is guaranteed to terminate. We now must show the truth of the proposed invariant is preserved by each iteration.

Assume the proposed invariant is true after the $(k - 1)$st iteration. We need to consider two cases for the kth iteration. If $In[k]$ is not less than min_{old}, then neither the value of *min* nor the value of *minindex* will be changed. By induction, at the beginning of the loop, $min = In[minindex]$, and $1 \le minindex \le k - 1$. So at the end, *min* still equals $In[minindex]$, and $1 \le minindex \le k - 1 < k$. Also by induction, $min \le In[j]$ for $1 \le j \le k - 1$. Since $In[k] \not< min$, we have $min \le In[j]$ for $1 \le j \le k$.

If $In[k]$ is less than the old value of *min*, then $minindex = k$ at the end of the kth iteration, so the first equation of the invariant is satisfied. Also $min = In[k]$ and $k = minindex$ at the end of the iteration, so the second equation is satisfied. Finally, min_{new} is less than min_{old}, which is less than $In[j]$ for $1 \le j \le k - 1$ by induction. We have already seen that it is equal to $In[k]$, so the third equation is established.

In either case the proposed invariant remains true; therefore, we have shown correctness of the minimization algorithm. ∎

Since there are n iterations each with time complexity $O(1)$, the overall time complexity is $\Theta(n)$.

3.3 Sequential Search in an Array

As a second example, we consider sequential search through an array for the first location of a desired array element *key*. A simple informal approach is to proceed from location 1 through location n, checking if each array element equals the desired key. If so, we return the index. If we reach location n without having found the element, we return unsuccessfully. An algorithm based on this approach is Algorithm 3.2, where we assume a returned value of 0 indicates an unsuccessful search.

One interesting feature of the algorithm is the use of the key as a *sentinel*, a dummy value not considered part of the data structure. Without a sentinel, the termination check for the loop must consider whether or not the element was found and whether or not the end of the sequence had been reached. With the desired key inserted as a sentinel at the end of the sequence, this second check needn't be made, since the search will terminate by finding the desired key in the sentinel position — if not sooner. The algorithm must, of course, distinguish between a search that only succeeded because of the sentinel and a true successful search.

Using the notation of the algorithm, a formal specification of the desired result is

return 0 if there is no j with $1 \leq j \leq n$ and $In[j] = key$; otherwise

return j, where $In[j] = key$ and $In[i] \neq key$ for $1 \leq i \leq j$.

```
 1. ArraySearch(key, In)          {looks for the element key in the array
 2.                                In and returns the first location where it is
 3.                                found (or 0 in the unsuccessful case).
 4.                                It is assumed that there is room at the end of
 5.                                the array to insert the key as a sentinel and
 6.                                that elements are stored between locations 1 and n.
 7.                                The size of the sequence is assumed to be
 8.                                available via size(In)}
 9. In[size(In) + 1] := key                                  {insert sentinel}
10. index := 1                                       {begin at first location}
11. while In[index] ≠ key                           {and search for element}
12.     index := index + 1
13. end while
14. if index = size(In) + 1 then return 0            {if sentinel reached}
15. {else} return index                         {then return unsuccessfully}
16.                                          {else return index where found}
```

Algorithm 3.2 Sequential Search in an Array

Theorem The Sequential search algorithm (3.2) is correct according to the preceding specification.

Proof The specification suggests the following proposal for an invariant to be true after iteration k:

If there is no j with $1 \leq j \leq k$ and $In[j] = key$, then the algorithm has not terminated.

Otherwise, the algorithm has terminated and returned j with $In[j] = key$ and $In[i] \neq key$ for $1 \leq i < j$.

Notice this invariant is trivially true before the first iteration, since there is no j less than k and the algorithm cannot have terminated. If the invariant is true after iteration n, then either the algorithm will not have terminated and a value of 0 will be returned at the next step, or a value j will be returned with the properties of the specification.

Suppose the proposed invariant is true after the $(k - 1)$st iteration. If the algorithm has terminated, then we have nothing to prove; otherwise, we must consider two cases.

If $In[k] \neq key$, then the algorithm will not terminate during iteration k, and the invariant will be true because there is no j with $In[j] = key$, and either $1 \leq j < k - 1$ or $j = k$.

Otherwise, k will be returned as the algorithm terminates. In this case, since the algorithm did not terminate before iteration k, there is no $j < k$ with $In[j] = key$. So we have the desired condition, which is $In[k] = key$, but $In[j] \neq key$ for $1 \leq j < k$. Thus the proposed invariant really is an invariant and the algorithm is correct. ■

The algorithm must terminate after at most n iterations. Since each iteration has time complexity $O(1)$, the overall time complexity is $O(n)$. Unlike the previous algorithm, this algorithm may terminate before all array locations have been visited, so we cannot say the time complexity is $\Theta(n)$. However, in the unsuccessful case we can make this claim. In the successful case, if all elements of the array are equally likely to be the desired key, then the average-case time complexity will be $\Theta(n/2)$, which is $\Theta(n)$.

3.4 Merging

As another example, we may consider merging two sorted sequences stored in arrays into a third array. The idea for an iterative merge algorithm is not much different than that for a recursive merge algorithm. At any point each input sequence has a smallest item not yet included in the output. At the next iteration, the smaller of these is added to the end of the output sequence. Once one input sequence has been exhausted, the other may be appended to the end of the output sequence.

```
 1. Merge(In, low1, upp1, low2, upp2)     {merges two adjacent runs
 2.                        from array In, passed by reference or value-result.
 3.                         Assumes Copy algorithm (see below). Uses local
 4.                        array Out. low1 and upp1 mark endpoints of first
 5.                         run; low2 and upp2 mark endpoints of 2nd run;
 6.                          local variables index1, index2, and k are used}
 7. index1 := low1                          {index1 points into first input}
 8. index2 := low2                        {index2 points into second input}
 9. k := low1                                        {k points into output}
10. while index1 ≤ upp1 and index2 ≤ upp2
11.    if In[index1] < In[index2] then                 {find smaller key and}
12.       Out[k] := In[index1]                       {put it in destination}
13.       index1 := index1 + 1
14.       k := k + 1                                              {  array}
15.    else
16.       Out[k] := In[index2]
17.       index2 := index2 + 1
18.       k := k + 1
19.    end if
20. end while                           {continue until one run is exhausted}
21. if index1 > upp1 then                          {if first run is exhausted}
22.    Copy(In, Out, index2, upp2, k)           {then copy rest of second}
23. else                                                        {otherwise}
24.    Copy(In, Out, index1, upp1, k)                    {copy rest of first}
25. end if
26. Copy(Out, In, low1, upp2, low1)            {copy back to input array}
```

```
1. Copy(Source, Dest, low, upp, start)               {copies the elements
2.                    from low to upp from array Source to array Dest,
3.                         beginning with location start. Both arrays are
4.                                 passed by value-result or reference}
5. offset := short-low
6. for i from low to upp
7.    Dest[i + offset] := Source[i]
8. end for
```

Algorithm 3.3 Merge Two Sorted Sequences Stored in Arrays (with Auxiliary Procedure Copy)

In order to make this merge algorithm the basis of a mergesort algorithm, we cannot assume the input sequences are stored beginning at location 1. This point has been reflected in the merge algorithm, given as Algorithm 3.3. Figure 3.1 shows a trace of a sample run of this algorithm.

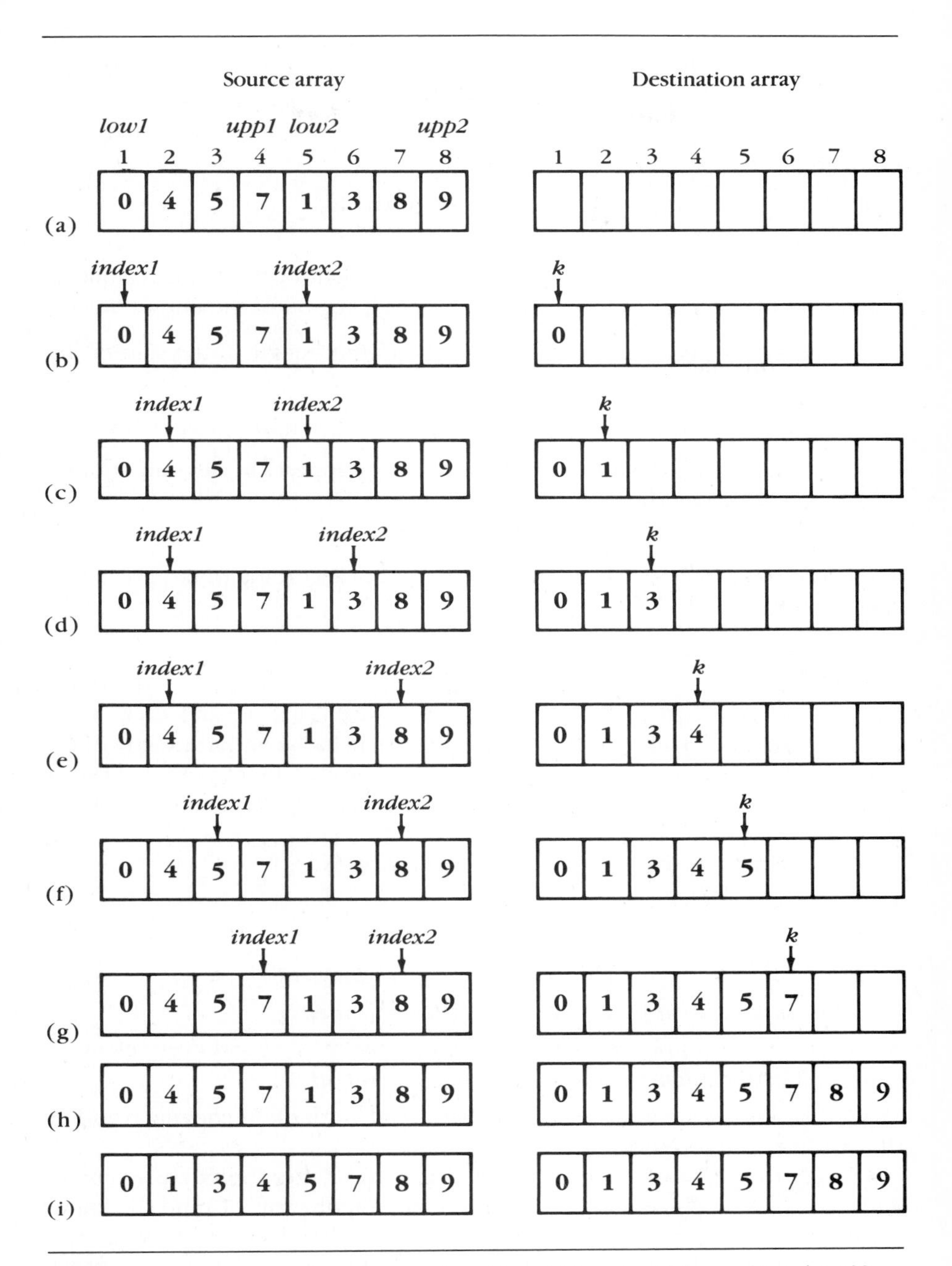

Figure 3.1 Stages in the Merge Algorithm (After Call Merge (A.1.4.5.8))

Although Algorithm 3.3 assumes that the two input sequences are in a single array, arguing for correctness of the algorithm will be simpler if we assume that the input sequences are in two different arrays called *In1* and *In2*. This assumption involves no loss of generality. With this terminology, we have the following specification for the algorithm:

If *In1* is originally sorted from *low1* to upp1 and

In2 is originally sorted from *low2* to *upp2* then

Out is sorted from *low* to *upp*, with each element of the given regions of input arrays appearing exactly once in the given region of the output array.

Notice the last condition implies that the output length is the sum of the input lengths, that is,

$$(upp1 - low1 + 1) + (upp2 - low2 + 1) = (upp - low + 1).$$

This equation implies that if every input element appears in the output, then every input element appears exactly once.

Theorem The Merge algorithm (3.3) is correct according to the preceding specification.

Proof Our proposed invariant must capture this information: the output array so far is sorted and contains only the smallest keys, and each input element already considered from the input appears in the output exactly once. In addition, if an input run is not exhausted, the next element from that run is as large as the largest output element. Thus we get the following proposed invariant:

After the mth assignment to the destination array,

1. k has value $low1 + m$
2. *Out* is sorted from $low1$ to $k - 1$
3. either $index1 > upp1$ or $Out[k - 1] \leq In1[index1]$
4. either $index2 > upp2$ or $Out[k - 1] \leq In2[index2]$
5. every element of *In* from location $low1$ to $index1 - 1$ and every element from location $low2$ to $index2 - 1$ appears in *Out*.

We also notice the input array is not changed, until the recopying phase in the last line of the algorithm.

Initially the proposed invariant is vacuously true. After the nth assignment, where n equals the size $upp - low + 1$ of the output array, (1) and (4) imply the desired result provided that $index1 = upp1 + 1$ and $index2 = upp2 + 1$. One of these will be true when the first **while** loop is exited, and the other becomes true after the appropriate call to Copy. Since neither *index1* nor *index2* can exceed the corresponding upper index by more than 1, and since each assignment to the destination array increments exactly one of the indices by 1, the correct number n of assignments will be performed.

We need to show the proposed invariant is preserved after the mth assignment. So suppose the statement is true after the $(m - 1)$st assignment. In particular, we assume that k has value $low1 + m - 1$.

Suppose that $index1 \leq upp1$ and $index2 \leq upp2$. Further suppose that $In[index1] < In[index2]$, so that lines 12–14 of the algorithm will be executed. Since k is incremented by 1 by the algorithm and m increases by 1, part 1 of the invariant remains true.

To show (2), we first note that under our assumptions, the new last element is $Out[k_{new} - 1] = Out[k_{old}] = In[index1_{old}]$, where the last equality follows by line 12. By (3), $In[index1_{old}] \geq Out[k_{old} - 1] = Out[k_{new} - 2]$. Also, by (2), *Old* is sorted up to $k_{old} - 1 = k_{new} - 2$. So *Out* is now sorted up to $k_{new} - 1$, and (2) remains true.

For (3), if $index1_{new} \leq upp1$, then $Out[k_{new} - 1] = In[index1_{old}] \leq In1[index1_{old} + 1] = In[index1_{new}]$, where the inequality follows from the initial sortedness of *In*. So (3) remains true.

For (4), we have $In[index2_{new}] = In[index2_{old}] > In[index1_{old}] = Out[k_{new} - 1]$, where the inequality follows from our assumption. So (4) remains true.

Part 5 remains true since the only element of *In* newly under consideration is $In[index1_{new} - 1] = In[index1_{old}] = Out[k_{new} - 1]$, and this latter element is the only new element of *Out*.

The arguments for the other three cases of the algorithm (lines 16–18 and the first two calls to Copy) are similar and are left as exercises. ■

Note that the algorithm will still work if the two input runs are taken from different arrays or if the values of k and $index1$ are initialized to different values. The algorithm can fail, however, if the same array locations are used for both input and output.

Since each assignment has time complexity 1, the overall time complexity is $\Theta(n)$, where n is the sum of the input lengths.

3.5 Partition (for Quicksort)

Our next example is the partition algorithm deferred from Section 2.11. Recall that given a sequence stored between locations *low* and *high* of an array *In*, the partition algorithm is supposed to return an index i and permute the array so that

1. for $low \leq j < i$, $In[j] \leq In[i]$ and
2. for $i < j \leq high$, $In[j] \geq In[i]$.

Intuitively, an algorithm to perform the partition should move small elements to the left and large elements to the right. To determine whether an element is small or large, we compare it to a *pivot* element. Several ways of choosing a pivot element are explored in the exercises; the simplest is to choose $In[low]$.

Partition Algorithm 3.4, which Bentley (1984) credits to Nico Lomuto, will traverse the array from left to right. For those elements already considered, those smaller than the pivot will be stored to the left and those larger to the right. We maintain the index *LastLow* of the rightmost small element. If the current element is larger than the pivot, then it is already to the right of position LastLow and nothing needs to be done. If the current element is smaller than the pivot, then *LastLow* is incremented, and the element at the new *LastLow* position is exchanged with the current element.

A trace of a sample partition instance is given in Figure 3.2.

Since the only loop executes a fixed number of times, Algorithm 3.4 will terminate.

Theorem The Partition Algorithm 3.4 is correct according to the specification.

Proof Our proposed invariant is just the formal counterpart of the observation that the part of the array to the left of the current position is properly partitioned. More precisely, after item k has been considered,

1. for $low < j \leq LastLow$, $In[j] < pivot$
2. for $LastLow < j \leq k$, $In[j] \geq pivot$.

The invariant is true for $k = low = LastLow$ at the beginning of the first iteration; for $k = high$ the invariant implies the desired result. Here k plays the role of the variable *current* in the algorithm.

```
1.  Partition(In, low, upp)          {the sequence to be partitioned is
2.                                    stored in array In with indices from low to
3.                                    upp. In[low] is used as the pivot element
4.                                    Variables pivot, current, and temp are used locally
5.                                    Partition returns an index LastLow that
6.                                    points to the end of the first subsequence}
7.  LastLow := low                    {LastLow is to point to rightmost small key}
8.  pivot := In[low]                  {choose the leftmost key as pivot element}
9.  for current from low + 1 to upp   {pass through sequence}
10.     if In[current] < pivot then   {if current key out of place}
11.        LastLow := LastLow + 1     {then include it among small}
12.        swap(In[LastLow], In[current])  {keys and update pointer}
13.     end if                        {to rightmost small key}
14. end for
15. swap(In[low], In[LastLow])        {swap pivot with last small key}
16. return LastLow
```

Algorithm 3.4 Partition

(a) Initial sequence	18	25	7	11	26	14	9	38	10
	LastLow	***Current***							
(b)	18	25	7	11	26	14	9	38	10
		LastLow	***Current***						
(c)	18	7	25	11	26	14	9	38	10
			LastLow	***Current***					
(d)	18	7	11	25	26	14	9	38	10
			LastLow		***Current***				
(e)	18	7	11	25	26	14	9	38	10
				LastLow		***Current***			
(f)	18	7	11	14	26	25	9	38	10
					LastLow		***Current***		
(g)	18	7	11	14	9	25	26	38	10
					LastLow			***Current***	
(h)	18	7	11	14	9	25	26	38	10
						LastLow			***Current***
(i)	18	7	11	14	9	10	26	38	25
(j) Final sequence	10	7	11	14	9	18	26	38	25

Figure 3.2 Trace of a Sample Instance of the Partition Algorithm

Suppose the invariant is true after index $k - 1$ has been considered. Then we must consider two cases for $In[k]$. If $In[\mathrm{k}] \geq pivot$, then *LastLow* and *In* remain unchanged and part 1 remains true. Part 2 is true for $LastLow < j \leq k - 1$ by induction and true for $j = k$, since by assumption $In[j] = In[k] > pivot$.

If $In[k] < pivot$, then part 1 remains true for $j \leq LastLow_{\mathrm{old}}$. Part 1 is also true for $j = LastLow_{\mathrm{new}} = LastLow_{\mathrm{old}} + 1$, since $In[LastLow]_{\mathrm{new}} = In[k]_{\mathrm{old}} < pivot$. Part 2 remains true for $LastLow_{\mathrm{new}} = LastLow_{\mathrm{old}} < j \leq k - 1$ by induction. In addition (2) is true for $j = k$, since $In[k]_{\mathrm{new}} = In[LastLow_{\mathrm{new}}]_{\mathrm{old}} = In[LastLow_{\mathrm{old}} + 1]_{\mathrm{old}} \geq pivot$. Here the last inequality follows by induction and (2), since $LastLow_{\mathrm{old}} + 1 \geq LastLow_{\mathrm{old}}$. ∎

Each of the n iterations in partition has time complexity $O(1)$, so the overall algorithm has time complexity $O(n)$. Thus we have justified the claims made in Section 2.11 for the Quicksort algorithm.

Floyd and Rivest (1975) have pointed out that the preceding partition algorithm may be used to select the kth largest element of an unsorted sequence. Assume Partition is called and returns j. If $j = k$ then the jth position of the array will contain the desired element. If $j < k$ then the element is sought to the right of the jth position. If $j > k$ then the element is sought to the left of the jth position. We get Algorithm 3.5.

This algorithm is similar to binary search, except that the size of pieces of the input sequence is not necessarily equal. Notice that the algorithm does *not* return the kth largest element between positions *low* and *upp,* and the parameter k is passed without change recursively. This is because the value j of the Partition function is an array index, not a pointer to the jth largest element in a subsequence.

A trace of the Select algorithm is given in Figure 3.3. Keys put in their correct place by Partition are in boldface. The time complexity of this version of Select is $\Theta(n^2)$, since there may be $\Theta(n)$ calls to Partition. Note that we only need to process one-half of the partitioned sequence, compared to two in

```
1. Select(k, In, low, upp)          {When called with low = 1, returns
2.                                      the kth largest element in the
3.                                      array In. In need not be sorted.
4.                                      upp is the size of the array.
5.                                      It is assumed that k < upp}
6. j := Partition(In, low, upp)
7. if j = k then return In[j]
8. {else} if j < k then return Select(k, In, j + 1, upp)
9. {else if j > k then} return Select(k, In, low, j − 1)
```

Algorithm 3.5 Select

Sequence												
1	*2*	*3*	*4*	*5*	*6*	*7*	*8*	*9*	*10*	*j*	*low*	*upp*
5403	4161	9899	6536	2836	9601	7539	1455	9111	8390		1	10
1455	4161	2836	**5403**	9899	9601	7539	6536	9111	8390	4	5	10
1455	4161	2836	5403	8390	9601	7539	6536	9111	**9899**	10	5	9
1455	4161	2836	5403	6536	7539	**8390**	9601	9111	9899	7	8	9
1455	4161	2836	5403	6536	7539	8390	9111	**9601**	9899	9		

Figure 3.3 Trace of a Select Algorithm

Quicksort. If the pivot element could be chosen closer to the median, then we might hope for a $\Theta(n)$ worst case. Horowitz and Sahni (1978) show how, with a great deal of overhead, this can be accomplished.

3.6 Selection Sort

We next consider a purely iterative sorting algorithm, *selection sort,* which constructs a sorted array by repeatedly selecting the smallest unselected element. When an element is selected, it should not be considered again, so that the element is swapped into the appropriate position into the input array rather than copied into a separate output array. An example of the result after each iteration of a typical instance of selection sort is given in Figure 3.4. In this figure, keys known to be in their correct position are in boldface.

A more formal version of the selection sort algorithm is Algorithm 3.6. Note the similarity of the inner loop to the loop of the minimization Algorithm 3.1.

Theorem Algorithm 3.6 sorts correctly.

Location	1	2	3	4	5	6	7	8
Initial sequence	2	1	7	3	6	8	9	4
After 1st iteration	**1**	2	7	3	6	8	9	4
After 2nd iteration	**1**	**2**	7	3	6	8	9	4
After 3rd iteration	**1**	**2**	**3**	7	6	8	9	4
After 4th iteration	**1**	**2**	**3**	**4**	6	8	9	7
After 5th iteration	**1**	**2**	**3**	**4**	**6**	8	9	7
After 6th iteration	**1**	**2**	**3**	**4**	**6**	**7**	9	8
After 7th iteration	**1**	**2**	**3**	**4**	**5**	**7**	**8**	**9**

Figure 3.4 Example of Selection Sort

Proof The preceding informal description suggests that after iteration k, where k corresponds to the index variable of the algorithm, the first k elements should be in their proper places and the remaining elements should all be larger. This corresponds to the following candidate for an invariant:

after the kth iteration,

1. for $1 \le i < j \le k$, $In[i] \le In[j]$
2. for $j > k$, $In[j] \ge In[k]$.

These two conditions together imply that the array is sorted after the $(n - 1)$st iteration of the outer loop. The proof that the inner loop correctly finds the minimum is essentially the same as that for the preceding minimization algorithm and is left as an exercise. After the first iteration, (1) is vacuously true and (2) is true, since the minimal element is moved to position $k = 1$.

Suppose the proposed invariant is true after iteration $k - 1$. Iteration k will move the smallest remaining element into position k, say from position m. By induction and (1), $In[i] \le In[j]$ for $1 \le i < j \le k - 1$.

$$\begin{aligned} \text{Now } In[k]_{\text{new}} &= In[m]_{\text{old}} \\ &\ge In[\mathrm{k} - 1]_{\text{old}} \text{ [by induction and (2)]} \\ &= In[k - 1)_{\text{new}} \end{aligned}$$

So $In[i] \le In[k - 1] \le In[k]$ whenever $1 \le i < j = k$. Thus (1) remains true after iteration k.

```
 1. Selsort(In)                   {assumes an array In with lowest index 1.
 2.                                   Size(In) is assumed to give the size
 3.                                              of the input sequence.
 4.                               Uses local variables j, index, m, and min}
 5. for index from 1 to size(In) − 1           {ith smallest to position i}
 6.    m := index                               {save position and value}
 7.    min := In[index]                     {  of smallest remaining key}
 8.       for j from index + 1 to size(In)            {look through rest}
 9.         if In[j] < min then                                {of input}
10.            m := j                               {update if a smaller}
11.            min := In[j]                                {key is found}
12.         end if
13.       end for
14.    swap(In[m], In[index])          {exchange key in position index}
15. end for                                      {with ith smallest key}
```

Algorithm 3.6 Selection Sort

For $j > k$, $In[j] \geq In[k]_{\text{new}}$, after iteration k, since $In[j]$ was one of the elements considered when $In[k]_{\text{new}}$ was determined to be the minimum. Note this is true even if $j = m$. Thus (2) remains true after iteration k, and we have shown correctness. ∎

Iteration k considers $n - k + 1$ elements, and thus has time complexity $\Theta(n - k + 1)$. Accordingly the entire algorithm has time complexity $\Theta\left(\sum_{k=1}^{n} n - k + 1\right)$. If we substitute j for $n - k + 1$, this becomes $\Theta\left(\sum_{j=1}^{n} j\right)$; by property (10) of Section 1.6 this time complexity is just $\Theta(n^2)$.

We have already seen asymptotically better algorithms in Mergesort and Heapsort, so selection sort is appropriate primarily for small sequences when low overhead is desired. Notice however that although selection sort makes a lot of comparisons, it makes few assignments, so selection sort might be appropriate for files in which records are awkward to move but keys are easy to compare.

3.7 Tail Recursion

Implementing recursion requires pushing information onto a stack for each procedure call (see Appendix D). Often equivalent nonrecursive algorithms can accomplish the same result without the overhead of stack manipulations or the space needed by a stack.

For example, consider the recursive definition of sequential search. In this definition, the subsequence beginning at the desired element is returned if the element is in the list; otherwise, the empty subsequence is returned.

Def 3.7 SeqSearch(x, L) =

1. L if L is empty or x = head(L)
2. SeqSearch(x, tail(L)) otherwise.

The straightforward recursive implementation of this definition would require n pushes before any pops for a list of length n. The space requirement would thus be $\Theta(n)$. Yet we have already seen a very natural algorithm for sequential search that doesn't require a stack at all.

This recursive algorithm has the property that the very last action performed is the recursive call, and there is no other recursive call. Such an algorithm is called *tail recursive.* During execution of a tail recursive algorithm, control passes from the last to the first statement in the definition, exactly as it would in a loop. We may therefore replace tail recursion by iteration.

The only wrinkle is that the arguments to the recursive call are different than those to the original call, so these arguments should be maintained as explicit variables that are initialized at the beginning of the loop and explicitly

```
1. SeqSearch(x, list)              {assumes a Boolean Empty function}
2. while not Empty(list) and head(list) ≠ x
3. list := tail(list)
4. end while
5. return list
```

Algorithm 3.8 Iterative Sequential Search (No Sentinel)

changed before control is transferred to the beginning of the loop. In the iterative version, the condition for returning to the beginning of the loop is the negation of the termination condition for recursion. An explicit output statement may also be required.

An equivalent iterative algorithm would therefore be Algorithm 3.8.[1]

The resulting algorithm is essentially the sequential search Algorithm 3.2, only without a sentinel.

This example suggests that some functions for which we have presented recursive definitions can be defined iteratively. The process of removing recursion is not a panacea; it need not simplify a definition. Even if it makes a procedure more efficient from the machine's point of view, the recursive definition may still be easier for humans to work with. Many situations in which removing recursion is worthwhile do exist. In fact, compilers and interpreters may be constructed to recognize some such situations and to remove recursion automatically.

The binary search Algorithm 2.63 is tail recursive. Even though a recursive call appears before the last line of the algorithm, no statements of the current call remain to be executed if this recursive call is made. Consequently the recursive binary search algorithm has the equivalent iterative version, Algorithm 3.9.

The *k*th function of Definition 2.15 is tail recursive; the tail recursion may be removed to get the iterative Algorithm 3.10.

A number of recursive algorithms don't qualify as tail recursive under the preceding definition, but are naturally implemented iteratively. These algorithms include the factorial function, the length function, and the exponential function of Sections 2.1 through 2.3. These recursive algorithms correspond to divide-and-conquer algorithms for which only one piece is nontrivial.

A typical divide-and-conquer function of this nature has the form for some C and F:

---

[1] Warning: in some language implementations, including many implementations of Pascal, the second conjunct in the while statement will be tested even if the first one fails. Thus in these languages, even if the list is empty, its head will be sought with a resulting run-time error. We are also assuming that parameters are called by value, so that assignments to them do not change their values in the calling routine.

```
1. BinarySearch(key, In, low, high)
2.    while low ≤ high
3.       mid := ⌊(low + high)/2⌋
4.       if key = In[mid] then return mid
5.       {else} if key < In[mid] then high := mid − 1
6.       else {if key > In[mid]} low := mid + 1
7.    end while
8. return unsuccessfully
```

Algorithm 3.9 Iterative Binary Search

```
1. Kth(k, S)
2.    while S not empty
3.       if k = 1 then return head(S)
4.       else
5.          k := k − 1
6.          S := tail(S)
7.       end if
8.    end while
9.    return unsuccessfully
```

Algorithm 3.10 Finding the *K*th Element of a Sequence

Def 3.11 DivideAndConquer(X) is

1. base(X)   if X is small
2. C(head(X), F(tail(X))) otherwise.

Here **head** and **tail** are not necessarily the functions defined in Section 2.3 for sequences, but instead functions appropriate for dividing elements with the data type of X. If X is not small, it is assumed to have a tail.

An equivalent iterative algorithm in general must initialize and maintain a variable to be returned, test X for smallness, apply C to head(X), and replace X by the tail of X. One attempt would be Algorithm 3.12.

One immediate problem with this approach is that the base function will be applied to the original input X, while in the recursive algorithm it is applied only to small X. We may solve this problem by restricting consideration to those functions for which base(X) returns a constant value which we shall call **base.** Note the preceding three functions all qualify. For no such function is a recursive call applied to the head of the input, so all such functions are of the first of the two divide-and-conquer types discussed in Section 2.2. If recursion

```
1. DivideAndConquer(X)
2.        result := base(X)
3.        while X not small
4.           result := C(head(X), result)
5.           X := tail(X)
6.        end while
7.        return result
```

Algorithm 3.12 Bad Iterative Divide-and-Conquer

is applied to two or more pieces of the input, then removing recursion is more troublesome.

Now consider a case where tail(tail(X)) is empty. The recursive algorithm will return

$$\mathrm{C(head}(X), \mathrm{C(head(tail}(X)), \mathrm{base})).$$

The iterative algorithm will return

$$\mathrm{C(head(tail}(X)), \mathrm{C(head}(X), \mathrm{base})).$$

These will not be the same unless C is commutative and associative; switching the order of the arguments to C in the definitions won't help. But if all small arguments get the same value and the function C is commutative and associative, then we can show that the preceding iterative algorithm works. We may restate this claim as follows:

If F(X) is defined by F(X) =

1. base if small(X)
2. C(head(X), F(tail(X))) otherwise,

where C is commutative and associative, then the following iterative algorithm (3.13) will compute F(X):

```
1. F(X)                         {calculates the function defined by
2.                                        F(X) = base   if X is small
3.                              C(head(X), F(tail(X))) otherwise}
4.     result := base
5.     while X not small
6.        result := C(head(X), result)
7.        X := tail(X)
8.     end while
9.     return result
```

Algorithm 3.13 Generic Divide-and-Conquer

```
1. Length(L)                {iterative function that returns the length
2.                                                    of a sequence L}
3.    result := 0
4.    while L not empty
5.       result := 1 + result
6.       L := tail(L)
7.    end while
8.    return result
```

Algorithm 3.14 Iterative Length

The argument proving the claim will be easier to believe after a few examples have been presented. For the length function, we get Algorithm 3.14.

For the factorial function, we have Algorithm 3.15.

And for the exponential function, we get Algorithm 3.16.

Note the iterative factorial function multiplies its arguments in the order from X down to 1, while the recursive function multiplies them in the order from 1 up to X.

More precisely, the values of k and *result* for successive iterations in the computation of factorial(X) and factorial($X - 1$) may be put into one-to-one correspondence. Figure 3.5 shows an example for $X = 6$.

Theorem The iterative DivideAndConquer algorithm calculates the function defined by the recursive DivideAndConquer definition.

Proof We first establish notation to generalize the example of Figure 3.5. For a given X, let $X^+(k)$ be the value of the variable X after the kth iteration of F(X). Let $result^+(k)$ be the value of the variable *result* after the kth iteration of F(X). Let $X^-(k)$ be the value of the variable X after the kth iteration of F(tail(X)). Let $result^-(k)$ be the value of the variable *result* after the kth iteration of F(tail(X)). These four terms correspond to four columns of Figure 3.5.

```
1. Factorial(X)                  {Iterative algorithm for the factorial
2.                            function assuming X a nonnegative integer}
3. result := 1
4. while X ≠ 0
5.    result := X × result
6.    X := X − 1
7. end while
8. return result
```

Algorithm 3.15 Iterative Factorial

```
1. Exponential(b, x)                {Iterative algorithm returning b^x}
2. result := 1
3. while x ≠ 0
4.     result := b × result
5.     x := x − 1
6. end while
7. return result
```

Algorithm 3.16 Iterative Exponential

For small X, both the iterative and recursive algorithms will return the value **base.** So we may assume that X has a tail, and that the computation of F(X) requires at least one iteration. We claim that for all k before computation halts,

1. $X^+(k+1) = X^-(k)$ and
2. $result^+(k+1) = \mathrm{C}(\mathrm{head}(X), result^-(k))$.

Note that (2), in the context of the example, simply means that the values attained by the variable *result* in the computation of 6! are just 6 times those in the computation of 5!

For $k = 0$ the claims are true, since

$$X^+(1) = \mathrm{tail}(X) = X^-(0) \text{ and}$$

$$result^+(1) = \mathrm{C}(\mathrm{head}(X), \mathrm{base}) = \mathrm{C}(\mathrm{head}(X), result^-(0)).$$

If the claims are true for $k - 1$, then

1. $X^+(k+1) = \mathrm{tail}(X^+(k)) = \mathrm{tail}(X^-(k-1)) = X^-(k)$
   where the middle equality is by induction and
2. $result^+(k+1) =$ (by the algorithm)
   $\mathrm{C}(\mathrm{head}(X^+(k)), result^+(k)) =$ (by induction)
   $\mathrm{C}(\mathrm{head}(X^+(k)), \mathrm{C}(\mathrm{head}(X), result^-(k-1))) =$ (by the properties of C)

Iteration k	$result^+(k)$	$X^+(k)$	Iteration k	$result^-(k)$	$X^-(k)$
1	6	5			
2	30	4	1	5	4
3	120	3	2	20	3
4	360	2	3	60	2
5	720	1	4	120	1
6	720	0	5	120	0

Figure 3.5 Stages in the Iterative Computations of 6! and 5!

C(head(X), C(head($X^+(k)$), $result^-(k-1)$))) = (by induction and (1))
C(head(X), C(head($X^-(k-1)$), $result^-(k-1)$))) = (by the algorithm)
C(head(X), $result^-(k)$).

Now since $X^+(k+1)$ and $X^-(k)$ have the same value for each k, they both become small at the same time. For the k that terminates the algorithm,

$$result^-(k) = \text{F(tail}(X)) \text{ and } result^+(k) = \text{F}(X),$$

where F refers to the value computed by the iterative algorithm. So by substitution into the equation of claim (2),

$$\text{F}(X) = \text{C(head}(X), \text{F(tail}(X))),$$

which is the recursive part of the recursive definition. Therefore, the iteratively defined function has the same value as the recursive on small X and satisfies the same recurrence. Consequently it must be the same function. ■

3.8 Bottom-Up Computation

The commutativity of the combine function C in the previous section was essential. For example, one useful function that lends itself to a recursive formulation is the function for the change of base for integers.

In this section we will write binary numbers with a subscript of 2, so that $1101_2 = 13$.

The usual representation of a string as a sequence suggests that the recursive formulation of the change of basis algorithm should give the value of 1101_2 in terms of the number 101_2. While $1101_2 = 10_2^3 + 101_2$ is true, there is no way to obtain the exponent 3 without first finding the end of the sequence. In other words, the recursive algorithm cannot examine the digits from left to right.

There is, nevertheless, a natural recursive formulation if the sequence is reversed. In our example, $1101_2 = 2 \times 110_2 + 1$, where the number 2 is independent of the length of the sequence. Thus if the nonempty sequence S is the reversed string of digits of a binary number, we have the following recursive function for base conversion:

Def 3.17 ConvertToDecimal(S) =

1. head(S) if S has length 1
2. head(S) + 2 × ConvertToDecimal((tail(S)) otherwise.

The number 2 can be replaced by any other radix; however, when a radix is greater than 10, digits greater than 9 must be converted first into decimal numbers.

Notice the function $C(x, y) = x + 2 \times y$ is not commutative, so this algorithm will not work if the sequence of binary digits has not been reversed.

```
1. ConvertToDecimal(a)     {The integer a is represented as an array}
2.                                its length is given by size(a)}
3.    result := 0
4.    for k := size(a) to 1 by −1
5.       result := a[k] + 2 × result
6.    end for
7. return result
```

Algorithm 3.18 Iterative ConvertToDecimal

On the other hand, if the sequence of digits is given in an array a, with the kth digit from the left in $a[k]$, we may compute the decimal equivalent using Algorithm 3.18. This corresponds to reversing the string and then removing recursion as in the previous section.

Reversing the input corresponds to the insight that the "conquer" phase of a recursively implemented divide-and-conquer algorithm processes the pieces of the data structure from smallest to largest, although the "divide" phase considers them from largest to smallest. Sometimes avoiding recursion and

```
 1. Mergesort(arr)              {assumes that the list to be sorted is stored
 2.                          between locations 1 and size(arr) of array arr.
 3.                       Uses locals low2, upp2 to mark endpoints of 2d run
 4.                             as well as local variables n, len, and index.
 5.                                 Uses the merge algorithm of Section 3.4}
 6. len := 1                                   {initialize size of input runs}
 7. n := size(arr)                                       {call the array size n}
 8. while len < n                            {if entire list is not a single run}
 9.    index := 1                                   {start at beginning of list}
10.    repeat
11.       low2 := index + len                  {calculate beginning of 2d run}
12.       if low2 ≤ n then                     {if there is a 2d run to merge}
13.          upp2 := low2 + len − 1                     {then calculate its end}
14.          if upp2 > n then upp2 = n;            {2d run may be truncated}
15.          merge(arr, index, low2 − 1, low2, upp2)           {see Sec. 3.4}
16.       end if
17.       index := index + 2 × len                 {point to next pair of runs}
18.    end repeat if index > n
19.    len := 2 × len                                   {prepare for next pass}
20. end while
```

Algorithm 3.19 Iterative Mergesort

beginning with the "conquer" phase, as in the dynamic programming algorithms of Section 2.19, is preferable. We have already seen examples of this task in the dynamic programming algorithms of Section 2.19.

Another example we have already seen is Mergesort. The natural top-down, divide-and-conquer algorithm for Mergesort was given in Section 2.15. The Mergesort algorithm given in Section 2.3 was a bottom-up algorithm, since it began with small pieces (runs of minimum size) and created successively larger runs. An iterative Mergesort algorithm for sequences stored in arrays is given as Algorithm 3.19. This bottom-up algorithm is closer to that of Section 2.3 than that of Section 2.15.

For Mergesort, the top-down approach does not lead to the same work as the bottom-up approach if the input size is not a power of 2. The tree structure (called a *merge tree*) defined by the divisions of the "divide" phase is given in Figure 3.6a. The resulting merges are shown in (b) of the figure. The somewhat different merges performed by the bottom-up algorithm are illustrated in Figure 3.7. Note in the top-down approach the highest-level merge is between two pieces of nearly equal size, which need not be true in the bottom-up case.

Radix Sort

Another example of a bottom-up version of a top-down algorithm is *radix sort,* the bottom-up version of bucket sort. In radix sort, the characters of a string are considered from right to left.[2] In order to finish each character at the same time, if necessary the strings are padded at the right end with blanks so that all have the same length. Numerical keys are padded with leading zeros, since numerical order is different than lexicographical order.

Theorem Radix sort sorts correctly.

Proof After iteration k of radix sort, the strings should have been sorted based on their last k characters. Clearly this is true initially, when $k = 0$; if it is true for $k = s$, the common string length, then the set of characters is sorted. Thus we need only show how to maintain this proposed invariant between iterations $k - 1$ and k.

The job of iteration k is to assign strings to buckets based on the kth character from their right. If this is done for each string and the buckets are concatenated in order, then the proposed invariant should be true—if the invariant was true before the iteration. Suppose that a_i occurs before a_j at the end of the kth iteration. We need to show that a_i comes before a_j when their last k characters are compared.

There are two cases: In the first case, a_i and a_j were sent to different buckets. Here the bucket containing a_i must come before the bucket containing a_j,

[2] These characters may be identified with digits in radix N, where there are N different possible characters. Hence the name "radix sort."

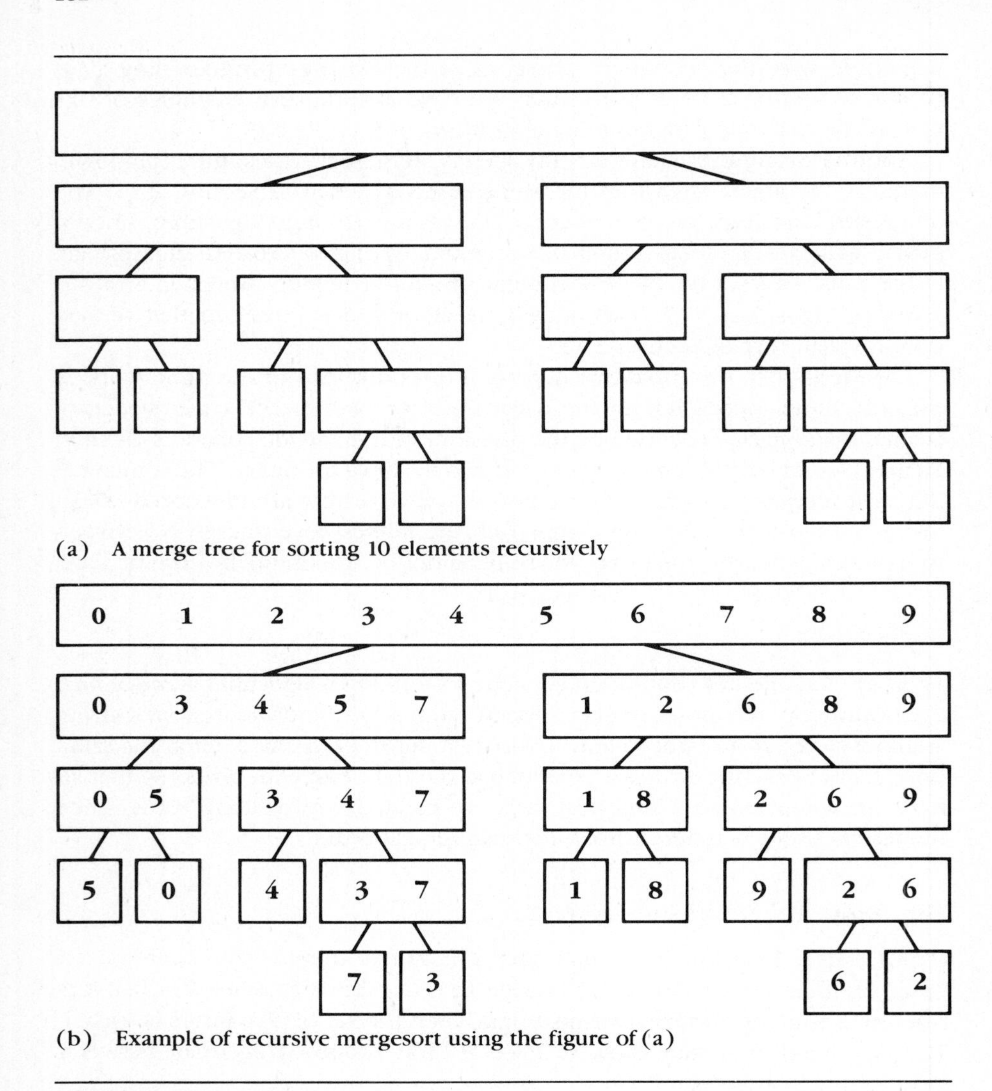

(a) A merge tree for sorting 10 elements recursively

(b) Example of recursive mergesort using the figure of (a)

Figure 3.6 Merge Trees

since the buckets were concatenated in order. But this means that a_i precedes a_j when their last k characters are compared.

In the second case, a_i and a_j were sent to the same bucket. So they agree on the kth character from the right. For a_i to precede a_j, it must be true that it precedes a_j based on the last $(k - 1)$ characters. By induction, a_i precedes a_j at the end of the $(k - 1)$st iteration. To force it to precede a_j at the end of the kth iteration, we need only make sure that characters in the same bucket appear in the same order as they are considered by iteration k (i.e., the same order as the output of iteration $k - 1$). This first-in, first-out behavior can be attained by maintaining each bucket as a queue, that is, adding new keys to the end of the

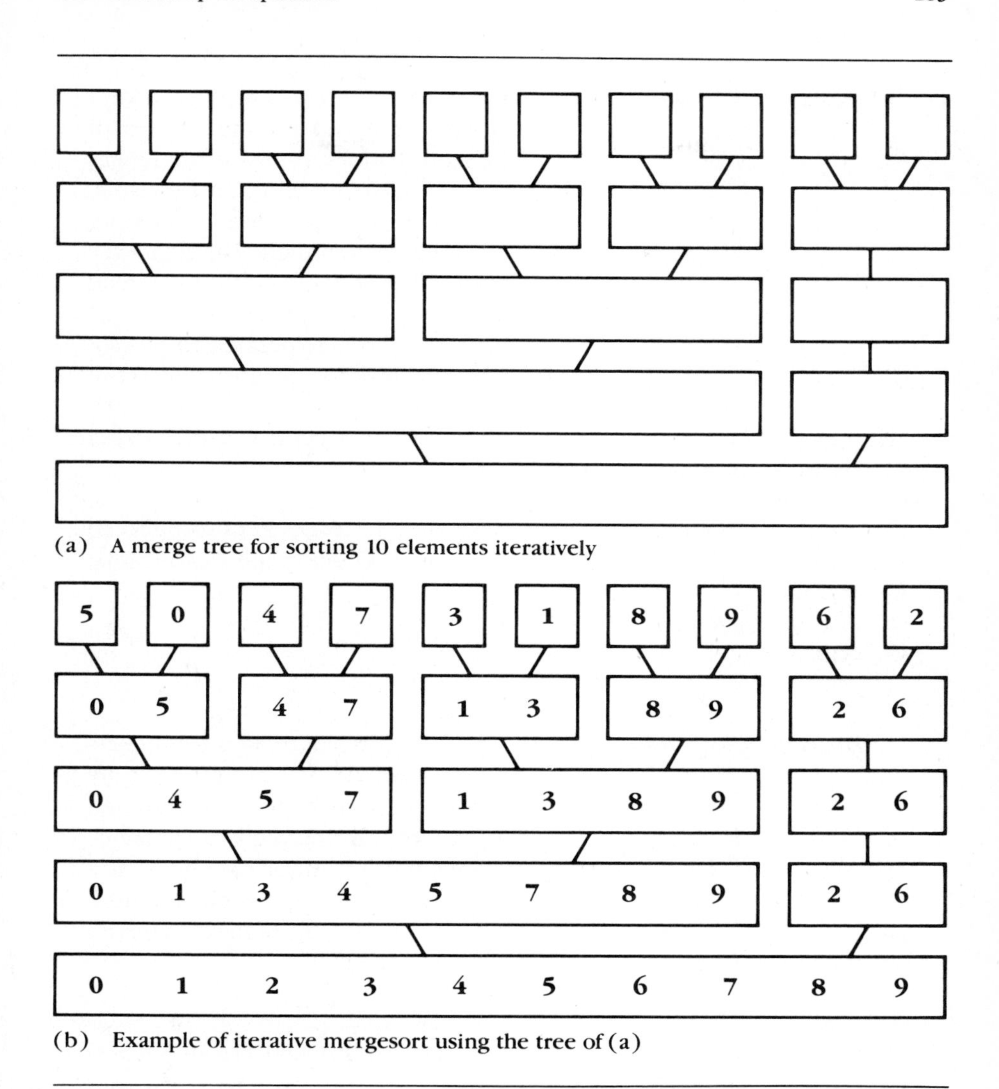

(a) A merge tree for sorting 10 elements iteratively

(b) Example of iterative mergesort using the tree of (a)

Figure 3.7 Merge Trees

bucket rather than to the beginning. This is easy to do if a pointer is maintained to the end of each bucket. Note that having a pointer to the end of each bucket also greatly simplifies concatenation in the linked representation. ■

Algorithm 3.20 presents a radix sort algorithm based on the preceding discussion. The algorithm has s iterations, each requiring time $O(n + b)$, where b is the number of buckets. If s and b are fixed, this gives a time complexity of $O(n)$—better than any sorting algorithm we have seen before. However, it is not always realistic to assume that s is independent of n. In fact, if the strings being sorted are distinct, then n grows exponentially with s, with

```
 1. Radixsort(seq, maxplace, radix)          {seq is a list of keys
 2.                                 to be sorted, radix is the number of
 3.                                 buckets, and maxplace is the length
 4.                                      of the longest input variable.
 5.                                     Place is a local variable pointing
 6.                                   to the desired digit. The digits are
 7.                                numbered from left to right beginning
 8.                                with 1. The algorithm returns a sorted
 9.                                               list pointed to by seq.
10.                                It assumes algorithms for distribute (see
11.                                     below) and collect (see text).
12.                              An array bucket of buckets is used locally}
13. create array bucket of buckets
14. for place from maxplace to 1 step − 1
15.     distribute(seq, place, bucket)          { assign keys to buckets}
16.     seq := collect(bucket)               { concatenate the buckets}
17. end for
18. destroy bucket
19. return seq
```

```
 1. Distribute(seq, place, bucket)         {distribute keys to buckets
 2.                               variables curr, i, and d are used locally
 3.                            A function Findchar(key, place) is assumed
 4.                               that extracts a given digit or character
 5.                                 from a given position to a given key
 6.                              seq, place, and bucket are as documented
 7.                                                  in radixsort above.}
 8. for i from 0 to radix − 1
 9.     bucket[i] := ( )
10. end for
11. while seq ≠ ( )                        {assign the next key to bucket}
12.     d := Findchar(head(seq), place)               {d, where d is the}
13.     bucket[d] := Concat(bucket[d], (head(seq))         {character in}
14. end while                                {position place of the key}
```

Algorithm 3.20 Radix Sort

the result that s grows logarithmically with n. The time complexity $O(s(n + b))$ thus becomes $O(n \log n)$, and radix sort is comparable to Mergesort and Heapsort in the worst case.

The function **collect** of the radix sort concatenates all the buckets. It is essentially the function Concatlist of Section 2.3. A trace of radix sort is shown in Figure 3.8.

Original sequence:	(211 312 123 321 111 232 221)
After distributing (*place* = 3)	1: (211 321 111 221) 2: (312 232) 3: (123)
After collecting (*place* = 3)	(211 321 111 221 312 232 123)
After distributing (*place* = 2)	1: (211 111 312) 2: (321 221 123) 3: (232)
After collecting (*place* = 2)	(211 111 312 321 221 123 232)
After distributing (*place* = 1)	1: (111 123) 2: (211 221 232) 3: (312 321)
After collecting (*place* = 1)	(111 123 211 221 232 312 321)

Figure 3.8 Example of Radix Sort

3.9 Iterative Quicksort

The general way to eliminate recursion is to maintain your own stack. This approach is necessary for algorithms with more than one recursive call, since only one recursive call can be eliminated by eliminating tail recursion.

We illustrate this approach by considering Quicksort. One recursive call to Quicksort does appear at the end of the algorithm, so it may be replaced by transferring control to the top of a loop. The other recursive call is replaced by pushing onto a stack the two indices bounding the section of the array to be sorted. These correspond to the only parameters of the recursive Quicksort algorithm that actually vary.

Before presenting the resulting Algorithm 3.21, we consider one more issue: the space requirements of Quicksort. (Before, the space requirements of the stack were hidden.)

A simple exercise in induction shows that if the pairs of indices on the stack are considered endpoints of unsorted intervals, then no two intervals on the stack can overlap. This means the stack size is minimized if the intervals on the stack are large. Thus an algorithm intended to minimize stack size should postpone the larger subinterval defined by the call to Partition and should immediately consider the smaller one.

This approach means that each interval on the stack is at most half as large as the interval pushed previously, so that the stack size is $O(\log n)$. If the larger interval is not postponed, then the lengths of successive intervals on the stack may differ by only a constant, giving a stack size of $O(n)$.

```
1.  Quicksort(arr, low, upp)      {assumes a sequence stored between
2.                                locations low and upp in an array arr.
3.                          A local stack stk and variable loc are used.
4.                  A test for triviality of a subsequence is assumed}
5.  if upp − low > 0 then
6.     create(stk)                     {stack operations are also assumed}
7.     push(low)
8.     push(upp)                                {initialize stack contents}
9.     while not empty(stk)         {if subsequences remain, keep at it}
10.       loc := partition(arr, low, upp)           {partition is assumed}
11.                                        {to find its own pivot element}
12.       if loc − low ≤ 1 then                {if left subsequence trivial}
13.          if upp − loc ≤ 1 then            {and also right subsequence}
14.             upp := pop(stk)
15.             low := pop(stk)        {get next subsequence from stack}
16.          else                   {only right subsequence needs sorting}
17.             low := loc + 1
18.          end if
19.       else
20.          if upp − loc ≤ 1 then           {only left piece needs sorting}
21.             upp := loc − 1
22.          else                                {else both are nontrivial}
23.             if loc − low > upp − loc then  {if the left is larger than}
24.                push(low, stk)
25.                push(loc, stk)             {push the larger subsequence}
26.                low := loc + 1      {proceed with smaller subsequence}
27.             else
28.                push(loc + 1, stk)         {push the larger subsequence}
29.                push(upp, stk)
30.                upp := loc − 1      {proceed with smaller subsequence}
31.             end if
32.          end if
33.       end if
34.    end while
35.    destroy stk
36. end if
```

Algorithm 3.21 Iterative Quicksort

3.10 Exercises

Solutions of exercises marked with an asterisk appear in Appendix C; references are given in that appendix for those marked with a double asterisk.

1. Trace the Partition algorithm about the key 7 for the sequence (7 2 9 8 1 3 5 4 6).
2. Trace the recursive and nonrecursive Quicksort algorithms of this chapter on the input sequence (23 24 11 56 16 57 78 31 43 7). How many comparisons and exchanges are required? What if you partition about the key in the middle position of each subsequence?
3. Trace the operation of radix sort on the sequence (142 212 234 443 241 211 314 112 342 144). Trace the operation of radix sort on the sequence (1234 1134 234 4 1123 2123 2222 2134 24 332).
4. Give an example to show that radix sort can fail if keys are added to the beginning of buckets.
5. Give an example to show that merge can fail if the same array is used as both source and destination.
6. What is the average-case number of key comparisons for the sequential search algorithm? Be exact.
7. Find the worst-case time complexity of each of the algorithms of Section 3.7.

*8. Find the worst-case time complexity of the algorithm for conversion to decimal, where n is the length of the input in bits. What is this time complexity when measured in terms of the integer returned by the algorithm?

9. Show that iterative Mergesort has $O(n \log n)$ time complexity.
10. What is the time complexity of radix sort if arrays are used instead of linked lists?

*11. Suppose a sequence of size n were sorted by radix sort using b buckets and each pass took time $5 \times b + 10 \times n$. If the keys range from 0 to $2^{20} - 1$, how much time would be required as a function of n if a) $b = 2$ b) $b = 2^2$ c) $b = 2^4$ d) $b = 2^5$ e) $b = 2^{10}$. Which choice of b would be best for $n = 25$?

*12. For the previous exercise, estimate the optimum choice of b by using the minimization techniques of calculus.

13. Is Quicksort a stable sorting algorithm? Can it be made stable?
14. Is radix sort a stable sorting algorithm? Is recursive bucket sort?
15. Implement the radix sort algorithm.
16. Design and implement a sequential search algorithm that uses no sentinel.

17. Implement the Quicksort algorithm.
18. Design versions of Quicksort in which the key used for partitioning is
    a) the key in the middle position of the unsorted sequence
    b) the mean of the first and last keys of the sequence
    c) the median of the first three keys of the sequence.
19. Design and implement a recursive version of Quicksort which minimizes the stack size.
20. Design and implement a recursive bucket sort algorithm
    a) if recursion terminates only for sequences of size 1
    b) if another sorting algorithm is used for sequences of size <5.
21. Implement a heap data type with purely iterative operations.
22. From which of the algorithms of Chapter 2 can you remove recursion?
23. Show without considering the equivalent recursive algorithms that each of the iterative algorithms in Sections 3.7 and 3.8 is correct.
24. Prove that overlapping intervals are never on the stack at the same time during Quicksort.
25. Show that the bubblesort algorithm you constructed in the exercises for Chapter 2 is correct.
26. Design and implement a partition algorithm along the traditional lines discussed in the text. Compare the number of comparisons and assignments required on sample input to the number required by the algorithm of the text. What if your sample input contains repeated keys?

*27. At some low level (e.g., assembly language or machine language level), 2-dimensional arrays are typically implemented in terms of 1-dimensional arrays. Two common choices exist: either the rows are stored one after another (*row-major order*) or the columns are stored one after another (*column-major order*). Assume each entry requires k words of storage, the first coordinate ranges from $lower_1$ to $upper_1$, the second coordinate ranges from $lower_2$ to $upper_2$, and the entire array is stored beginning at location α. Construct a formula for determining the storage location of the entry with coordinates (i, j).

*28. Generalize the formula of the above problem to arrays of more than 2 dimensions.

*29. Can you write the formula of the previous problem so that for an r-dimensional array, the formula involves only r additions and $r - 1$ multiplications?

30. A *symmetric* matrix is an array A of 2 dimensions such that $A[i, j] = A[j, i]$ for all i and j. In the representation of the previous problem, about half the information stored is redundant. If only those values of A are stored for which $i \leq j$, construct a formula using the assumptions of Exercise 27 for the storage location of element with coordinates (i, j).

31. Can you generalize your formula of the previous problem to more than 2 dimensions?
32. Can you represent sparse arrays of 1 dimension as sequences? Can these sequences in turn be represented in terms of arrays?
33. A bit string can be used to represent each row in a sparse array. Nonzero entries in the array would be changed to 1. The actual nonzero values could be stored in row-major order somewhere else. We would also store the matrix of bits in row-major order. What would be the advantages and disadvantages of such a representation? Can you think of any improvements to it?
34. Suppose the representation of the previous problem is modified to replace nonzero entries to pointers to the actual values. How does this new representation compare to others for sparse arrays?
35. Suppose only a small number of keys, say the integers from 1 to M, occur in a given file to be sorted (here we allow duplicate keys). Consider the following sorting procedure: using an array A indexed from 1 to M, count the number of occurrences of each key. From this information we know exactly where in the output each key will appear. In one pass through the input we can thus put each key in the appropriate position in the output array. What is the time complexity of this sorting procedure? How much space is required? Under what circumstances would it be preferable to the sorting algorithms we have already discussed?

**36. A sorting procedure similar to both Quicksort and recursive bucket sort may be described as follows: Given a set of keys (which we think of as binary numbers), use the partition algorithm of Quicksort to put all keys with the most significant bit of 0 before all keys with the most significant bit of 1. These two subsets of the keys may now be sorted, perhaps recursively, in the same way. How much time would this procedure require? How much space? Write and test a program to implement this method.

Additional Reading—Chapter 3

Many algorithms of this chapter are covered, not necessarily with proofs, in the data structures books mentioned in the reading for Chapter 1. Tail recursion and the relationship between recursion and iteration is treated in Abelson and Sussman (1985). Different versions of the Partition algorithm are presented in Wirth (1976), 76–84. Representation of arrays is covered in Knuth (1973), volume 1, Section 2.2.6; Standish (1980), Chapter 8; and Smith (1987), Section 2.2.

Data Structures and Applications

4.1 Trees and Algebraic Expressions

In this chapter we first consider several applications of trees, and then we consider applications in which both trees and graphs are useful. After looking at a few purely numerical applications, we finish the chapter by considering string and text processing and its relation to computing as a whole.

A tree structure is a natural way of capturing hierarchical information, as in the examples of Figure 4.1. One common sort of hierarchy involves precedence of operations. Thus trees may be used to represent algebraic expressions, Boolean expressions, and parse trees.

We begin by considering algebraic expressions. First assume each operator has exactly two operands, which may themselves be algebraic expressions. Then the operator may be identified with the root of a binary tree—the first operand with the left subtree, and the second operand with the right subtree. Operands that are not algebraic expressions (e.g., that are numbers or variables) correspond to leaves. Several examples are given in Figure 4.2. The parse tree of Figure 2.30 also serves as an example.

Unary operators, those with one operand, may be represented as nodes with a right subtree but no left subtree, although it is more natural to represent expressions with nonbinary operators as ordered trees rather than as binary trees. Except where noted, our observations in this section generalize to ordered trees.

The three recursive tree traversal methods discussed in Section 2.4 give three distinct ways of expressing an algebraic expression as a sequence of symbols.

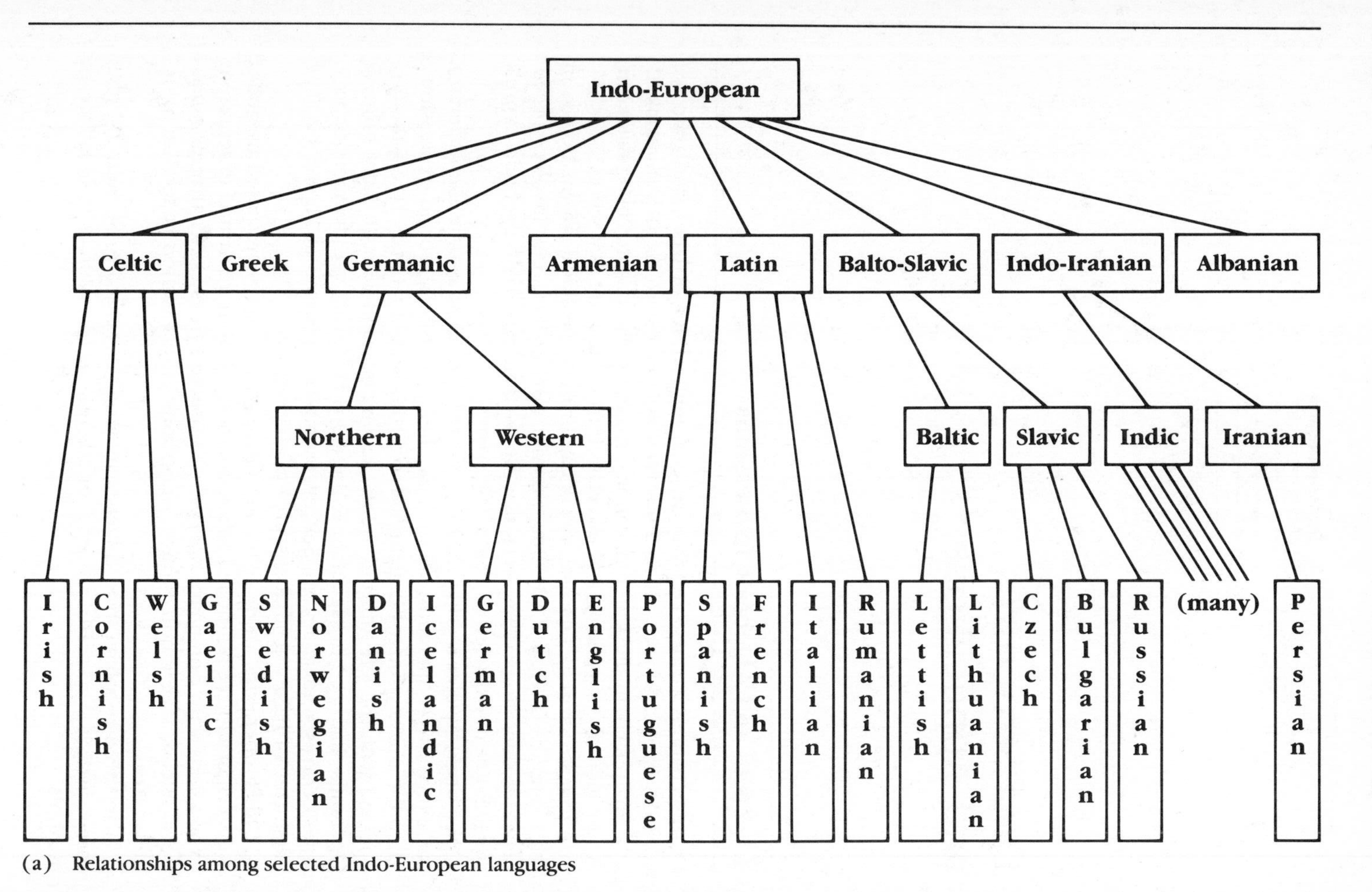

(a) Relationships among selected Indo-European languages

Figure 4.1 Hierarchies Represented as Trees

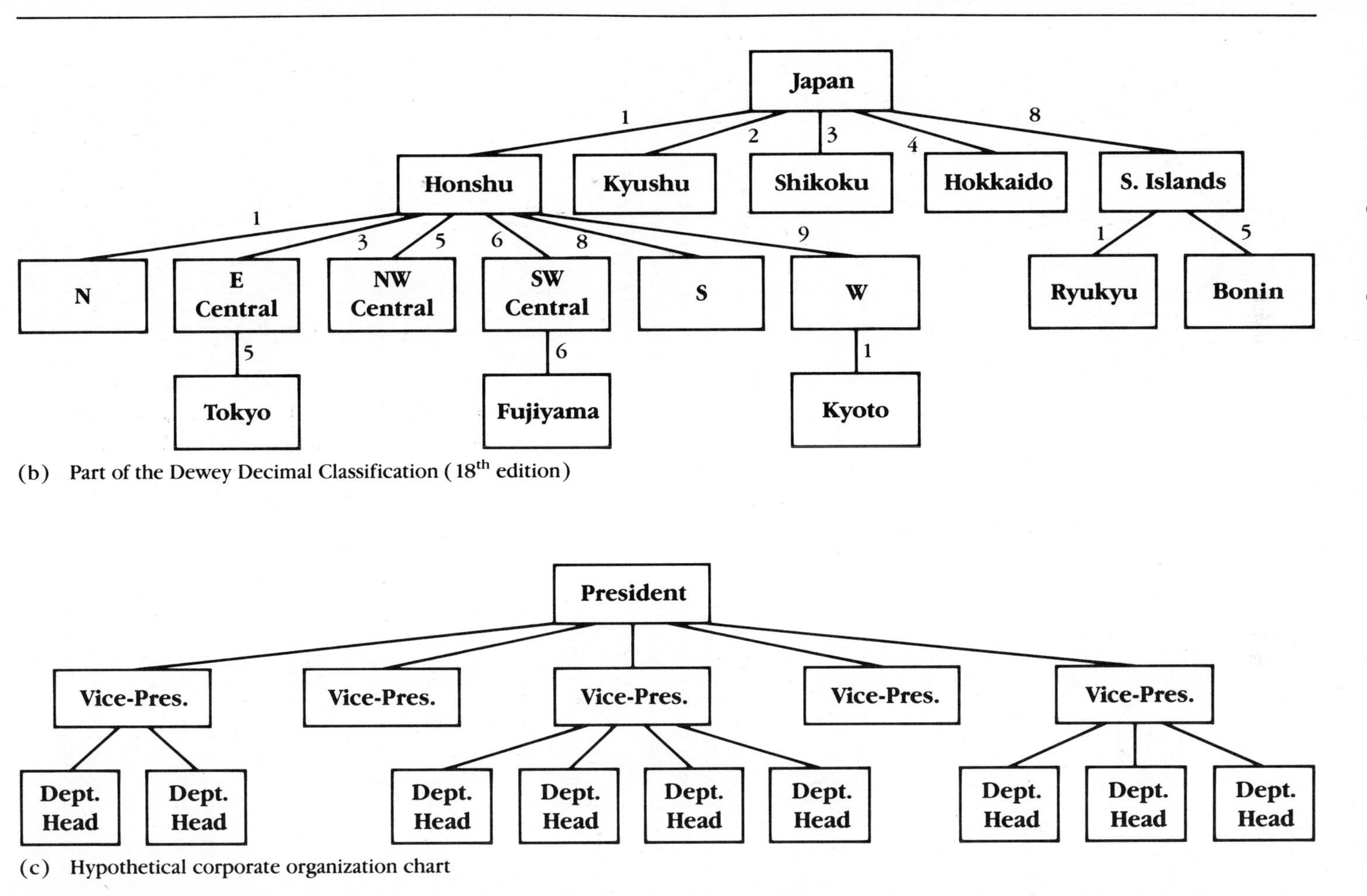

(b) Part of the Dewey Decimal Classification (18th edition)

(c) Hypothetical corporate organization chart

Figure 4.1 *(continued)*

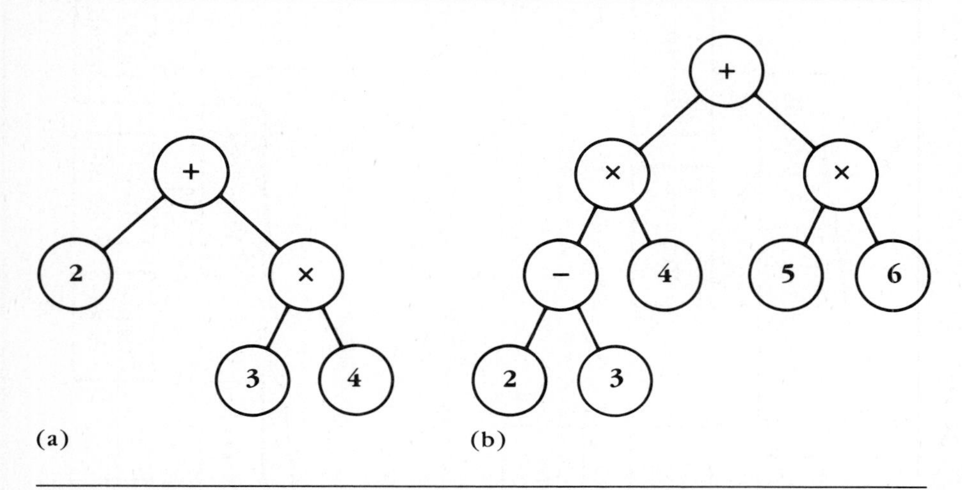

Figure 4.2 Representation of Algebraic Expressions by Trees

The notation corresponding to the preorder traversal is called *prefix* notation. That corresponding to the postorder traversal is *postfix* notation, and that corresponding to symmetric order is *infix* notation. In (solved) Exercise 16 you are asked to show that a prefix or postfix expression corresponds to a unique tree, given a knowledge of the number of operands required by each operator.

For the tree of Figure 4.2b,

the prefix expression is	$+ \times - 2\ 3\ 4 \times 5\ 6$
the postfix expression is	$2\ 3 - 4 \times 5\ 6 \times +$ and
the infix expression is	$2 - 3 \times 4 + 5 \times 6$

Notice that the operands appear in the same order in each expression (see Exercise 67 of Chapter 2), and there is one more operand than operator (see Exercise 77 of Chapter 2).

The infix expression is probably not the one you would have chosen for the algebraic expression represented by the tree. In general, there are many ways of interpreting a given infix expression. Equivalently, there are many different trees with the same inorder traversal. An example of two trees each representing the infix expression $2 + 3 \times 4$ is given in Figure 4.3. This ambiguity gives rise to the well-known conventions about precedence and parentheses in infix expressions.

Another problem with the infix notation is that it does not generalize well to operators requiring more than two operands. The infix notation does have the advantage that the operands are written near the corresponding operator.

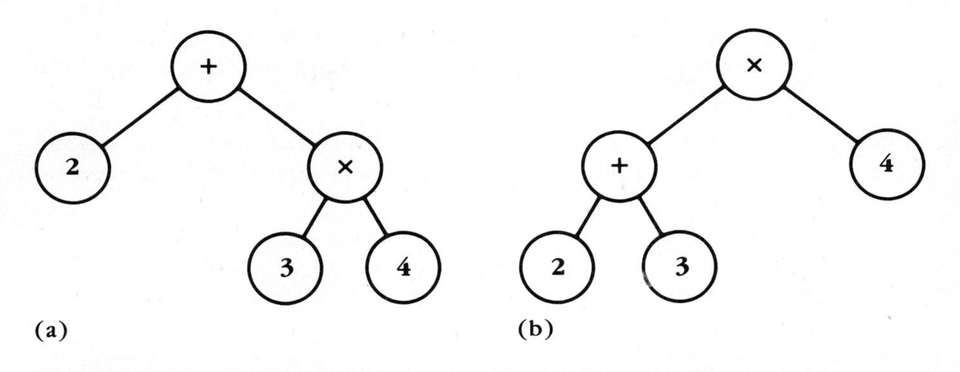

Figure 4.3 Ambiguity in the Infix Notation for Algebraic Expressions

The operators in this application need not be algebraic. The preceding remarks apply to Boolean operators. Figure 4.4 shows the tree representation of the Boolean expression

$$[x_1 \vee (x_2 \wedge x_3)] \wedge [(x_1' \wedge x_4) \vee x_3]$$

A recursive definition for the value of an algebraic expression corresponding to a binary tree is

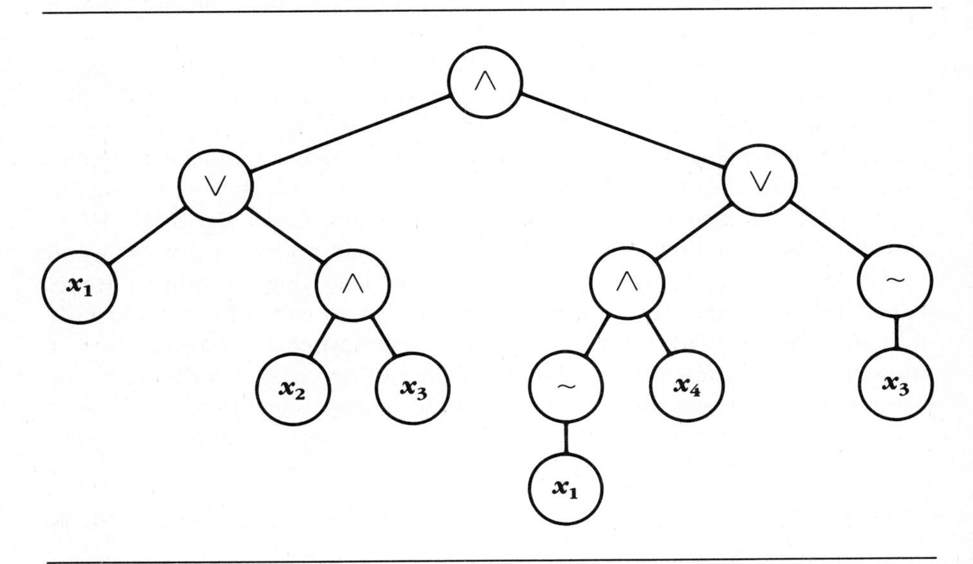

Figure 4.4 Tree Representation of a Boolean Expression

Def 4.1 *Value*(T) is

1. X if T consists of a single node representing the operand X
2. f(Value(left(T)), Value(right(T))) if the root of T represents the operator f.

Here we assume that the value of each operand is known. Since each subtree must be evaluated before the operator in the root is applied, evaluation corresponds to postorder traversal.

4.2 Decision Trees

Decomposition of a problem using the divide-and-conquer strategy also implies a hierarchy and thus a tree structure. These trees are called *decision trees.* We have seen one example for binary search in Section 2.10. Each internal node corresponds to a decision, and each subtree to a possible outcome of that decision. Leaves correspond to possible outcomes of the entire problem. Figure 4.5a represents a decision tree for a simple instance of ZOK. This is not a complete tree, because some sequences of decisions correspond to a knapsack overflow. Figure 4.5b corresponds to a sorting algorithm for 3 items. Here the 6 leaves each correspond to a possible permutation of the input.

Decision trees can be used to find lower bounds for decision-based algorithms for solving particular problems. This is the topic of the next section.

Trees used to represent the progress of athletic tournaments may be considered decision trees, since each match corresponds to a decision as to which player is better.

We may abstract from this to get an algorithm for finding the minimum element of a sequence. This algorithm, with only minor changes, will also find the maximum. The details of the algorithm are left as an exercise, but an example may be seen in Figure 4.6. Since each element except for the minimum has to "lose" at least one comparison,[1] there are at least $n - 1$ comparisons—no fewer than for the straightforward minimization algorithm. However, once the minimum is found, the second smallest element may be found merely by replaying the $O(\log n)$ comparisons involving the minimum element. The computation of the next few smallest elements for the example of Figure 4.6 is shown in Figure 4.7. If this latter process is repeated $n - 1$ times, we have sorted the original set in time $O(n) + (n - 1)O(\log n)$, or $O(n \log n)$.[2] The resulting *tournament sort* algorithm is highly reminiscent of heapsort. The details of termination and correctness are left as an exercise.

[1] Note that this observation applies not just to a tournament, but to every algorithm that finds the maximum of n elements.

[2] In the next section we use the decision tree model to show that this algorithm must be $\Theta(n \log n)$.

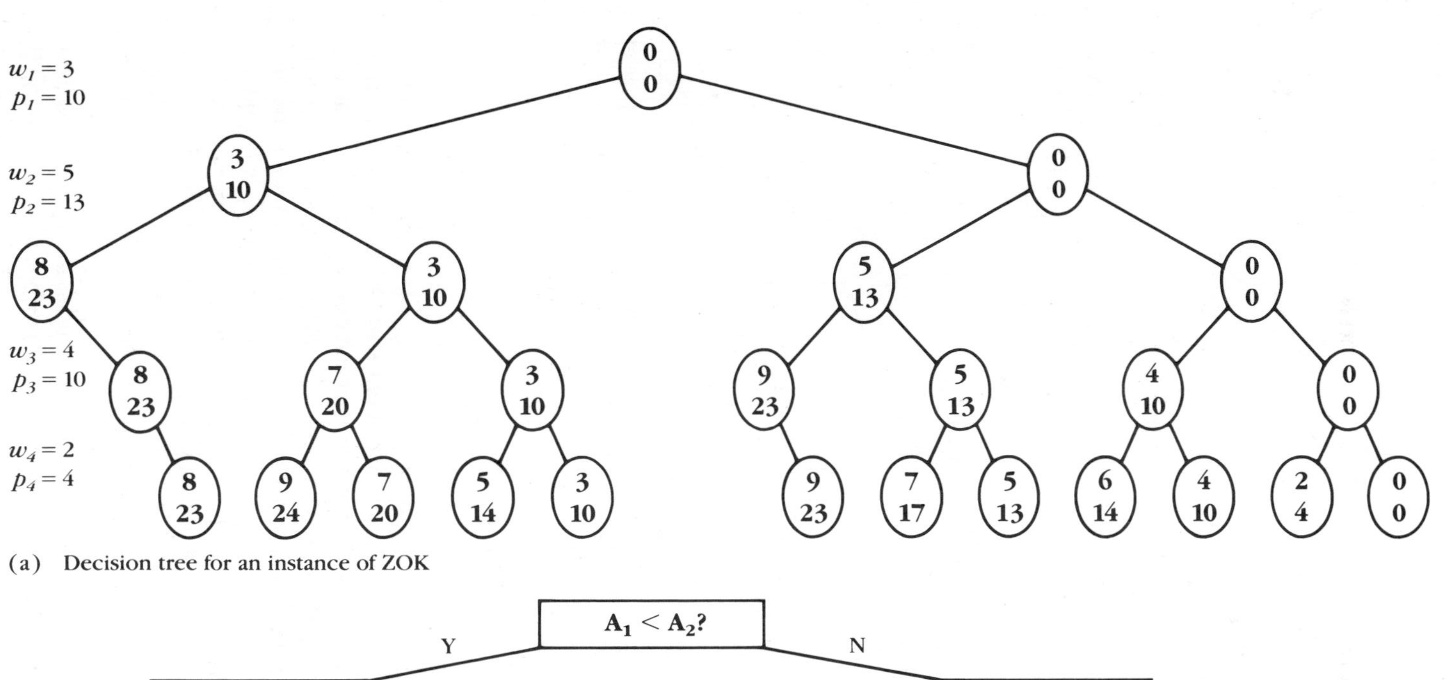

(a) Decision tree for an instance of ZOK

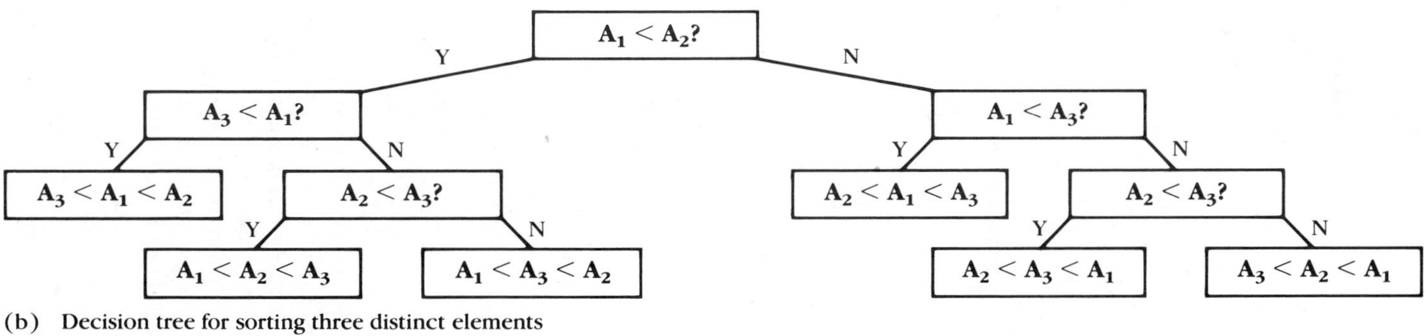

(b) Decision tree for sorting three distinct elements

Figure 4.5 Decision Trees

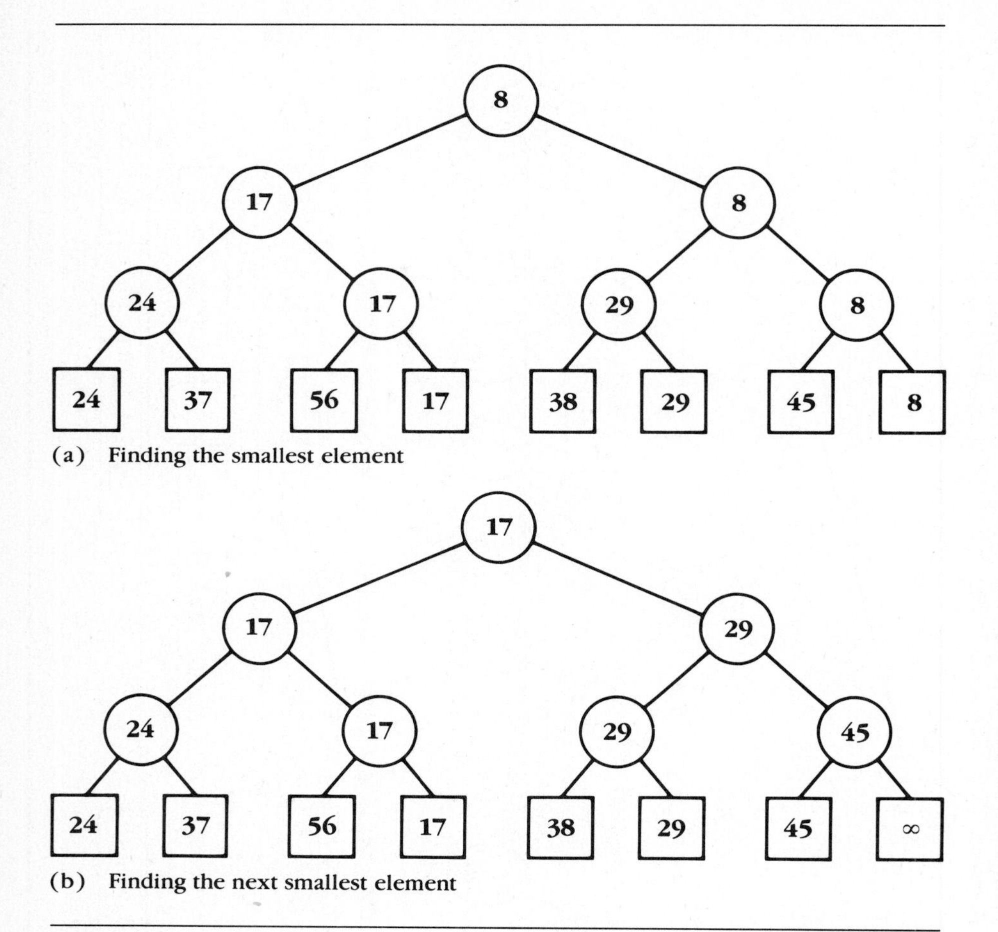

Figure 4.6 Tournament Sort

Tournament sort can be converted easily into an *m-way* merge algorithm. To merge m runs, one may set up a tournament sort based on a binary tree with m nodes. When an element wins the tournament, it is replaced at the lowest level by its successor in its run, and the comparisons involving that element are redone in time $O(\log m)$. We use ∞ as a sentinel to mark the end of each run. Once the sentinel wins the tournament, the merge is complete. Like 2-way merge, this merge algorithm may be the basis of a mergesort.

Figure 4.8 shows an example of the beginning and end of a merge for $m = 4$ and input runs (1 8 9 12), (3 4 6 14), (5 10 15 16), and (2 7 11 13). Each element in the tree needs to have a pointer to its run; we have not shown this in the figure.

The merge pattern of an m-way mergesort may itself be represented as a *merge tree* (see Figure 2.41). The parent node corresponds to the merged

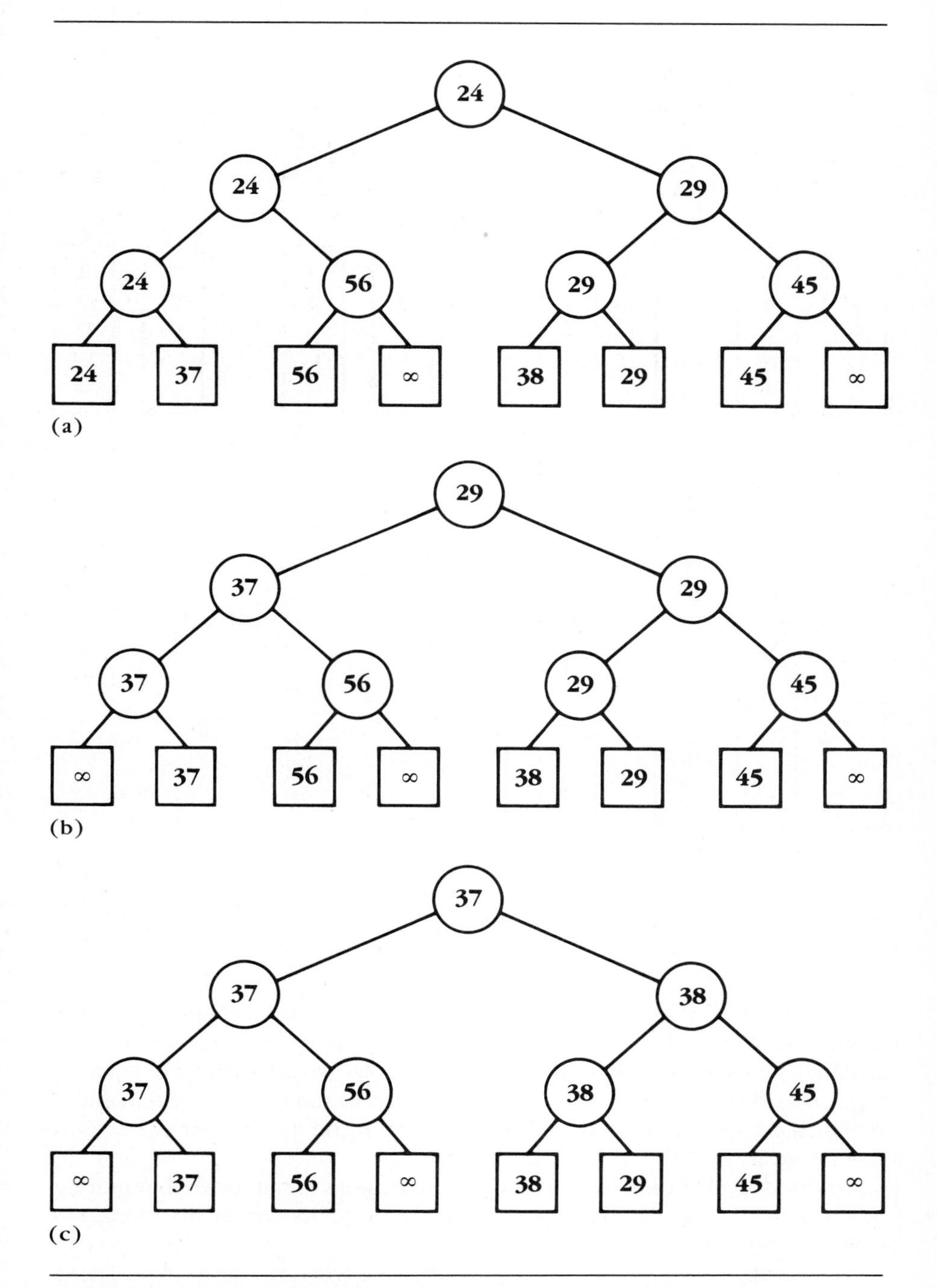

Figure 4.7 Continuation of the Tournament Sort of Figure 4.6.

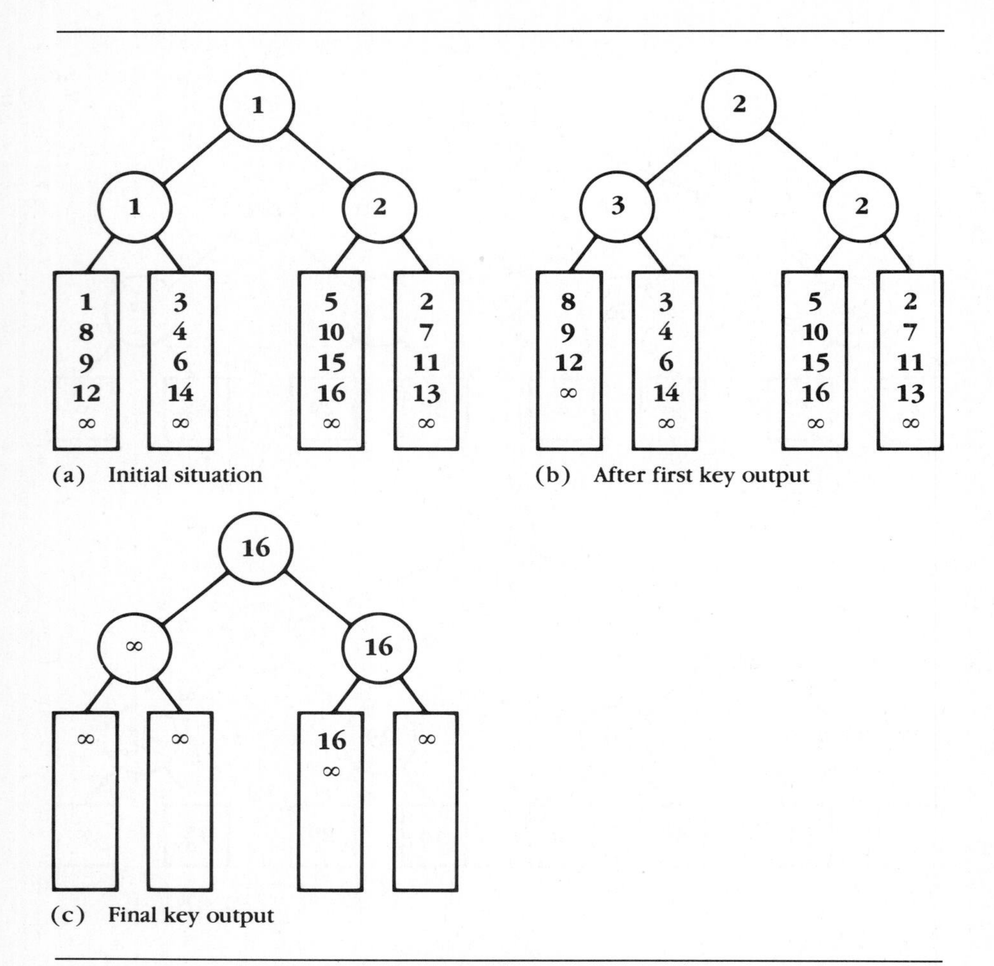

(a) Initial situation

(b) After first key output

(c) Final key output

Figure 4.8 Stages in a 4-Way Merge

sequence and the m children correspond to the runs combined into the merged sequence. Since each merge in an m-way merge tree increases the run size by a factor of m, the resulting merge tree has height essentially equal to $\log_m (n/r) = \log_m n$. Here we assume r, the average run size, is independent of n, which equals both the original number of runs and the number of leaves in the merge tree.

For each level of the merge tree, there are n nodes to output, each requiring time essentially $\log_2 m$. Thus m-way mergesort requires very nearly $n(\log_2 m)(\log_m n) = n \log_2 n$ comparisons. Note this number is independent of m.

Yet it may matter how m is chosen. In the case when the items to be sorted are stored in external memory, to which access is very slow, it is important to minimize the number of times that the average element must be accessed. This number is just the height of the merge tree.[3] Thus m should be taken to be quite large. The only upper bound on the size of m is given by this observation: For a fixed-size internal memory, many runs mean short runs, and getting a large number of different short runs from external memory may be inefficient. In practice, Mergesort appears to be the most efficient sorting algorithm for data stored externally.

4.3 Lower Bounds from Decision Trees

We have seen that the structures of many algorithms can be modeled by decision trees. In those cases where the number of decisions is an adequate measure of the time complexity of the algorithm, we may find the worst-case time complexity merely by looking at the height of the decision tree.

This technique can be applied to problems instead of to particular algorithms if the number of children per node and the number of leaves in the decision tree are independent of the algorithm. If each internal node has at most b children and there are m distinct outcomes that need to be distinguished, then there are at least m leaves in the decision tree. Its height is then at least $\log_b m$. We consider some examples.

Sorting: A sorted sequence of n distinct elements can arise from any of $n!$ possible permutations of the input. Having sorted an input sequence implies a knowledge of which of the $n!$ permutations corresponds to the input sequence. In the case when all elements are distinct, a *3-way* comparison of elements (i.e., one that when comparing x and y can distinguish among $x < y$, $x = y$, and $x > y$ in a single operation) will give no more information than a *2-way* comparison (i.e., one that can only distinguish $x \leq y$ from $x > y$ or $x \geq y$ from $x < y$ or $x = y$ from $x \neq y$).

So each node in the decision tree has at most 2 children, and there are at least $n!$ leaves. This implies a height, and thus a time complexity, of at least $\log_2 n!$. This value is $\Theta(n \log n)$, by an argument from Section 2.16. We have already seen algorithms (Heapsort, Mergesort) with this time complexity, so this lower bound is the best possible. In short, $\Theta(n \log n)$ is the best possible asymptotic worst-case time complexity for a sorting algorithm that depends on comparing pairs of elements.

Recall from Section 3.8 that the special case of radix sort, in which we assume elements of bounded length, has $\Theta(n)$ time complexity. Radix sort does not compare pairs of elements. If we don't assume bounded length, we get $\Theta(n \log n)$ time complexity.

Search in a sorted sequence: For a sequence of n elements, we must consider only n cases for a successful search. If all comparisons are 2-way comparisons, then the corresponding decision tree will have height at least $\log_2 n$. A

[3] Note the similarity to the application of B-trees to external files discussed in Section 2.6.

sequence of n distinct elements will define $n + 1$ intervals, corresponding to external nodes, into which an unsuccessful search could be directed. Together with the n successful cases, this gives $2n + 1$ cases to distinguish, so that the height of a decision tree will be at least $\log_2(2n + 1) = \Theta(\log n)$. Again, we have seen an algorithm (binary search) that achieves this bound, so the bound is the best possible. Note that asymptotically it makes no difference whether a 2-way or a 3-way comparison is used. In fact a 3-way comparison won't help much even nonasymptotically, since the subtree corresponding to equality will have only one element.

Finding the maximum of a sequence: Here there are n outcomes to be distinguished. In the worst case all elements will be distinct and an equality test will always fail. Thus a 2-way comparison is as good as a 3-way comparison, and we may assume that the branching factor of the decision tree is 2. This implies a height of at least $\log_2 n$ for the decision tree. So the worst-case time complexity of an algorithm to find the maximum of a sequence of n elements is $\Omega(\log n)$. We saw in the previous section that a different argument gives a better lower bound of $\Omega(n)$.

ZOK: Here there are 2^n subsets to be distinguished. Since each item can only be included or excluded, each internal node in the decision tree has 2 children. This gives a height of $\log_2 2^n = n$ for the decision tree and a time complexity of $\Omega(n)$. We have not seen a polynomial-time algorithm for ZOK, much less an $O(n)$ algorithm for it.

The 12 coins problem: An old problem asks whether or not a counterfeit coin can be identified from a group of 12 coins in 3 weighing operations on a balance scale. The assumptions are that there is exactly one counterfeit coin, and that it is either heavier or lighter than the rest. Here there are 24 outcomes—12 cases where the bad coin is light and 12 where it is heavy—to be distinguished. Each nonleaf has 3 children, since the left side of the balance may be heavier, lighter, or the same as the right side. Because the 24 outcomes require $\log_3 24$ comparisons, and this number is less than 3, the decision tree model does not rule out a successful algorithm. The construction of the successful algorithm is left as an exercise. Another exercise: What if the number 12 is replaced by 13? And another: What if it is possible that no coin is counterfeit?

4.4 Inorder Predecessors and Successors

In Section 2.5 we saw sorted lists represented by binary search trees. However, the operation of finding the predecessor or successor in a sorted list corresponds to finding the inorder predecessor or successor in a binary search tree. This operation cannot be implemented efficiently in the standard linked representation for binary search trees.

Note that an efficient nonrecursive inorder successor operation would give an efficient stackless inorder traversal algorithm. We only need to find the first node in inorder (if a header node as described in Appendix D didn't point directly to it) and then repeatedly find inorder successors until no successor could be found. Such an operation for databases would also make it easy to find all records with keys in a given interval—once the first was found, the inorder successor algorithm could be used repeatedly to find the others. Similar remarks apply to inorder predecessors.

Before introducing a new representation in which these operations can be performed more efficiently, we examine them in the usual linked representation of binary search trees. We concentrate on inorder successors; there is a symmetric analysis for inorder predecessors.

Given a node in a binary search tree, inorder traversal of the subtree rooted there works by traversing its left subtree, visiting the node itself, and traversing the right subtree. If the given node has a right subtree, then the first node in inorder in this right subtree is the desired inorder successor.

The inorder traversal algorithm implies that to find the first inorder node in any subtree, we descend left as far as possible, that is, until we find a node with no left child.

In short, to find the inorder successor of a node with a right child, descend to that child and then descend left as far as possible. If there is no right child, then the linked representation will have an empty link in the right link field. This link field can be used to point directly to the inorder successor, provided this new pointer can be distinguished from an ordinary tree link.

In the example of Figure 4.9, to find the inorder successor of 35, we descend right to 70, then left as far as possible past 68 to 60. The inorder

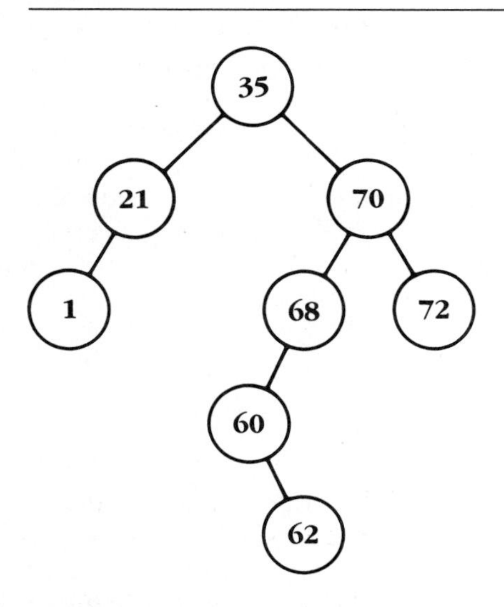

Figure 4.9 A Binary Tree for Finding Inorder Successors

successor of 62 is 68, which cannot be found given only a pointer to 62 and the child pointers in the tree.

4.5 Deletion in a Binary Search Tree

To delete from a binary search tree, it is not always enough to mimic the case of deletion from a linked list and point around the node being deleted. In the example of the previous section, to delete 70 we need to retain pointers to its subtrees rooted at 68 and 72. The pointer from 35 to 70, although now free, cannot be pointed to either the left child 68 or the right child 72 without losing access to the other child and the subtree of which it is the root.

On the other hand, if the key to be deleted is in a leaf (e.g., 62 in the previous case), we can delete it simply by giving the pointer to the leaf from its parent to the empty value NIL (assuming the parent pointer is available). And if a node (e.g., 21 in the previous case) has just a single child, we may delete the key in the node simply by changing the pointer from the parent to the node so that the pointer points to this single child (again this parent pointer must be available). Showing this type of deletion preserves the binary search property is left as an exercise.

To distinguish deletion of a key from deletion of a node is an important task. In order to delete a key from a node with two children, we can replace the key with a second key and delete the original occurrence of the second key. If the binary search property is to be maintained, the replacing key must be the inorder successor (or the inorder predecessor). For a node with two children, we may use the inorder successor algorithm of the previous section. And since the node containing the inorder successor has no left child, it may be deleted as suggested earlier in this section.

So we have Algorithm 4.2 for deletion of a given key from a given binary search tree.

Using the same tree as in Figure 4.9, the examples in Figure 4.10 show the resulting tree after deletion of various nodes in the original tree. We can easily verify each new tree is a binary search tree.

4.6 Threaded Trees

We have already suggested that any node without a right child in a binary search tree could be given instead a pointer to its inorder successor. This type of pointer is usually called a *thread*. The right-child pointer field can be used for threads if there is some indication (e.g., a related Boolean field) whether it

```
 1. Delete(key, bst)            {deletes a given key from a given binary
 2.                              search tree bst. Call by value-result is to be
 3.                              used for the tree variable bst. Returns true
 4.                              or false depending on whether key was found.
 5.                              An auxiliary function DeleteSucc is given below}
 6. if bst = nil then
 7.    return false                                        {key not found}
 8. else
 9.    if key < data(bst) then                        {if key small, then}
10.       return delete(key, left(bst))                 {delete from LST}
11.    else
12.       if key > data(bst) then                     {if key large, then}
13.          return delete(key, right(bst))             {delete from RST}
14.       else                                        {if key found, then}
15.          if left(bst) = nil then                  {check possibility}
16.             bst := right(bst)
17.          else                                     {of just pointing}
18.             if right(bst) = nil then              {around the node}
19.                bst := left(bst)                     {to be deleted}
20.             else
21.                data(bst) := DeleteFirst(right(bst))     {otherwise}
22.             end if                       {get key from successor}
23.          end if                  {and delete original occurrence}
24.       end if                                         {of that key}
25.    end if
26. end if
27. return true
```

```
 1. DeleteFirst(ptr)            {deletes first node in inorder in a binary
 2.                              search tree ptr with nonempty left child
 3.                              and returns the key in the deleted node.
 4.                              Call by value-result is to be used for ptr}
 5. if left(ptr) = nil then           {if no left child, then have found}
 6.    temp := data(ptr)                          {node to return, so}
 7.    ptr := right(ptr)                {delete node (it has ≤1 child)}
 8.    return temp                          {and return deleted node}
 9. else                           {otherwise continue descending left}
10.    return DeleteFirst(left(ptr))
11. end if
```

Algorithm 4.2 Delete from a Binary Search Tree (with Auxiliary Function DeleteFirst)

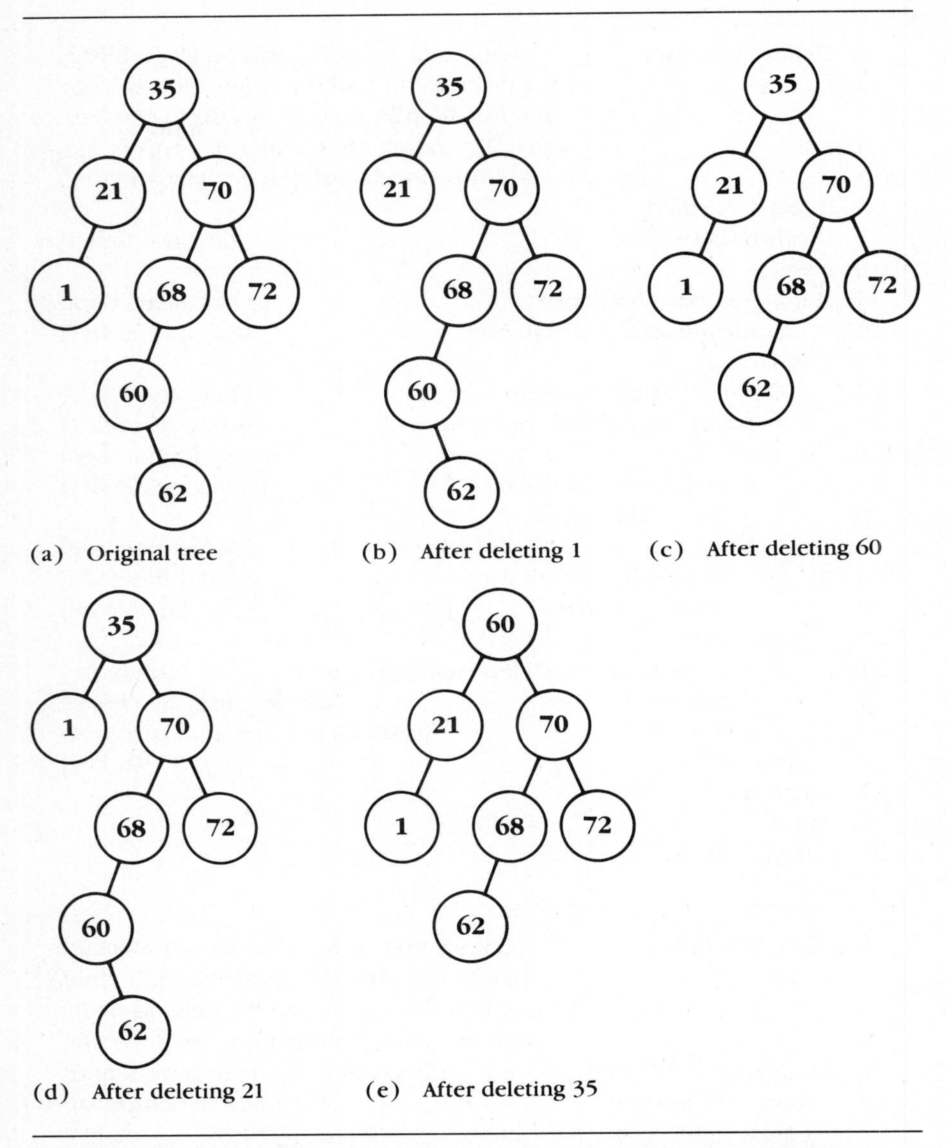

Figure 4.10 A Binary Search Tree and the Result After Various Deletions

contains a thread or a pointer field.[4] Similarly, a left thread may be used for inorder predecessors. A tree with threads for inorder predecessors and successors is called *symmetrically threaded.* In the algorithms of this section, *lthread*

[4] Note that variables of Boolean type can be stored in as little as one bit, so these fields need not consume much space. Also, these fields remove the need for a parent field in finding inorder successors and predecessors.

and *rthread* of a node have the value "true" if the corresponding pointer field contains a thread and not a pointer. Threads were first discussed by Perlis and Thornton (1960).

Even without threads, we can find the inorder successor of nodes with right children according to Section 4.4. In addition, nodes without right children

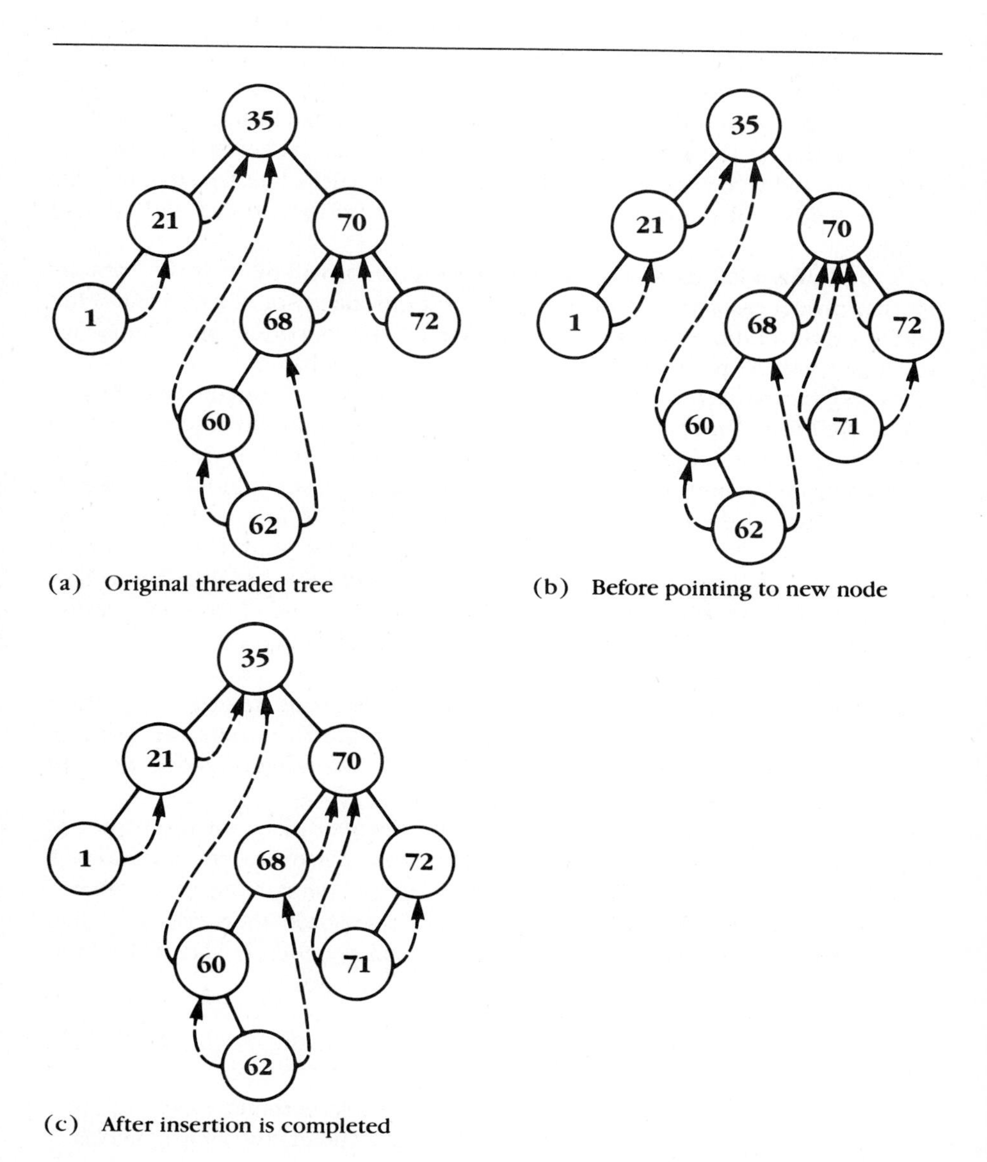

Figure 4.11 Insertion of the Key 71 into the Threaded Tree of Figure 4.9

have right pointer fields available to point directly to the successor. The analogous observation holds for left-child pointer fields and inorder predecessors.

For example, Figure 4.11a shows the tree of Figure 4.9 with threads added. As is customary, the thread pointers are indicated by dotted lines.

The search algorithm for binary search trees will work as before, as long as we check before following a pointer to make sure the pointer is really a pointer and not a thread. Insertion and deletion are slightly more complicated, since they must maintain the threaded structure. We only discuss insertion here; deletion is left as an exercise.

Assuming that both left and right threads are to be maintained, the new insertion algorithm is reminiscent of that for doubly linked lists (see Appendix D). In particular the successor and predecessor must be reachable from the new node, and vice versa. Of course the new node will be a leaf, so that both its pointer fields will contain threads (as indicated by the lthread and rthread fields).

If the leaf is a left child, then its inorder successor will be its parent, and its inorder predecessor will be the former inorder predecessor of its parent. If the leaf is a right child, its inorder predecessor will be its parent, and its inorder successor will be the former inorder successor of its parent.

As an example, let us consider inserting 71 into the threaded tree of Figure 4.11a. We locate the spot for insertion in the usual way. In this case we want to insert 71 as the left child of 72. We do not want to insert 71 as the left child of 72 immediately, since we would lose the thread to 70, the old inorder predecessor of 72 and the new inorder predecessor of 71. So first we create a new node and insert 71 as the key. Since it is a left child, we point the right thread to its parent and the left thread to the inorder predecessor of the parent. This inorder predecessor may be found by following a left thread from the parent. We mustn't forget to initialize the lthread and rthread fields of the new node to "true." We are in the situation of Figure 4.11b.

Now we are ready to point to the new node. We create an ordinary pointer from the parent to the child as usual, remembering to change the value of lthread for the parent node to "false." Part (c) of the figure shows the end result of the insertion.

The insertion Algorithm 4.3a searches for a proper place for insertion and then calls Doinsert to do the insertion. Duplicate insertions are not allowed; this algorithm can easily be modified to allow them. Algorithm 4.3b for traversing a threaded tree is also presented; it calls a function InorderSucc to find inorder successors. Since this function may be independently useful, the InorderSucc function is listed separately as Algorithm 4.3c.

---

1. InsertThread(*key, tree*) {inserts the key *key* into a}
2. threaded tree *tree,* which is
3. passed by value-result.
4. A value of *true* is used to distinguish

```
 5.                                  pointers from threads. The "repeat" loop
 6.                              will be terminated by a "return" statement.
 7.                        DoInsertThread (defined below) does insertion.
 8.                                  curr and par are used as local variables
 9. if tree = nil then
10.    DoInsertThread(key, nil, tree, false)
11. else                                {insert as root—parent pointer is nil}
12.    curr := tree                        {we can't use value-result here}
13.    repeat
14.       if key = data(curr) then            {can't do duplicate insertion}
15.          return error
16.       else
17.          if key < data(curr) then                 {insert in left subtree}
18.             par := curr                          {update parent pointer}
19.             if lthread(curr) then                  {if at a leaf, insert}
20.                DoInsertThread(key, par, curr, false)           {as left}
21.                return                        {child since 4th arg. false}
22.             else
23.                curr := left(curr)                  {follow left pointer}
24.             end if
25.          else                                    {insert in right subtree}
26.             par := curr
27.             if rthread(curr) then             {if at a leaf, then insert}
28.                DoInsertThread(key, par, curr, true)           {as right}
29.                return                         {child since 4th arg. true}
30.             else
31.                curr := right(curr)                {follow right pointer}
32.             end if
33.          end if
34.       end if
35.    end repeat
36. end if
```

```
 1. DoInsertThread(key, parent, ptr, right)        {inserts new key key
 2.                                          to be pointed to by ptr, which
 3.                                               is passed by value-result.
 4.                                         parent is parent pointer (which may
 5.                                          be empty) and right is true if ptr is
 6.                                  to be right child of parent, false otherwise}
 7. ptr := Get                               {get new node and point ptr to it}
 8. lthread(ptr) := true
 9. rthread(ptr) := true                                     {new node is leaf}
10. data(ptr) := key
11. if parent = nil then
12.    left(ptr) := nil                                        {insert as root}
```

```
13.     right(ptr) := nil
14.     return
15. else
16. if right then
17.     left(ptr) := parent                      {insert as right child}
18.     right(ptr) := right(parent)      {initialize threads from node}
19.     right(parent) := ptr                             {and pointer}
20.     rthread(parent) := false                         {to new node}
21. else
22.     right(ptr) := parent                      {insert as left child}
23.     left(ptr) := left(parent)        {initialize threads from node}
24.     left(parent) := ptr                              {and pointer}
25.     lthread(parent) := false                         {to new node}
27. end if
```

Algorithm 4.3a Insert into Threaded Tree (with Auxiliary Procedure DoInsertThread)

```
 1. TraverseThread(tree)         {traverses nonrecursively a threaded
 2.                   tree pointed to by tree, passed by value. No header
 3.                  node is assumed. A function InorderSucc is defined
 4.                           below. It returns an inorder successor}
 5. if tree = nil then
 6.     return error                                      {empty tree}
 7. else                                       {find first node in inorder}
 8.     while left(tree) ≠ nil
 9.         tree := left(tree)
10.     end while
11.     repeat
12.         visit tree
13.         tree := InorderSucc(tree)          {find inorder successor}
14.     end repeat if tree = nil              {until last inorder node}
15. end if
```

Algorithm 4.3b Traverse a Threaded Tree

```
1. InorderSucc(tree)         {returns pointer to inorder successor of}
2.                     the node pointed to by tree, passed by value,
3.                                                in a threaded tree}
4. if not rthread(tree) then               {if there is a right subtree}
```

```
5.     tree := right(tree)                        {then go right once}
6.     while (not lthread(tree))         {[notice left(tree) can't be nil]}
7.        tree := left(tree)                {and left as far as possible}
8.     end while
9.     return tree
10. else return right(tree)                    {if no right subtree then}
11. end if                                         {follow right thread}
```

Algorithm 4.3c Find Inorder Successor in a Threaded Tree

4.7 AVL Operations

Recall that a binary search tree is an AVL tree iff for every node in the tree, the left subtree and the right subtree of the node have heights differing by at most 1. In this section we describe an insertion algorithm for AVL trees and show that the algorithm preserves the AVL property. Preservation of the AVL property implies that search has O(log n) time complexity.

We can easily see that all trees with 1 or 2 nodes are AVL trees. Consider the possibilities for a binary tree with 3 nodes. There are 5 different cases, 1 completely full and 4 corresponding to the trees of Figure 4.12. Only the completely full tree is an AVL tree, so there must be some way of transforming each of the others to this tree. In fact, if the binary search property is to be preserved, then the transformation for each of the other 4 trees is completely determined.

These 4 cases are completely general — every AVL tree can be restructured after an insertion by an operation similar to 1 of these 4. In order to formalize this assertion, we need more precise definitions of the four types of imbalance of the figure.

For each node, we may define the *balance* as the height of the right subtree less the height of the left subtree.[5] Each node in an AVL tree must therefore have a balance of 0, 1, or -1. A subtree whose root has balance of $+1$ will become unbalanced if an insertion is made into the right subtree that changes this subtree's height. We may distinguish RL imbalance, in which the insertion is made into the right subtree's left subtree, from RR imbalance, in which the insertion is made into the right subtree's right subtree. LL and LR imbalance are defined symmetrically. Note these 4 types of imbalance are precisely the 4 types of Figure 4.12. Also, any insertion that changes the height of the subtree rooted at a node will change heights all the way down the path from this node to the point of insertion.

[5] Some authors call this a *balance factor,* but it is not really a factor.

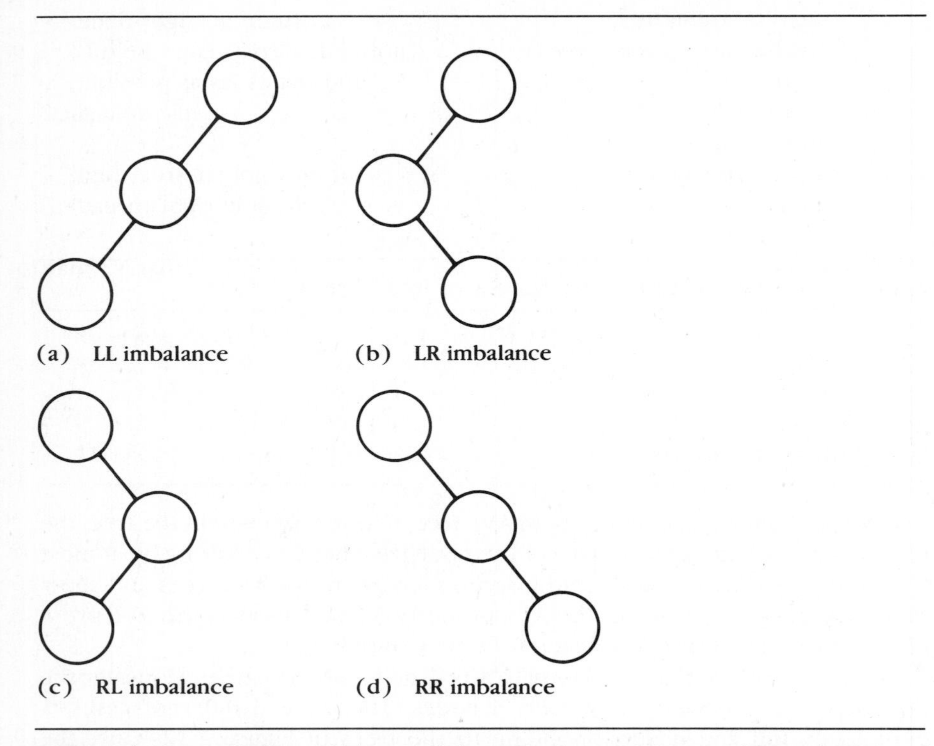

Figure 4.12 Unbalanced Trees of Three Nodes, with Names for Each Type of Imbalance

Consider the lowest node *lastbad* on the path from the root to the newly inserted node with a nonzero balance before insertion. There are three cases:

1. no such node exists
2. the new node was inserted as a child of *lastbad*
3. there are one or more nodes between *lastbad* and the new node. All these nodes have balance 0 before insertion, by definition of *lastbad*.

Only in case 3 can a balanced tree become an unbalanced tree. In case 1, all balances on the path to the new node were 0 and can only change by 1. No balances off this path can change. In case 2, the new node inserted into the empty subtree must give its parent a balance of 0. This insertion does not change the height of the subtree rooted at the parent, and therefore does not change the height of the subtree rooted anywhere on the path from the root to the new node.

Assume the insertion unbalances the tree; that is, we are in case 3. Then if *lastbad* had balance $+1$, the new balance must be $+2$, and the insertion must have been into the right subtree. Since we are not in case 2, the insertion must have been made into a nonempty left or right subtree of this subtree, corresponding respectively to RL or RR imbalance. By symmetry, a balance of -1 implies either the LL or the LR case.

Additionally, if *lastbad* originally has balance 1, then any insertion into the right subtree must increase the height of the right subtree and thus unbalance the tree. This can be seen by a simple induction on the length of the path from *lastbad* to the new node, since all nodes on the path must have balance 0. A similar remark applies to an original balance of -1.

In each of the cases LL, LR, RL, and RR, the two nodes labeled $N2$ and $N3$ in Figure 4.13 must have had balance 0 before the insertion (except that $N3$ may be the new node), since they are on the path to the new node. From this information and the definition of height, we get the heights before insertion of the four subtrees $S1$, $S2$, $S3$, and $S4$. Note that after rebalancing, these subtrees may have new parents, but the binary search property is still preserved.

We can verify in each case after the rebalancing operation given in Figure 4.13, the height of the tree rooted at *lastbad* is unchanged from the height before the insertion.

For example, in the RR case, suppose the tree rooted at $N1$ had height h before insertion. We have seen that the balance of $N1$, which equals *lastbad* here, must have been $+1$ before insertion, while those of $N2$ and $N3$ were 0. So the right subtree of $N1$, rooted at $N2$, had height $h - 1$ and the left subtree $S1$ had height $h - 2$. Because $N2$ had balance 0 and height $h - 1$, both $S2$ and its right subtree had height $h - 2$. Similarly the subtrees $S3$ and $S4$ of $N3$ had height $h - 3$. The new node must have gone to $S3$ or $S4$, so after rebalancing, one of these subtrees has height $h - 2$. Thus in the rebalanced tree the subtrees rooted at $N1$ and $N3$ have height $h - 1$ and that subtree rooted at $N2$ has height h. Thus the height of the tree rooted at *lastbad* remains unchanged. The arguments for the other cases are similar.

Since the height of the subtree rooted at *lastbad* doesn't change, no node above *lastbad* can have its balance changed by the insertion and rebalancing. Therefore, at most one rebalancing, involving only a few assignment statements, is necessary for each insertion.

An AVL insertion algorithm then may have the following form:

1. perform an ordinary binary search tree insertion
2. determine if the tree is unbalanced and rebalance if necessary
3. update balances. This need only be done below the point of restructuring, if there is one (i.e., below *lastbad* if it exists).

The time complexity of AVL insertion is therefore proportional to the height of the tree, or $O(\log n)$.

In Algorithm 4.4, some of this work is done in parallel. For example, determination of the type, if any, of imbalance is done while making the original insertion. When descending to the spot of insertion, the variable *lastbad* is kept pointing to the last node with balance ± 1. If insertion is made into the subtree of greater height, then the variable *howbad* is set to "?" to indicate there will be imbalance of some unknown type at this point if *lastbad* is not later changed. At the next level, the type of imbalance is determined. If

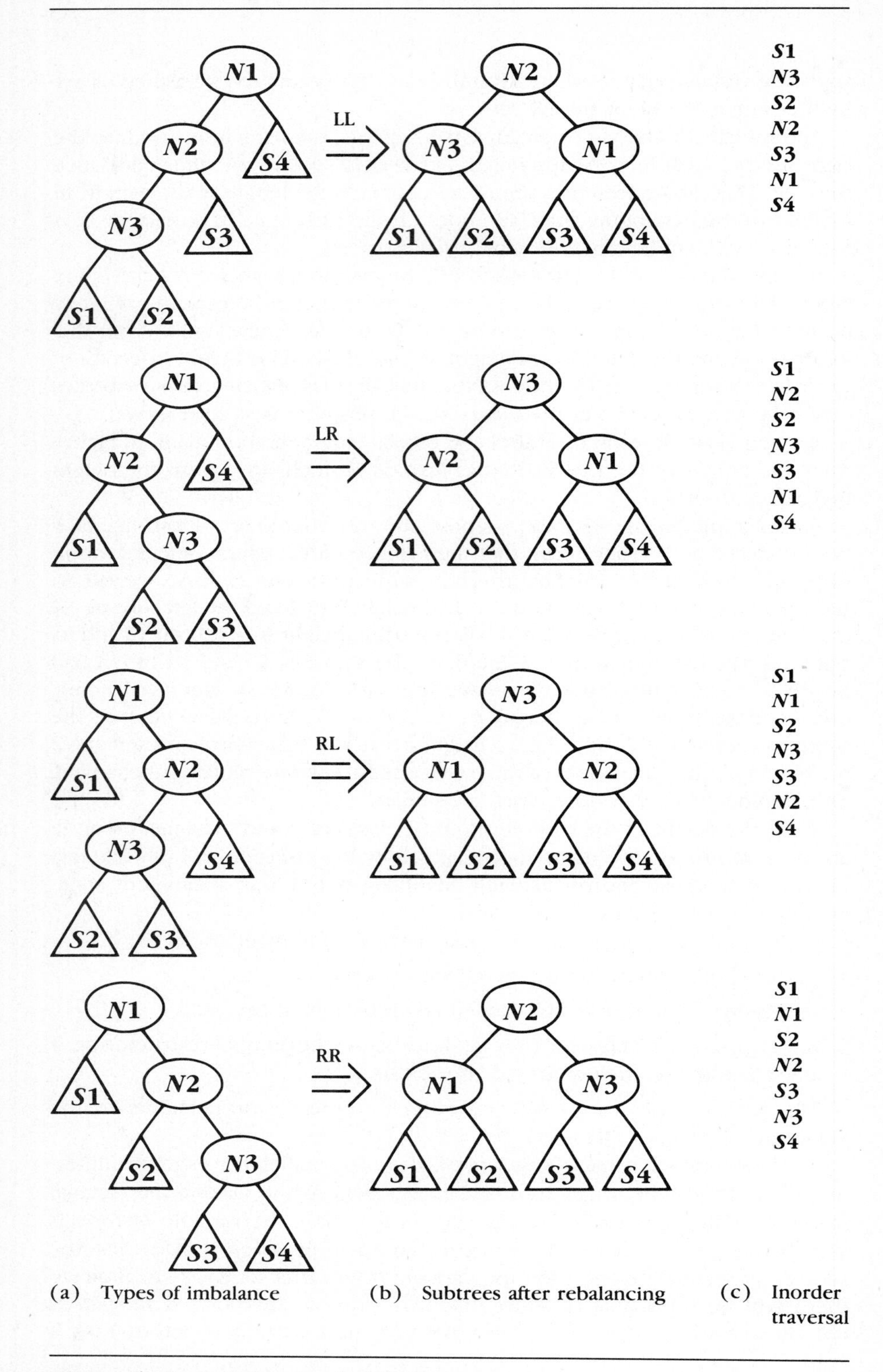

Figure 4.13 Rebalancing in AVL Trees

```
1. Insert(key, avl)                              {inserts a given key into a given
2.                                                    AVL tree. The avl parameter
3.                                                  is passed by value-result.
4.                                              Duplicate insertions are allowed.
5.                                    Uses local variables lastbad and howbad
6.                                               that are set by RecursiveInsert}
7. RecursiveInsert(key, avl, lastbad, howbad)                  {insert normally}
8. Rebalance(key, avl, lastbad, howbad)                          {and rebalance}
```

```
1. RecursiveInsert(key, avl, lastbad, howbad)                   {inserts key into
2.                          an AVL tree recursively. The last 3 parameters are
3.                          value-reference; lastbad and howbad are as in text}
4. if avl = nil then                                         {upon reaching a leaf}
5.    avl := NewNode(key)                                        {insert new node}
6. else
7.    if key < key(avl) then                        {if key small, then insert to L}
8.    if balance(avl) = −1 then                          {if was imbalanced to L}
9.         lastbad := avl                                 {mark as last imbalance}
10.        howbad := "?"                               {of type to be determined}
11.   else
12.      if balance(avl) = 1 then                            {if was imbalanced to R}
13.        lastbad := avl                             {then mark as last imbalance}
14.        howbad := "notbad"                         {but new imbalance possible}
15.      else                                      {else no new imbalance, but old}
16.        if (lastbad ≠ nil) and (howbad = "?") then howbad := "L"
17.      end if                            {unknown imbalance may need resolving}
18.   end if                                         {in any case, continue down}
19.   RecursiveInsert(key, left(avl), lastbad, howbad)
20. else                                                     {else insert into RST}
21.   if balance(avl) = 1 then                             {if was imbalance to R}
22.      lastbad := avl                          {then mark last spot of imbalance}
23.      howbad := "?"                                           {of unknown type}
24.   else
25.      if balance(avl) = −1 then                            {if was imbalance to L}
26.        lastbad := avl                        {then mark last spot of imbalance}
27.        howbad := "notbad"                         {but new imbalance possible}
28.      else                                      {else no new imbalance, but old}
29.        if (lastbad ≠ nil) and (howbad = "?") then howbad := "R"
30.      end if                            {unknown imbalance may need resolving}
31.   end if                                          {in any case continue down}
32.   RecursiveInsert(key, right(avl), lastbad, howbad)
33. end if
```

```
1. procedure Rebalance(key, avl, lastbad, howbad)
2.                         {rebalances a tree after insertion given the key
3.                    inserted, the tree, the last point of imbalance, and
4.                          the type of imbalance. The last 3 parameters are
```

```
 5.                                               passed by value-result}
 6. if lastbad = nil then                   {if no imbalance then merely}
 7.    Update(avl, key)                    {update balances of all nodes}
 8. else if howbad = "?" then               {if imbalance removed then}
 9.    balance(lastbad) := 0                          {correct balance}
10. else if howbad = "notbad" then           {if no new imbalance then}
11.    Update(lastbad, key)                 {only some balances change}
12. else
13.    if balance(lastbad) = 1 then              {else handle imbalance}
14.       if howbad = "L" then RL(key, avl, lastbad)
15.                                  {RL: +1 imbalance, howbad = "L"}
16.       else RR(key, avl, lastbad){RR: +1 imbalance, howbad = "R"}
17. else                                             {balance = -1}
18.    if howbad = -1 then           {LL: -1 imbalance, howbad = "L"}
19.       LL(key, avl, lastbad)      {symmetric to RR, left as exercise}
20.    else                          {LR: -1 imbalance, howbad = "R"}
21.       LR(key, avl, lastbad)      {symmetric to RL, left as exercise}
22.    end if
23. end if
```

```
 1. procedure RL(key, avl, lastbad)        {removes RL imbalance after
 2.                            insertion of key into AVL tree avl where
 3.                          lastbad is point of lowest imbalance. Last
 4.                                two parameters passed by value-result.
 5.                 The local variable ptr is used for temporary storage}
 6. temp := key(lastbad)
 7. key(lastbad) := key(left(right(lastbad)))
 8. key(left(right(lastbad))) := temp
 9. ptr := left(right(lastbad))
10. left(right(lastbad)) := right(ptr)
11. right(ptr) := left(ptr)
12. left(ptr) := left(lastbad)
13. left(lastbad) := ptr
14. if left(left(lastbad)) ≠ nil then                             {leaf}
15.    balance(left(lastbad)) := -1
16.    balance(right(lastbad)) := 1
17.    Update(lastbad, key)
18. end if
19. balance(lastbad) := 0
```

```
 1. procedure RR(key, avl, lastbad)        {removes RR imbalance after
 2.                            insertion of key into AVL tree avl where
 3.                          lastbad is point of lowest imbalance. Last
 4.                                two parameters passed by value-result.
 5.                 The local variable ptr is used for temporary storage}
```

```
 6. temp := key(lastbad)
 7. key(lastbad) := key(right(lastbad))
 8. key(right(lastbad)) := temp
 9. ptr := right(lastbad)
10. right(lastbad) := right(right(lastbad))
11. right(ptr) := left(ptr)
12. left(ptr) := left(lastbad)
13. left(lastbad) := ptr
14. balance(lastbad) := 0
15. Update(right(lastbad), key)
```

```
1. NewNode(key)                     {returns a node initialized to}
2.                                        {contain the given key}
3.    temp := Get                  {get a new node from storage}
4.    left(temp) := nil                     {and fill its fields}
5.    right(temp) := nil
6.    key(temp) := key
7.    balance(temp) := 0
8.    return temp
```

```
 1. Update(ptr, key)                  {changes all balances beginning
 2.                                    with ptr after insertion of key}
 3.
 4. while left(ptr) ≠ nil or right(ptr) ≠ nil        {don't change leaf}
 5.    if key < key(ptr) then                      {insertion into LST}
 6.       balance(ptr) := balance(ptr) − 1       {decrements balance}
 7.       ptr := left(ptr)
 8.    else                                        {insertion into RST}
 9.       balance(ptr) := balance(ptr) + 1       {increments balance}
10.       ptr := right(ptr)
11.    end if
12. end while
```

Algorithm 4.4 Insertion into an AVL Tree (with Auxiliary Operations)

insertion is not made into the subtree of greater height, then *howbad* is set to "notbad" to indicate there is no imbalance, but that the balances will have to be changed from *lastbad* on down.

To summarize the interpretation of *lastbad:*

if balance(*lastbad*) = 1 and *howbad* = "L", there is an RL imbalance;
if balance(*lastbad*) = 1 and *howbad* = "R", there is an RR imbalance;
if balance(*lastbad*) = −1 and *howbad* = "L", there is an LL imbalance;

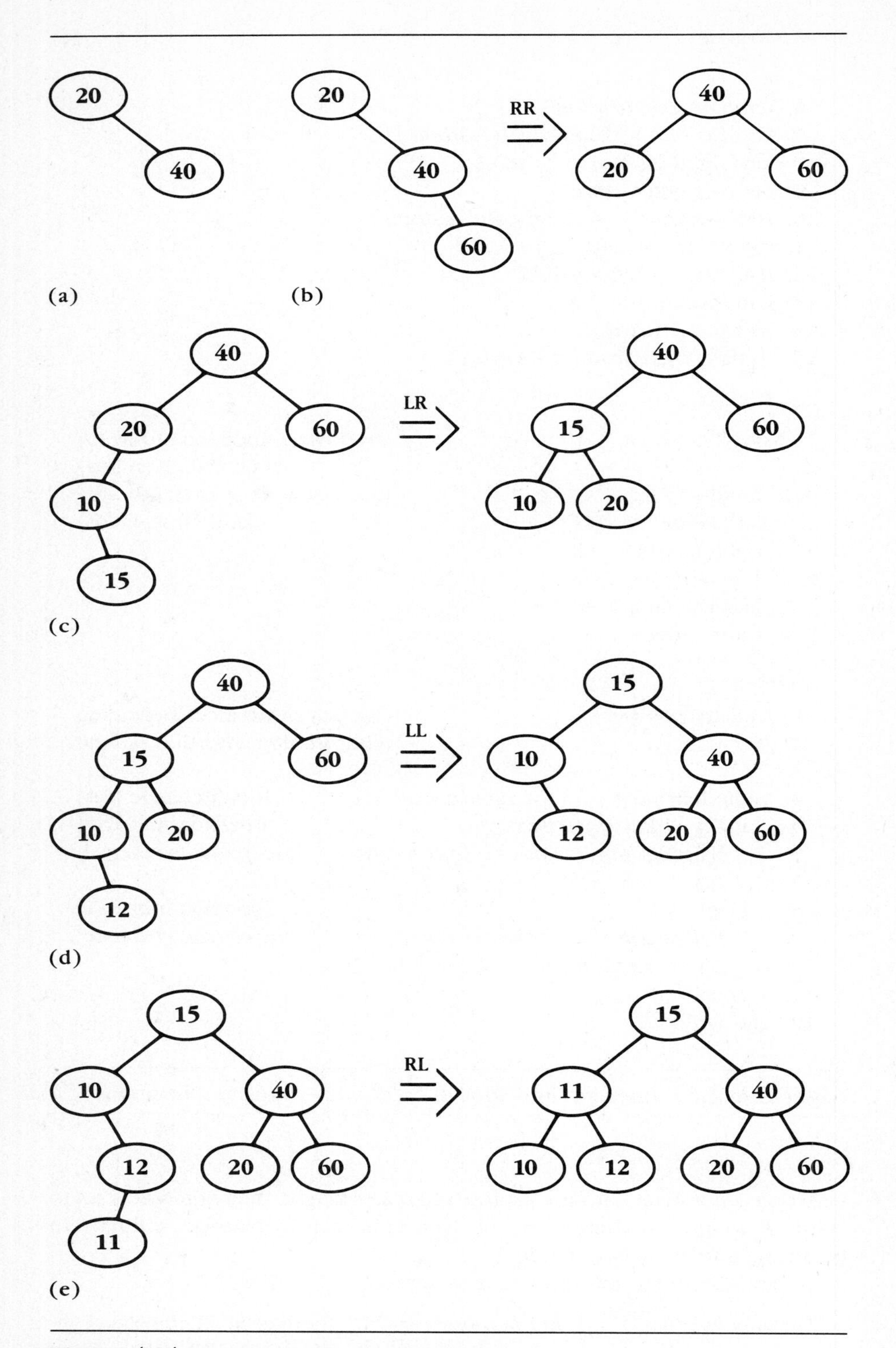

Figure 4.14 Insertions into an AVL Tree

if balance(*lastbad*) $= -1$ and *howbad* = "R", there is an LR imbalance;
if *lastbad* = nil or *howbad* = "notbad," there will be no imbalance.

We have already seen that the correct type of imbalance can be determined at the first pass through the tree. The steps making up the restructuring are easy to determine by comparing with Figure 4.13. They are given as part of Algorithm 4.4 for RR and RL. The cases LL and LR are symmetric and are left as exercises.

We need to show the balances are updated correctly if there is no restructuring. We have already shown no further updating of the balances is necessary if there is a restructuring. There are several cases.

If the final value of *lastbad* is "nil," then all balances on the path to the new node are 0. A simple induction on the length of the path shows that all nodes on the path require new balances of $+1$ or -1, depending on whether the insertion is into the LST or the RST. The call to the Update function accomplishes this, beginning at the root.

If the final value of *howbad* is "?," then *lastbad* must be the parent of the new node; otherwise, the value of "?" would have been changed at the next level. This means the insertion was done into the empty subtree of *lastbad* rather than into the subtree of greater height. Thus the subtree rooted at *lastbad* keeps the same height, and no node above it has a changed balance. The balance of *lastbad* itself becomes 0. In this case the algorithm does the correct thing.

If the final value of *howbad* is "notbad," then all lower nodes have original balances 0, but no restructuring should take place. As in the previous case, these lower nodes need to have their balances changed to $+1$ or -1 as appropriate. Again this is accomplished by a call to the Update algorithm, beginning at *lastbad*.

Figure 4.14 shows a sequence of insertions together with the necessary rebalancings.

Design of an $O(\log n)$ deletion algorithm for AVL trees is left as an exercise.

4.8 B-Tree Operations

In this section we consider insertion and deletion algorithms for B-trees of order m. We first consider insertion.

As in the case of insertion into a binary search tree, we begin with a search. For any key k the search algorithm of Section 2.6 will lead to a particular leaf node. If fewer than $m - 1$ keys are in this node, then the key may be inserted into this node. If the node is implemented as a sorted array, then $\Theta(m)$ keys and $\Theta(m)$ pointers must be shifted in the worst case to make room for the new key. In the case of external files, however, this internal manipulation is still much cheaper than even a single reference to external memory.

If the node found by the search algorithm already contains the maximum number of keys, then there are two possible strategies: Keys may be sent to a neighboring sibling node or the current node may be split in two, with the middle key passed up to the parent node.

The first strategy is illustrated for the B-tree of order 5 of Figure 4.15b, based on an attempted insertion as shown in (a) of the figure. Note that one key of the parent node must also be shifted. The problem with this strategy is that the sibling may be full as well. In general the keys may be evenly distributed between the two siblings to minimize future overflows.

The alternative strategy of splitting a node has the advantage that it always works. The disadvantage, as we shall see, is that this splitting may propagate up the tree.

Splitting a node assumes that an attempt has been made to insert an mth key. It requires that a new key be installed in the parent node in order to separate the keys in the two new nodes [see Figure 4.15c, based on the tree of (a)]. If these keys are split as evenly as possible, then of the $m - 1$ remaining keys, $\lceil (m - 1)/2 \rceil$ will go into one node and $\lfloor (m - 1)/2 \rfloor$ into another. These values satisfy the constraints on the number of keys in nodes of a B-tree.

The addition of a new key to the parent node may lead to overflow in this node as well. In the worst case, this overflow may propagate to the root. Here the root will be split into two nodes, each to become a subtree of a new root node (see Figure 4.16, also representing a B-tree of order 5). For this reason the definition of a B-tree allows as few as two children for the root.

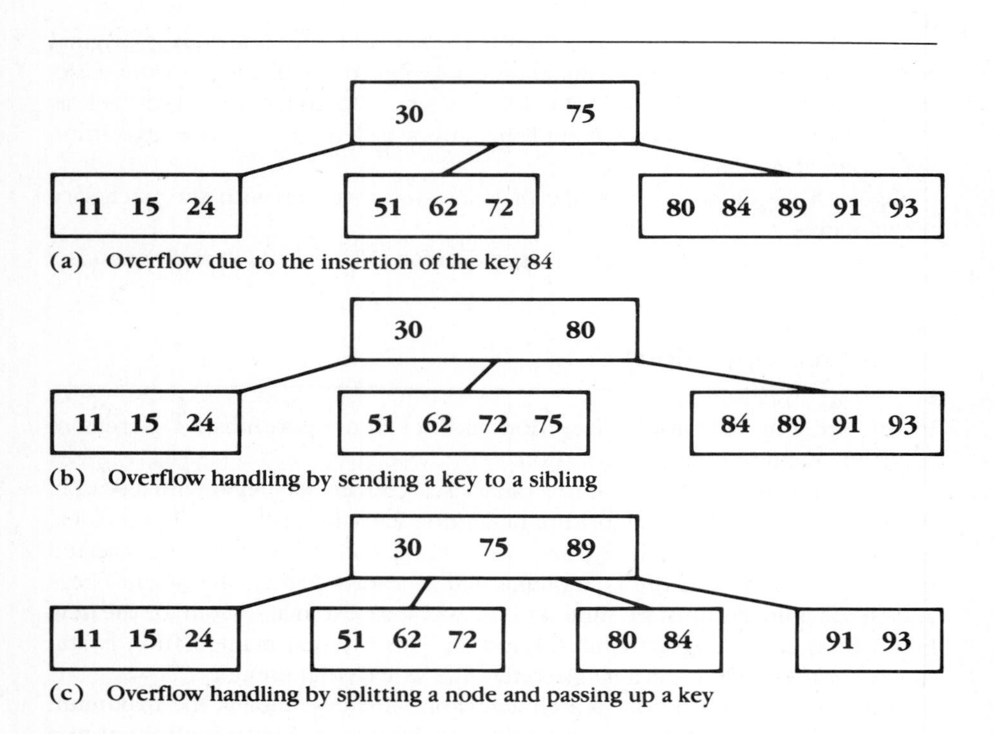

(a) Overflow due to the insertion of the key 84

(b) Overflow handling by sending a key to a sibling

(c) Overflow handling by splitting a node and passing up a key

Figure 4.15 Overflow Handling in B-Tree Insertion

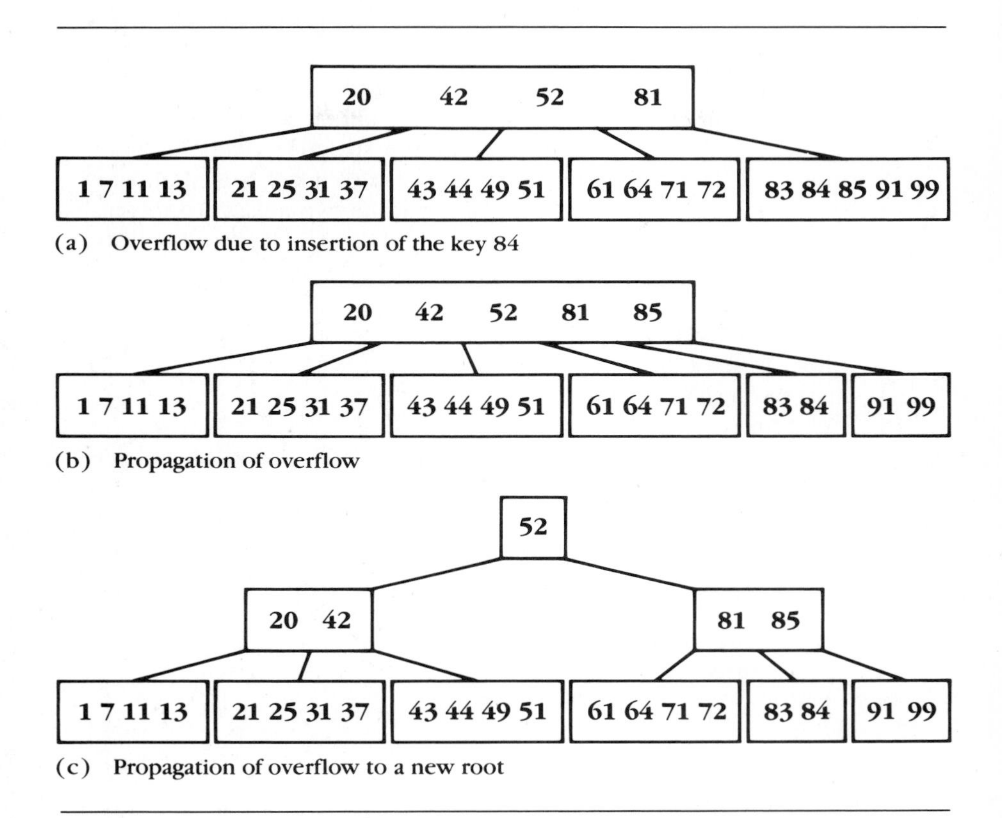

Figure 4.16 Increase in B-Tree Height After an Insertion

In the B-tree Algorithms 4.5 and 4.6, we assume that each node has a size field containing the number of keys in the node, a key field that is an array indexed from 1 through m, and a pointer field that is an array indexed from 0 through m. The value of m is assumed to be a constant.

Most of the work of the insertion Algorithm 4.5 is done by the recursive function **RecInsert**, which will pass a key up along with two subtrees that are to flank it in the new node. A single overflow structure is used to combine these three components. If such a structure is returned to the top-level **Insert** operation, it will be installed as the new root. For example, when the key 52 is passed up between stages (b) and (c) of Figure 4.16, two subtrees (one containing 20 and 42 in its root, and the other 81 and 85) must be passed up as well.

Several low-level pointer-moving operations have been assumed; filling in these blanks has been left as an exercise.

When deleting a key, there is the possibility the key is not stored in a leaf, although this would not happen in B^+-trees. Here, as for binary search trees, we may replace by an inorder successor or predecessor and delete the original occurrence of this successor, which will be in a leaf. Recall that B-tree heights

```
 1. Insert(key, btree, m) {inserts a key into a btree of order m, passed
 2.                          as a value-result parameter. If the btree is
 3.                          empty, creates a new one, with the structure
 4.                          described in the text. Otherwise uses RecInsert
 5.                          (defined below) to do the actual insertion.
 6.                          RecInsert returns a structure overflow that is
 7.                          either empty or contains a key passed up from
 8.                          below and its two adjacent subtrees}
 9. if btree = nil then                              {if btree is empty}
10.    btree := Get                          {assign it to a new node and}
11.    size(btree) := 1                            {initialize its fields}
12.    key(btree)[1] := key
13.    for i from 0 to 1
14.       ptr(btree)[i] := nil
15.    end for
16. else                      {otherwise do a recursive insertion and see}
17.    overflow := RecInsert(key, m, btree, nil, dummy)
18.    if overflow ≠ nil then                {if a key is passed up. If so,}
19.       new(btree)                               {get a new root node}
20.       size(btree) := 1                     {and initialize its fields}
21.       key(btree)[1] := key(overflow)
22.       ptr(btree)[0] := left(overflow)
23.       ptr(btree)[1] := right(overflow)
24.    end if
25. end if
```

```
 1. RecInsert(key, m, nodeptr, parent, parindex)
 2.       {inserts a given key into the node pointed to by nodeptr in a
 3.          B-tree of order m. Also takes a pointer parent to its parent
 4.             node and the index parindex of the pointer from the
 5.                parent to the node. Returns an overflow structure
 6.           as described previously for Insert. Assumes operations
 7.             (involving mainly pointer manipulations and so left as
 8.          exercises) Search, which searches a node and returns the
 9.          index of the subtree to contain the key, InsertIntoNode,
10.               which puts a key and adjacent pointers into a node,
11.    CanBorrow, which determines whether one node can borrow
12.              from its right neighbor, Borrow, which does the
13.           actual borrowing, and SplitNode, which given a node,
14.             the index of the new key, and an overflow passed up,
15.             splits the node and returns a new overflow structure.
16.   Assumes the number of keys of node x is available via size(x).}


17. index := Search(key, nodeptr, m)       {find subtree — if not a leaf}
18. if ptr(nodeptr)[index] ≠ nil then               {continue recursively}
19. overflow := RecInsert(key, m, ptr(nodeptr)[index], nodeptr, index)
```

```
20. else                                  {if at leaf, prepare new insertion to be}
21.     overflow := Get                        {handled by InsertInto Node by}
22.     key(overflow) := key                          {getting and initializing}
23.     left(overflow) := nil                          {overflow structure}
24.     right(overflow) := nil
25. end if
26. if overflow = nil then                          {no key passed up, so}
27.     return nil                                  {nothing to propagate}
28. else
29.     if size(nodeptr) = m − 1 then          {make room for new node}
30.         if CanBorrow(m, parent, parindex) then       {if can borrow}
31.             left := ptr(parent)[parindex − 1]      {from right, do so}
32.             Borrow(nodeptr, m, left, parent, index, parindex, overflow)
33.             return nil                {borrowing means no key to pass up}
34.         else                            {could also try to borrow from left}
35.             return SplitNode(nodeptr, m, index, overflow)
                                                            {gets overflow}
36.         end if                               {if can't borrow, split node}
37.     else                                  {node has room for key passed up}
38.         InsertIntoNode(nodeptr, m, index, size(nodeptr), overflow)
39.         return nil                               {no overflow to propagate}
40.     end if
41. end if
```

Algorithm 4.5 Insert (into B-Tree) (with Auxiliary Function RecInsert)

```
1. Delete(key, btree, m)                   {deletes the given key from the
2.                 given btree of order m. The parameter btree is passed by
3.                              value-result. The recursive delete function
4.                              RecDelete is given and documented below}
5. if btree = nil then
6.     return unsuccessfully
7. else                         {value returned is irrelevant but will be 0}
8.     return RecDelete(key, btree, m, nil, dummy)
9. end if                      {value of fourth parameter is also irrelevant}
```

```
1. RecDelete(key, btree, m, parent, parindex)      {deletes a given key
2.  from a non-empty btree of order m. The parameter btree is passed
3.                        by value-result. Takes as additional parameters a
4.               pointer to the parent node and the index of the pointer
5.               to the btree from the parent node. Returns either 0 or
6.                   the index of a key and pointer to be deleted from the
7.             parent node. Assumes Search algorithm is as documented
8.                        in text and DeleteFromNode as given below.}
```

```
 9. index := Search(key, btree, m)        {look for key in current node}
10. if ptr(btree)[0] = nil then                             {if at a leaf}
11.    if key = key(btree)[index] then          {if node found, delete}
12.       return DeleteFromNode(btree, m, index, parent, parindex)
13.    else
14.       return unsuccessfully          {if key not found — no deletion}
15.    end if
16. else                                                       {nonleaf}
17.    if key(btree)[index] = key then               {if found, then find}
18.       temp := ptr(btree)[index]
19.       while ptr(temp)[0] ≠ nil                   {inorder successor}
20.          temp := ptr(temp)[0]
21.       end while
22.       key(btree)[index] := key(temp)[1]             {replace with it}
23.       key := key(temp)[1]                  {and delete replacement}
24.       wheredelete := RecDelete(key, ptr(btree)[index],
                                                   m, btree, index)
25.    else                {if not in internal node, check correct child}
26.       wheredelete := RecDelete(key, ptr(btree)[index],
                                                   m, btree, index)
27.       if wheredelete > 0 then           {if key taken down, delete it}
28.          return DeleteFromNode(btree, m, wheredelete,
                                                   parent, parindex)
29.       else
30.          return 0              {otherwise done, no more propagation}
31.       end if
32.    end if
33. end if
```

```
 1. DeleteFromNode(btree, m, index, parent, parindex)
 2.                     {deletes the key at the given index from the node
 3.                   btree, which is passed by value-result. Takes as extra
 4.                 parameters a pointer to the parent node and the index
 5.                       of the pointer from this node to btree so that the
 6.              Underflow function (given below) may use these values.
 7.                      Returns 0 or the index of the key and pointer to be
 8.     deleted from the parent node. The roles of m and size are as in
 9.        Algorithm 4.5. The MoveLeft operation, which removes a key
10.                                 from a node, is left as an exercise.}
11. if parent = nil then                            {if at root node with}
12.    if size(btree) > 1 then                    {more than 1 key then}
13.       MoveLeft(btree, index, size(btree))                  {delete}
14.    else                                   {else new root is lone child}
15.       if index = 1 then
16.          btree := ptr(btree)[0]
```

```
17.       else
18.          btree := ptr(btree)[1]
19.       end if
20.       return 0                              {and no propagation to do}
21.    end if                               {since root has no minimum size}
22. else                                                  {if not at root}
23.    if size(btree) > (m − 1) div 2 then                {if enough keys}
24.       MoveLeft(btree, index, size(btree))
25.       return 0                              {then delete without propagating}
26.    else         {otherwise, Underflow decides how to share nodes}
27.       return Underflow(btree, m, index, parent, parindex)
28.    end if
29. end if
```

```
 1. Underflow(btree, m, index, parent, parindex)
 2.            {given a node btree with a minimum number of keys, an
 3.               index of a key in that node to be deleted, a pointer to
 4.               the parent node and the index of the pointer from the
 5.            parent to the node, performs the deletion and returns 0
 6.               or the index in the parent of a key and pointer taken
 7.               down. The btree parameter is passed by value-result.
 8.                               The order of the B-tree is given by m.
 9.            Algorithms for BorrowFromLeft, BorrowFromRight,
10.         MergeWithLeft and MergeWithRight are left as exercises}
11. if parindex < size(parent) then              {if right sibling exists}
12.    if size(ptr(parent)[parindex + 1)] > (m − 1) div 2 then
13.       BorrowFromRight(btree, m, index, parent, parindex)
14.       return 0                                  {no key taken down}
15.    else                             {right sibling has no extra nodes}
16.       MergeWithRight(btree, m, index, parent, parindex)
17.       return parindex + 1         {must delete pointer and key there}
18.    end if
19. else                                 {node may have no right sibling}
20.       if size(ptr(parent)[parindex − 1)] > (m − 1) div 2 then
21.          BorrowFromLeft(btree, m, index, parent, parindex)
22.          return 0                               {no key taken down}
23.       else
24.          MergeWithLeft(btree, m, index, parent, parindex)
25.          return parindex             {must delete pointer and key there}
26.       end if
27. end if
```

Algorithm 4.6 Delete from a B-Tree (with Auxiliary Function Operations)

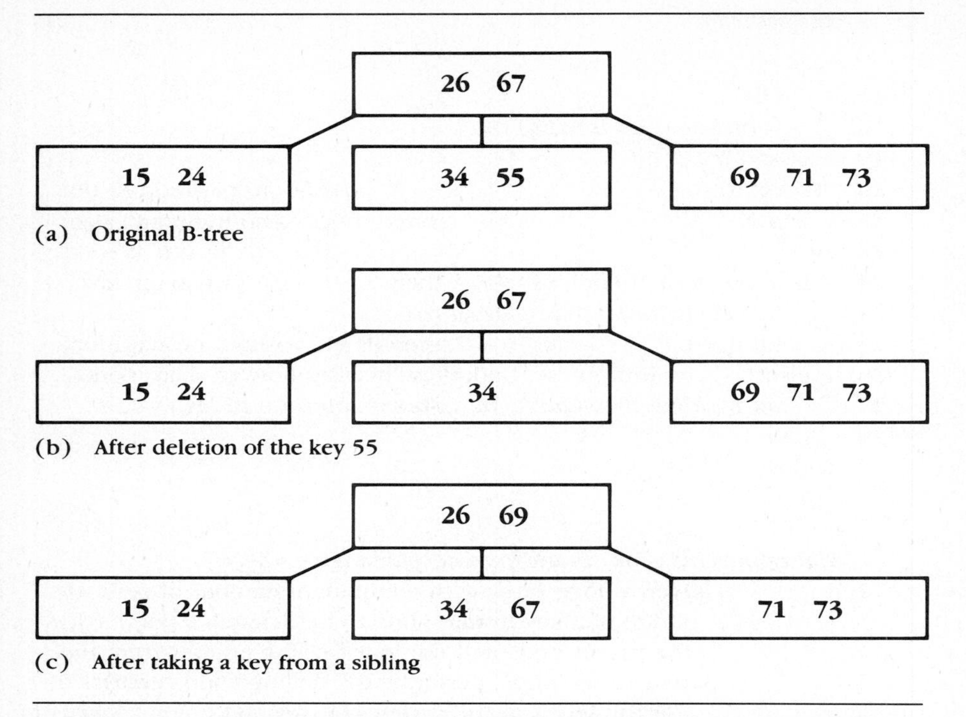

(a) Original B-tree

(b) After deletion of the key 55

(c) After taking a key from a sibling

Figure 4.17 Deletion in a B-Tree of Order 5 by Taking a Key from a Sibling

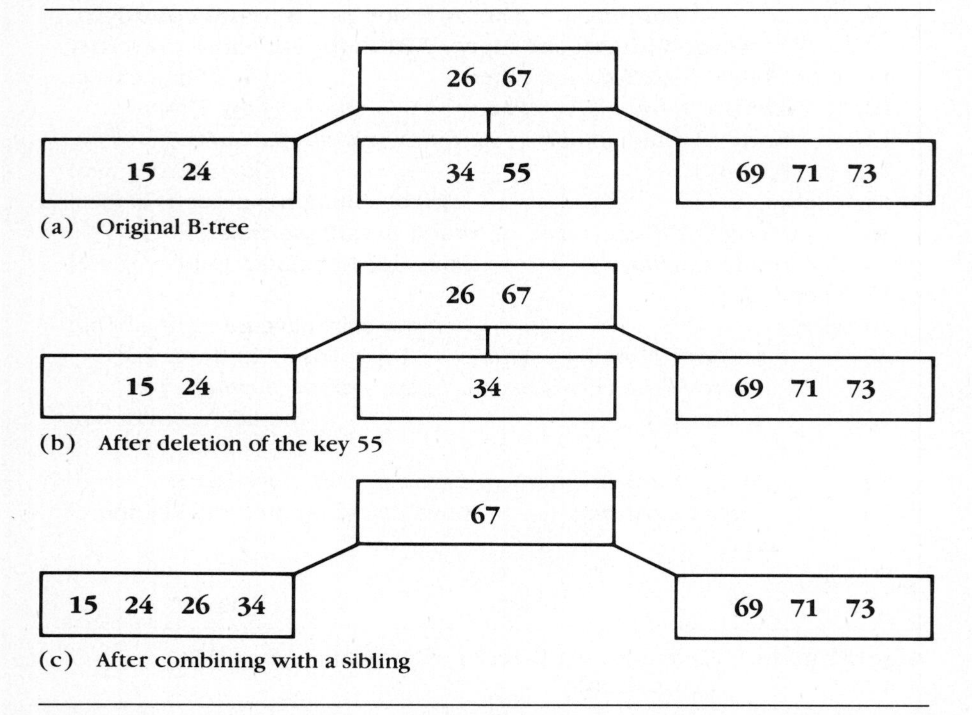

(a) Original B-tree

(b) After deletion of the key 55

(c) After combining with a sibling

Figure 4.18 Deletion in a B-Tree of Order 5 by Combining with a Sibling

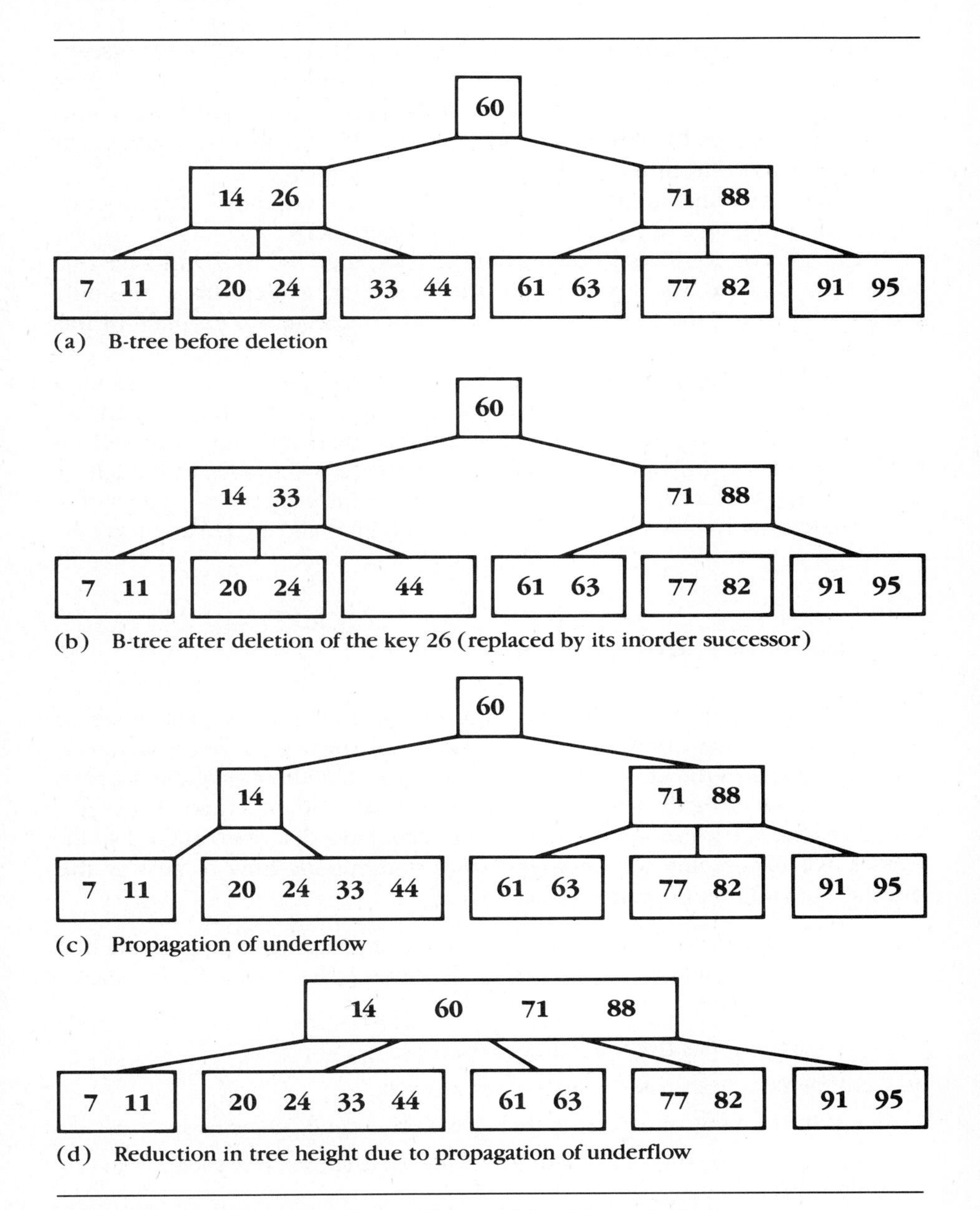

Figure 4.19 Reduction in a B-Tree Height After a Deletion

are quite small and keys are usually in leaves, so the extra search for the inorder successor is not that inefficient.

To delete from a leaf, we reverse the process of insertion. This reversal may involve taking keys from a sibling node or combining two nodes and bringing a key down from a parent. If a node and its sibling each have the minimum number $\lceil m/2 \rceil - 1$ of keys before deletion, then if the nodes are combined and a key from the parent node replaces the deleted key, $2\lceil m/2 \rceil - 2$ keys will be in the combined node. This is a legal number of keys for a B-tree, since $2\lceil m/2 \rceil - 2 \leq (m + 1) - 2 = m - 1$. If the node that is short has a sibling key with more than the minimum number of nodes, then that short node may take one or more from the sibling.

Sharing keys is exemplified in Figure 4.17, while an example of combining nodes and bringing a key down is shown in Figure 4.18.

If the last key is brought down from the root, then the root will have one child left. This child should become the new root. Notice this is the only situation in which the height of the B-tree will decrease. An example of this behavior is shown in Figure 4.19.

The deletion Algorithm 4.6 assumes a MoveLeft algorithm, which given a node, an index, and a size, will delete the key and pointer at that index from the node. The MoveLeft algorithm also assumes that Underflow can call on functions to borrow from the left or right or merge with nodes to the left or right. Of course the appropriate key from the parent node must be included. These functions involve mostly pointer manipulations and are left as exercises.

4.9 Union-Find Structures

In the minimum spanning tree discussion of Section 2.19 we had a set of vertices divided into disjoint classes and needed to answer the question, when are two vertices in the same class? Two vertices of an undirected graph were in the same class iff there was a path from one to the other. These classes are called *connected components.* Since many problems deal with such disjoint classes, we need some terminology before determining how to answer the question for minimum spanning trees.

Def 4.7 A *partition* of a set S is a collection of subsets, each pair of which is disjoint and whose union is S. This implies each element is in exactly one subset.

Every partition has the following properties:

1. Each element x is in the same subset as itself
2. If x is in the same subset as y, then y is in the same subset as x
3. If x is in the same subset as y and y is in the same subset as z, then x is in the same subset as z.

Def 4.8 Subsets with properties 1, 2, and 3 — called respectively *reflexivity, symmetry,* and *transitivity* — are called *equivalence classes.* An *equivalence relation* is defined by saying that x is related to y (notation: $\mathbf{x} \sim \mathbf{y}$) if and only if they are in the same equivalence class.

An equivalence relation $\sim$ may also be defined independently of equivalence classes as follows:

Def 4.9 A relation $\sim$ is called an *equivalence relation* iff

1. $x \sim x$
2. if $x \sim y$ then $y \sim x$ and
3. if $x \sim y$ and $y \sim z$, then $x \sim z$.

The *equivalence class* containing x is just $\{z|z \sim x\}$.

Thus any partition defines an equivalence relation; any equivalence relation defines a partition into equivalence classes.

Equivalence classes occur frequently in mathematics as well; two fractions are equivalent (e.g., 2/4 and 3/6) if they represent the same rational number. The relations of congruence mod n for integers and similarity for triangles are easily seen to be equivalence relations.

In the minimum spanning tree application, there are two important operations. It is important to determine whether or not two vertices are in the same component. One way to do so is to maintain a fixed representative for each class and then determine whether the two vertices have the same representative, as is done for fractions and congruence classes mod n. We call the operation of determining the representative the *Find* operation. Two classes can also be combined, which is called the *Union* operation. After this operation the classes that are combined should have the same representative. Note in the minimum spanning tree application there will be only $n - 1$ Union operations if there are n vertices.

These two operations are just as important when considering general equivalence classes. We would like a representation that allows them to be performed efficiently.

Unfortunately, we cannot simply maintain a table giving the representative for each element. This implementation would make the Union operation unacceptably slow. Suppose that two sets are represented by elements x and y. If their union is to be represented by x, all elements in set y must have their set representative changed from y to x. In the given representation the only way to find the elements in set y is to perform a sequential search through all elements.

We introduce the best representation known for this problem by tracing a sample problem. The data structure is called a *Union-Find* structure. We consider an example where, as in the minimum spanning tree problem, we begin with n classes of size 1. We assume the n elements are numbered 1 through n. Element k is initially its own representative. The class with representative k has size 1 for all k. The situation is that of Figure 4.20a.

Suppose that classes 1 and 2 are to be combined by a Union operation. Without loss of generality we assume the new class is to be represented by 1. In order to reflect that element 2 is represented by 1, we construct a pointer[6] from element 2 to element 1. We are now in the situation of Figure 4.20b.

[6] In typed languages this pointer could be stored in a variable of integer type. In our example, the pointer associated with element 2 would have integer value 1. These pointers will need to be initialized to zero or to the null pointer.

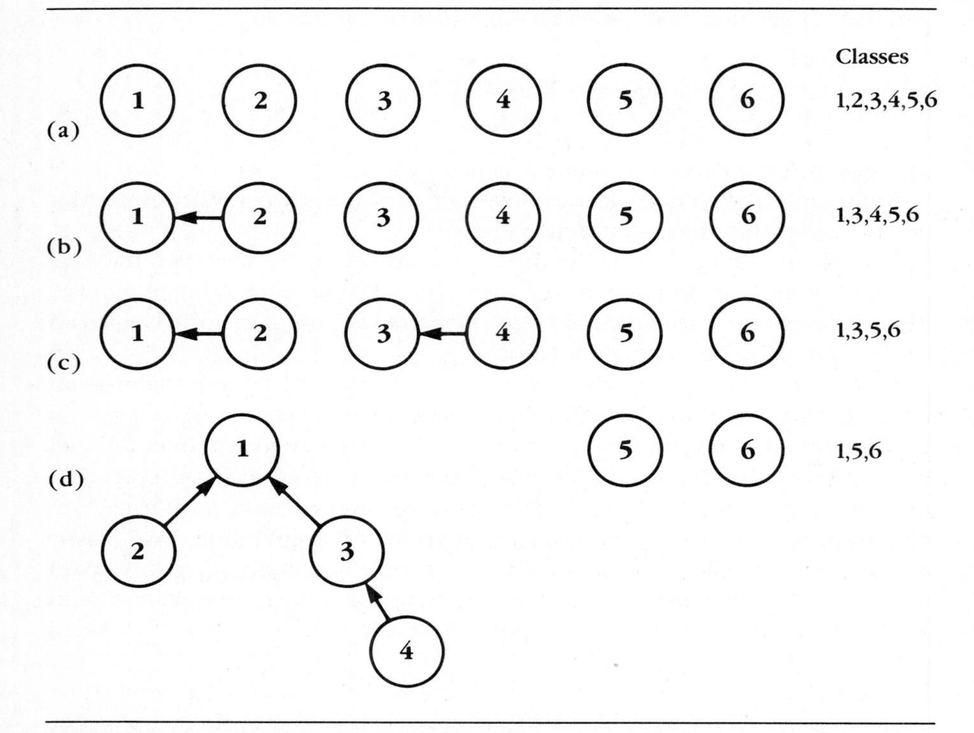

Figure 4.20 The Union Operation in a Union-Find Structure

In the same way, constructing an edge between elements 3 and 4 and calling the new class 3 requires a pointer from element 4 to element 3. This corresponds to (c) in the figure.

In this representation, any element without a pointer — or with a null pointer, for example 1 and 3 in (c) — is its own representative.

Now suppose that the classes containing 2 and 4 are to be combined. Constructing a pointer from element 2 to element 4 is not enough, since then element 3 would still have a null pointer and it would be impossible to recognize that element 3 was now in the new class. For a similar reason a pointer from 4 to 2 would not solve the problem. We need to construct a pointer from element 3 to element 1. In other words the Union operation must have classes, given by representatives, as operands and not arbitrary elements. If the representatives of the classes are not available, the Find function may be applied.

After construction of the pointer from 3 to 1, we see in Figure 4.20d that element 1 is accessible from any of the elements 1, 2, 3, and 4 by following pointers as far as possible. Element 1 is thus the only element with a null pointer. This suggests the property of the root of a tree, and as (d) of the figure suggests, our data structure is precisely a forest of trees in which each node points to its parent and leaves are accessible directly. The name of the root element is the name of the class.

Now suppose an edge is created between elements 3 and 5. By calling the Find operation, we see the respective representatives are elements 1 and 5. These two nodes should be connected with a pointer. The two possible resulting trees are given in Figure 4.21. The first alternative is better, since on the average fewer pointers follow in the Find operation, assuming all possible arguments to Find are equally likely.

For this reason, in a general Union operation, the smaller tree always points to the larger one; that this improves the average-case behavior of Find is left as Exercise 59. It thus becomes important to keep track of the number of elements in each class, which requires a field associated with each root. Note that it is precisely the roots that have empty pointer fields. Thus the same field may be used for pointer and size counter. (Even in typed languages integer variables may be used to store the pointers.)

Since a general element need not have a pointer directly to its representative, the Find operation will not have time complexity $O(1)$. It is left as Exercise 60 to show the tree height, and thus the time complexity, is $O(\log n)$.

For example, finding the class containing 4 in the situation of Figure 4.20d requires checking 3 nodes. But after the representative is found, we may point from 4 directly to it. In fact, we may do so from every ancestor of 4.

In general, for any Find operation with parameter x (see Figure 4.22) we can point directly to the root from each node on the path from x to the root. This requires at worst following the path to the root a second time, which doesn't worsen the asymptotic time complexity of the Find algorithm. It does, however, make subsequent Finds more efficient. This process is called ***path compression.***

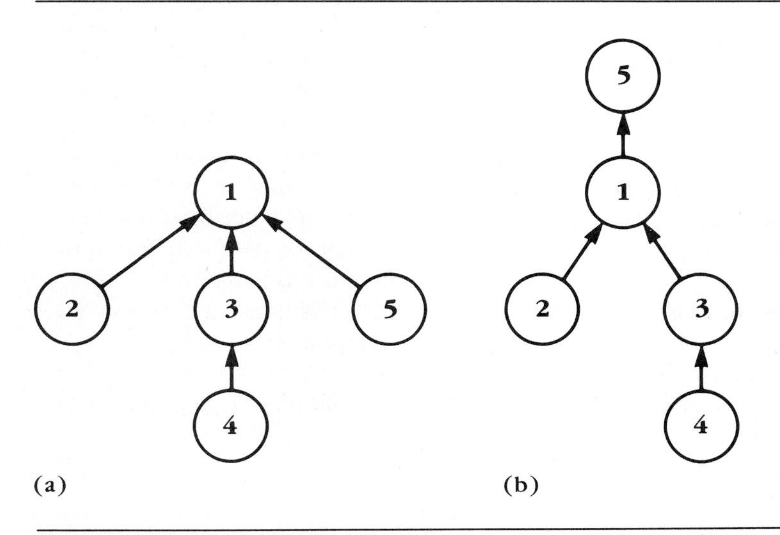

Figure 4.21 Two Ways of Taking the Union of Union-Find Structures

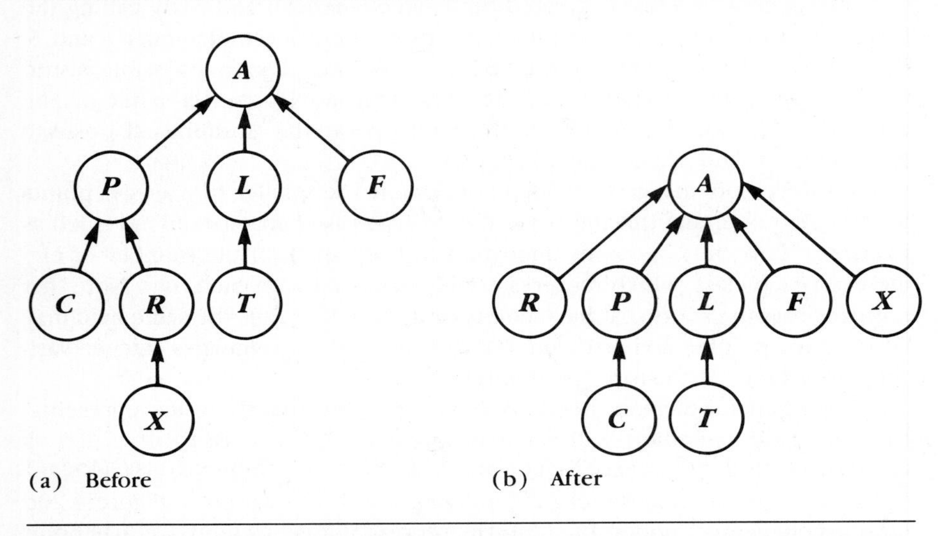

Figure 4.22 Union-Find Structures Before and After a Find(X) Operation

```
 1. Find(x, par)                  {returns index rep of the root of the tree
 2.                                 containing element x. par is an array of integers
 3.                                 containing indices of parents. Assumes a
 4.                                 function empty to determine when a parent field
 5.                                 is empty. Variable temp is used locally}
 6. curr := x                     {curr is used locally as a pointer in the tree}
 7. while par[curr] ≠ curr                   {ascend until root reached}
 8.      curr := par[curr]                   {parent field of root}
 9. end while                                {contains a negative number}
10. rep := curr                              {save result for path compression}
11. curr := x                                {begin retracing path to root}
12. while not empty(par[curr])               {perform path compression}
13.    temp := par[curr]                     {by saving pointer to next node up}
14.    par[curr] := rep                      {pointing current node to rep}
15.    curr := temp                          {and following pointer up}
16. end while
17. return rep
```

Algorithm 4.10a Find (for an Equivalence Class)

```
 1. Union(x, y, par)          {of the trees containing x and y, makes the
 2.                               root of the smaller tree a child of the
 3.                                          larger tree. Par is defined
 4.                              as for Find. The size of the tree rooted
 5.                            at each representative z is assumed to be
 6.                                               accessible as size(z)}
 7. if x = y or x ≠ par[x] or y ≠ par[y] then
 8.    return error         {need x and y different trees and both roots}
 9. else
10.    if size(x) > size(y) then
11.       size(x) := size(x) + size(y)                {update size of x}
12.       par[y] := x                                   {point y to x}
13.    else
14.       size(y) := size(x) + size(y)                {update size of y}
15.       par[x] := y                                   {point x to y}
16.    end if
17. end if
```

Algorithm 4.10b Union (of Equivalence Classes)

Algorithms 4.10a and 4.10b for Find and Union reflect the discussion above. The initialization algorithm necessary to use these two algorithms has been left as an exercise.

In order to analyze the time required by Union and Find, one assumption and a few observations are useful. Let us assume that m Finds and $n - 1$ Unions are performed in an arbitrary order, with $m \geq n$ and beginning with classes of size 1. The inequality is not an unreasonable assumption, since we have seen that typically Unions require Finds beforehand. Also at most $n - 1$ Unions can be performed on n original classes.

Under these assumptions we can conclude 1) any tree has at most n nodes, since it can have at most $n - 1$ pointers; 2) the height of each tree is at most $\log_2 n$, (we have already left this as Exercise 60); 3) each Union takes $O(1)$ time; and 4) each Find takes time proportional to the height of the tree, which is at worst $O(\log n)$.

From the preceding assumptions we may conclude that the $n - 1$ Unions take $O(n)$ time and that the m Finds take $O(m \log n)$ time. Since $m \geq n$, the entire sequence of operations takes time $O(n) + O(m \log n) = O(m \log n)$.

This analysis does not give the best possible bound, since we have ignored the path compression that is part of the Find operation. The optimal bound is attributed to Tarjan (1975), who has shown that the worst case time complexity to process m Finds and $n - 1$ Unions in arbitrary order under our assumptions is $\Theta(m\alpha(m, n))$, where α is the spectacularly slow-growing function defined in Exercise 36 of Chapter 2.

Because of the extremely slow growth of α, in practice we may oversimplify and say that under our assumptions the Union and Find operations take total time $O(m)$.

4.10 State Spaces

In this section and the next we expand on the notion of generate and test as discussed in Section 2.1. In the case of optimization problems, we would like to generate all possible solutions for a problem to see which are feasible. For this to make sense, we need some notion of possible solution different from the notion of feasible solution.

We have already seen some examples. In the case of ZOK and the Boolean satisfiability problem, all subsets of a set of n variables are possible solutions. Feasible solutions are those that satisfy the capacity restriction. In the case of TSP, the permutations of the vertices correspond to possible solutions. Feasible solutions are those in which each vertex is adjacent to its predecessor in the permutation.

In both cases we may generate these possible solutions by a sequence of guesses. To generate subsets of a set we may order the set items arbitrarily, and for each element consider whether or not to include it in the subset. To generate permutations of a set we may order the set items arbitrarily, and for each position in the permutation consider all possibilities for the element to occupy that position.

In the second case, there are intermediate states that do not correspond to permutations. For example, if the original set is $\{1, 2, 3, 4\}$ and we decide to include 3 first and then 1, the sequence $(3\ 1)$ is not a permutation of the original set. Yet the sequence $(3\ 1)$ is an intermediate state on the way to generating a permutation such as $(3\ 1\ 2\ 4)$. In order to handle such intermediate states, we make the following definition.

Def 4.11 A *state space* model consists of a possibly infinite set of states. One of these states is an *initial state,* or *start state.* For each state there is a finite, perhaps empty, set of its *successor states.* The set of states has a subset of *goal states.* Goal states have no successors.

For instance, in considering permutations of the set $\{1, 2, 3, 4\}$, each state may be identified with a sequence of elements with no repetitions. The start state corresponds to the empty permutation $(\ )$. Each successor of a state corresponds to adding an unused element to the end of the sequence. For example, the successors of the state $(1\ 3)$ are $(1\ 3\ 2)$ and $(1\ 3\ 4)$. Goal states are those sequences with exactly 4 elements.

In the case of subsets of a set, the start state corresponds to the empty set, in which no choices have been made. For the preceding set, a successor of the start state may be obtained either by including or excluding item 1, giving states corresponding to $\{1\}$ and to $\varnothing$. Notice this latter state is not the same state as the start state, because none of its successors will correspond to a state containing the element 1.

For decision problems, we may identify goal states with solutions. But for optimization problems, more commonly we identify goal states with feasible solutions. Of course finding a feasible solution is not the same as finding an optimal solution, so we eventually need to say more about the use of the state space model in optimization problems.

Often the set of states is not specified in advance, especially if the set is large or infinite. In this case the set of successors of a given state may be implicitly defined by *rules* by which a state can be transformed into a successor state. The productions of context-free grammars are examples of such rules. For context-free grammars, the start symbol corresponds to the initial state and a string over $V \cup T$ to a general state. A string over T corresponds to a goal state.

A second class of examples is given by games. In many board games, for example, an initial board position corresponds to the initial state, and rules describe how to proceed from one board position to another. Certain board positions correspond to wins for one player or another.

A third sort of example arises in artificial intelligence when attempting to automate the process of deduction. We begin with a set of axioms, or assumptions. This set corresponds to the initial state and is sometimes called a *knowledge base.* Then based on rules of inference, we can add new deductions to the knowledge base. If the deduction is intended to prove some hypothesis, then the goal states are those in which the hypothesis appears as a member of the knowledge base (see Exercise 76).

The examples of parse trees and decision trees suggest that tree-based models are useful for understanding state spaces. We make the following definition:

Def 4.12 A *state-space tree* is a tree corresponding to a state space where the root of the tree corresponds to the initial state, the successors of a state correspond to its children in the tree, and the goal states correspond to a subset of the leaves of the tree.

Examples of state-space trees are pictured in Figure 4.23. Figure 4.24 corresponds to an instance of ZOK. Except for the omission of nodes in which the capacity has been exceeded, its shape is typical of state-space trees for generating subsets of a given set. Figure 4.25b corresponds to an instance of TSP with the cost matrix of Figure 4.25a. Its shape is typical of state-space trees for generating permutations of a given set. Note that a TSP solution may begin at any vertex, so we need to consider only $(n - 1)!$ different tours for n vertices. In the case of undirected graphs, half of these are reflections of the other half.

In many cases some of the nodes at lower levels of the tree are indistinguishable except for their position in the tree from states at earlier levels. For example, in many board games, moves may be undone so that a node's successor may represent the same board position as its predecessor. Identifying these

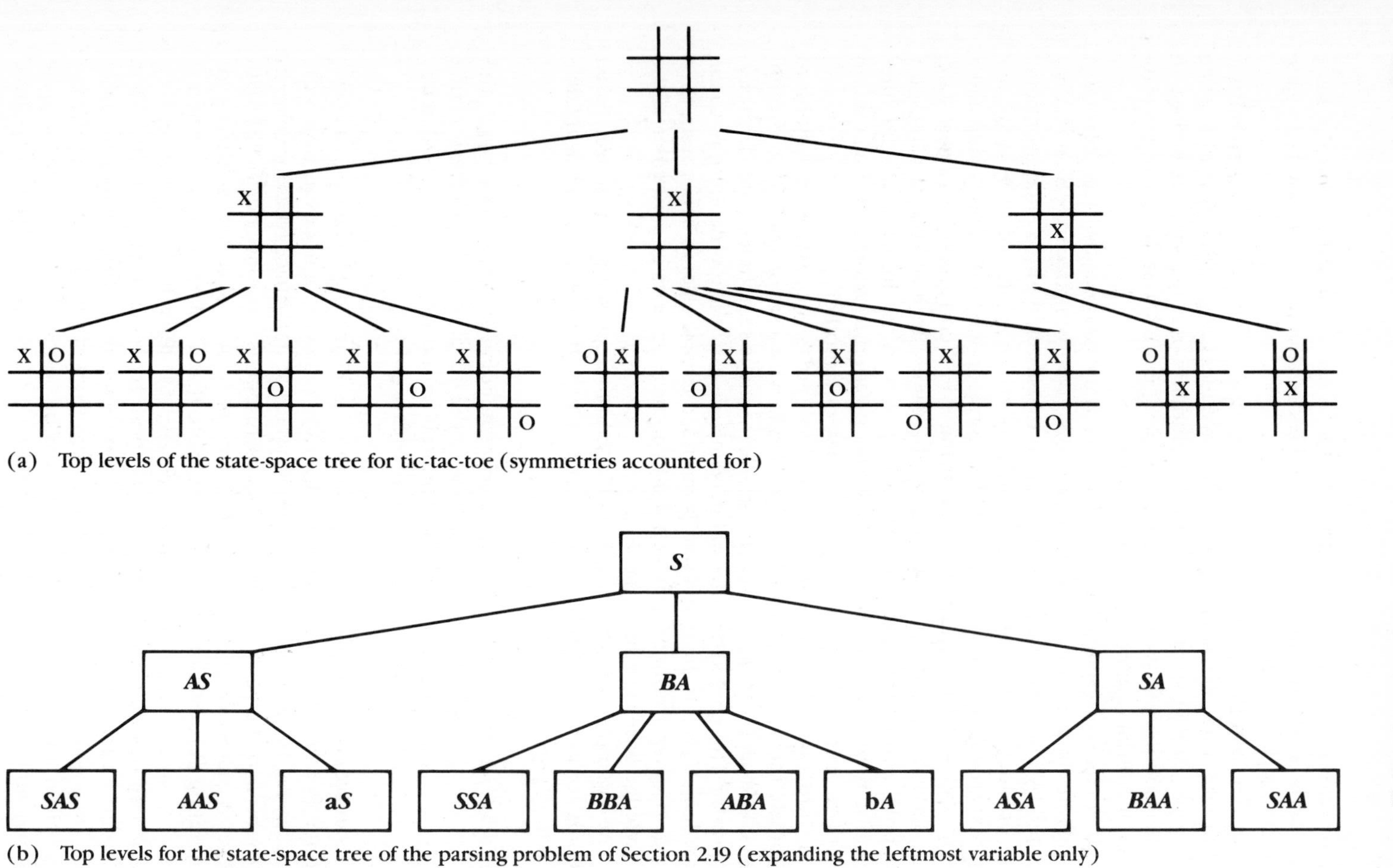

(a) Top levels of the state-space tree for tic-tac-toe (symmetries accounted for)

(b) Top levels for the state-space tree of the parsing problem of Section 2.19 (expanding the leftmost variable only)

Figure 4.23 State-Space Trees

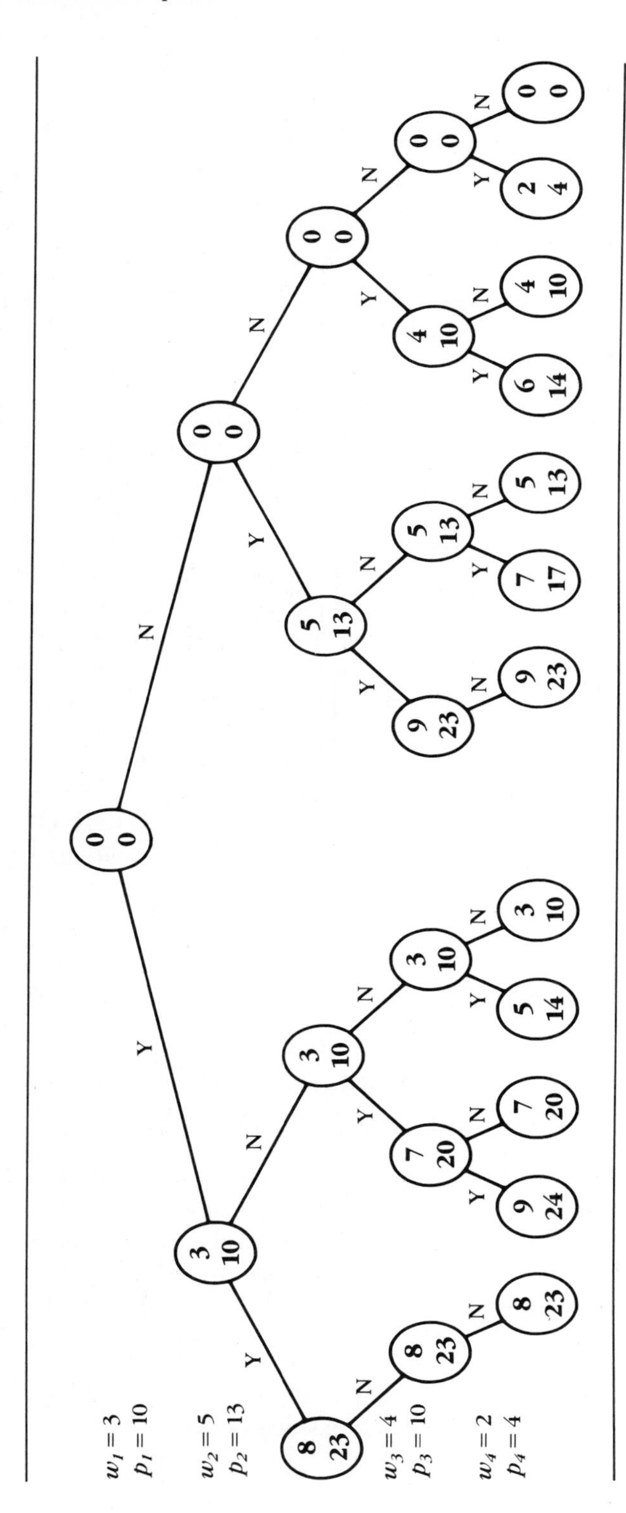

Figure 4.24 State-Space Tree for an Instance of ZOK

$$\begin{pmatrix} 0 & 3 & 5 & 12 \\ 3 & 0 & 6 & 4 \\ 5 & 7 & 0 & 2 \\ 8 & 1 & 4 & 0 \end{pmatrix}$$

(a) Cost matrix for a TSP instance

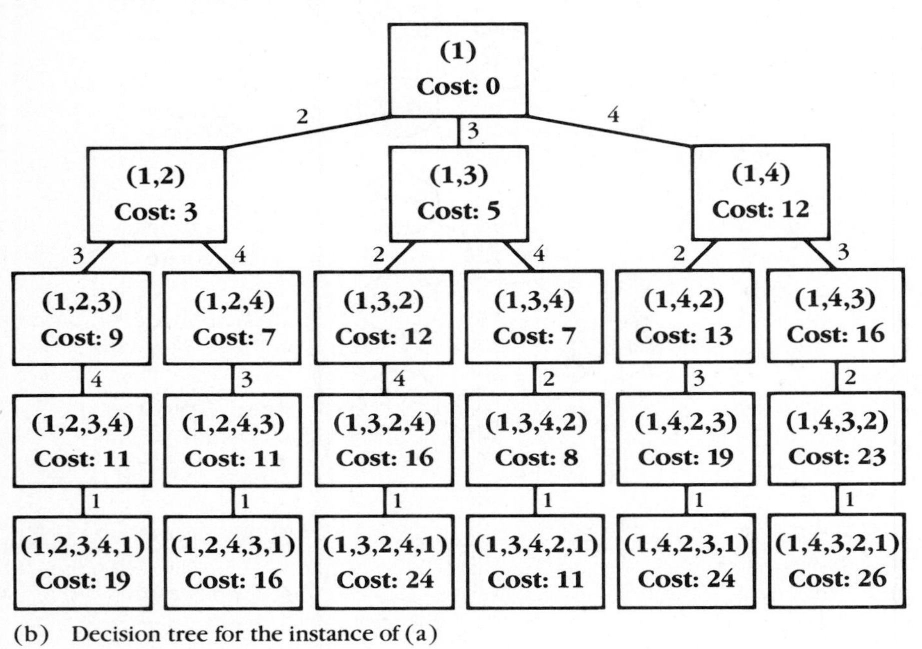

(b) Decision tree for the instance of (a)

Figure 4.25 A Decision Tree for the TSP

indistinguishable states turns the state-space tree into a *state-space graph* that may have cycles. We use the phrase *state space* to include both trees and nontrees.

One solution to the preceding problem is simply to ignore it and treat nodes generated differently as different nodes. However, indistinguishable nodes would be processed in indistinguishable ways, with the result that the same computations could be repeated many times over. Even simply storing multiple copies of indistinguishable nodes can be awkward, since state-space graphs are often very large if not infinite.

On the other hand, determining whether or not two nodes correspond to the same state can be costly. Imagine, for example, the cost of determining whether or not a newly generated chess position has been seen before.

4.11 Searching and State Spaces

We now further refine the generate-and-test strategy with which we began the previous section. Given an explicit or implicit state space, we may begin at the start state and systematically generate new nodes as successors of nodes already generated. That new nodes are being generated systematically is important to insure that no goal node is missed.

We must consider several issues. One issue is whether the problem is a decision problem or an optimization problem. Actually the distinction between the two is not clear cut in practice, since even for decision problems we may want to say that one goal node is better than another. For example, we may want the shortest parse of a string or the shortest path between two vertices of all the cheapest paths between the vertices.

Even for decision problems, finding a single goal node may be sufficient, or we may need to find all goal nodes. There are in fact intermediate cases, such as the "and-or" problems we shall encounter in Section 4.13. If there are many goal nodes, finding all of them may require a traversal of the entire state space.

For optimization problems, in which a goal node corresponds to a feasible solution, finding a goal node in general doesn't end the search. The question becomes, under what circumstances is it possible to be sure that the optimal solution has been found?

In any case we hope to reach a feasible solution quickly. Search strategies that do so are more desirable than those strategies that do not.

One systematic way of generating nodes is to generate and store the successors of the node currently being considered, and next consider some stored node. This node may be a node previously stored; it will not necessarily be one of the new successors. This strategy will also work for graphs, given a way to mark each node as already visited. Then nodes already visited need not be considered or saved. The following terms will be useful:

Def 4.13 To *expand* a node is to generate its successors. An *open* node is one that has been generated but not expanded.

If the successors of the current node are stored in a queue, then we have an analog of the level-order traversal of Section 2.7. In this context the technique is called *breadth-first search.* If we define the level of a node to be the shortest distance to the node from the start state, than each node has a finite level. If each node also has a finite number of successors, then only a finite number of nodes can exist on each level. Thus each node in the state space will eventually be reached even if the state space is infinite. The first goal node reached will be the one at the lowest-numbered level.

During breadth-first search of a finite state-space graph, a *breadth-first spanning tree* may be generated by including in the tree each node when it is first generated along with an edge from its predecessor. A graph and its breadth-first spanning tree rooted at A are given in Figure 4.26. Nodes are numbered in

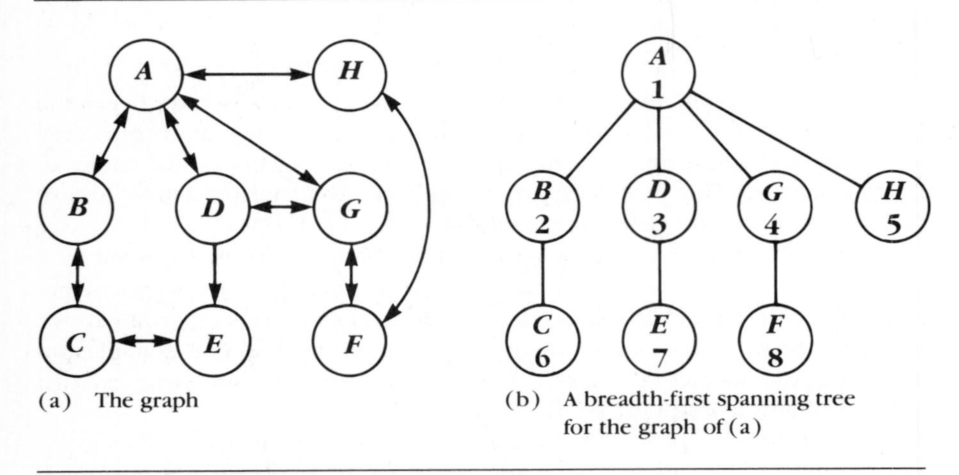

(a) The graph

(b) A breadth-first spanning tree for the graph of (a)

Figure 4.26 A Directed Graph and a Breadth-First Spanning Tree

the order expanded during breadth-first search, that is, the same order in which they are generated. In the figure we assume that child nodes are stored in alphabetical order.

A disadvantage of breadth-first search is that the storage requirements are high. For example, in a completely full binary tree, immediately after consideration of the rightmost node on level k, the queue will contain all 2^{k+1} nodes on level $k + 1$.

We do not present here an algorithm for breadth-first search, since breadth-first search is a special case of a more general searching technique to be discussed later.

If breadth-first traversal is used with a stack instead of a queue the result is not quite analogous to the recursive traversals of Section 2.4. The reason lies in our current search method in which all children are generated at once. If they are immediately pushed onto a stack, they will be popped in reverse order. A recursive traversal such as preorder will process the children one at a time in the correct order. In Figure 4.27a, nodes are numbered in order of generation as they would be if all children were immediately pushed. Nodes of (b) are numbered in the order that they are generated in preorder.

We consider both techniques to be examples of *depth-first search.* In depth-first search a successor of the current node will always be visited as the next node if possible. A version of depth-first search that finds a single solution in the tree rooted at N by generalizing recursive traversal is given as Algorithm 4.14.

In contrast with breadth-first search, depth-first search has small storage requirements; the stack essentially contains only a path from the start node to the current node. On the other hand, a search proceeding forever without visiting a given node is a possibility. This is true even if each node has only a

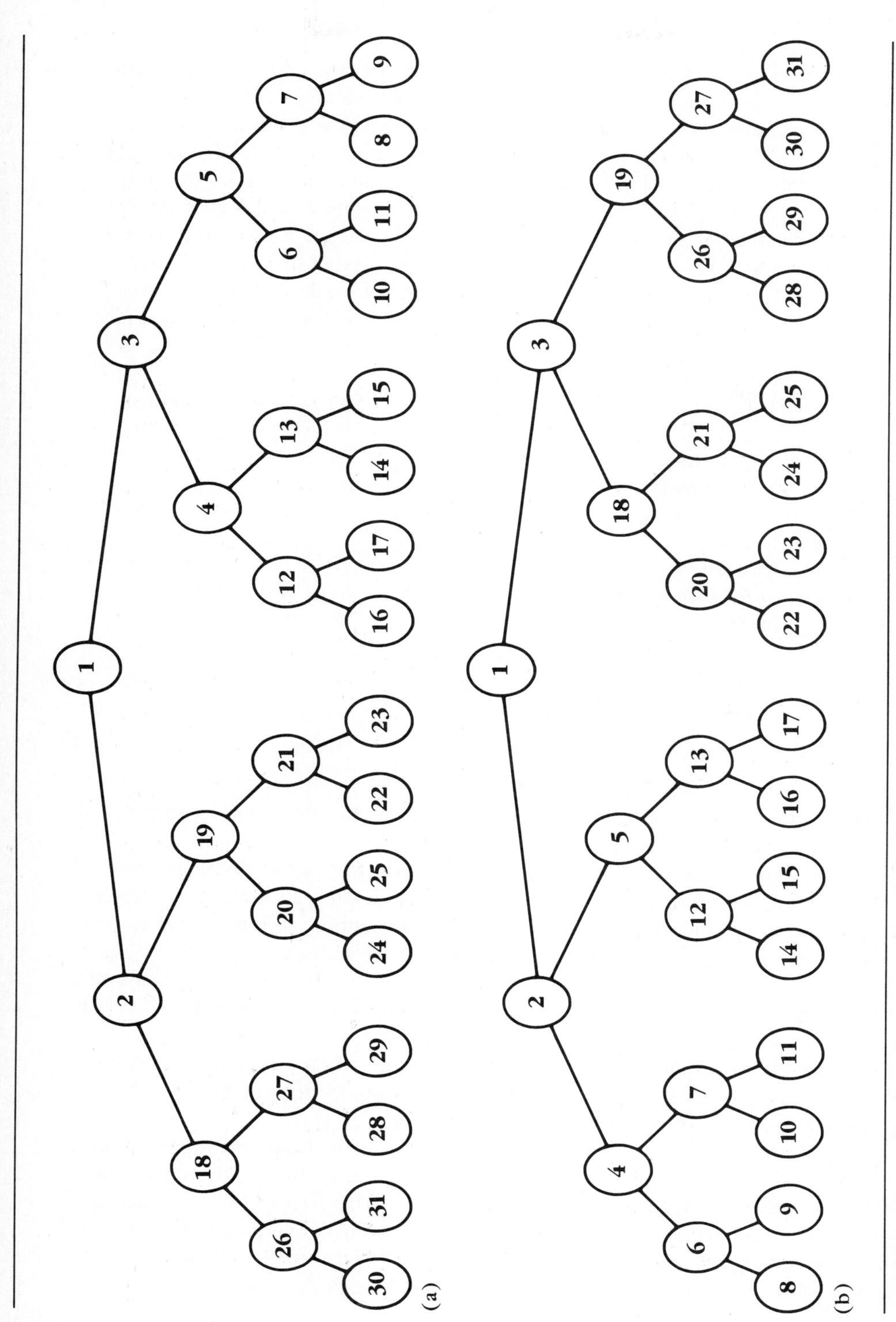

Figure 4.27 Two Methods of Depth-First Search

```
1. RecursiveDepthFirstSearch(N)        {returns the first node in depth-
2.     first order beginning at node N for which Goal is true. Assumes
3.     functions giving the number of successors and the ith successor
4.         for each node as well as a Goal function. Assumes marking
5.   and mark testing functions and initially unmarked nodes. Assumes
6.       a function returning the number of children of a given node}
7. answer := dummy                {dummy value means no solution so far}
8. mark(N)                     {make sure not to visit or expand N again}
9. if Goal(N) then                     {if root is goal, don't try children}
10.    return N                          {could set answer to N instead}
11. else
12.    for i from 1 to NumberOfChildren(N)
13.       if answer = dummy then               {don't look for 2d solution}
14.          M := Child(N, i)
15.          if not marked(M) then
16.             answer := RecursiveDepthFirstSearch(M)
17.          end if
18.       end if
19.    end for
20.    return answer
21. end if
```

Algorithm 4.14 Recursive Depth-First Search

finite number of successors. For example, the node marked N in Figure 4.28 will never be visited.

Depth-first search will give a *depth-first spanning tree* in the same way that breadth-first search will give a spanning tree. In Figure 4.29 we show the state-space graph of Figure 4.26a and its depth-first spanning tree rooted at node A. Here we assume children are generated one at a time, as shown in Figure 4.27a. Again we show the order in which nodes are expanded and assume that children are generated in alphabetical order. In this case the order of expansion is not the same as the order of generation.

One characteristic of depth-first search is that when a node has no successors, the next node is obtained by unmaking one or more of the most recent decisions. This process is called *backtracking*.

So far we have not considered the possibility that the nodes might contain information to guide the search. Searches in which this happens are called *informed searches;* searches in which this doesn't happen are called *blind searches.* Such information might suggest, for example, that a particular successor is most likely to lead to an optimal solution or most likely to lead quickly to a feasible solution. Following the most promising path is not likely to give a

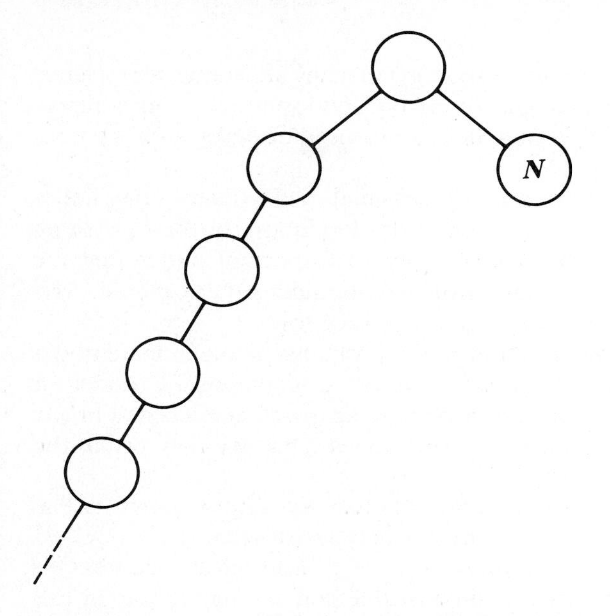

Figure 4.28 A Tree That Cannot be Traversed in a Depth-First Manner

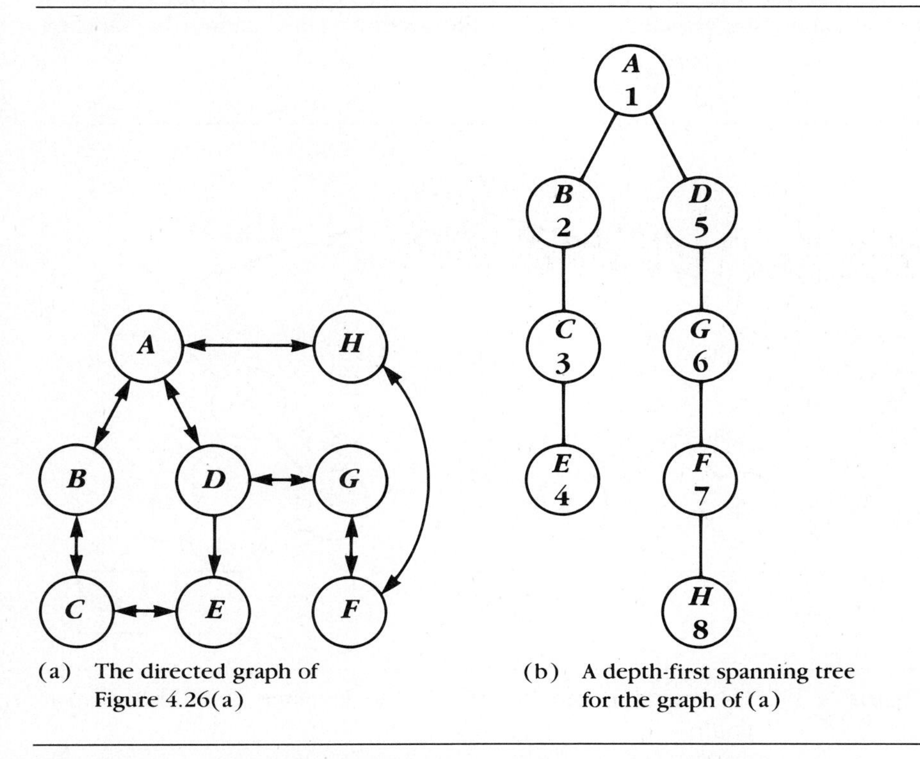

Figure 4.29 A Directed Graph and a Depth-First Spanning Tree

pure depth-first or a pure breadth-first strategy. In many situations, for instance in games like chess, quickly finding a reasonably good solution is more important than finding the optimal solution. In these cases informed search is particularly important.

The informed strategies we present are designed to avoid inspecting nodes that cannot lead to a solution (or to an optimal solution, in the case of an optimization problem) and give low priority to inspecting nodes that are unlikely to lead to a solution. Ignoring or deleting nodes that cannot correspond to solutions is called *pruning* the state-space tree.

One simple way of pruning in optimization problems is to eliminate nodes that cannot correspond to a feasible solution. For example, in the model for ZOK discussed previously, if the capacity of the knapsack is exceeded in any node, it will be exceeded in all descendant nodes. Thus we may prune the entire subtree rooted at the node.

An example of pruning in conjunction with backtracking is shown in Figure 4.30. The problem is Boolean satisfiability and the instance is $(x_1' \vee x_2') \wedge (x_2 \vee x_3) \wedge x_3' \wedge x_4'$. Once one conjunct (called a *clause* in this context) is unsatisfiable, the entire expression is unsatisfiable and we may prune. In this example an assignment of "true" to the next variable corresponds to the first child, and an assignment of "false" corresponds to the second child. Nodes in the figure are numbered in depth-first order.

At the node 3 the first clause cannot be satisfied; at nodes 5 and 9 the third clause cannot be satisfied; at node 6 the second clause cannot be satisfied.

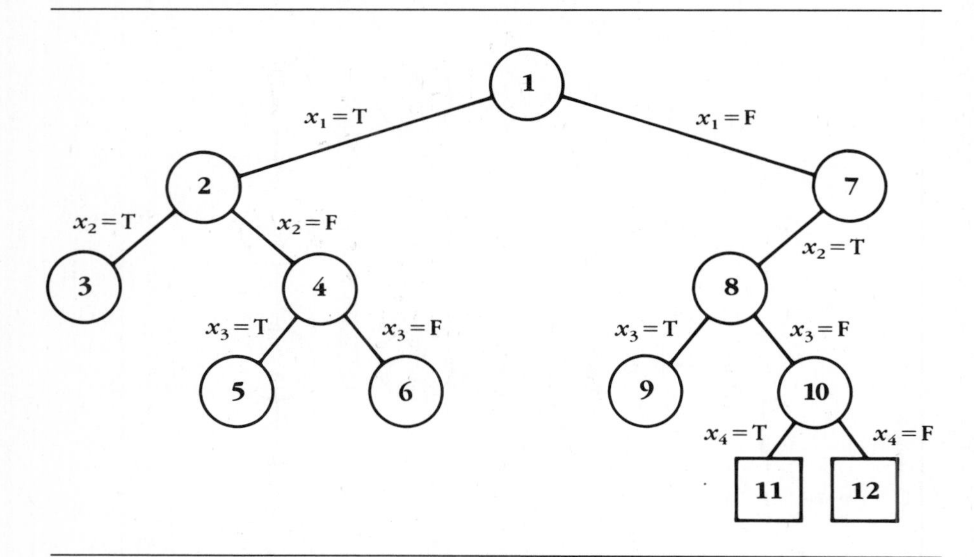

Figure 4.30 Backtracking and Pruning for an Instance of Boolean Satisfiability

Therefore we may prune at all of these nodes. Node 11 is not a goal node, but node 12 is, and that is the last node visited.

A second way of pruning in optimization problems is to eliminate nodes that cannot correspond to an optimal solution. We can sometimes accomplish this by comparison with the value of a feasible solution already reached. Methods that prune based on values calculated from partial solutions are often called *branch-and-bound* methods. Depth-first search is valuable in this regard because early on it tends to reach goal nodes, or at least nodes with no successors. Even if no feasible solutions have yet been reached, as is common in breadth-first search, finding a lower bound for the value of a solution to a maximization problem or an upper bound for a minimization problem may be possible. If an upper bound for solutions in the subtree rooted at a given node can be found that is lower than the given lower bound, then once again the subtree can be pruned.

Finally, the order of the traversal can be chosen to maximize the chance that a good solution can be found quickly. We now consider these issues in the context of ZOK.

An instance of ZOK has a useful property: there is a natural upper bound for the value of the solution, an upper bound likely to be very close to the actual value. This upper bound is simply the value of the solution of the continuous knapsack instance with the same capacity, weights, and profits. Since any feasible ZOK solution is also a feasible continuous solution, the value of the optimal continuous solution must be at least as large. Yet the optimal ZOK solution can be expected to contain most of the items with the highest profit densities, so we may expect the upper bound to be very good.

By analogy with the greedy algorithm for the continuous case, we would expect to reach a good, if not optimal, feasible solution for ZOK very quickly by considering the items in order of nonincreasing profit density and by considering inclusion of each item before exclusion using depth-first search (see Exercise 157 of Chapter 2).

We may also use this approach to find a good, but not necessarily optimal, continuation of a partial solution (i.e., a descendant of an internal node). Starting with the remaining items and remaining capacity, consider the items in order of profit density. If the next item fits, include it; otherwise omit it. This continuation of the partial solution will automatically be feasible and therefore will be a lower bound on the optimal solution. The best such lower bound seen so far may be used for the second method of pruning discussed under branch-and-bound. A possible disadvantage of this approach is that $O(n)$ additional work is required per node.

Thus we may proceed as follows for ZOK:

1. Order the items by nonincreasing profit density
2. Order the children of each node so that including an item is considered before excluding it
3. Maintain as a lower bound the value of the best solution seen so far or the maximum of the lower bounds corresponding to each node

4. Before expanding a node, get an upper bound for the best extension of the associated partial solution by solving the continuous knapsack instance for the remaining items and capacity and add the value of this solution to the profit achieved so far. If this sum is less than the value of the current lower bound from part 3 on an optimal solution, or if the capacity has already been exceeded, then do not consider its children.

Now the depth-first search algorithm may be used with a good chance that the optimal solution may be found quickly. (Notice that the state-space tree has height n, where n is the number of items to be considered.) A search technique that can be better is to expand the unexpanded node with the largest upper bound as defined in part 4. This technique, called ***best-first*** search, is likely to require expanding fewer nodes, but it does require a heap or some other dynamic data structure for repeatedly finding the maximum. The operation of finding the maximum cannot be expected to have time complexity $O(1)$ even in the average case.

Note that the greedy algorithms for finding the upper and lower bounds have time complexity $O(k)$ if k items remain. To be worthwhile then, they need to prune a relatively large fraction of the nodes. On the other hand, we shall see that these estimates need not be recomputed for every node.

As an example, consider the ZOK instance with $n = 6, M = 18$, weights 6, 10, 3, 8, 5, and 4, and profits 20, 31, 9, 21, 13, and 10. These items have already been sorted by profit density. If the only nodes pruned are those not leading to feasible solutions, then 80 nodes will be considered, including 34 leaves of the possible 64.

Depth-first search, with pruning by comparing the local upper bound of part 4 with either of the local bounds of part 3, requires considering the 27 nodes and 5 leaves of Figure 4.31. Each node in the figure is marked on the outside with the order in which it was generated, on the inside with the weight and profit so far, and (except for leaves) with the lower and upper bound associated with the node.

Best-first search would consider just 17 nodes and 2 leaves, as shown in Figure 4.32 (with the same labeling as the previous figure). Best-first search terminates when we find a better solution than the upper bound associated with any unexpanded node. In the example, a solution with profit 53 is found, and at that time no unexpanded node can lead to a solution more profitable than 52.6.

This performance can even be improved based on three observations. First, if the greedy continuation of a partial solution does not include any new nodes, then the lower bound it has found is also an upper bound. In our example, this occurs at the node numbered 4 of Figure 4.32; its actual profit is 51. Since no other item will fit into the knapsack, we may conclude at this point that the best possible continuation of this partial solution also has profit 51, as can also be seen from the figure. Consequently, the subtree rooted at this node can be pruned.

Second, if the upper bound found by the continuous knapsack algorithm involves including each item entirely or not at all, then the upper bound is also

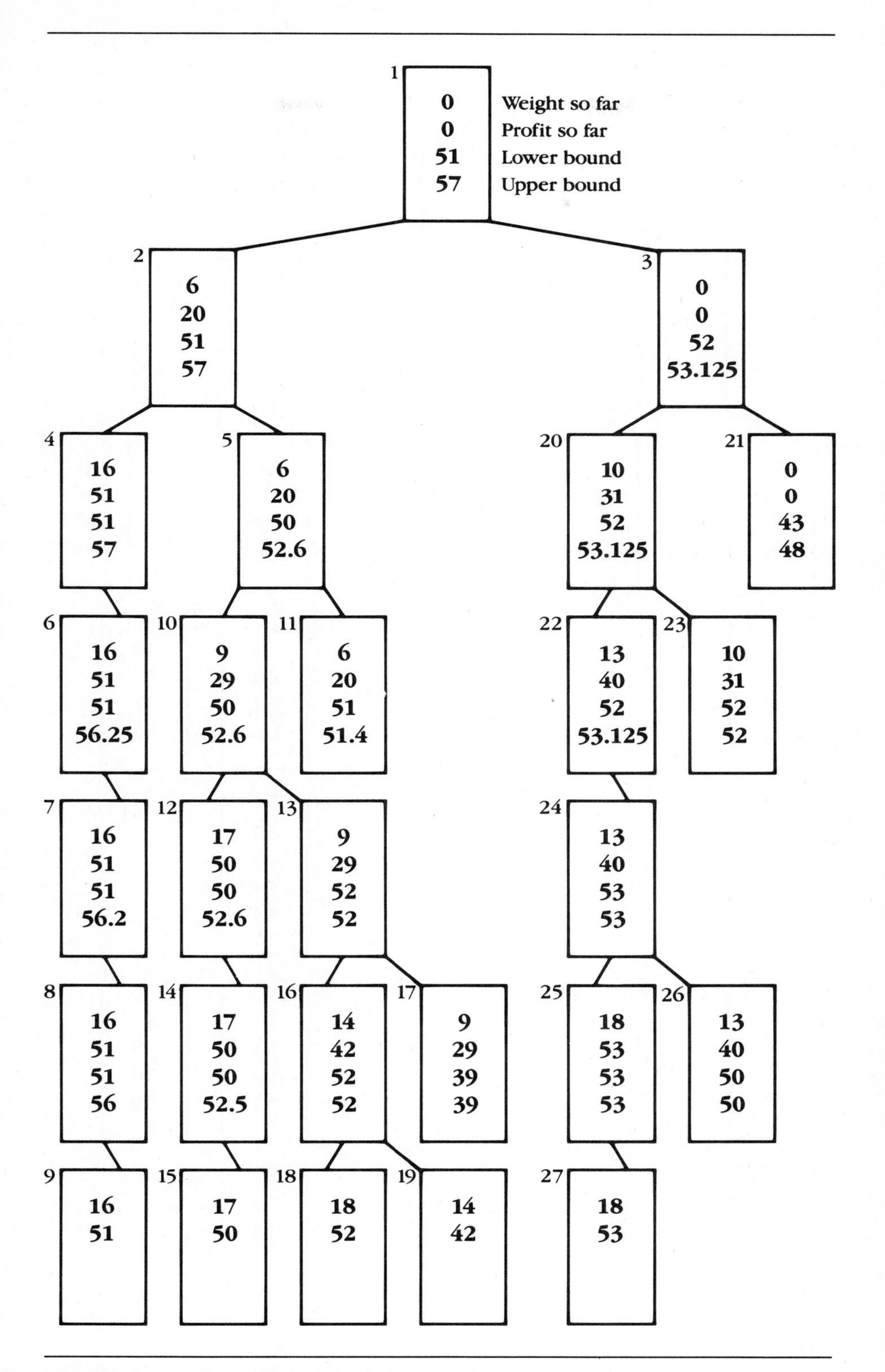

Figure 4.31 Depth-First Search with Pruning

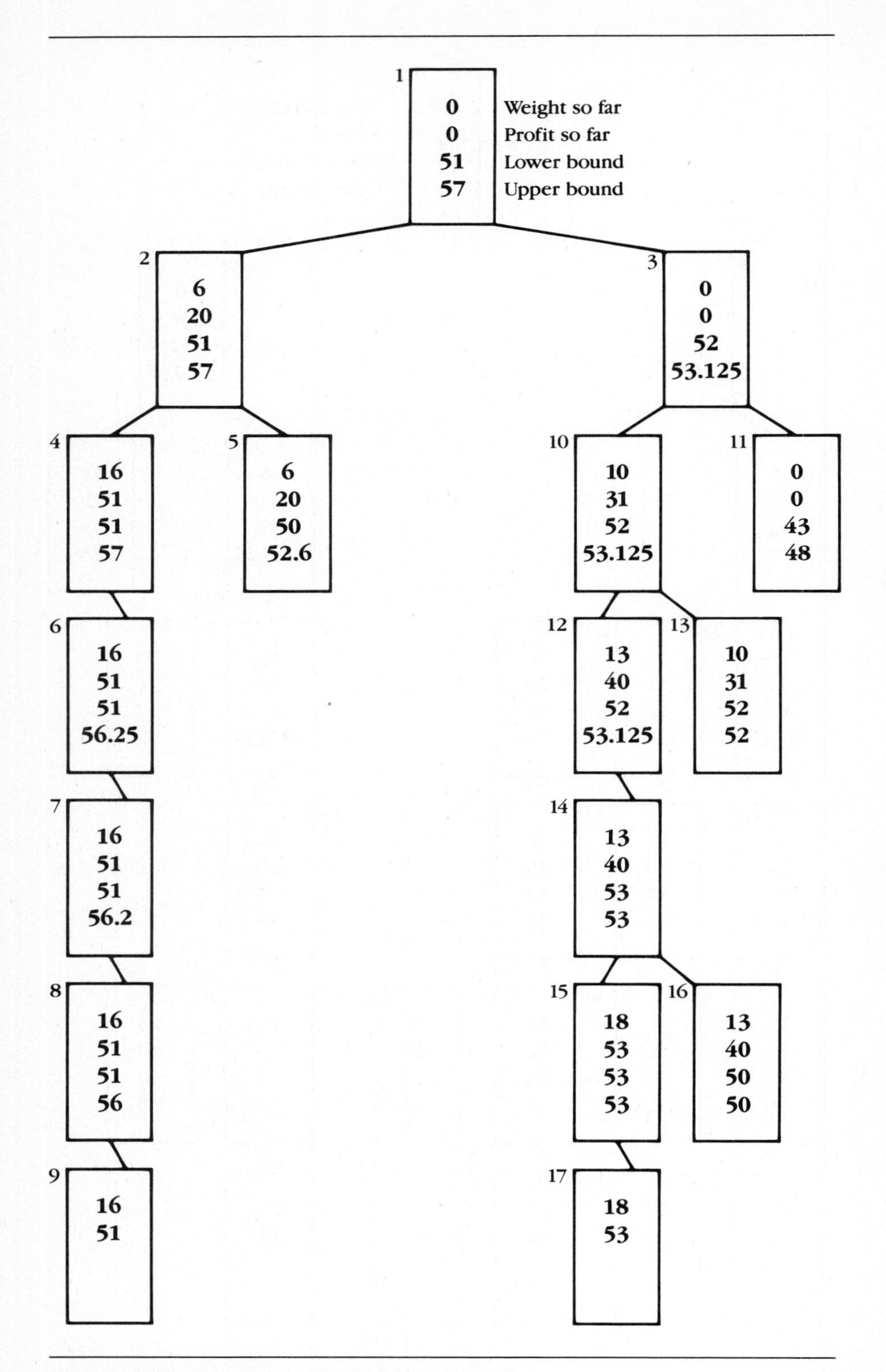

Figure 4.32 Best-First Search with Pruning

a lower bound, and the tree may be pruned. In Figure 4.32, this applies to node number 12 with upper bound 53 and lower bound 40. The upper bound of 53 corresponds to including all of the item with profit 13 and none of the item with profit 10, which is a feasible solution to ZOK as well.

Finally, if only one solution is found and the profits are known to be integers, pruning can take place even if the lower bound is below the upper bound. For example, if when considering expansion of node 11 with weight 51.4 of Figure 4.31, the best solution so far was 51 instead of 52, there could not be a better solution accessible from the node.

In the preceding example it made no difference which lower bound was used in part 3. This is not always true. If the greatest lower bound of all the nodes is used instead of the value of the best solution seen so far, it is possible to prune even before reaching a single leaf. Figure 4.33 corresponds to the ZOK instance $n = 8, M = 21$, weights 15, 10, 9, 5, 5, 5, 5, and 5, and profits 29,

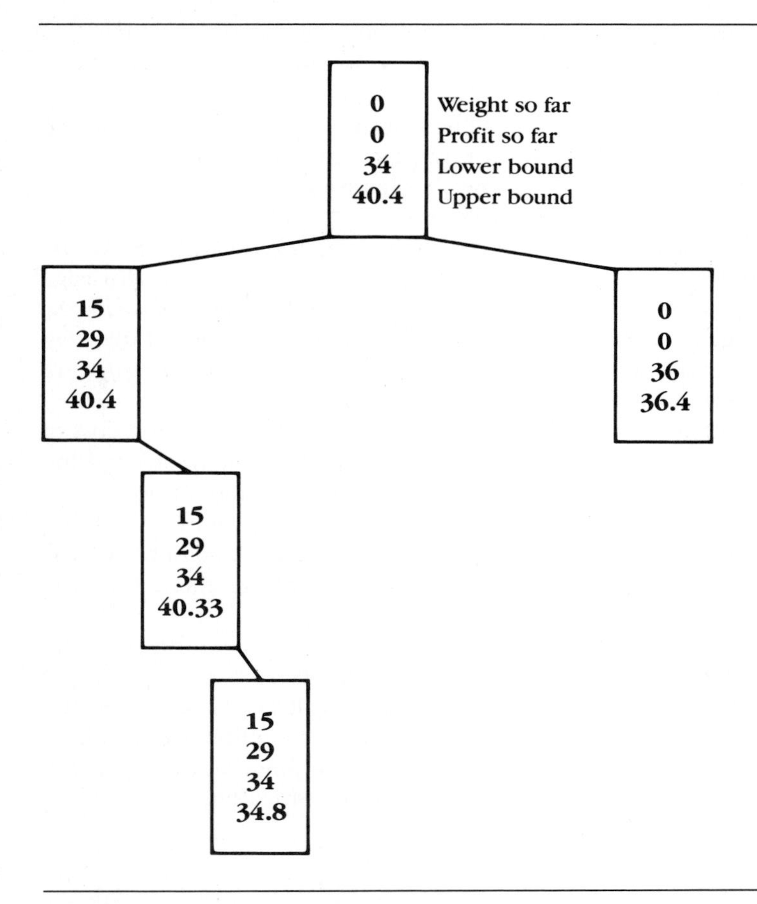

Figure 4.33 Early Pruning in Best-First Search

19, 17, 5, 4, 3, 2, 1. Here if the item of weight 15 is included, then the algorithm discovers that the items of weights 10 and 9 must be excluded. Now the current path may be abandoned, since the best continuation has profit 34.8, and the right subtree guarantees a solution of profit 36.

General Search

We have seen informally that both depth-first search and breadth-first search are special cases of an algorithm that repeatedly stores successors and removes single nodes from storage. In order to use a neutral term for the data structure that supports this insertion and deletion, we will call it simply the *set of open nodes.* The information implied by the term *informed search* is used to remove the most promising nodes from the set of open nodes.

More precisely, we assume an estimating function **h** defined on the set of nodes. The general search algorithm that uses such an estimating function is called the A* algorithm, attributable to Hart, Nilsson, and Raphael (1968).[7] The A* algorithm applies to both decision problems and optimization problems. (A fairly but not completely general version of this algorithm, given as Algorithm 4.15, will be discussed later.)

By symmetry, we may assume that all optimization problems are minimization problems; we therefore speak in terms of minimizing a cost function. Maximization problems like ZOK can be handled by subtracting the function to be maximized from a suitably large constant.

The A* algorithm assumes that the cost to be minimized is the cost of the path to the goal node. This may seem unduly restrictive, but generally edge costs can be defined in the state-space graph to make this true. For example, the cost of an edge in the state-space tree for ZOK would be zero if the item was not included and the item's profit otherwise (or actually the negative of this, since ZOK is a maximization problem).

With this assumption, the cost of a node has two components: the cost of the decisions made to get to the node (i.e., of the path to the node) and the cost of the decisions needed to reach a goal node (i.e., of the cheapest path from the node to a goal node). Using Nilsson's notation, we let $f^*(N)$ be the actual cheapest cost of a solution reachable from N. It is the sum of $g^*(N)$, the cheapest cost of reaching N from the start node, and $h^*(N)$, the cost of the cheapest path from N to a goal node.

When N is being expanded, the value of $h^*(N)$ is not known but must be estimated, say, by a function h. We may assume that $h(N) = 0$ if N is a goal node. The function h will depend on the problem; it is not a part of the A* algorithm. Altogether, the algorithm assumes the following problem-dependent information: 1) a start state, 2) a successor function, 3) a function to determine whether a state is a goal state, and 4) an estimating function h for all nongoal nodes.

---

[7] See also Hart, Nilsson, and Raphael (1972), Gelperin (1977), and Nilsson (1980).

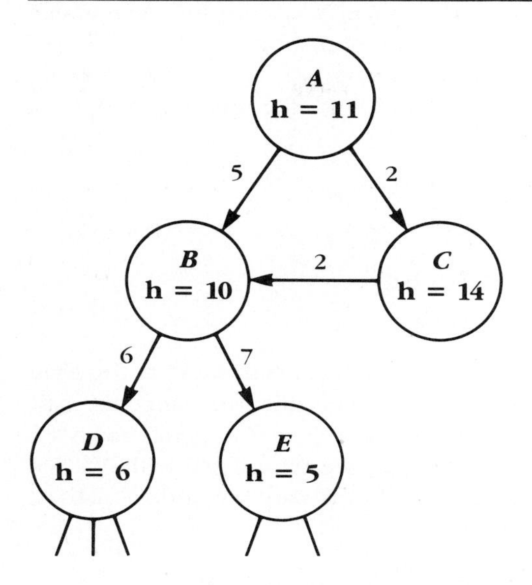

Figure 4.34 Portion of a State-Space Graph for Which g* Cannot Be Estimated Exactly

If h is equal to 0 and g is equal to the distance from the root, then the nodes will be expanded in order of distance from the root. This is simply breadth-first search. To simulate depth-first search, f would need to decrease as the distance from the root increased. For example, $g(N)$ may be 0 and $h(N)$ may be some large constant less the level of N.

One further complication is that although in general we may estimate $g^*(N)$ by $g(N)$, this estimate may not be exact when N is being expanded. An example is shown in Figure 4.34. Here $f(B) = 5 + 10 = 15$ and $f(C) = 2 + 14 = 16$, so B will be expanded first. Its children will be expanded after C, since $f(D) = 5 + 6 + 6 = 17$ and $f(E) = 5 + 7 + 5 = 17$. When C is expanded, a cheaper path will be discovered to B, although B is no longer open. In other words $g^*(B) \leq 4$, although $g(B)$ was 5 when B was expanded.

Cases exist when we may be sure the estimate of $g^*(N)$ by $g(N)$ is exact. One case would be when the state-space graph is a tree, so that there is only one path to N.

The algorithm will certainly terminate if the state space is finite. It will also terminate for infinite state spaces if for every M there are only a finite number of nodes with f value less than M. This case could happen if there was a constant c with $g^*(N)$ greater than the constant times the level of N, since $f \geq g \geq g^*$. This in turn could happen if the edge costs were bounded below by a constant.

We do not yet clearly know how the algorithm could work correctly on optimization problems. We hope that for optimization problems the first feasible solution found would be optimal. This hope is actually realized under certain reasonable assumptions for appropriate estimating functions h.

How could the first goal node found be nonoptimal? If N^* is an optimal solution node and N the first feasible solution found, the nonoptimality of N means

$$f^*(N^*) < f^*(N)$$

Since N is a goal node, we have

$$f^*(N) = f(N) \text{ or}$$

$$f^*(N^*) < f(N)$$

Now let us see what is happening along the optimal path to N^* at the time when N is removed. Some open node M must lie on this path, since the first node (i.e., the root) on the path is initially open and since if all nodes had been considered and removed, the algorithm would have terminated with N^*. We may assume that M is the first such open node. In this case the optimal path to M has already been found, which means

$$g(M) = g^*(M)$$

Since N was removed before M, we have

$$f(N) \leq f(M)$$

So by the last inequality of the previous paragraph, we have

$$f^*(N^*) < f(M)$$

Now all nodes on the path to N^* have the optimal value $f^*(N^*)$ of f^*, so

$$f^*(M) = f^*(N^*)$$

Combining with the previous inequality, we have

$$f^*(M) < f(M)$$

And since $g^*(M) = g(M)$, we may subtract this expression from both sides of the inequality to get

$$h^*(M) < h(M)$$

Only when an inequality of this form holds will we get a nonoptimal node as the first goal node.

In other words, if we insist that $h(M) \leq h^*(M)$ for all nodes M, then the A* algorithm will work for optimization problems as well. In fact, this condition on h is usually made part of the definition of the algorithm. Since for maximization problems h must be an overestimate, in general we may say h should be an optimistic estimate. (See Figure 4.32.)

There is no provision in the A* algorithm, like that used in the example of Figure 4.31, for discarding newly generated nodes before they are inserted into the set of open nodes. Adding such a provision is left as an exercise.

```
 1. AStar(start)                    {searches a state space rooted at start.
 2.             Assumes an optimistic evaluator h, a goal-testing function
 3.                goal and functions children and child that return the
 4.                number of children and the ith child respectively for a
 5.             given node. Assumes create, destroy, insert, and delete
 6.                        functions for sets of open nodes. Assumes that
 7.             predecessors of nodes can be assigned by pred. Assumes
 8.                   that any pointer reassignment is done at a low level;
 9.                        see text. Returns a goal node if there is one.
10.             The path to it may be recovered through the pred fields}
11. create(opennodes)                              {create a set of open nodes}
12. insert(start, h(start), opennodes)             {make the start node open}
13. while not empty(opennodes)                     {quit if no node is open}
14.    nextnode := delete(opennodes)               {else get the next node}
15.    if goal(nextnode) then                      {if it's a goal}
16.       destroy(opennodes)
17.       return nextnode                          {we're done}
18.    else                                        {otherwise}
19.       for i := 1 to children(nextnode)         {put the children}
20.          nextchild := child(nextnode, i)       {in the openlist}
21.          pred(nextchild) := nextnode           {with predecessor links}
22.          hval := h(nextchild)
23.          insert(nextchild, hval, opennodes)
24.       end for                                  {and continue}
25. end while                           {if while loop ends with no solution}
26. destroy(opennodes)
27. return unsuccessfully                          {then return empty-handed}
```

Algorithm 4.15 A* (See Assumptions in Text)

Algorithm 4.15 returns not just the first goal node found, which is necessarily the best, but also links from each node to its predecessor. By tracing these links when the goal node is found, the path to the goal node may be recovered in reverse.

If the state-space graph is not a tree, then finding a better path to a node is possible (see Figure 4.34). If the node is an open node, the predecessor pointer must be redirected. Algorithm 4.15 does not explicitly address this issue or the issue of how to recognize whether or not a successor has already been generated. If the node is no longer open, then not only would the predecessor pointers need to be redirected, but the node may have successors, and the predecessor pointers of these nodes may need to be updated. These modifications are left as an exercise.

Fortunately there is a common special case in which nodes which are no longer open need not be considered. To describe this case, we need a definition.

Def 4.16 Let c be the cost function for edges in a state-space graph. The function h is *monotone* iff $h(M) \leq h(N) + c(M, N)$ whenever N is a successor of M.

Monotonicity of h means that as we move away from goal nodes, where $h = 0$, the function h doesn't grow too large too fast. The following theorem deals with the general version of A*. The theorem implies that only open nodes need be reconsidered if better paths are found to them. Its statement and proof are essentially those of Nilsson (1980).

Theorem If h is monotone, then at the time a node N is being expanded, the best path to N has been found.

Proof Consider a lowest-cost path to N. Let M be a node on this optimal path and P its successor on the path. Then by definition of g* we have

$$g^*(M) + c(M, P) = g^*(P)$$

By monotonicity, we have

$$g^*(M) + h(M) \leq g^*(M) + h(P) + c(M, P)$$

Combining these two inequalities, we get

$$g^*(M) + h(M) \leq g^*(P) + h(P)$$

Since M and P are arbitrary on the path, we get that $g^* + h$ is a nondecreasing function along the path. In particular, for any M on the path,

$$g^*(M) + h(M) \leq g^*(N) + h(N)$$

Being even more particular, we may consider the time when N was expanded and choose M to be the first open node along the optimal path at that time. Since M's predecessors along the optimal path have all been expanded, we have

$$g(M) = g^*(M)$$

But since $M = N$ or N was selected before M, we have

$$f(N) \leq f(M)$$

or

$$g(N) + h(N) \leq g(M) + h(M)$$

or, using the previous equation,

$$g(N) + h(N) \leq g^*(M) + h(M)$$

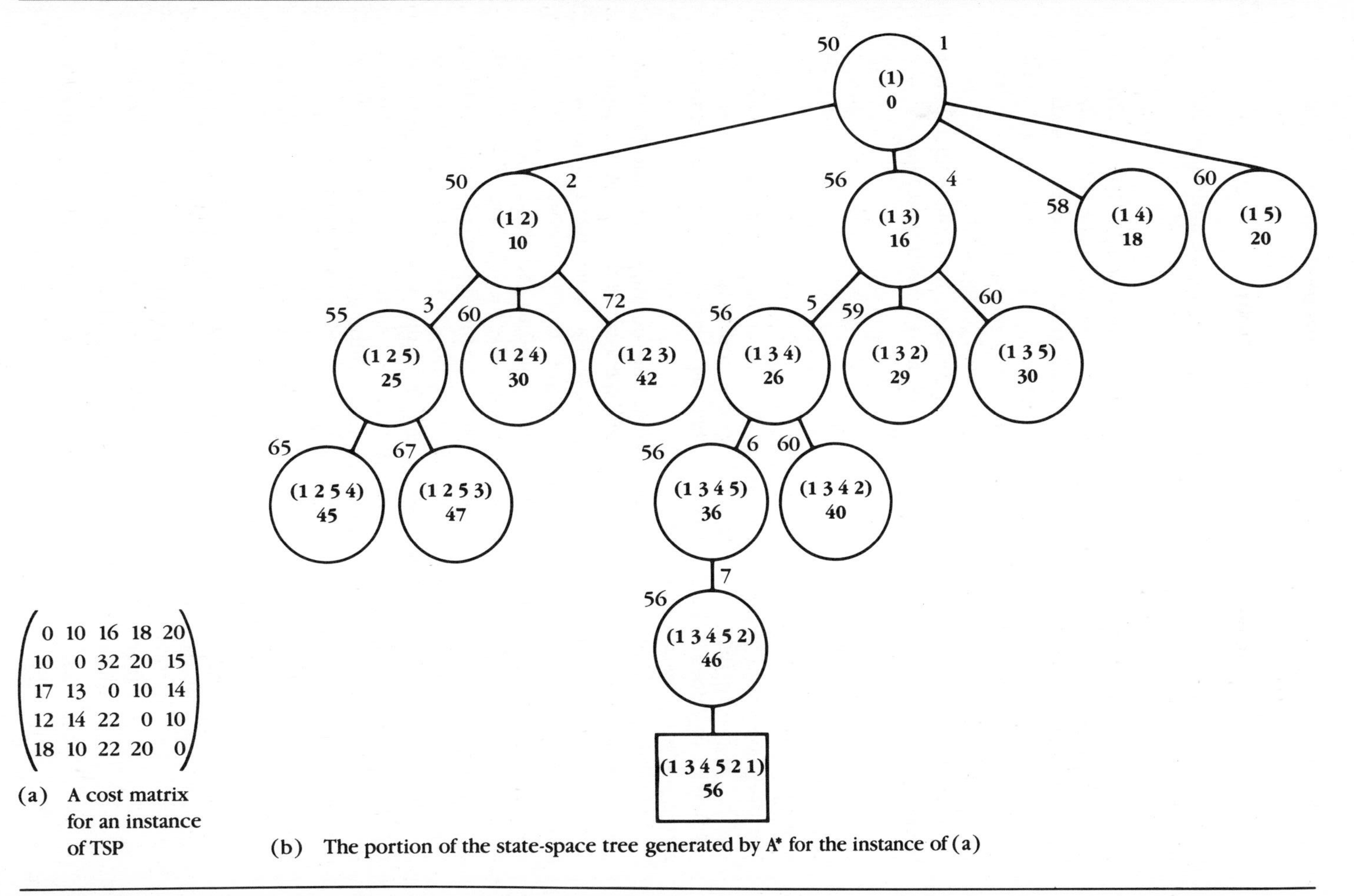

(a) A cost matrix for an instance of TSP

(b) The portion of the state-space tree generated by A* for the instance of (a)

Figure 4.35 Trace of the A* Algorithm on an Instance of TSP

We already have an inequality involving this right-hand side; using the inequality and transitivity, we get

$$g(N) + h(N) \leq g^*(N) + h(N)$$

Thus $g(N) \leq g^*(N)$. Since we always have $g(N) \geq g^*(N)$, we must have $g(N) = g^*(N)$, which is what we want. ■

A trace of the A* algorithm for TSP is shown in Figure 4.35. The instance is given by the cost matrix of (a). The portion of the state-space tree generated by A* is given in (b). Each node is labeled on the left with its f value; on the right with the order in which it is expanded; and inside with the corresponding path and the value of g, which here equals g*. The cheapest tour is (1 3 4 5 2 1), the first one completely generated. It has cost 56.

Rich (1983) covers A* and other search techniques from the artificial intelligence perspective, and Pearl (1984) covers many search techniques.

4.12 Games and Game Trees

We have suggested in previous sections that games like tic-tac-toe, checkers, chess, or go can be fit into the state-space model. However, some special properties of games and *game trees* (i.e., state-space trees representing games) deserve some special discussion.

Although games may be more general, in this section we assume that each game has two players who alternate moves. We assume that when the game is over, each player receives a payoff depending on the final state of the game. We also assume that the game is a *zero-sum* game, that is, the sum of the two payoffs is zero. We need only consider the payoff to the player moving first, since the payoff to the second player will simply be the negative of this payoff.

The root of the game tree corresponds to the initial state of a game. Other nodes correspond to other states; leaves correspond to final states. Children of a node correspond to those states resulting from the parent state by performing a legal move. (Notice the analogy with parse trees—it's the rules that define the game.) We want to know the value of the game (i.e., the payoff) to the first player if both players play according to their best strategy.

Def 4.17 The *value* of a node in a game tree to the first player in a two-person, zero-sum game is

1. the given value of the node, if the node is a leaf,
2. the maximum of the values of its children, if the node corresponds to a move by the first (maximizing) player,

2′. the minimum of the values of its children, if the node corresponds to a move by the second (minimizing) player.

The *value* of the game to the first player is the value of the root.

An example of a game tree is shown in Figure 4.36a. In this tree, only the leaves have known values. In (b) of the figure, the other nodes have been shown with the values they are given by the definition above. The value of the game to the maximizing player is $+1$, the value in the root. Note that this solution corresponds to one of the leaves in the tree. All nodes on the path from the root to this node have value $+1$.

If the game tree is infinite or even simply very large, then generation and inspection of the complete game tree is impossible. In this case an approximate evaluation may be done with the use of an evaluation function. Sometimes this function is called a *static* evaluation function, since it does not depend on other nodes in the tree. Contrast this with the *dynamic* evaluation function of Definition 4.17.

Nodes that are impractical to evaluate dynamically may have their values estimated statically. Part 1 of Definition 4.17 may be augmented to return the static evaluation function in these situations.

For example, depth-first search may be performed down to a fixed level. Any nodes on this level would be evaluated statically. It is possible that the estimate would be inexact enough to lead to the selection of an inferior move. This selection need not be fatal, though, since the opponent is laboring under a similar handicap (although with the same amount of work the opponent, starting one level deeper, can look one move further ahead).

Estimating functions may also be used to control the search as in best-first search.

Pruning a game tree is also possible. Consider for example a depth-first evaluation of the game tree of Figure 4.37a. The minimizer will choose 5 as the value of node B. The maximizer will choose 6 and 3 as the respective values of nodes D and E. Given 3 as a choice for the value of C, the minimizer will choose a value at least that small. But a value for the right subtree of 3 or smaller will not survive the maximizing stage at node A, since the maximizer already has the value of 5 to choose from. Thus there is no point to evaluating F or any additional siblings of the node with value 3.

To make this argument systematic, it helps to take the view of the current player rather than the maximizing player. From this point of view, the tree of Figure 4.37a becomes that of (b) and all levels are now maximizing levels. At any point in the evaluation of a node, there are upper and lower bounds on its final value. The lower bound is given by the value of its subtrees; the upper bound is imposed by the earlier choices and strategy of the opponent.

Historically these bounds were called *alpha* and *beta*, so this sort of pruning is called *alpha-beta pruning*. Algorithm 4.18 is an alpha-beta pruning algorithm taking the point of view of the current player. Initially the lower and upper bounds are $-\infty$ and $+\infty$. The value of a leaf is assumed to be available as in the basis step of Definition 4.17. Otherwise any improvement in the lower bound due to higher values of subtrees is taken, but if the lower bound becomes greater than the upper bound, the evaluation of the node halts. The value of the lower bound is returned merely as a convenience; since it is too

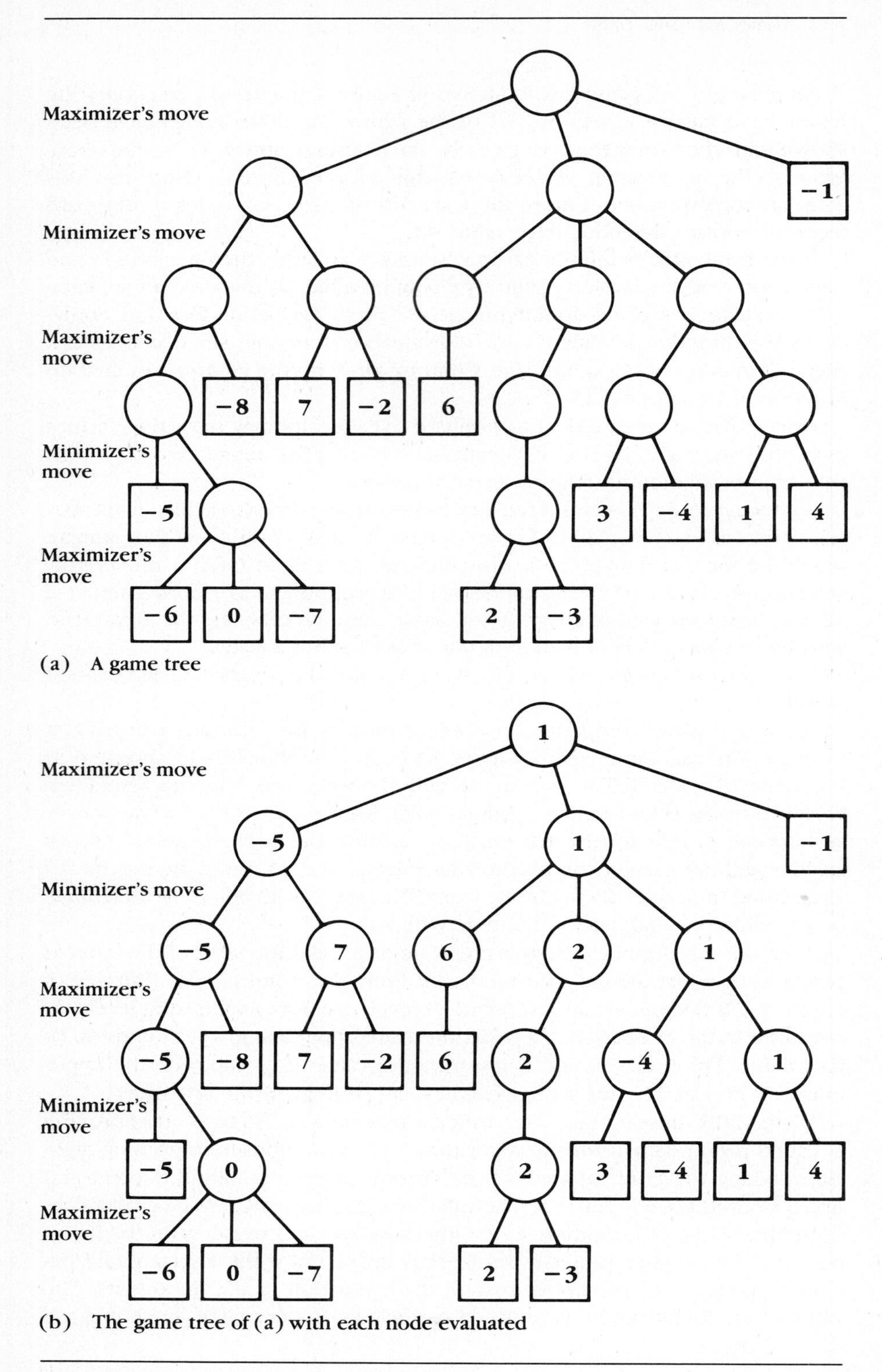

(a) A game tree

(b) The game tree of (a) with each node evaluated

Figure 4.36 Evaluation of a Game Tree

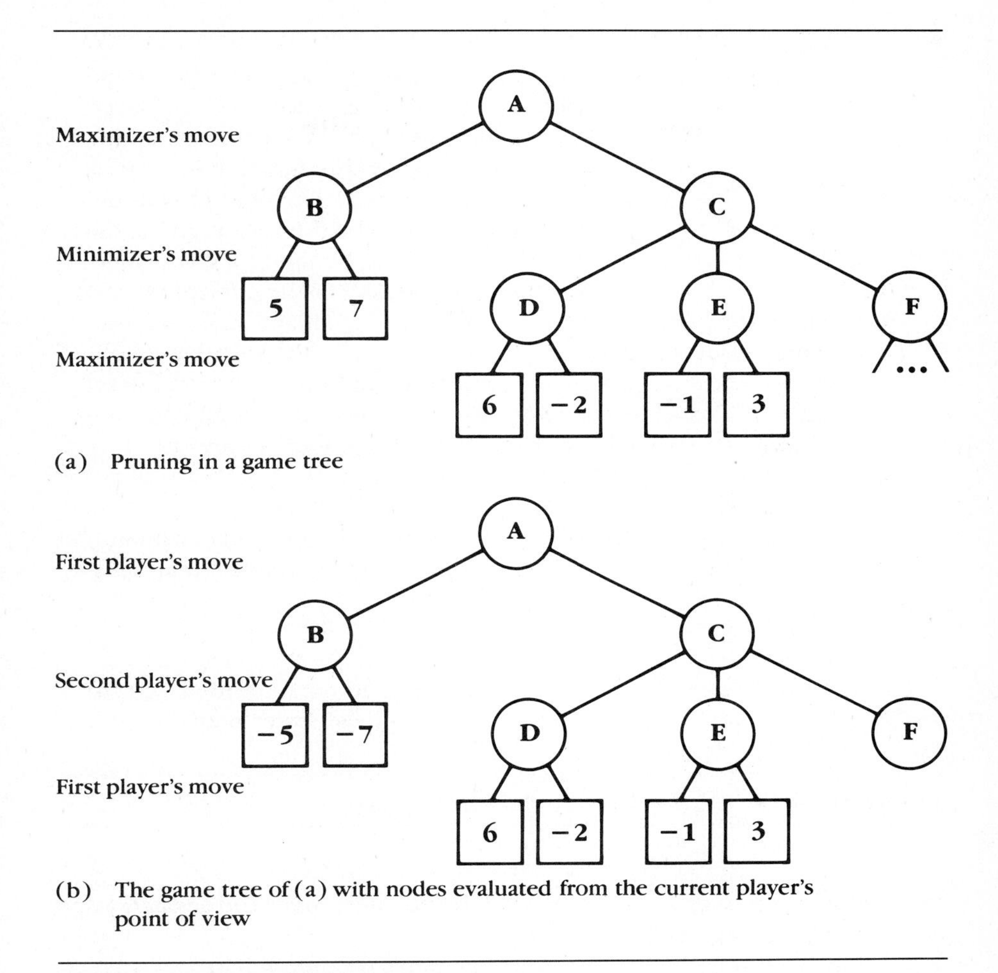

Figure 4.37 Alpha-Beta Pruning

large, the value of the lower bound will and should play no role in evaluating the tree.

Note in the recursive call the lower bound and upper bound are negated and swapped as the point of view switches from the current player to the opponent. Similarly the value returned by the recursive call must be negated.

For example, in the case of Figure 4.37b, the interval defined by the lower and upper bounds of the root is $(-\infty, +\infty)$. Evaluation of the left subtree shrinks the interval to $(5, +\infty)$. This interval is passed as $(-\infty, -5)$ to the right subtree. The value of $+6$ obtained for node D by the maximizer is converted to -6 and used to shrink the interval to $(-6, -5)$. The value of $+3$ for node E is converted to -3. The lower bound improves to -3, but now the interval for node C has become $(-3, -5)$, which is impossible. So node C cannot figure in

```
 1. AlphaBeta(node, lower, upper)       {given a node and a lower and
 2.                          upper bound for its value, returns an improved
 3.                               lower bound based on a search of the node's
 4.                              subtrees. Assumes a value function for leaves,
 5.                        a children function giving the number of children
 6.                             of each node, and a child function giving the
 7.                                                  kth child of a given node.
 8.                            Takes the point of view of the current player}
 9. if children(node) = 0 then                          {if at a leaf, then use}
10.    return value(node)                                 {the value function}
11. else                                      {otherwise find the maximum lower}
12.    for i := 1 to children(node) do               {bound over all children}
13.       temp := -AlphaBeta(child(node, i), -upper, -lower)
14.       if temp > lower then
15.          lower := temp
16.          if lower > upper then              {subject to given upper bound}
17.             return lower               {if upper bound exceeded, then quit}
18.          end if
19.       end if
20.    end for
21.    return lower                    {if upper bound not exceeded, return}
22. end if                                             {best lower bound found}
```

Algorithm 4.18 Alpha-Beta Search of a Game Tree

the solution. Returning a value of -3 will be harmless, since the negated value of $+3$ will be disregarded at node A.

If a game tree is not finite, then the opponent's strategy will be less predictable. Specifically, the particular goal node to be reached during the play of the game cannot be predicted. The strategy implicit in the evaluation of the game tree will not be a completely determined series of moves, but a complicated strategy with many conditionals like "if my opponent makes this move, I will make that move." The relation of this sort of strategy to state spaces is an important topic of the next section.

4.13 And-Or Trees

Some differences exist among the decision tree examples we have seen in this section. In the case of ZOK, we only had to find a single leaf, or equivalently, the path to the leaf; so for each node beginning with the root, we need to choose only one of its successors.

By contrast, the tournament-based algorithm for finding the minimum of a set needs to consider all leaves. In order to find a solution graph including the paths from the root to all these leaves, we must consider the whole tree.

Game trees represent an intermediate case. In the cases where the tree is too large for play to be completely determined, the first player need only choose one of his moves but must consider all possible responses of his opponent.[8] Thus at some nodes only one child needs to be chosen, while at other nodes all children need to be chosen. This will lead to some fraction of the goal nodes being possible outcomes of the game.

We may describe a framework to cover such cases. We allow each node to be an AND *node,* in which case all successors must be included in a solution, or an OR *node,* in which case only one child needs to be considered. Goal nodes are considered to have value "true," and other nodes with no successors to have value "false." The resulting tree or graph is called an *and-or* tree or graph, and we may determine whether it has a solution by defining the value of each of its nodes as follows:

Def 4.19 The *value* of a node of an and-or tree is

1. the value of the node, if it is a leaf;
2. the value of **and** applied to all its children, if the node is an AND node;
3. the value of **or** applied to all its children, if the node is an OR node.

The *solution* to an and-or tree is a subtree such that

1. whenever an AND node is included in the solution, all its children are included;
2. whenever an OR node is included in the solution, one of its children is included.

In other words, an AND node is feasible iff all its children are feasible. An OR node is feasible iff some child is feasible.

Figure 4.38 shows a type of and-or tree you may be familiar with—graduation requirements (simplified!) from a mythical department of mathematics and computer science. Here, as is customary for and-or trees, links to the children of AND nodes are connected by an arc. The heavy lines in the figure indicate one possible solution to the and-or tree. This solution requires the student to take data structures, Lisp, algorithms, and artificial intelligence to graduate with a computer science degree. There are of course other equally good solutions for this particular and-or tree.

Figure 4.39 shows the game tree of Figure 4.36 as an and-or tree. As before, the tree is from the point of view of the first (maximizing) player. For alternate

---

[8] So far we have avoided using the pronouns "he" and "she." However, in game trees the players are often named after their strategies, so that the first player is called Max and the second player Min. Thus there are as many male as female players.

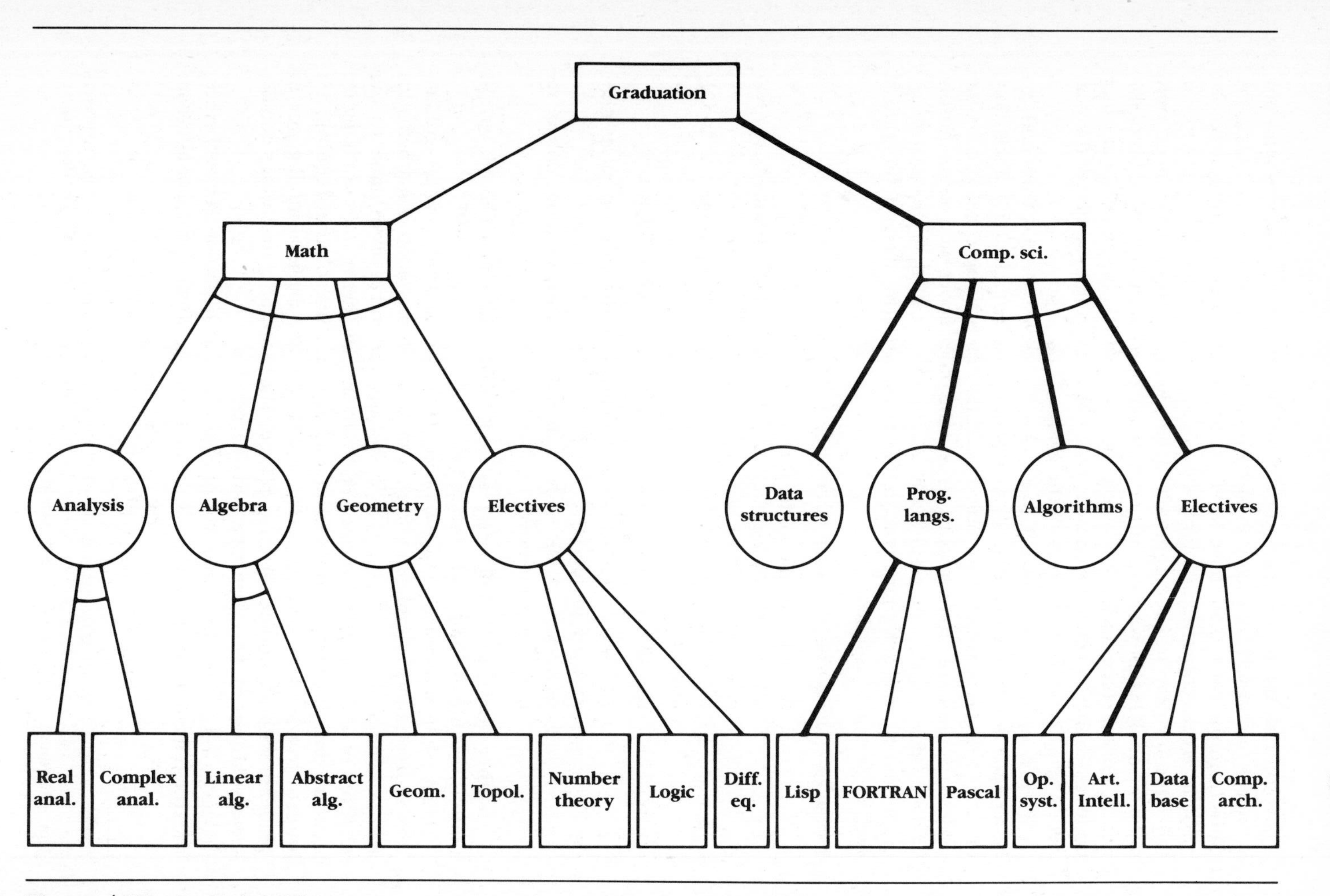

Figure 4.38 An And-Or Tree

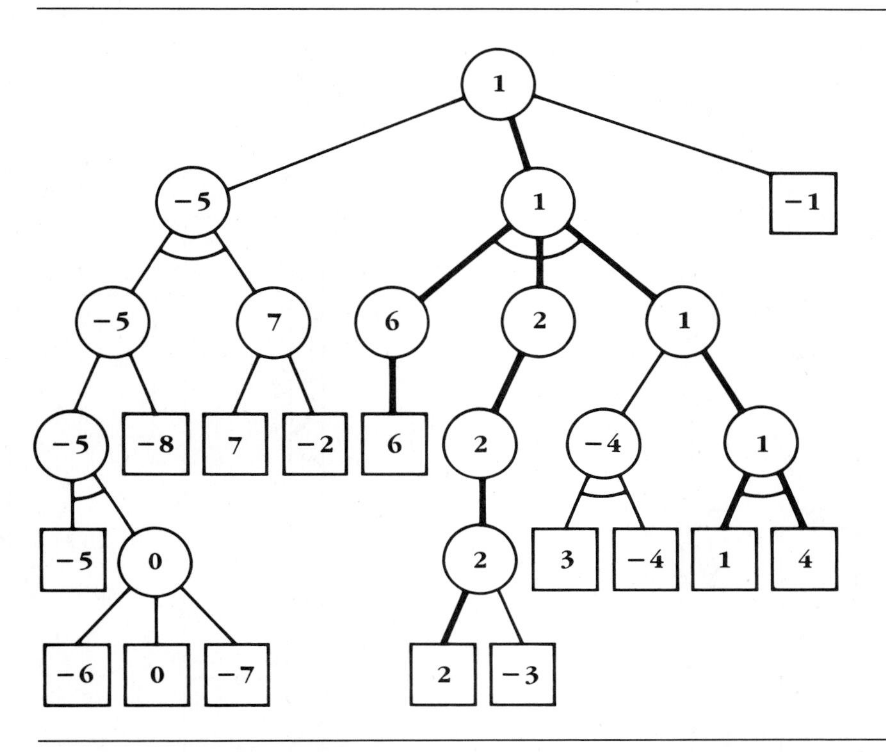

Figure 4.39 Solution of the Game Tree of Figure 4.36 Considered as an And-Or Tree

moves beginning with the first, the maximizing player is choosing the move so needs only to deal with one of the choices. These moves correspond to OR nodes. For the other moves, he needs to have a reply ready for all possible responses of his opponent. These nodes correspond to AND nodes.

The strategy indicated with heavy lines in Figure 4.39 is a solution to the and-or tree in the sense of Definition 4.19. Under the assumption that the second player might not use her optimal strategy, the game may end in any leaf of the solution. The value of the game may thus be either 6, 2, 1, or 4 to the first player. The value 1 of the game tree found in the previous section is just the minimum of these values.

As suggested at the beginning of this section, if all nodes in an and-or tree are OR nodes (as in ZOK), the solution is simply a path. If all nodes are AND nodes (as in tournaments), the solution is the entire tree; that is, solving the problem requires a tree traversal.

Figure 4.40a shows part of an and-or tree for the problem of flying from Los Angeles to New York. Nodes crossed out correspond to subtasks that are not feasible. If a solution is to be found quickly, depth-first traversal with backtracking is appropriate.

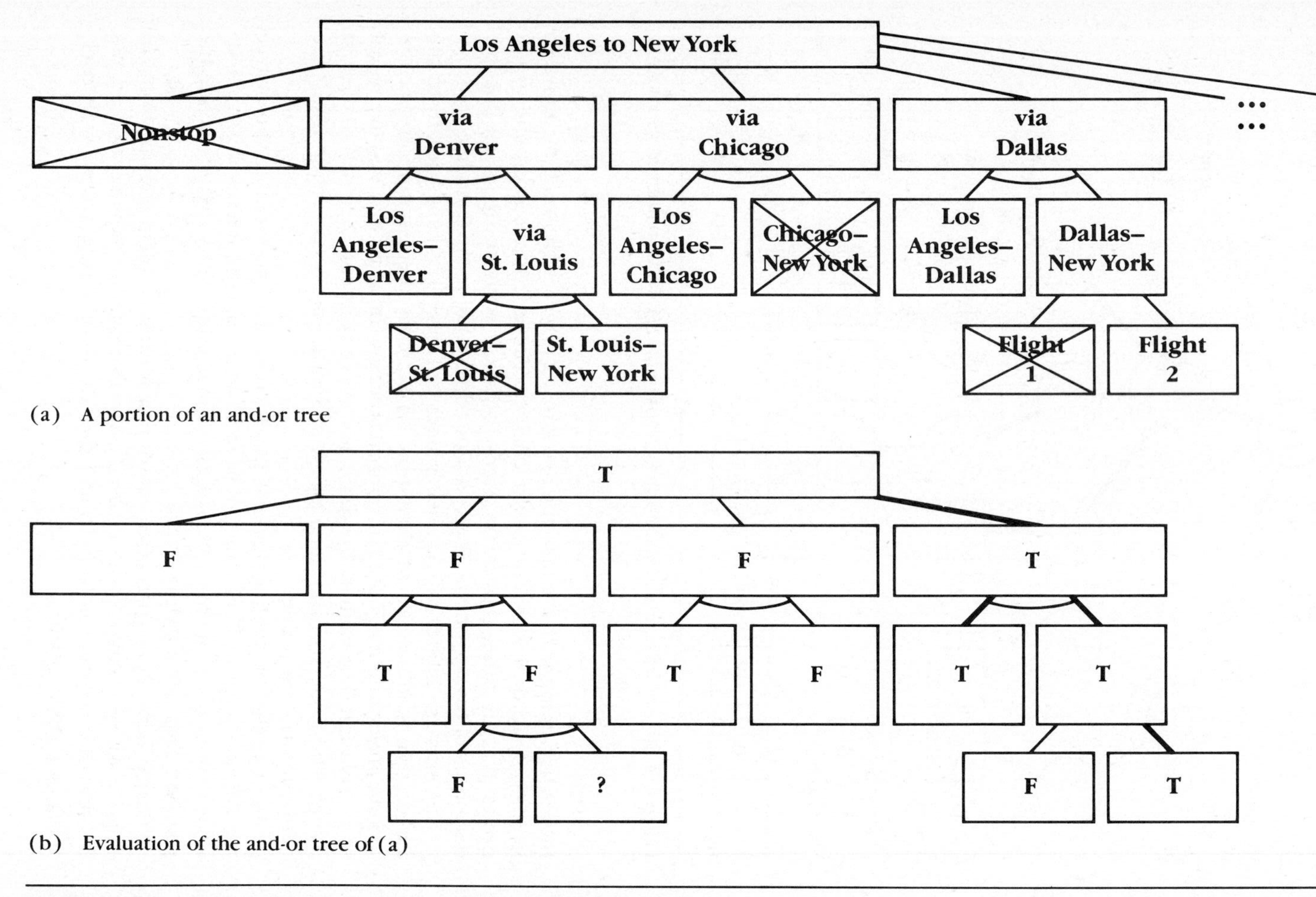

(a) A portion of an and-or tree

(b) Evaluation of the and-or tree of (a)

Figure 4.40 Evaluation of an And-Or Tree Using Backtracking

The first hope is for a nonstop flight; this is not feasible. The second hope is for a flight through Denver. A flight is available through Denver, but the trip from Denver to New York requires a stop in St. Louis. No space is available on the flight from Denver to St. Louis, so the leg from St. Louis to New York is not investigated.

Backtracking leads to investigating a flight through Chicago. A flight is available to Chicago, but not from Chicago to New York. The next intermediate city tried is Dallas. There is a flight to Dallas and a choice of flights from Dallas to New York. The first flight is full, but the second is feasible, completing a solution. The portion of the state-space tree generated is shown in Figure 4.40b. The solution found for the and-or tree has been highlighted. Note that other intermediate cities were possible but were not investigated.

In and-or trees, leaf nodes may be marked with constraints instead of simply as feasible or not feasible. Other nodes correspond to combinations of these constraints. For example, we might have had a budget constraint of $200 for the flight. Each leg would have a minimum price, and each AND node would correspond to adding these minima to give a minimum price for the combined journey. A minimum price for the whole trip of $200 or more would not be compatible with the budget constraint. See Charniak and McDermott (1985) for more detail on constraints and and-or trees.

Like general state-space trees, and-or trees are not completely general. And-or graphs may be defined in the obvious way. There is an analog of the A* algorithm for and-or graphs, usually called the AO* algorithm. See Nilsson (1980) or Rich (1983) for more information on the AO* algorithm.

4.14 Applications of Graph Traversal

We have seen that a depth-first spanning algorithm can be applied to graphs as well as trees. However if the underlying graph is not connected, the algorithm will terminate without having visited all of the vertices. This property can be used to determine the connected components of a graph. One repeatedly selects an unvisited vertex and traverses, beginning at this vertex, until the traversal terminates.

It is possible, given a depth-first spanning tree and the other edges of a graph, called ***back edges*** in this context, to determine whether it will remain connected if a vertex v and all edges of the form $\{v, w\}$ are deleted. If not, v is called an ***articulation point.*** The details are left as an exercise, but a look at Figure 4.41 is instructive.

Part (a) of the figure is a graph. A depth-first spanning tree with its back edges shown as dotted lines is illustrated in (b). First note that the root A is an articulation point, since it has more than one child and its deletion would separate the children. None of the leaves D, E, F, and H is an articulation point. Nodes B and C may be deleted without harm because back edges bypass them from all their subtrees. Therefore nodes B and C are not articulation points. This is not true for node G; its removal would isolate F and F's children, if there were any. So A and G are the articulation points.

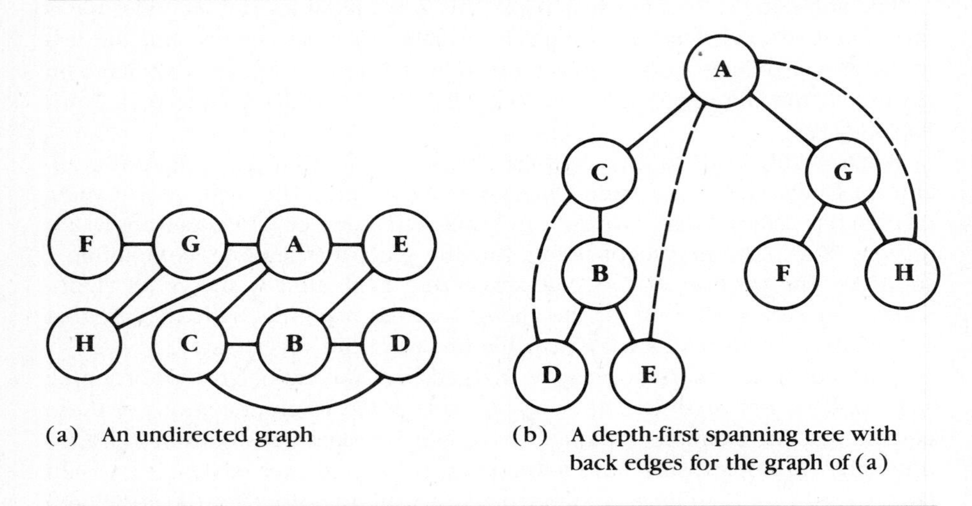

Figure 4.41 Depth-First Search and Articulation Points

One important application of graph traversal is *garbage collection.* In general, memory in a computer will be allocated dynamically, by operations similar to our **Get** operation, but not necessarily explicitly returned to available memory. Memory locations currently in use will presumably have access paths from named variables or from registers or other locations known to the system. Memory locations to which there is no access path are collectively called *garbage.* If garbage is not made available to the memory allocation process, requests for memory could fail even though not all memory is in use.

Rather than depend on programmers to free all memory no longer needed, many systems implement garbage collection algorithms that first mark all memory locations in use, then physically sweep through memory and return garbage to a *free space area* available to satisfy requests for memory. The marking phase corresponds to a general graph traversal algorithm. Suggested organizations for the free space area are explored in the exercises.

We mention but cannot explore here two important issues involving garbage collection. First, garbage collection, as we have described it, is very time-consuming, since it involves an exhaustive search through memory (although only groups of memory locations and not individual memory locations must be visited). Garbage collection could therefore be extremely awkward in real-time systems.

Second, we have suggested that a conventional breadth-first or depth-first search traversal could be used in garbage collection. These traversal methods require considerable additional memory in the first case; however, automatic garbage collection algorithms are invoked precisely when memory is about to be exhausted. A number of less memory-intensive garbage collection algorithms are discussed in Standish (1980).

4.15 Topological Sort and Critical Paths

Suppose that a set of tasks must be accomplished and that some of the tasks are prerequisites for others. Assume that any given task has a set of prerequisites, all of which must be completed before the given task can be completed (i.e., that there are no alternate prerequisites as in Figure 4.38).

We may find it important to know 1) in what order, if any, the tasks can be accomplished if only one can be done at a time and 2) if each task requires a given amount of time, how much time will be required to complete the entire set of tasks if different tasks may be done at the same time.

The prerequisite structure induces a graph structure in a natural way. Let the set of vertices be the set of tasks, and let there be a directed edge from x to y iff x is a prerequisite for y. If the resulting graph contains a cycle, then clearly no solution is possible.

Consider the first of the two problems, called the *topological sorting* problem. If there is some task with no prerequisite (i.e., with no incoming edge), then this task may be performed first. By removing the task and all edges from it, we have reduced it to a smaller problem of the same type.

We need to show that unless there is a cycle in the graph, some vertex has no incoming edges.

Theorem In a prerequisite graph with a finite number of tasks, either there is a cycle or some vertex has no incoming edges.

Proof by minimal counterexample. Of all graphs with no cycle in which each vertex has an incoming edge, consider the graph with the minimal number of edges.

For an arbitrary edge (v, w) in the minimal counterexample, removing the edge certainly gives a graph with no cycles. So by minimality, some vertex in the resulting graph has no incoming edge. This vertex must be w. Since the choice of edge was arbitrary, all vertices in the original graph have exactly one incoming edge.

Choose a vertex x in this graph. Then there is an edge (w, x) in the graph and also an incoming edge (v, w) for w. Since there is no cycle in the graph, all three vertices are distinct. Consider the graph obtained by removing w and all edges to or from w, but adding (v, x). Each vertex in the new graph has exactly one incoming edge.

Additionally, this new graph can have no cycle, since any cycle not including (v, x) would be a cycle in the original graph, and any cycle including (v, x) would correspond to a cycle containing (v, w) and (w, x) in the original graph. Since the new graph has fewer edges — we have deleted at least the two edges (v, w) and (w, x) — we have violated minimality, and every directed graph with no cycle has a vertex with no incoming edge. ∎

A topological sorting algorithm may thus be constructed to work by repeatedly 1) finding a vertex with no incoming edge and adding it to the output

Task	A	B	C	D	E	F	G
Prerequisites	C, D	F	—	—	A, C	E, G	D
Time required	3	2	1	6	4	5	12

(a) An example for topological sort and critical paths

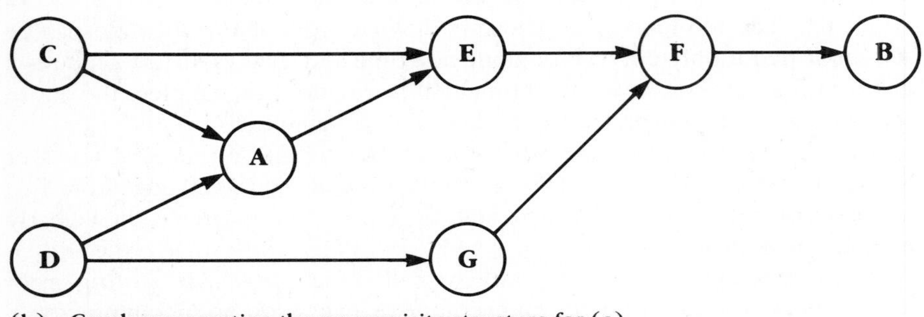

(b) Graph representing the prerequisite structure for (a)

Figure 4.42 An Instance of the Topological Sort Problem

sequence and 2) deleting the vertex and all its outgoing edges from the graph, as long as there are vertices not yet output.

Finding a vertex in (1) needn't require a sequential search if the vertices are stored in order of the number of their incoming edges, say, in a heap. If adjacency lists are used, it is easy when a vertex v is deleted to find all vertices (v, w) and to decrement the count of incoming vertices for each w. The entry for w in the heap will have to be deleted and reinserted with its new count. If this count becomes 0, the vertex w can be put on a queue of vertices to be output rather than back into the heap.

Figure 4.42a gives an instance, including tasks and prerequisites, of the topological sort problem for seven tasks. Part (b) of the figure represents the prerequisite structure as a directed graph. Figure 4.43 shows the construction

Task			A	B	C	D	E	F	G
Unmet prerequisites			C, D	F	—	—	A, C	E, G	D
"	after including	C	D	F	—	—	A	E, G	D
"	"	D	—	F	—	—	A	E, G	—
"	"	A	—	F	—	—	—	E, G	—
"	"	E	—	F	—	—	—	G	—
"	"	G	—	F	—	—	—	—	—
"	"	F	—	—	—	—	—	—	—
"	"	B	—	—	—	—	—	—	—

Figure 4.43 Operation of the Topological Sort Algorithm

```
 1. TopologicalSort(n, pre, post, count){takes a number n of tasks and
 2.                 arrays pre and post (indexed from 1 to n) of pointers to
 3.                         linked lists giving respectively the list of
 4.                     prerequisites and the list of postrequisites to each
 5.                     task. The data in each node in the list is contained
 6.                     in a task field. Also takes an array count giving the
 7.                       number of prerequisites for each task. Assumes a
 8.                 function findmin that will give the number of the task
 9.                   with the fewest unsatisfied prerequisites. Returns an
10.                    array order containing a sequence of tasks such that
11.                   each task appears in the sequence only after all of its
12.                                                          prerequisites}
13. for i := 1 to n                        {until all tasks have been ordered}
14.    task := FindMin(pre)                {find a task with no prerequisites}
15.    if count[task] > 0 then return unsuccessfully              {if one exists}
16.    else                                                            {if so,}
17.       order[i] := task                {include it in the output sequence}
18.       ptr := post[task]                       {and traverse the list of its}
19.       while ptr ≠ nil                                      {postrequisites}
20.          count[task(ptr)] := count[task(ptr)] − 1           {decrementing}
21.          ptr := next(ptr)              {the number of unmet prerequisites}
22.       end while                                                  {of each}
23.    end if
24. end for
25. return order
```

Algorithm 4.20 Topological Sort

of one order (C, D, A, E, G, F, B) in which the tasks may be included in the topologically sorted output. After each task the figure shows all unmet prerequisites for each task.

A topological sorting algorithm as described previously is given as Algorithm 4.20.

The solution to the second problem turns out to be relatively simple, given a topological sorting algorithm. The divide-and-conquer approach suggests that the earliest possible finishing time $ET(x)$ for a given task x is just $T(x) + \max\{ET(y)\}$, where $T(x)$ is the time required to finish x and the minimum is taken over all prerequisites y of x. Ordinarily it would not be clear that we have reduced the problem of finding $ET(x)$ to a simpler problem, but now we know that all such y must precede x in topological order. So we may evaluate $ET(x)$ in topological order of the tasks, letting the maximum over an empty set of prerequisites be 0. The time required for the entire task is just the maximum of $ET(x)$ over all tasks x. Let's call this time FT, for "finish time."

With just a little extra work, we may derive a result of great practical importance. Once the best possible finish time FT has been determined, we may define LT(x) informally to be the latest time that x may be completed consistent with a finish time of FT. Formally, we have the definitions

Def 4.21 (a) *ET*(x) =

1. T(x) if x has no prerequisites,
2. T(x) + max{ET(y)}, where the maximum is taken over all prerequisites y of x.

(b) *FT* = max{ET(x)}

(c) *LT*(x) =

1. FT if x is not a prerequisite for any task;
2. $\min_y$\{LT(y) − T(y)\} otherwise, where the minimum is taken over all tasks x for which x is a prerequisite for y.

Note that LT may be evaluated for all tasks by considering them in reverse topological order. Also note that the latest possible starting time for task x is LT(x) − T(x), if the overall finish time FT is not to be changed.

In general, a task x may finish at any time between ET(x) and LT(x) if the overall finish time is to be FT. There will always be some tasks x for which ET(x) = LT(x); otherwise, FT could be reduced. We can easily see such tasks lie on one or more paths from a task with no prerequisites to a task that is

Task	**C**	**D**	**A**	**E**	**G**	**F**	**B**
Prerequisites	—	—	C, D	A, C	D	E, G	F
Postrequisites	A, E	A, G	E	F	F	B	—
Time	1	6	3	4	12	5	2
ET	1	6	9	13	18	23	25
LT	5	6	14	18	18	23	25

(a) Determination of critical tasks

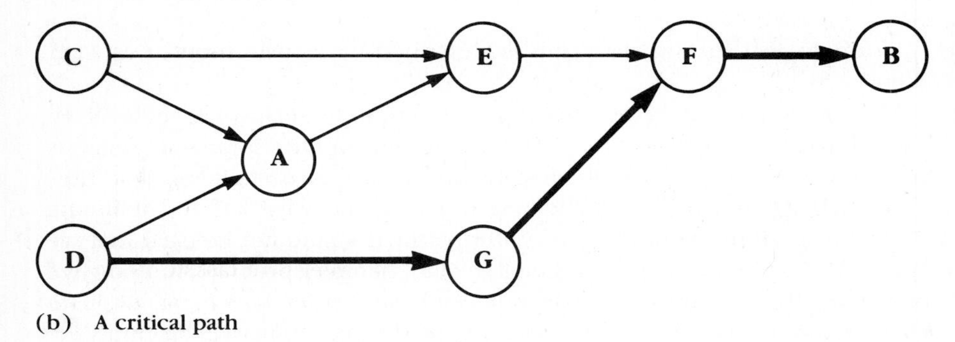

(b) A critical path

Figure 4.44 Critical Tasks and Critical Paths

```
1. CriticalTasks(order, pre, post, time)                {takes an array
2.                 order produced by the topological sort algorithm,
3.                            arrays pre and post containing lists of the
4.                   prerequisites and postrequisites of each task, and an
5.               array time containing the time required to perform each
6.                     task. Returns an overall finish time ft and arrays et
7.                            giving the earliest time that each task can be
8.                  completed and lt giving the latest time that each task
9.                                 can be completed consistently with ft}
10. for i := 1 to n                                      {for each task}
11.    task := order[i]                            {in topological order}
12.    ptr := pre[task]                                {find the earliest}
13.    earliest := 0                                {possible finish time}
14.    while ptr ≠ nil
15.       if et[task(ptr)] > earliest then earliest := et[task(ptr)]
16.       ptr := next(ptr)                          {consistent with all}
17.    end while                                {prerequisites, and enter}
18.    et[task] := earliest + time[task]             {in output array et}
19. end for
20. ft := 0                                         {find latest finish time}
21. for i := 1 to n                            {for all tasks, and assign}
22.    if et[i] > ft then ft := et[i]                    {this value to ft}
23. end for
24. for i := n to 1 by −1                      {for each task in reverse}
25.    task := order[i]                           {topological order, find}
26.    ptr := post[task]                             {the latest possible}
27.    latest := ft                               {finish time consistent}
28.    while ptr ≠ nil
29.       if lt[task(ptr)] − time[task(ptr)] < latest then
30.          latest := lt[task(ptr)] − time[task(ptr)]
31.       end if                                  {with an overall finish}
32.       ptr := next(ptr)                                    {time of ft}
33.    end while                                   {and put this value in}
34.    lt[task] := latest                               {the output array lt}
35. end for
36. return lt, et, ft
```

Algorithm 4.22 Critical Tasks

finished at time FT. Such a path is called a *critical path* and tasks on a critical path are *critical tasks.* If a noncritical task is delayed, the overall finish time may not change. But any delay in a critical task will delay the entire project. Thus management resources might reasonably be concentrated on the critical tasks.

A trace of the topological sorting algorithm is shown in Figure 4.43.

Figure 4.44a, where the tasks are listed in topological order, shows the values of ET and LT for the example of Figure 4.42. The overall finish time FT is 25. The critical path for this example is shown in (b) of the figure.

Algorithms for topological sorting and for finding the critical tasks are given as Algorithms 4.20 and 4.22 respectively. Finding ET for all vertices requires considering n vertices. For each vertex there is an inner loop that requires checking the prerequisites for the corresponding task. The total time complexity for all n inner loops is $\Theta(e)$, where e is the number of prerequisites (i.e., the number of edges in the prerequisite graph). Thus so far we have a time complexity of $O(n + e)$. The same argument applies to finding all values of LT.

For determining the critical tasks, the only part of Algorithm 4.22 not considered is the loop that finds the finish time FT. This has time complexity $\Theta(n)$, so the overall time complexity remains $O(n + e)$. Since e is $O(n^2)$, the worst-case time complexity in terms of n is $O(n^2)$.

For topological sort itself, we need to consider the time complexity of finding the minimum n times. Sequential search for the n minima would have time complexity $\Theta(n^2)$; using a heap cuts this to $\Theta(n \log n)$. So the overall time complexity of topological sort if a heap is used is $O(e + n \log n)$. Again in terms of n, the worst case is $O(n^2)$. If the minimum is sought sequentially, the time complexity becomes $O(e + n^2) = O(n^2)$.

4.16 Greatest Common Divisors and Inverses Mod n

If a and b are positive integers, then there are a number of situations when it is useful to know the greatest number d that divides evenly both a and b. This number d is called the *greatest common divisor* (abbreviation: *GCD*) of a and b. The greatest common divisor of a and b is often abbreviated (a, b). For one application, the fraction a/b will be in lowest terms if both numerator and denominator are divided by their greatest common divisor. A second application, encryption, will be discussed in Section 5.5.

We may assume that $a > b > 0$. If b divides a, then clearly b is the greatest common divisor. Otherwise, if d is a divisor of a and b, then by definition $a \bmod b = a - kb$ for some k. Since d divides a and b, d divides $a - kb$ and therefore divides $a \bmod b$. Thus $(a, b) = (b, a \bmod b)$, and we have reduced to a smaller problem of the same type. We are led to the following recursive definition:

Def 4.23 If $a > b > 0$, then $(a, b) =$

1. b if $a \bmod b = 0$;
2. $(b, a \bmod b)$ otherwise.

For example, $(140, 77) = (77, 63) = (63, 14) = (14, 7) = 7$.

```
 1. GCD(a, b)              {given a > b, find and return d = (a, b) and
 2.                         numbers m and n such that m × a + n × b = d}
 3. oldm := 1                                     {initialize the variables}
 4. oldn := 0                            {used in the discussion of the text}
 5. newm := 0
 6. newn := 1
 7. q := ⌊a/b⌋
 8. r := a − q × b                                         {r = a mod b}
 9. while r > 0
10.    temp := a × (oldm − q × newm)
11.    oldm := newm                             {find new value of m and}
12.    newm := temp                            {update old value of m}
13.    temp := b × (oldn − q × newn)                {do the same for n}
14.    oldn := newn
15.    newn := temp
16.    a := b
17.    b := r                         {replace (a, b) by (b, a mod b)}
18. end while                                   {if r = a mod b = 0, done}
19. return (b, newm, newn)                              {and GCD is b}
```

Algorithm 4.24 Euclidean Algorithm for GCD and Inverses

This definition leads in a straightforward manner to a tail-recursive algorithm, which may be converted to an iterative algorithm by the methods of Section 3.7. The resulting algorithm is called the *Euclidean* algorithm for finding the GCD. Algorithm 4.24 does a little more work than just finding the GCD, as we shall now explain.

If d is the greatest common divisor of a and b, we can show that there are integers m and n such that $am + bn = d$. If $d = 1$, then m is the reciprocal of a mod b, and n is the reciprocal of b mod a. Thus we may find it useful (e.g., in the encryption application) to be able to find m and n.

Theorem If d is the greatest common divisor of a and b, then there are integers m and n with $am + bn = d$.

Proof Suppose (a, b) is a pair appearing in the computation of (A, B). The inductive hypothesis (or invariant) is that both a and b can be written in the form $m \times A + n \times B$. Certainly the claim is true for $(a, b) = (A, B)$, since $1 \times A + 0 \times B = A$ and $0 \times A + 1 \times B = B$.

Let (a, b) be the last pair arising in the calculation of (A, B) for which the claim is true. If $a \bmod b = 0$, then $b = (A, B)$ and there is nothing to prove. Let $q = \lfloor a/b \rfloor$, so that $a \bmod b = a - q \times b$. By induction, a and b can be written in terms of A and B; recall that if $(a, b) \neq (A, B)$, then a appears as the second element of a pair (c, a). To be definite, say

$$a = m_{\text{old}} \times A + n_{\text{old}} \times B \text{ and}$$

$$b = m_{\text{new}} \times A + n_{\text{new}} \times B$$

Then

$$a \bmod b = a - q \times b = (m_{\text{old}} \times A + n_{\text{old}} \times B) - q \times (m_{\text{new}} \times A + n_{\text{new}} \times B)$$
$$= (m_{\text{old}} - q \times m_{\text{new}}) \times A + (n_{\text{old}} - q \times n_{\text{new}}) \times B$$

Thus the claim is true for both b and a mod b. Since $(b, a \bmod b)$ is a pair arising later than (a, b) in the computation of (A, B), we get a contradiction. ∎

Therefore we may find m and n, by the algorithm suggested previously, so that $m \times A + n \times B = (A, B)$. This algorithm may be performed in parallel with the GCD algorithm, as in Algorithm 4.24.

As an example to be used in the discussion of encryption of Section 5.5, we may find the inverse of 17 mod 40. Initially we have $a = 40$, $b = 17$, and

$$40 = 1 \times 40 + 0 \times 17$$

$$17 = 0 \times 40 + 1 \times 17$$

Dividing 40 by 17, we get $q = 2$ and $r = 6$, so

$$6 = (1 - 2 \times 0) \times 40 + (0 - 2 \times 1) \times 17 = 1 \times 40 - 2 \times 17$$

Dividing 17 by 6, we get $q = 2$ and $r = 5$, so

$$5 = (0 - 2 \times 1) \times 40 + (1 - 2 \times (-2)) \times 17 = -2 \times 40 + 5 \times 17$$

Dividing 6 by 5 we get $q = 1$ and $r = 1$, so

$$1 = (1 - 1 \times (-2) \times 40 + (-2 - 1 \times (5)) \times 17 = 3 \times 40 - 7 \times 17$$

Thus the inverse of 17 mod 40 is -7. To get a positive value, we note that $-7 = 33 \bmod 40$, so $d = 33$.

4.17 Fast Fourier Transform

Often arithmetic operations are easier in one representation than another. For example, the straightforward algorithm of Section 2.14 for multiplication of two polynomials of degree $n - 1$ requires $O(n^2)$ multiplications, and even if the asymptotically improved algorithm of that section is used, $O(n^r)$ multiplications are required, where $r = \log_2 3$. On the other hand, if each polynomial is represented by its values on n distinct points, then only n multiplications are necessary.

The technique of transforming one representation to another is very useful in mathematical problem solving. The continuous analog (using an integral instead of a sum) of the discrete Fourier transform discussed in this section is useful in differential equations, analyzing periodic motion, and elsewhere.

If an operation would be simplified in another representation, it may be worthwhile to change the representation, perform the operation efficiently in the other representation, and then change the result back into the original representation. In the example of multiplying polynomials, if in time better than $O(n^2)$ we could find the value of each polynomial on n points (the *evaluation* problem), and from the values on n points we could find the coefficients of the product polynomial (the *interpolation* problem), then this approach would be reasonable. Note that n here must stand for the number of coefficients in the product polynomial, not the factors, since we are only allowing n points to represent it.

Unfortunately the most natural algorithms for the evaluation and interpolation problems don't help in this regard. For evaluation of one polynomial $\mathrm{p}(x) = \sum_{k=0}^{m-1} \mathrm{a}_k x^k$ at $x = \mathrm{a}$ is going to take time $\Omega(n)$, since all coefficients must be inspected. A simple and efficient $O(n)$ recursive algorithm for evaluation follows from observing that

$$\mathrm{p}(x) = a_0 + x \times \left(\sum_{k=1}^{m-2} \mathrm{a}_k x^{k-1} \right)$$

and the problem of evaluating a polynomial of degree $m - 1$, given m coefficients, has been reduced to one of evaluating a polynomial of degree $m - 2$, given $m - 1$ coefficients. The resulting algorithm can of course be written iteratively. However, if evaluation using this algorithm is to be done for n points, time $O(n^2)$ will be necessary.

The simplest way to construct a polynomial p satisfying n equations of the form $\mathrm{p}(x_i) = y_i$ is to let $\mathrm{p}(x) = \sum_{i=1}^{n} y_i \mathrm{p}_i(x)$, where $\mathrm{p}_i(x_j) = 0$ for $i \neq j$ and 1 for $i = j$. To have the given zeros, $\mathrm{p}_i(x_j)$ must have as factors $(x - x_j)$ for all $j \neq i$. If this product is divided by its (nonzero) value at $x = x_i$, we may assume that $\mathrm{p}_i(x_i) = 1$.

Thus if we define a product notation by analogy with our notation for sums, replacing Σ by Π, we get

$$\mathrm{p}(x) = \sum_{i=1}^{n} y_i(\Pi_{j \neq i, 1 \leq j \leq n}(x - x_j)/(x_i - x_j))$$

The time complexity for constructing a single p_i is $\sum_{k=1}^{n} k = \Theta(n^2)$, leading to an overall time complexity of $\Theta(n^3)$ for the interpolation problem. There are other, more efficient interpolation algorithms, but their time complexity is still not better than $O(n^2)$. We need a new approach. We start by reconsidering the evaluation problem, keeping in mind that the n points at which to evaluate may be chosen arbitrarily.

Recall from elementary algebra that the value p(a) of the polynomial p when x = a is the same as the remainder when p(x) is divided by x − a.[9] If p has degree n, then it will also take n multiplications to perform the division by x − a.

However, only n multiplications of coefficients are needed to find p(x) mod $x^2 - a^2$. This remainder will have degree less than 2 and the same value as p(x) when x = a and when x = −a, since $x^2 - a^2 = (x - a)(x + a)$. Dividing this remainder by both x − a and x + a requires only two more multiplications, so we get two evaluations with only $n + 2$ multiplications.

This observation may be generalized to a divide-and-conquer strategy with only one drawback. The proper generalization would be to first divide p(x) by $x^n - b$, where b has n nth roots. But over the real numbers, b can have at most 2 nth roots. Thus we need to introduce a number system in which any number b has n nth roots. We need only consider b = 1.

Def 4.25 A *complex number* is a number of the form $a + b$i, where a and b are reals and $i^2 = -1$. The arithmetic operations on complex numbers are the same as those for polynomials in i with real coefficients, except that i^2 may be replaced by −1.

This latter condition means that the polynomials in i are linear or constant. For example, $(a + bi)(c + di) = (ac - bd) + (ad + bc)i$, since both sides are equal to $ac + (ad + bc)i + bdi^2$. In other words, the product of two complex numbers is a complex number.

Note that $(-1)^2 = i^2 = -1$.

Def 4.26 A *primitive* n*th root of 1*, or *of unity*, is a complex number ω such that $\omega^n = 1$ and $\omega^k \neq 1$ whenever $k < n$.

For example, the numbers i, −i, 1, and −1 are fourth roots of 1, but only i and −i are primitive 4th roots. Primitive nth roots of 1 exist for all n. If ω is a primitive nth root of 1, then ω^k is an nth root of 1 for $0 \leq k \leq n$.

One primitive nth root of 1 is $\omega = \cos(2\pi/n) + i\sin(2\pi/n)$. It is easy to verify that $\omega^k = \cos(2\pi k/n) + i\sin(2\pi k/n)$.

A few simple algebraic identities will greatly simplify our calculations with complex numbers.

1. If ω is a primitive nth root of 1, then $\omega^{n/2} = -1$ (since $\omega^{n/2}$ is not 1 by primitivity, but its square is 1, it must be −1).
2. If ω is a primitive nth root of 1, then $\omega^{k+n} = \omega^k$ and $\omega^{k+n/2} = -\omega^k$. This follows from the rule $x^{a+b} = x^a x^b$, $\omega^n = 1$, and the previous rule.

[9] The proof is not difficult. By the same argument as for integers, there are polynomials q and r such that $p(x) = q(x)(x - a) + r(x)$. Here the degree of r is less than that of x − a, so r must be constant. When x = a, p(a) = r(a).

3. If ω is an nth root of 1, then $\sum_{k=0}^{n-1} \omega^k = 0$, unless $\omega = 1$. This follows from Example 4 of Section 2.1, since $(\omega^n - 1)/(\omega - 1) = 0$, as long as the denominator is nonzero. If $\omega = 1$ then the sum is just n.

A divide-and-conquer algorithm, corresponding to evaluation at the n nth roots of 1, now may be formulated as follows: Let n be a power of 2 and ω a primitive nth root of 1. To find the remainder with $p(x)$ is divided by $x - \omega^r$ or $x + \omega^r = x - \omega^{r+n/2}$, we first find the remainder $r(x)$ when $p(x)$ is divided by $(x^2 - \omega^{2r}) = (x + \omega^r)(x - \omega^r)$ and then find the remainders mod $x + \omega^r$ and $x - \omega^r$ of $r(x)$.

In this approach, the linear polynomials $x - \omega^r$ and $x + \omega^r = x - \omega^{r+n/2}$ are related to their product $x^2 - \omega^{2r}$. This relationship is diagramed in the tree of Figure 4.45a. Part b of the figure is the special case $n = 4$ of (a); part c is the special case $n = 8$. For a given r, the divisors corresponding to the divide-and-conquer reduction are just the ancestors of $x - \omega^r$ in the tree. We can see that every r requires eventual consideration of the divisor $x^n - 1$. Also note that the 2^k divisors at level k of the tree correspond to the 2^k 2^kth roots of 1.

Computation takes place from the root downward with intermediate results saved. An example for the case $n = 4$ is shown in Figure 4.45d for the polynomial $p(x) = x^3 + 3x^2 + x - 1$. The tree of divisors corresponds to the tree of remainders in (b) of the figure. It follows that $p(x)$ has value 4 at $x = 1$, 0 at $x = -1$, and -4 at $x = i$ and $x = -i$.

It is an easy induction to show that level k of the tree requires 2^k divisions of polynomials of bounded degree into polynomials of degree $n/2^k$. Thus each level requires $O(n)$ multiplications of coefficients. Since there are $\log_2 n$ levels, this divide-and-conquer algorithm for the discrete Fourier transformation problem has time complexity $O(n \log n)$. So it may indeed be called the fast Fourier transform algorithm (FFT).

We still need to handle the problem of interpolation. Amazingly enough, the algorithm just described will work with only minor changes. Suppose that $y^k = p(\omega^k)$ for each root ω^k of 1. If we evaluate with $\{y_k\}$ replacing $\{a_k\}$, we get for ω^i,

$$\sum_{k=0}^{n-1} y^k(\omega^i)^k =$$

$$\sum_{k=0}^{n-1} y^k \omega^{ik} =$$

$$\sum_{k=0}^{n-1} \left(\sum_{j=0}^{n-1} a_j \omega^{kj} \right) \omega^{ik} =$$

$$\sum_{j=0}^{n-1} a_j \left(\sum_{k=0}^{n-1} \omega^{(i+j)k} \right)$$

But the inner sum is 0 unless $\omega^{i+j} = 1$, that is, unless $i + j = 0 \bmod n$, in which case the sum is n. Thus the outer sum is just a_{-i}, where negation is taken mod n.

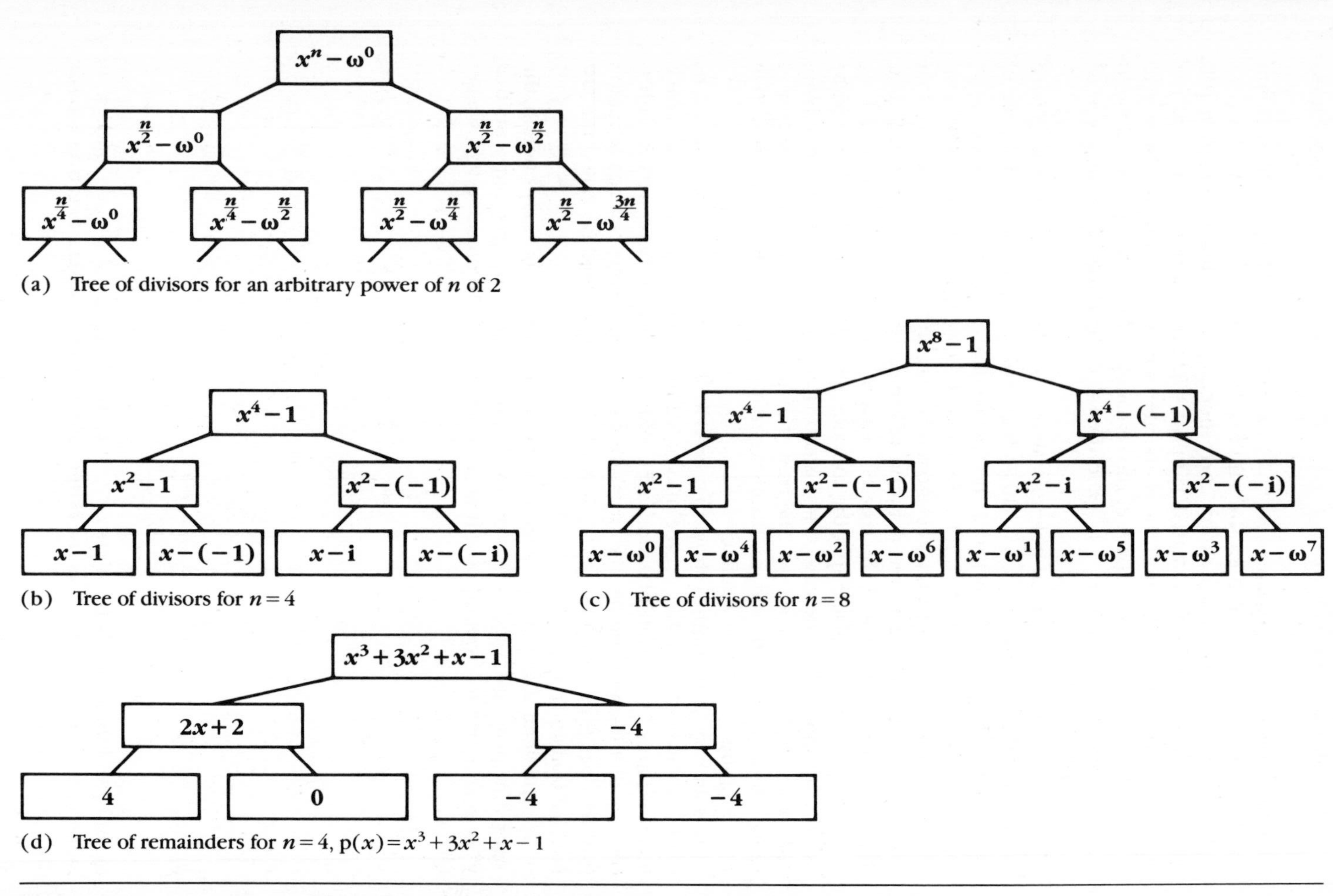

Figure 4.45 Trees of Divisors and Remainders for the Fast Fourier Transform

If we consider the input polynomial as a sequence $(a_k)_{k=0}^{n-1}$ of coefficients and the Fourier transform as evaluating the corresponding polynomial at (ω^k) to get a new sequence $(y_k)_{k=0}^{n-1}$, we have shown that a second application of the Fourier transform will recover the (a_k) except that 1) each term is multiplied by n and 2) the terms are in the order $0, n-1, n-2, n-3, \ldots, 1$. In other words, we have shown that

$$a_k = (1/n) \sum_{j=0}^{n-1} y_j \omega^{-jk}$$

where a_k is associated with the $-k$th power of ω, or equivalently, the $(n-k)$th power instead of the kth power.

Figure 4.46 shows the use of the fast Fourier transform algorithm to find the product of $x^3 + x - 2$ and $x^2 - 1$. Here $n = 8$, since the product will have more than 4 coefficients. The trees of remainders correspond to the tree of divisors of Figure 4.45c. You may want to use the identities $\omega + \omega^3 = \sqrt{2}i$, $\omega + \omega^5 = 0$, $\omega^5 + \omega^7 = -\sqrt{2}i$, $\omega^{n+8} = \omega^n$, $\omega^{n+4} = -\omega^n$, $\omega^2 = i$, and $\omega^7 = \frac{\sqrt{2}}{2} - \left[\frac{\sqrt{2}}{2}\right] i$ to verify the steps of the computation.

The leaves are interpreted in the order $(16\ -8\ -16\ 0\ 0\ 8\ 0\ 0)$. When these coefficients are divided by $n = 8$, we get $(2\ -1\ -2\ 0\ 0\ 1\ 0\ 0)$, corresponding to the correct product $2x^0 - 1x^1 - 2x^2 + 1x^5 = x^5 - 2x^2 - x + 2$.

4.18 Maxima, Minima, and Adversary Arguments

For the problem of finding the maximum element in a sequence of size n, the decision tree model gives a lower bound of $\Theta(\log n)$ comparisons. We already have an algorithm requiring $n - 1$ comparisons. Thus a considerable gap exists between the best time complexity known so far and the best time complexity we can hope for.

In this section we show by a new technique, the use of *adversary arguments,* that $n - 1$ comparisons is the best possible. We also apply an adversary argument to the problem of finding both the maximum and the minimum.

The lower bound given by the decision tree model for the problem of finding the maximum decision tree is $O(\log n)$ and is not optimal. In fact there are two easy ways to see this. One way is simply to note that any algorithm that purports to find the minimum must actually inspect all the elements of the sequence. Since only 2 elements can be inspected at once, at least $n/2$ comparisons are required.

This perfectly correct argument gives an asymptotic lower bound equal to the time complexity of a known algorithm for the problem. However, if we are concerned with actual numbers of comparisons rather than just asymptotic results, the argument still leaves open the possibility of an algorithm requiring fewer than the $n - 1$ comparisons of the algorithm in Section 2.2.

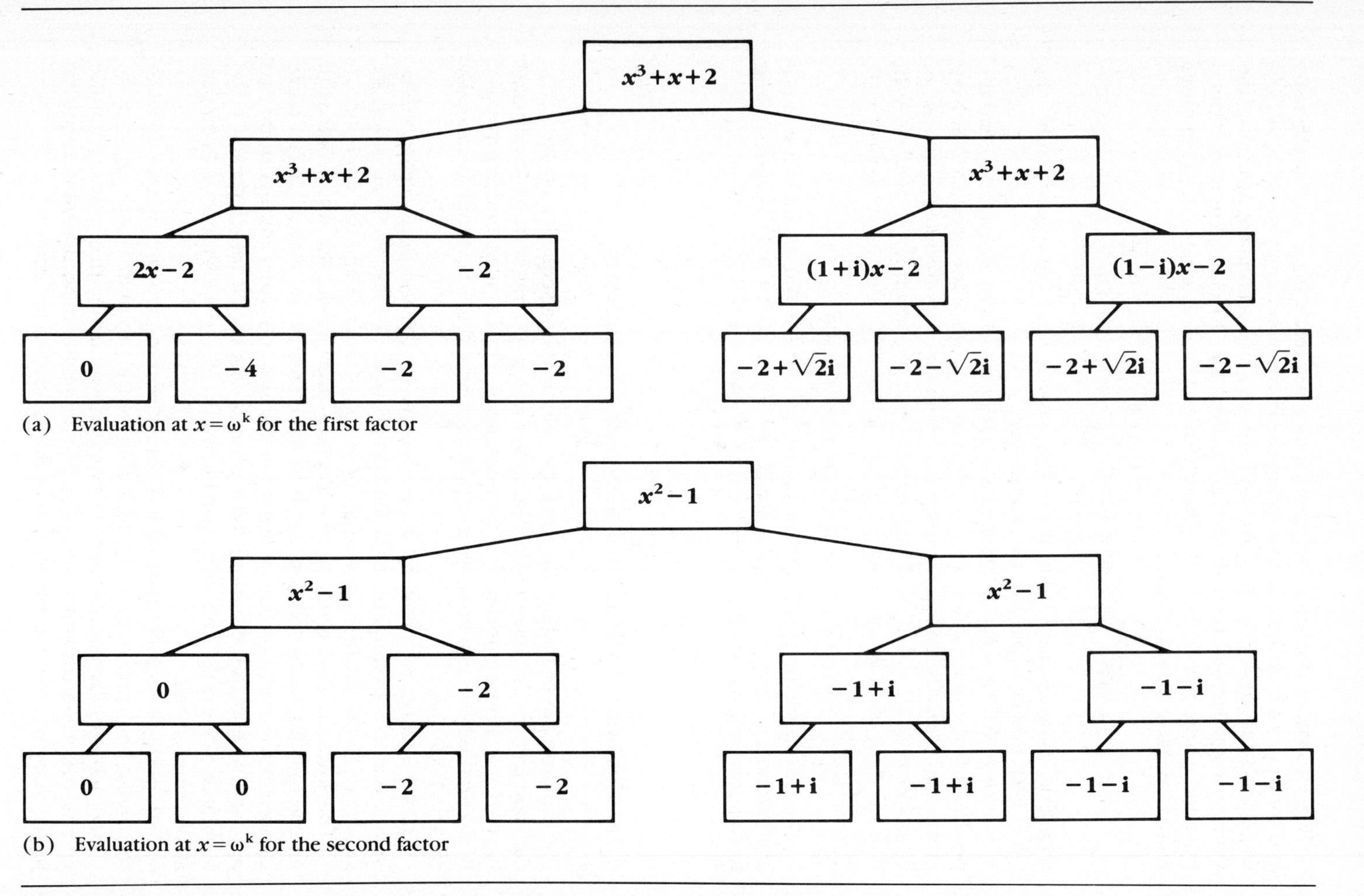

Figure 4.46 Polynomial Multiplication Using the Fast Fourier Transform

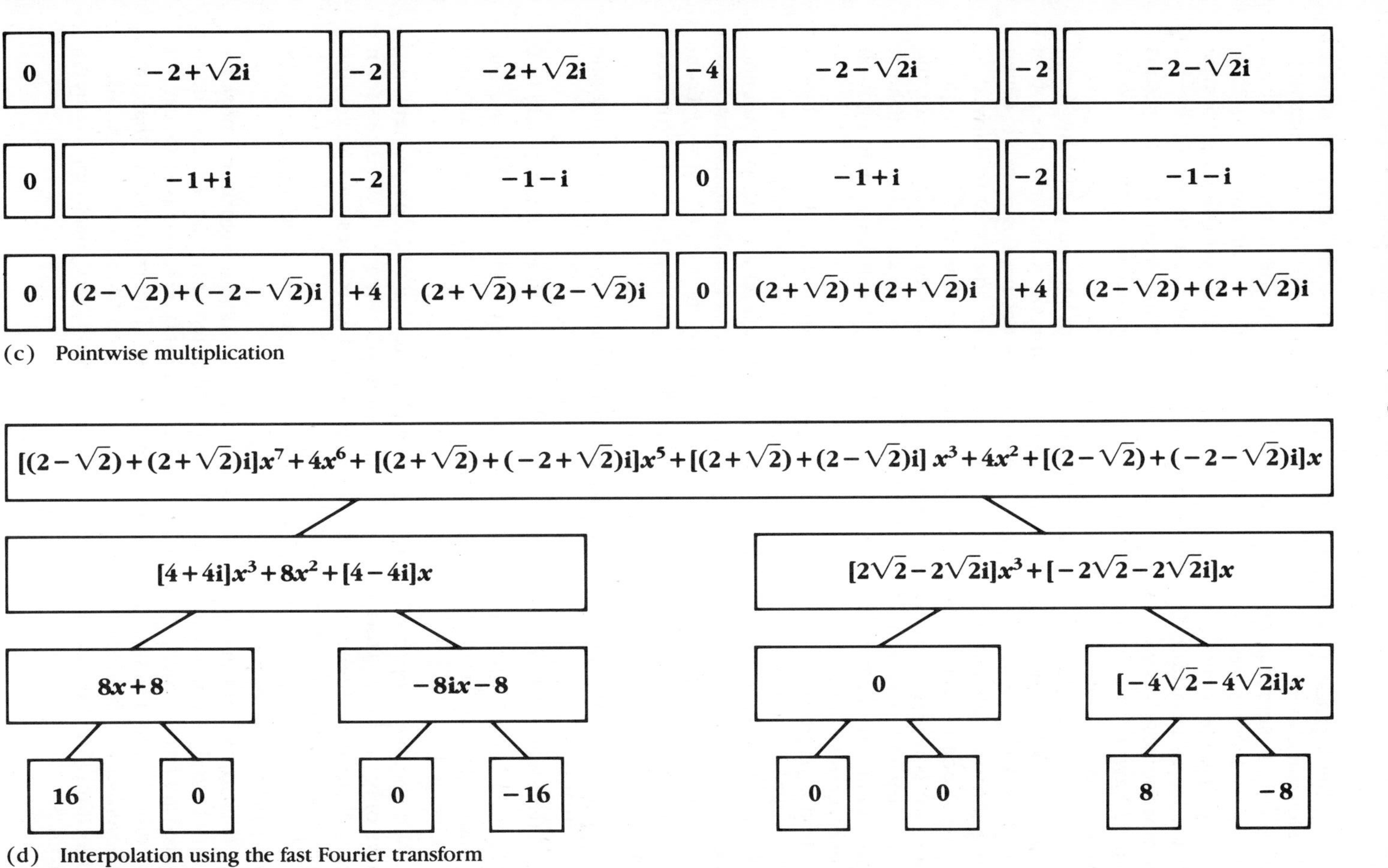

Figure 4.46 *(continued)*

Another argument was given in the language of tournaments in Section 4.2: No element can be ruled out as the maximum unless it has lost a comparison. Since $n - 1$ elements must each lose a different comparison, there must be at least $n - 1$ comparisons in any algorithm finding the maximum.

This argument is correct but hard to generalize. For example, it is not true in a sorting algorithm that the kth largest element must have lost $k - 1$ comparisons. If this were so, then any sorting algorithm based on comparisons would have time complexity $\Omega(n^2)$. In general one may show that $a < b$ by transitivity, without ever comparing a and b.

We may formalize and generalize the argument of Section 4.2 for the maximization problem by means of an *adversary argument.* Intuitively, the job of an adversary is to make a proposed algorithm work as hard as possible. Given an algorithm defined as a sequence of decisions, an adversary is allowed to answer each decision in any way consistent with previous decisions. In practice, we construct an adversary argument by constructing a strategy for the adversary and proving that this strategy will require the algorithm to make at least a certain number of decisions. Often this proof will shed some light on how to construct an efficient algorithm.

In the maximization problem, the adversary's strategy could be to declare as the loser an element that has already lost a comparison, if such an element exists. Primarily in preparation for a more complicated adversary argument, we formalize this adversary's strategy. Let L be the set of elements that have lost a comparison and N the set that has not. Initially $|L| = 0$ and $|N| = n$. From our point of view, we would like to make $|N| = 1$ and $|L| = n - 1$. From the adversary's point of view, there are 3 cases:

1. Two elements from N are compared; in this case, one goes to L and one remains in N
2. Two elements from L are compared; in this case both elements remain in L
3. An element from L is compared with an element from N; in this case the adversary declares the element from L the loser, and the sets are unchanged.

We see that progress is made by increasing the size of L one element at a time only in case 1. Thus finding the maximum will take $n - 1$ decisions from case 1. In case 3 the adversary has a choice of outcomes and makes the decision that gives us the least help.

We have shown that any algorithm for finding the maximum must take at least $n - 1$ comparisons.

This formal argument also suggests how an efficient algorithm must be constructed. We need to arrange that the only elements that are ever compared are two elements from N. In general, the best type of comparison cannot be done as often as we want; however in this case, if there are fewer than two elements in N, then the maximum has already been found. Note in the usual algorithm for this problem the only comparisons are between elements from N.

We may generalize this argument to the problem of finding the simultaneous maximum and minimum. Remember that for this problem, there was an obvious algorithm that did not take the minimum number of comparisons. So far the best algorithm we have seen was discussed in Section 2.16, and required $\lceil (3/2)n - 2 \rceil$ comparisons.

For this problem, the adversary ought to distinguish not only between losers and nonlosers, but between winners and nonwinners. Thus the adversary should maintain four sets:

W: the elements that have won but not lost

L: the elements that have lost but not won

N: the elements that have neither won nor lost

B: the elements that have both won and lost

Initially $|N| = n$ and $|W| = |L| = |B| = 0$. Ultimately we want $|N| = 0$, $|W| = 1$, $|L| = 1$, and $|B| = n - 2$. Notice that we care only about W and L, but if N is not empty, then we have not found the maximum or the minimum.

Our goal is to move elements from N to either W or L to B. If the adversary is to minimize this movement, the following strategy is appropriate for elements from the two sets given in each case (refer to Figure 4.47 to check that we are covering all the cases):

1. *N vs. N:* Here the adversary has no choice. One element goes to W, the other to L
2. *W vs. W:* No choice again. One element goes to B, the other stays in W
3. *L vs. L:* No choice again. One element goes to B, the other stays in L
4. *B vs. B:* No choice. Both elements remain in B

	N	W	L	B
N	(1) +2	(5) +1	(6) +1	(10) +1
W	(5) +1	(2) +1	(9) 0	(7) 0
L	(6) +1	(9) 0	(3) +1	(8) 0
B	(10) +1	(7) 0	(8) 0	(4) 0

Figure 4.47 Strategy for an Adversary

5. *W vs. N:* The element in N must move, but the element in W need not. Thus the adversary declares the element in W the winner, keeping it in W and moving the element from N to L
6. *L vs. N:* By symmetry with the previous case, the adversary declares the element from L the loser, keeping it in L and moving the element from N to W
7. *W vs. B:* The element in W is declared the winner, and neither element moves
8. *L vs. B:* The element in B is declared the winner, and neither element moves
9. *W vs. L:* Again the element in W is declared the winner, and neither element moves
10. *B vs. N:* It doesn't really matter what the adversary does. Let us say the element in B is the winner, staying in B, and the element in N moves to L.

To summarize, we may say that in case 1 two good things happen; in cases 2, 3, 5, 6, and 10 one good thing happens; and in the other cases nothing good happens. We may formalize this notion of goodness by saying that an element in N has goodness 0, an element in W or L has goodness 1, and an element in B has goodness 2. If the goodness of a state is the sum of the goodnesses of the n elements, then the goodness will increase by 2 in case 1; by 1 in cases 2, 3, 5, 6, and 10; and by 0 in the other cases. This information is summarized in Figure 4.47. Overall, the goodness starts at 0 and should increase to $2(n - 2) + 1(1) + 1(1) = 2n - 2$.

We may conclude immediately that at least $n - 1$ comparisons are required, since the goodness can increase by at most 2 per comparison. Unfortunately, we see that on the one hand this isn't news, because we already know $n - 1$ comparisons are required to find the maximum alone; on the other hand the bound of $n - 1$ is not achievable, because we can't remain in case 1 forever. In fact, since each comparison from case 1 removes 2 items from N, there are at most $\lfloor n/2 \rfloor$ possible comparisons from case 1. Thus the goodness can only increase by $2\lfloor n/2 \rfloor$ using case 1.

With this established, our best hope is to increase the goodness by 1 at each remaining comparison. This best hope requires $(2n - 2) - 2\lfloor n/2 \rfloor$ comparisons in addition to the $\lfloor n/2 \rfloor$ already accounted for. The best possible total is therefore $(2n - 2) - 2\lfloor n/2 \rfloor + \lfloor n/2 \rfloor$

$$= 2n - 2 - \lfloor n/2 \rfloor$$

$$= 2n - 2 - (n - \lceil n/2 \rceil)$$

$$= n - 2 + \lceil n/2 \rceil$$

$$= 2n/2 + \lceil n/2 \rceil - 2$$

$$= \lceil 3n/2 - 2 \rceil \text{ comparisons}$$

Now let us see if this total can actually be achieved. The only cases besides N *vs.* N that are permitted are W *vs.* W, L *vs.* L, W *vs.* N, L *vs.* N, and B *vs.* N. Assume the N *vs.* N comparisons are done first. Then there is at most one element in N that we may compare, say, with an element from W. Now every element is either in W or L, and we may make W *vs.* W comparisons until W has size 1, and L *vs.* L comparisons until L has size 1. After the N *vs.* N comparisons, only those comparisons that increase the goodness by one have been performed, so the lower bound we have found is in fact achievable (and is also achieved by the divide-and-conquer algorithm we have already discussed). A good exercise is counting the number of comparisons of each type, in order to double-check that the suggested algorithm does use the minimum number of comparisons.

4.19 Finite Automata and Pattern Matching

Section 4.19 has several functions: It presents a family of text-processing functions. It also contains an example of a useful type of argument called an *accounting argument.* Finally it provides a background for models of computation, both deterministic and nondeterministic, and gives some practice in their use. We need such models in order to answer the questions, what can be computed, and what can be computed efficiently? We discuss the first of these questions in Section 4.20 and the second in Chapter 5.

The text-processing issues relate to the problem of *pattern matching.* An instance of this problem is a string P called the *pattern* and a string S called the *subject string.* The problem is to find the first occurrence, if any, of the pattern string as a substring of the subject string.[10] Here we assume that the length n of S is greater than the length m of P.

A natural nondeterministic algorithm exists for this problem. One merely guesses the point in S where the occurrence of the pattern begins and then verifies in $\Theta(m)$ time that the pattern does occur in S. A deterministic version of this algorithm would require $O(n)$ guesses in the worst case to find the starting place and $O(m)$ steps to test each guess. It would therefore have $O(mn)$ worst-case behavior.

The nondeterministic algorithm can be modeled by a diagram. Suppose that P = "ssesse." Say that the algorithm is in *state k* if the first k characters of the pattern have been checked. In Figure 4.48 state k is represented by a circle containing k. A labeled arrow between two states means that the algorithm moves from the first state to the second by reading and consuming the character serving as the label. Sometimes there is more than one character allowing a

[10] Natural generalizations of this problem would require all occurrences to be found or allowance of more complicated patterns than single strings. For example, one might want to check a subject string for an occurrence of either of the substrings "June" or "July."

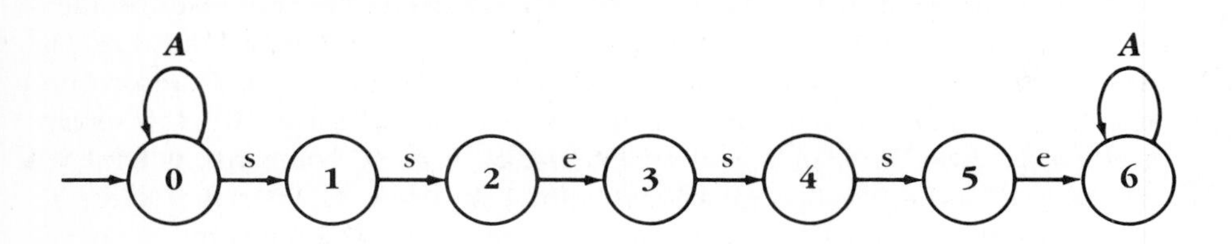

Figure 4.48 An NFA Used in Pattern Matching

move between two states; in this case arrows are labeled by sets of characters. In the figure two arrows are labeled A, representing the entire alphabet or character set. The first A means that the algorithm can consume one input symbol at a time while remaining in state 0 (where it starts). Once the algorithm sees an "s" and guesses that it is time to move to state 1, it reads the rest of P and enters state 6. At that point the algorithm may read any number of characters and remain in state 6. Any string S for which there is a series of legal moves ending in state 6 (or in general, state m) contains a copy of P.

For example, the string "assesses" of length $n = 8$ would visit in order the sequence (0 0 1 2 3 4 5 6 6) of states. The state 0 at the beginning of the sequence represents the state before any symbols are consumed. So the sequence has length $n + 1 = 9$. The string "reassessed" would visit the sequence (0 0 0 0 1 2 3 4 5 6 6).

There is no way of getting to state 6 if S = "assessor." A deterministic algorithm would need to provide a unique choice of moves for every possible character for every state. In particular, a way of handling mismatches is necessary. For example, if S had the prefix "asset," the algorithm could visit states 0, 1, 2, and 3, but then could not proceed to state 4, since that move requires a "t" rather than an "s." The algorithm could not simply halt, since a later occurrence of P is possible.

Fortunately, at least for this restricted class of problems, there is a way of determinizing every nondeterministic algorithm. More precisely, for every NFA there is a new DFA that does the same thing and for which the set of strings reaching the final state is the same as for the NFA. Describing the process in terms of diagrams like Figure 4.48 is easier than doing so in terms of algorithms, but to do so we need some notation and terminology.

Def 4.27 An *alphabet* is a finite set of characters.

Def 4.28 A *deterministic finite automaton* or *DFA* consists of a finite set of *states,* a fixed alphabet, a special state called the *start state,* a set of states called *final states,* and a *transition function d* giving a destination state for every state and character. A *nondeterministic finite automaton* or *NFA* has the same definition except that the transition function returns a set of destination states.

In our discussion, states of NFAs will always be numbered 0 through m. The start state will always be state 0 and state m will be the only final state.

Figure 4.48 represents an NFA. Here d(0, s) = {0, 1}, d(2, e) = {3}, and d(1, e) = ∅.

In both DFAs and NFAs it is useful to define precisely what states are reached by a given string. For DFAs, any string will reach a unique state. We let "_" represent the empty string.

Def 4.29 For any DFA, the function s is defined for all strings over its alphabet by

1) s(_) = 0, and 2) s(xa) = d(s(x), a) where $|a| = 1$.

For an NFA, s is defined by

1) s(_) = {0}, and 2) s(xa) = the union of d(p, a) over all p in s(x).

Def 4.30 A DFA *accepts* x iff s(x) is a final state. An NFA *accepts* x iff s(x) contains a final state. In either case, the *language accepted* by a finite automaton M is just $\{x|M \text{ accepts } x\}$. Two finite automata accepting the same language are *equivalent*.

For the time being, we will consider our pattern-matching algorithm as a decision algorithm that merely needs to decide whether S contains a substring of P. Eventually we will consider how to find the position in S where the substring occurs.

The determinization algorithm is quite straightforward. We merely allow the possibility of being in several states at once and treat the set of states that a computation is in as a single state. For the NFA above, the determinization algorithm would treat {0, 1} as a single state, so that d(0, a) would be a single state. Of course, it would then have to define d on all characters for the state {0, 1}. Intuitively, the meaning of the state {0, 1} is that the new DFA is in states 0 and 1 at the same time.

The general approach is as follows: Given a set Q of states for an NFA, let 2^Q be the set of states of the new DFA. If 0 is the NFA's start state, then {0} will be the DFA's start state. The final states of the DFA are those containing a final state of the NFA. The DFA's transition function is defined by saying that if W is a set of states, then state r is in d'(W, a) iff it is in d(t, a) for some t in W.

For example, for the DFA constructed from the preceding NFA, d'({0, 2}, a) = {1, 3} since 1 is in d(0, a) and 3 is in d(2, a).

This new DFA is illustrated in Figure 4.49. Part (a) of that figure is the original NFA. In (b) we have labeled the states of the DFA with sets without brackets or commas, so that for instance {0, 1, 4} appears as 014. In (c) we have given each state a label W_k. If we let $P_{i,j}$ be the substring of P from position i to position j, then W_k corresponds to the prefix $P_{1,k}$ of P. In (d) we have used these prefixes to label states. In (e) we have labeled each state W_r with $P_{1,s}$ for each s in W_r. In (b) through (e) all transitions not shown are to state 0.

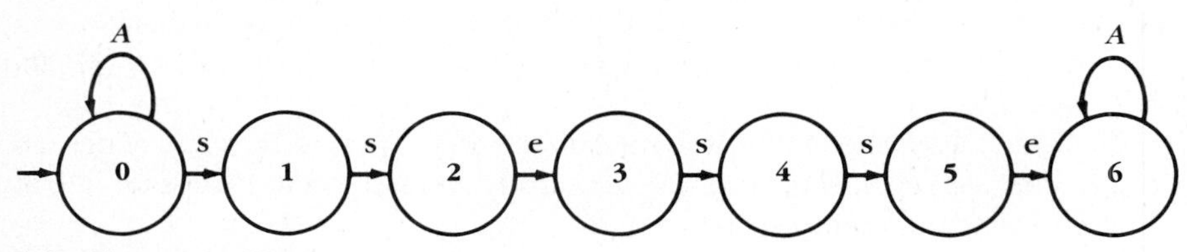

(a) The original NFA

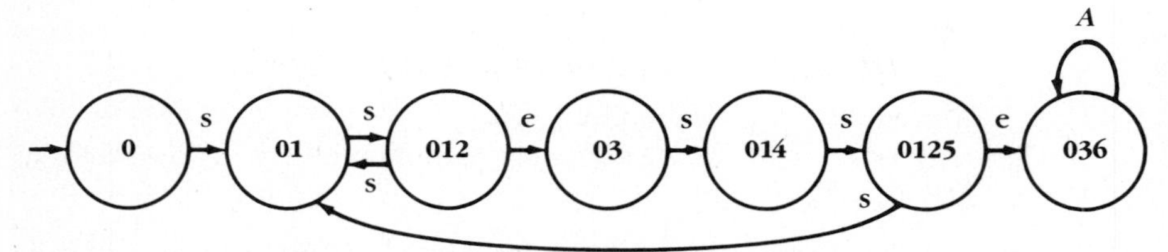

(b) An equivalent DFA (transitions not shown are to state 0)

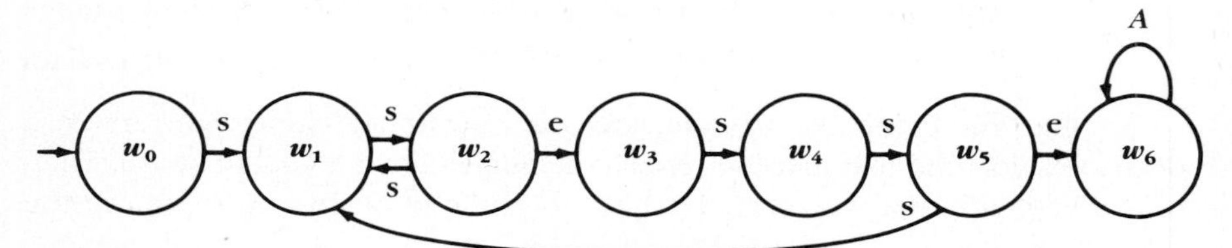

(c) The DFA with names given to sets of states (transitions not shown are to state w_0)

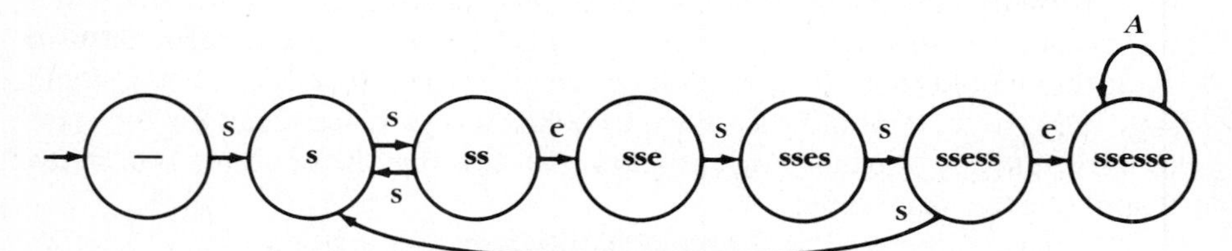

(d) The DFA with states labeled by longest prefixes matching suffixes (transitions not shown are to state—)

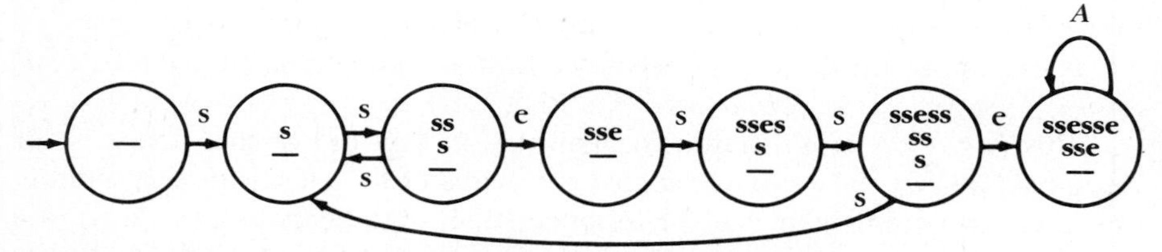

(e) The DFA with states labeled by all prefixes matching suffixes (compare (b)) (transitions not shown are to state—)

Figure 4.49 An NFA and an Equivalent DFA

Notice that s is W_r iff the prefix of length s is a suffix of the prefix of length r. Since the length of the suffix must also be s, this means $P_{1,s} = P_{r-s+1,r}$. Shortly we will show that this holds in general for DFAs constructed from NFAs.

We also use the notation that S_i is the ith character of S, and that P_j is the jth character of P.

The new transition function d′ is given in tabular form in Figure 4.50. Note that all characters not appearing in the pattern are equivalent. We have used the character "z" to represent these characters.

Remember that our motivation for constructing DFAs is to determinize nondeterministic algorithms like the pattern-matching algorithm previously described. At this point it may seem that we have not gained much, since the new DFA will have 2^{m+1} states. We can show, however, that in the case of NFAs constructed from particular pattern strings, enough states can be combined and discarded so that the DFA has no more states than the NFA. We establish this as the consequence of several other results. Referring to the example of Figure 4.49 and 4.50 when reading these theorems is a good idea.

Theorem Suppose that in an NFA state 0 is the only start state, state m is the only final state, and $d(r, c) = \emptyset$ for all r and c except that $d(0, c)$ contains 0 for all $\{c\}$, $d(m, c) = \{m\}$ for all c, and $d(r-1, P_r)$ contains r for all $1 \leq r \leq m$. Then in the DFA constructed from the NFA, we may combine all final states without changing the language accepted.

Proof It is enough to show that in the DFA all transitions from final states go to final states. Let W be a final state of the DFA. Then W contains m, the only final state of the NFA. Let c be in the alphabet and $X = d'(W, c)$. By definition of d′, since m is in $d(m, c)$ and m is in W, $X = d'(W, c)$ contains m and is therefore a final state. ■

Theorem In an NFA of the form of the previous theorem, s is in W_r iff $P_{1,s} = P_{r-s+1,r}$. (We assume this condition holds vacuously if $s = 0$.)

Proof by induction on r. The induction hypothesis is the statement of the theorem. The proof is mostly a matter of unraveling definitions and showing that the match given by the induction hypothesis is extended by one more character.

Basis: $W_1 = d'(\{0\}, P_1) = \{d(0, P_1)\} = \{0, 1\}$. And $P_{1,s} = P_{2-s,1}$ iff $s = 0$ or 1.

Induction step ($r > 1$): By definition, $W_r = d'(W_{r-1}, P_r)$. By definition of d′, s is in $d'(W_{r-1}, P_r)$ iff $s = d(t, P_r)$ for some t in W_{r-1}. By induction, t is in W_{r-1} iff $P_{1,t} = P_{r-t,r-1}$. By checking the transitions of the NFA we see that $s = d(t, P_r)$ is true iff $s = 0$ or both $t = s - 1$ and $P_r = P_s$.

If $s = 0$ then the condition holds and s is in W_r. Otherwise, we can replace t by $s - 1$ to get $P_{1,s-1} = P_{r-s+1,r-1}$ and $P_s = P_r$. This holds iff $P_{1,s} = P_{r-s+1,r}$, so we are just extending the match by one more character. ■

Corollary The highest numbered state in W_r is state r.

Proof The longest prefix of $P_{1,r}$ matching a suffix of $P_{1,r}$ has length r. ■

d′	s	e	z
{0}	{0, 1}	{0}	{0}
{0, 1}	{0, 1, 2}	{0}	{0}
{0, 1, 2}	{0, 1}	{0, 3}	{0}
{0, 3}	{0, 1, 4}	{0}	{0}
{0, 1, 4}	{0, 1, 2, 5}	{0}	{0}
{0, 1, 2, 5}	{0, 1}	{0, 3, 6}	{0}
{0, 3, 6}	{0, 3, 6}	{0, 3, 6}	{0, 3, 6}

Figure 4.50 Table for the Transition Function d′ for the DFA of Figure 4.49b

Theorem Given an NFA as in the previous two theorems, the W_r for $0 \le r < m$ are all the accessible nonfinal states of the DFA.

Proof We will show by induction on r that all states accessible from W_r are of the form W_s. Again checking Figure 4.49 for examples may help you follow this proof.

Basis: $W_0 = \{0\}$ is the start state, so is accessible.

Induction: We need to find $X = d'(W_r, c)$ for each c. By definition, $X = \{t | t = d(s, c)$ for some s in $W_r\}$. For a fixed c, let q be the largest such t. Since q is in X, $q = d(s, c)$ and s is in W_r. If $q = 0$ then $X = W_0$ and we're done. Otherwise, by the assumed properties of d, $q - 1 = s$ and $c = P_q$, so $q - 1 = s$ is in W_r. We claim $X = W_q$.

To show that X is contained in W_q, we choose an arbitrary t in X. By definition of X, $t = d(s, c)$ for some s in W_r. If $t = 0$ then the prefix matches a suffix trivially. Otherwise, the transitions of the NFA require $s = t - 1$ and $P_t = c = P_q$. Since $t - 1 = s$ is in W_r, we have that $P_{1,t-1}$ matches a suffix of $P_{1,r}$. We may assume $t < q$. Since $q - 1$ is in W_r, we have that $P_{1,q-1}$ matches a longer suffix of $P_{1,r}$. Thus $P_{1,t-1}$ matches the suffix of $P_{1,q-1}$ of length $t - 1$, that is, $P_{q-t+1,q-1}$. Finally, since $P_t = P_q$, we can extend the match one more character to get $P_{1,t} = P_{q-t+1,q}$. Thus t is in W_q.

This proof may be reversed to show that W_q is contained in X. Suppose that t is in W_q. Then $P_{1,t} = P_{q-t+1,q}$ and so $P_t = P_q$. Deleting the last character, we get $P_{1,t-1} = P_{q-t+1,q-1}$. And since $q - 1$ is in W_r, $P_{1,q-1} = P_{r-q+1,r}$. Since $t \le q$, $P_{q-t+1,q-1}$ equals both $P_{1,t-1}$ and $P_{r-t+1,r}$. So $P_{1,t-1} = P_{r-t+1,r}$, and therefore $t - 1$ is in W_r. Since $P_t = P_q = c$, $d(t - 1, c) = d(t - 1, P_t) = t$. Thus t satisfies the conditions for being in X. ■

In the example of Figure 4.49, each set W_r was obtained from some earlier W_k by adding the single state r. The state k will just be the second largest state in W_r. This condition is true, happens in general, and will be important for pattern matching. We have the following corollary:

Corollary For NFAs as in the previous three theorems, there is a function f such that for $r > 0$, $W_r = \{r\} \cup W_{f(r)}$.

Proof by induction on r. Recall that $W_r = d'(W_{r-1}, P_r)$.

Basis: $W_1 = \{0, 1\}$ so we may let $f(1) = 0$.

Induction step: By induction, $W_{r-1} = \{r - 1\} \cup W_{f(r-1)}$. Thus $d'(W_{r-1}, P_r) = d(r - 1, P_r) \cup d'(W_{f(r-1)}, P_r)$. But $d(r - 1, P_r) = \{r\}$, and by the proof of the previous theorem, $d'(W_{f(r-1)}, P_r) = W_q$, where q is the largest t such that $t = d(s, P_r)$ for s in $W_{f(r-1)}$. By checking the transitions of the DFA, we see $s = 0$ or $q = t = d(s, P_r) = s + 1$. The second case is true only if $P_r = P_{s+1}$. So q is one more than the largest s in $W_{f(r-1)}$ such that $P_{s+1} = P_r$ (or 0 if there is no such s). We are done if we let $f(r) = q$. ∎

Construction of the DFA leads to a pattern-matching algorithm requiring only $O(n)$ character comparisons. Unfortunately, this does not include the time required to construct the DFA. Given the NFA, we need to construct the values of d′ for each of $m + 1$ states and for each character in the alphabet. Even realizing any characters not in the pattern may be considered equivalent won't help a great deal.[11] There are still $\Theta(m)$ inequivalent characters in the worst case, so the table will have $\Theta(m^2)$ entries. Additionally, the method of the previous theorem for constructing each table entry might itself require $\Theta(m)$ time.

We do have two advantages: A DFA for P can be used for any number of subject strings; in addition, the technique of DFAs generalizes to other types of patterns than single strings. Nevertheless, in many cases the time to construct the DFA (analogous to the time required for compilation in more general programming) is unacceptable.

One way to simplify the DFA is to note that what we care about most is whether or not the next character matches. If so, we proceed to the next state. If not, then we back up to some previous state. In some cases the previous state is independent of the character seen; in other cases this state depends on the character seen. Our point of view in the following example is that we will only consider the two alternatives of match or mismatch; in the case of mismatch we will make the most conservative assumption possible. If we think of a mismatch as allowing the advance of P rightward in S, then we advance P as little as possible.

For example in Figure 4.51, we assume that the first two characters of the pattern P have found matches. If the next character of S is "e," we have a match for P_3 and continue with P_4. If the next character is "s," then we can only advance P one position. If the next character is "z," then we may advance P three positions. If we only know that there is a mismatch and not what the next character is, the only safe move is to advance only one position.

---

[11] This is true partly because when using the DFA we'd need to determine the equivalence class of each character of S.

```
P:        s s [e] s s e       -->   P:          s s e [s]
S: . . .  s s [e] . . .             S: . . .    s s e [.] . .
```

(a) If next character of S is "e"

```
P:        s s [e] s s e       -->   P:          s s [e] s s e
S: . . .  s s [s] . . .             S: . . .  s s s [.] . .
```

(b) If next character of S is "s"

```
P:        s s [e] s s e       -->   P:                [s] s e s s e
S: . . .  s s [z] . . .             S: . . .  s s z [.] . .
```

(c) If next character of S is "z"

Figure 4.51 Action Taken in Pattern Matching Based on Different Possibilities for the Next Character of the Subject String

The state corresponding to the least advance of P from state W_r is the successor of the second-highest-numbered state of W_r. This second-highest-numbered state corresponds to the second-longest prefix of $P_{1,r}$ matching a suffix, that is, the next possible match to try to extend. If $r = 2$ in our sample DFA, then we have matched "ss." If we cannot match the next character with $P_{r+1} =$ "e," then our next-best hope is to assume that the second-longest prefix "s" has matched and try and extend this match. This is another way of describing the conclusion of the previous paragraph.

For arbitrary W_r, the second-longest prefix previously described corresponds to the function $f(r)$ of the last corollary. The corollary gives a dynamic programming algorithm for computing f. We showed there that $f(1) = 0$, and in general $f(r)$ is the successor of the largest s in $W_{f(r-1)}$ such that $P_{s+1} = P_r$ (or 0 if there is no such s). Note that successive states in W_t are just $f(t)$, $f(f(t))$, $f(f(f(t)))$, and so on; therefore, all the states in $W_{f(r-1)}$ can be found just by knowing f.

For the actual pattern matching, we need not know exactly the function f but instead what to do in case of a mismatch. If P_j fails to match S_k, we need to know which two characters to compare next or, equivalently, how far to advance P and where in P to begin comparing. This can best be done by means of a *TryNext* function, which tells which character of the P to check next if there is a mismatch.

If there is a mismatch at P_j, then the longest successful match goes through P_{j-1}. The next-longest successful match goes through $P_{f(j-1)}$, so the next character of P to compare is $P_{1+f(j-1)}$. Thus $\text{TryNext}(j) = 1 + f(j - 1)$.

Note that the character of S to be compared with P is the same character resulting in the mismatch, unless the mismatch was with P_1. In this case P_1 should be compared with the next character of S.

```
 1. ProcessPattern(P)             {given a pattern P, returns a table
 2.                     TryNext for the pattern-matching Algorithm 4.32}
 3. TryNext(1) := 0                                            {basis}
 4. for j from 2 to m              {now find TryNext(j) by searching}
 5.    k := TryNext(j − 1)             {through W_{j−1} to find first}
 6.    while k > 0 and P_k ≠ P_{j−1}
 7.       k := TryNext(k)              {character that extends match}
 8.    end while                                {i.e., to find f(j − 1)}
 9.    TryNext(j) := k + 1              {TryNext(j) = 1 + f(j − 1)}
10. end for
11. return TryNext
```

Algorithm 4.31 Preprocessing Algorithm for Pattern Matching

The iterative Algorithm 4.31 calculates TryNext using the bottom-up calculation typical of dynamic programming.

For example, for the pattern "ananas" we have the table of Figure 4.52.

Now it is simple to present the actual pattern-matching Algorithm 4.32. The two algorithms of this section are essentially those of Baase (1978) and attributed to Knuth, Morris, and Pratt (1977).

An example of the behavior of the pattern-matching algorithm is given in Figure 4.53. For each position of the pattern, we show which of its characters are compared to characters of the subject string. One comparison, corresponding to the underlined character, can be assumed to give a match and is not actually made. In this case the algorithm "knows" that since P_1 matches P_3 and P_3 matched S_4, P_1 must match S_4. Note that the last character of S involved in a comparison is S_{18}, but that only 22 comparisons were allowed altogether.

The problem with analyzing Algorithm 4.31 is that k varies unpredictably. But we may find a bound on the number of times the loop k controls is executed by an *accounting argument*.

The argument goes as follows: The initial value of k is TryNext(1) = 0. Any time k is incremented, it is incremented by 1 (by the last statement of the outer loop and the first statement of its next iteration, given that j increases by 1 in between). Since k can only be incremented when j is, k is incremented by $m - 1$ in total. This corresponds to a household that begins with no wealth and

k	1	2	3	4	5	6
P_k	a	n	a	n	a	s
TryNext(k)	0	1	1	2	3	4

Figure 4.52 TryNext Function for the String "ananas"

```
 1. KMP_PatternMatching(P, S, TryNext) {takes a pattern P, a function
 2.                                     TryNext as constructed in Algorithm 4.31,
 3.                                     and a subject string S and returns the
 4.                                     location of the first occurrence of P as
 5.                                     a substring of S}
 6. j := 1                              {initialize pointer into pattern P}
 7. i := 1                              {and pointer into subject string S}
 8. while i ≤ n
 9.    while j > 0 and P_j ≠ S_i        {shift pattern right until}
10.       j := TryNext(j)               {a partial match can be extended}
11.    end while                        {or a match of P and S is found}
12.    if j = m then                    {if loop ends with complete match}
13.       return i − (m − 1)            {S_i matches P_m, so S_{i−(m−1)} matches P_1}
14.    else                             {if a partial match can be extended}
15.       i := i + 1                    {then continue with next}
16.       j := j + 1                    {characters of P and S}
17.    end if
18. end while
```

Algorithm 4.32 Knuth-Morris-Pratt Pattern Matching (Given Failure Function TryNext)

gets $m - 1$ units of income. Just as the household can only spend $m - 1$ units, so can k be decremented at most $m - 1$ times. (This happens in the inner loop; recall that TryNext[j] $< j$.) Accordingly the inner loop can be executed at most $m - 1$ times throughout the entire algorithm. The time complexity is therefore $O(m)$. In fact, there can be at most $2m - 2$ character comparisons.

P = "ananas"

```
                      S: "Panamanian bananas are good"
                          a
                           anan
                             an
                              a
                               ana
                                a
                                 ana
                                  a
                                   a
Successive shifts of P              ananas
```

Figure 4.53 Trace of the Knuth-Morris-Pratt Pattern-Matching Algorithm

The analysis of Algorithm 4.32 is similar. The variable j can only be incremented by 1 when i is, and i is incremented at most $n - 1$ times. Thus j can be decremented (in the inner loop) at most $n - 1$ times, and the algorithm is $O(n)$. In fact, there can be at most $2n$ character comparisons.

Clearly this algorithm is an improvement over the straightforward algorithm suggested at the beginning of this section.

Amazingly enough, there is another pattern-matching algorithm that in some cases does not have to look at all characters of the subject string. A hint how this might happen is given by Figure 4.51. Assume that it is the rightmost character P_m of P that is compared first ["e" in (c) of the figure], and that enough preprocessing has taken place to know that the character ("z" in the figure) P_m is compared to appears nowhere in the pattern P. Then the pattern can be advanced a full m characters (as in the figure), based on only one comparison.

The algorithm does more work than the Knuth-Morris-Pratt algorithm in the case of a character mismatch, determining what type of mismatch occurred. This algorithm is attributed to Boyer and Moore (1977).

4.20 Languages and Models of Computation

The DFAs of the previous section are not a sufficiently general model of computation. In other words, there are decision problems that have solution algorithms but that can't be solved by DFAs. And of course there are optimization problems that are not decision problems. A third objection, that DFAs only concern problems of language acceptance, does not hold water, since instances of a problem may always be represented as strings in some language.

We first consider the second objection.

It turns out to be easy to reformulate optimization problems as decision problems. We simply add a target parameter T to the problem and ask whether or not there is a solution as good as T. For example in ZOK, the question becomes whether or not there is a subset of items whose total weight does not exceed the capacity of the knapsack and whose total value is at least T. For the TSP, the question becomes whether or not there is a permutation of the vertices such that each vertex is adjacent to its successor in the permutation, the last is adjacent to the first, and the sum of the edge costs involved is at most T.

You may imagine this restriction represents a loss of generality. However, if we have an efficient algorithm for solving the related decision problem, we may efficiently solve an optimization problem, at least if the value of the optimal solution is known to be an integer.

Assume for definiteness that we have a maximization problem. We first find a value M of T so large that the answer to the decision problem for this M is no. Such a value of T may often be found from the problem instance. For example in ZOK, any value greater than the sum of all the profits will suffice. In general increasing powers of 2 may be tested until a sufficiently large value is found for T. Then a binary search can be used to find the boundary between values of T for which there is a solution and values of T for which none exists. Clearly this approach will work for minimization problems as well.[12]

For example, if the optimal solution is 43, the related decision problem could be tried for 1, 2, 4, 8, 16, 32, 64, 48, 40, 44, 42, and 43.

The first objection posed at the beginning of this section is more serious. One question for which there is a straightforward algorithm but which can't be solved by a DFA is whether or not a string of bits has an equal number of 0's and 1's. We can easily see that this question can't be solved by a DFA.

Define 0^n in the natural way (i.e., 0^0 is the empty string, $0^n = 0^{n-1}0$). If, as in the previous section, we identify strings with the states to which they are taken, then each different 0^n will be taken to a separate state. This is true because if 0^m and 0^n are taken to the same state, where $m \neq n$, the string 1^n will have to go from that state to a final state (so that 0^n1^n is accepted). But then 0^m1^n would be accepted also, and it shouldn't be. Therefore an infinite number of states are required, which is not possible in a DFA.

Intuitively, the problem with DFAs is they have no memory, except for that required to keep track of the current state. The *Turing machine* model of computation in the following definition will have an infinite tape for a memory.[13]

You might object to a model of computation containing an infinite memory, since all real-world machines have finite memory. However, limiting the theoretical computing power of algorithms based on contemporary technology seems even more inappropriate.

Def 4.33 A (deterministic) *Turing machine*, or *TM*, consists of an input alphabet and a set of *states*, including a *start state* 0, an *accepting state* y, and a *rejecting state* n. A TM also has a tape alphabet containing both the input alphabet and a ***blank*** symbol not in the input alphabet. Finally, a TM has a *transition function d* that, given a tape symbol and a state not in $\{y, n\}$, returns a triple consisting of a new tape symbol, a new state, and one of the directions $+1$ or -1.

The idea of the transition function is that the tape is divided into squares, and there is always a current tape square containing the current tape symbol. A

---

[12] It will also work in cases where the value of a solution is known to be a terminating decimal (or the binary equivalent). Note that the numbers appearing in the problem instance must have a finite representation, so that irrational numbers cannot be represented as infinite decimals.

[13] There is an intermediate class of abstract machines called *pushdown automata*, whose nondeterministic versions accept precisely the same languages as are generated by the context-free grammars of Section 2.19—except that CFGs of the special type defined there could not generate the empty string.

TM transition will, depending on this symbol and the current state, replace the current tape symbol with a new symbol, change state, and move one square either to the left (for the direction -1) or to the right (for the direction $+1$).

We make a few conventions. As in the previous section, states will be numbered and the start state will be state 0. We consider the tape as being infinite in both directions and of the current tape square at the beginning of the computation as being square 0. An input string of length n is stored in order in squares 0 through $n - 1$. Initially, the rest of the tape contains the blank symbol.

The progress of the computation at any time may be given by specifying an *instantaneous description*, or *ID*, consisting of the tape contents and the current state and tape square. We will describe an ID by giving the tape contents and inserting the state, underlined, before the current symbol. Strings of blanks at the beginning and end of the tape will not be written.

For example, at the beginning of the computation on string x, the ID is $\underline{0}x$. A rightward move, say, given by the value $d(m, a) = (n, b, +1)$ of the transition function, changes the ID $y\underline{m}ax$ to $yb\underline{n}x$. A leftward move, say, given by the value $d(m, a) = (n, b, -1)$, changes the ID $yc\underline{m}ax$ to $y\underline{n}cbx$. Note that the computation halts upon entering state y or state n.

Def 4.34 A TM *accepts* the string x iff there is a sequence of IDs beginning at the ID $\underline{0}x$ and ending at some ID corresponding to the final state y, with each ID resulting from its predecessor by the use of the transition function. The *language accepted* by a TM M is $\{x|M \text{ accepts } x\}$.

Note that there is no notion of consuming input symbols, so that the TM need not look at all the input. Also, there is no requirement that the TM halt, although it can only accept if it halts in state y. The TM may fail to accept by halting in state n or by computing forever.

Figure 4.54a shows an example of a Turing machine. Each edge of the underlying graph is labeled by a triple corresponding to the input tape symbol, the new tape symbol, and the new direction. The blank symbol is denoted by "_" and transitions not shown are to the rejecting state n.

This TM is designed to accept the language consisting of those strings with the same number of 0's as 1's. Recall this language could not be accepted by a DFA, so TMs are more powerful than DFAs. The underlying algorithm is as follows: in state 0 we move to the right looking for a 0 or 1. If a 0 is seen, it is crossed off and state 1 is entered, in which case we move right and look for a 1. Similarly seeing a 1 in state 0 causes a move to state 2 after the 1 is checked off. In state 2 we move right looking for a 1. In state 1 or 2 if the desired symbol is found we enter state 3. If the desired symbol is not found before reaching a blank at the end of the input, the input is rejected (due to too many 1's or 0's). State 3 moves left looking for a blank to the left of the input; when one is found, state 0 is entered to look for more pairs of 0's and 1's.

The computation for input string "110001" in (b) of the figure consists of a

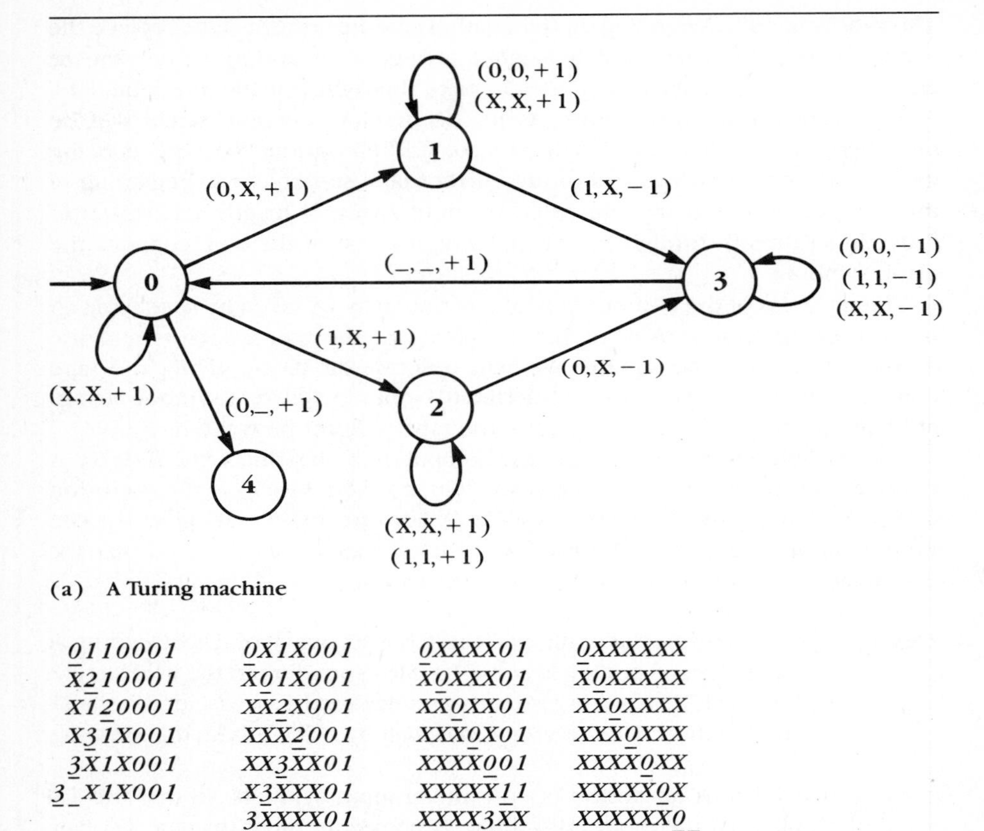

(b) A computation of the Turing machine of (a) on input 110001

Figure 4.54 A Turing Machine and One of Its Computations

sequence of IDs, to be read downward. An input string of "1100010" would be processed the same way up to the ID XXXXXX00, where it would check off the 0, enter state 1, and immediately find a blank. Thus this input would be rejected.

The given TM is not the most efficient possible.

It turns out that the notion of a Turing machine, or more precisely a Turing machine that halts on all inputs, appears to correspond to our intuitive notion of an algorithm. Since the intuitive notion of an algorithm can't be formalized, this statement, known as *Church's thesis,* can never be proved. A number of alternate formalisms exist, including models of real-world machines, that are equivalent in the sense that they accept the same class of languages as TMs. (Here we do not require TMs to halt on all input.) No formalism has ever been shown to accept a larger class of languages.

Showing that a real-world machine, say, is no more powerful than a TM is a long and uninstructive process. But we can imagine implementing arithmetic operations in terms of a TM, simulating random-access memory operations by search along the TM tape, and so on.

The time complexity of a Turing machine computation on a particular string is just the number of transitions needed to either accept or halt in a nonaccepting state. This time complexity need not be the same as for equivalent models, but polynomial-time computations in other models turn out with one important exception to be polynomial-time computations in TMs.[14] The exception is that the natural way of simulating nondeterministic TMs by deterministic TMs may require exponential time to simulate polynomial-time computations in the nondeterministic TMs. Since nondeterministic algorithms will be important in Chapter 5, we discuss this issue next.

We may define nondeterministic TMs as we did NFAs. We simply require the transition function to take on sets of triples instead of single triples. However, this is not convenient for the discussion of nondeterministic algorithms in Sections 5.1 and 5.2.

This version of a nondeterministic TM allows a choice of transitions at any time. We find it easier to work with nondeterministic TMs where the nondeterminism is isolated into a separate guessing stage at the beginning of the algorithm. Our definition will correspond to the intuitive notion of solving a decision problem by making a sequence of guesses and then verifying that the guesses lead to a solution of the problem. The point is that this verification stage will be deterministic.

The guessing stage will correspond to the start state g of a nondeterministic TM. All legal guesses will be tape symbols different from the input symbols. All transitions from state g will place one symbol in the current tape square, move left, and either remain in state g or move to a new state r. The role of state r is to move right until the first input symbol is seen. When this happens, a transition is made to the start state of the deterministic verification phase. This phase is allowed to consult the guesses made in the guessing phase.

As for deterministic TMs, we assume a unique accepting state and a unique rejecting state with no transitions defined from either.

---

[14] The definition of polynomial time complexity depends on the definition of the size of a problem instance. We assume, however, that each integer k has size $\Theta(\log k)$. This means, for example, that the naive deterministic algorithm for determining whether or not an integer k is prime (by testing all smaller integers as factors) has time complexity proportional to k but exponential in the size of k. Consequently it is not a polynomial-time algorithm. If we were allowed to assume the size of k was just k, this would be a polynomial-time algorithm.

Def 4.35 A *nondeterministic Turing machine,* or *NDTM,* consists of

1. a set S of states containing states g, r, 0, y, and n,
2. a guessing alphabet G and an input alphabet, both subsets of a tape alphabet T,
3. a blank symbol _ that is an element of the tape alphabet but not the guessing or input alphabets, and
4. a transition function d defined on $(S - \{g\}) \times T$ such that
   a. $d(g, \_) = \{g, r\} \times G \times \{-1\}$
   b. $d(r, b) = (r, b, +1)$ for all b in $\{\_\} \cup G$
   c. $d(r, b) = (0, b, 0)$ for all b in the input alphabet
   d. $d(y, b) = (y, b, 0)$ for all b in T
   e. $d(n, b) = (n, b, 0)$ for all b in T
   f. $d(s, b)$ is in $S \times T \times \{+1, -1\}$ for all b in T and all s in $S - \{g, r, y, n\}$.

The initial ID is gx for input x. Legal successor IDs of $\underline{g}y$ are $\underline{g}cy$ or $\underline{r}cy$ for some c in G. If the state s is not equal to g, then successors are defined as for DFAs.

Def 4.36 An NDTM *accepts x* iff there is a sequence of IDs beginning with the initial ID gx, ending with an ID with state component y, and with each ID a legal successor of its predecessor in the sequence. The *language accepted* by an NDTM M is $\{x|M \text{ accepts } x\}$. The *(nondeterministic) time complexity* of the computation on x is the length of the shortest such sequence—or 1 if there is no such sequence.

We may compare the two possible definitions of nondeterminism with the concept of decision trees. Consider an instance of ZOK (perhaps the one of Figure 4.24). All decisions are between just two choices; let's use the choice Y to mean including an item and the choice N to mean rejecting it. The optimal solution of Figure 4.24 is obtainable from the sequence of decisions (Y N Y Y). In our model of nondeterminism, we would make these choices in advance and refer to them as needed. In other models the choices could be made throughout the algorithm. The only problem in showing the equivalence of these two models occurs because our model may not know how many guesses to make or what the possible guesses are in each case. But we may handle this simply by saying that ill-formed sequences of guesses like (Y N Y) or (Y N Y Y N) or (Y X X Y) do not lead to a solution.

Note that nondeterminism is asymmetric between answers of yes and no. An ordinary deterministic algorithm for a problem A would give a deterministic algorithm for the complementary problem A'. For example an algorithm for deciding whether or not a number n is prime is also an algorithm for determining whether it is composite. In the nondeterministic case, there is a natural nondeterministic algorithm for the question of whether or not n is composite—just guess a factor and verify it is a divisor of n. There is no corresponding algorithm for the question of whether or not n is prime.

As a final hint of the power of the Turing machine model of computation, we sketch the proof that there are *undecidable* decision problems, that is, decision problems having no algorithms for their solution.

We may assume all input strings to TMs are expressed as a string of 0's and 1's, using an encoding if necessary. Since each Turing machine has a finite description for its transition function, we may also encode the TM as a string of 0's and 1's. This may require an encoding of the states and tape symbols of the TM. There may be different encodings for the same TM.

Since each TM has a string of bits that represents it, by interpreting this string as a binary integer we may assign an integer to each TM. Adding a "1" to the beginning of each TM encoding if necessary, we have that each integer represents at most one TM.

In the same way, if we add a "1" to the beginning of each input string and interpret the resulting string as an integer in binary, each integer represents a unique input string.

Thus we may talk about the ith Turing machine and the jth input string. We claim that the following question is undecidable: Does the ith TM M_i accept the ith string w_i?[15] In fact we will show something stronger—that no TM accepts $L = \{w_i | M_i \text{ does not accept } w_i\}$. This is a stronger result because we are not merely claiming that no *halting* TM accepts L; we are claiming that L cannot be accepted by any TM at all. Note that a halting TM for accepting L corresponds to an algorithm for deciding the negative of the given question. Such an algorithm would necessarily decide the given question.

So suppose that there is a TM M accepting L. We have already seen that $M = M_j$ for some j. Now we may ask whether M_j accepts w_j. If so, then w_j is not in L, by the definition of L. But $M_j = M$ accepts L, and hence cannot accept w_j, a contradiction.

On the other hand, if M_j does not accept w_j, then by the definition of L, w_j is in L. This time, since $M_j = M$ accepts L, M_j must accept w_j, and we get another contradiction.

Since M can neither accept nor not accept w_j, we get a contradiction to the existence of M. Therefore no TM accepts L, and the question of whether M_i accepts w_i is undecidable.

A more formal proof of the existence of undecidable questions would need to explore the details of the encodings we have assumed. The original proof is attributable to Turing (1936). For more information on Turing machines and undecidability, see Hopcroft and Ullman (1979) or Lewis and Papadimitriou (1981).

[15] If the binary representation for i does not correspond to a TM, then we simply assume the answer is no.

4.21 Exercises

Solutions of exercises marked with an asterisk appear in Appendix C; references are given in that appendix for those marked with a double asterisk.

Algebraic Expressions and Decision Trees

*1. Which of the following are legal prefix expressions?

a) X
b) 2
c) + X 2
d) − X 3 Y
e) + × 3 4 − 6 7
f) − + × 3 4 5 6
g) − 3 4 × 5 + 6
h) 4 − 3

*2. Which of the following are postfix expressions?

a) X
b) 2
c) −
d) X + 2
e) X Y −
f) A B C D + × −
g) A B − C D + ×
h) A B − + C D ×

*3. For each of the trees of Figure 4.55,

a) Evaluate the corresponding algebraic expression
b) Give the corresponding prefix and postfix expressions.

*4. Which binary trees represent each of the following expressions?

a) 3 − 4 × 5 + 6 / 2
b) (3 − 4) × 5 + 6 / 2
c) 3 − (4 × (5 + 6) / 2)
d) + × − 3 4 5 6
e) + × 3 − 4 5 6
f) + × 3 4 − 5 6
g) + 3 × 4 − 5 6

5. In parts (d) through (g) of the previous problem, convert the expressions to postfix without looking at the corresponding tree. Check that your answer is correct by evaluating the postfix expression.

*6. Suppose that an algebraic expression has n operators, all of which are binary. How many operands does it have?

*7. If you're familiar with the language APL, try this one: Why does the language allow a maximum of two arguments to user-defined functions?

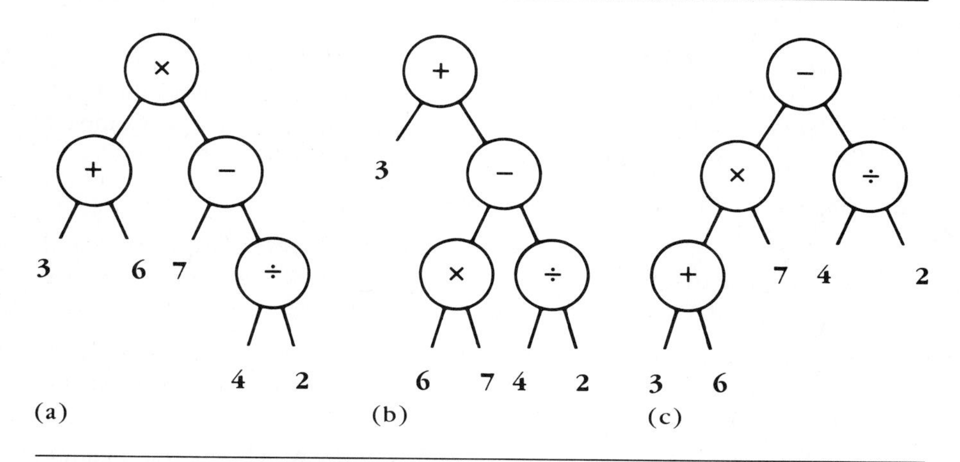

Figure 4.55 Binary Trees Representing Algebraic Expressions

8. Give an algorithm for the 12-coins problem using the minimum number of comparisons. What if it is possible that no coin is counterfeit? What if there are 13 coins? What if there are 13 coins and no coin is counterfeit?
9. Find a sorting algorithm using the minimum number of comparisons for 5 elements.
10. Suppose you have 3 elements to sort, and you also want to know exactly which elements are equal. What is the lower bound on the number of comparisons given by the decision tree model for a) 2-way comparisons? b) 3-way comparisons? Give an algorithm for achieving the minimum in each case.

*11. A sorting algorithm that for many values of n gives the fewest possible number of key comparisons is attributable to Ford and Johnson (1959). It begins by comparing elements in pairs and sorting the winners recursively. (Any element not in a pair is a loser.) The losers are then inserted one at a time into the sequence of winners using binary search to find the spot of insertion. Since binary search is most efficient when the sequence searched has size $2^k - 1$, the order of the insertions is important. Let the sorted sequence of winners be (w_i) and let l_i be the element that lost to w_i. In general the element l_i can be inserted into the sequence consisting of w_j for $j \leq i$ and any l_j already inserted. Note that l_1 is smaller than any w_i, so it may be inserted first.

   a) Show that for $n = 22$, if insertions are done in the order l_3, l_2, l_5, l_4, l_{11}, l_{10}, l_9, l_8, l_7, l_6, l_5, then each element is inserted into a sequence of size one less than a power of 2.

   b) Note that this sequence is constructed by concatenating the reversals of certain subsequences where the ith subsequence is of the form (l_j) where j ranges from $t_{i-1} + 1$ to t_i. Find a recurrence for the sequence (t_i).

c) Solve your recurrence.

d) For n ranging from 1 to 13, how many comparisons does the Ford-Johnson algorithm require to sort a sequence of size n? Compare this value with the lower bound found in Section 4.3.

12. Implement the Ford-Johnson algorithm.

13. Design a recursive algorithm that evaluates a prefix expression and another one that evaluates a postfix expression. Would it help in either case to be able to read the input from right to left?

14. Improve the algorithms of the previous problem so that they handle ill-formed input. Can you recognize different types of errors and handle them differently?

15. Design algorithms for recognition of well-formed a) prefix expressions, b) postfix expressions, and c) infix expressions. You will need to be able to determine the number of arguments each operator can take.

16. Design and implement an algorithm to construct the tree representation of an expression in a) prefix, b) postfix, c) fully parenthesized infix, and d) unparenthesized infix (using the normal precedence of operators) given knowledge of how many operands each operator requires. Show that in cases (a) and (b) the tree is unique.

17. Design and implement algorithms to construct from the tree representation an algebraic expression in a) prefix, b) infix, and c) postfix. Eliminate redundant parentheses from the infix expression if you can.

**18. Design algorithms to translate algebraic expressions from each of the three representations into either of the other two. Is there an easier way than building and then traversing a tree? Can your algorithm recognize which representation the input is using—so that the only input necessary is the expression and the form to be used for its output?

19. Design an algorithm to initialize the tree in tournament sort.

20. Design and implement a tournament sort algorithm. Prove that your algorithm is correct.

21. In the tournament sort algorithm, the losing key rather than the winning key is sometimes assigned to the node at which the comparison is made. This would give the tree of Figure 4.56b rather than that of Figure 4.56a. What might be the advantage of this approach? Implement a tournament sort algorithm using such a *tree of losers.* Can you do the same for k-way merge?

22. Design and implement an algorithm that, given an algebraic expression in infix, mimics a compiler. The algorithm should return (instead of the

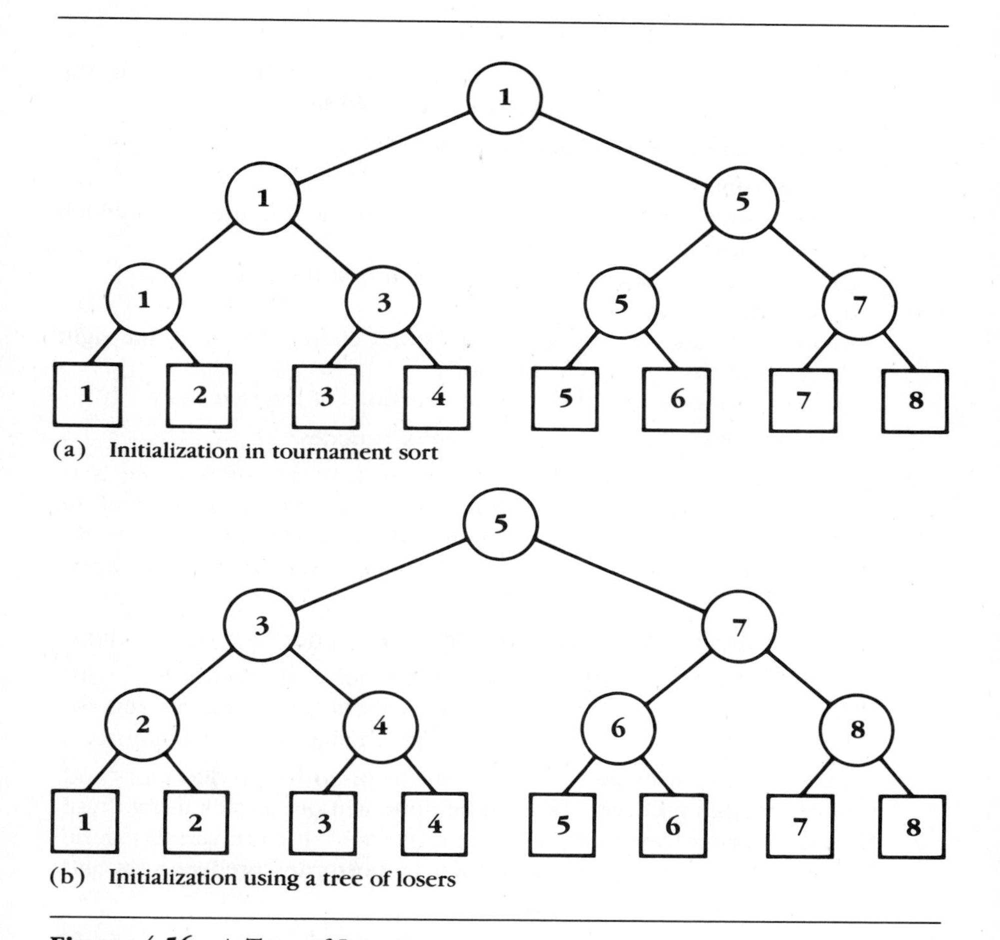

(a) Initialization in tournament sort

(b) Initialization using a tree of losers

Figure 4.56 A Tree of Losers

value of the expression) a list of two-operand instructions of the form A = B % C, where % stands for a binary operator, such that if done in the correct order, the new operations will evaluate the original expression. For example, the output for the expression $2 \times 3 + 4$ would be

$$A = 2 \times 3$$

$$B = A + 4$$

23. What lower bound is given by the decision tree model for the number of comparisons necessary to a) sort 10 elements? b) find *k* largest elements in order? c) find all connected components in a graph? d) eliminate all duplicates in list? Give an algorithm that achieves the lower bound in each case if you can.

Successors, Deletion, and Threaded Trees

24. In the trees of Figure 4.14d, use the algorithms of the text to find the inorder predecessor and successor of each node.
25. For the binary trees of Figures 2.7 and 4.9.
    *a) insert inorder threads,
    *b) construct the external nodes and determine the interval to which each belongs,
    c) construct the sequential representation of the tree,
    d) insert the keys 12, 34, 51,
    e) delete (from the original tree) the left child of the root, the right child of the root, then the root,
    f) do (d) and (e) for the threaded versions of the two trees.

*26. How many threads are there in a tree of n nodes?

*27. One possible way to delete a node from a binary search tree is to replace the node by its right subtree. In general, the left subtree of the node would become the left subtree of the first node in inorder of the old right subtree (see Figure 4.57). Why might this approach be undesirable?

28. Why does the inorder successor of an internal node have no left child?
29. Can preorder or postorder traversal be done nonrecursively with threads as described in this chapter? What about traversal in the reverse of inorder? Why or why not? If so, design an algorithm for doing so.
30. Suppose that threads are used to point to preorder predecessors and successors. Can preorder traversal be done without a stack using these threads? inorder traversal? postorder traversal? Which traversals can be done without a stack if threads point to postorder predecessors and successors?

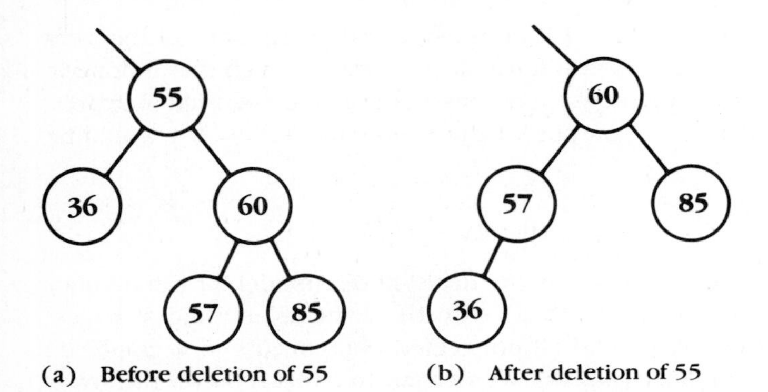

(a) Before deletion of 55 (b) After deletion of 55

Figure 4.57 A Possible Method of Deletion from a Binary Search Tree

31. Design an algorithm for finding the inorder predecessor and successor in a binary tree given a) a parent pointer from each node, and b) the consecutive storage representation.
32. Implement each of the algorithms of Section 4.6.
33. Can one delete from a binary tree using the inorder predecessor instead of the inorder successor? If so, design an algorithm to do this.

**34. Design an algorithm to find the parent of a node in a threaded tree. What is its time complexity?

35. Design an algorithm to convert a binary search tree to a threaded tree.

*36. Show that during threaded traversal, assuming no header node, each pointer and right thread is followed exactly once and no left thread is followed at all.

*37. How much time is required for tree traversal in a threaded tree if we start at the first node in inorder and repeatedly find the inorder successor?

38. Do the previous problem for the case in which we must start at the root in order to find the inorder successor.
39. Show the deletion algorithm for binary search trees preserves the binary search property in all cases.
40. Show that when deleting from a node with two children in a binary search tree, if a key is replaced, it must be replaced by an inorder predecessor or successor.

*41. Show that each node in a symmetrically threaded tree has as many threads pointing to it as nonempty pointers pointing from it.

42. Show that after insertion of a new node in a threaded tree, the new node is reachable from its inorder predecessor and successor by the usual algorithms.
43. Prove that each algorithm of Section 4.6 is correct.

*44. Show that every nonempty thread in a symmetrically threaded tree points to an ancestor.

45. Can trees share subtrees in the linked representation? What about threaded trees?
46. Implement a simple database with insertion and deletion operations and two find operations—one that returns the record with a given key and another that returns a sequence of records with keys in a given range.

AVL Trees and B-Trees

47. a) Trace the successive insertion of keys 40, 1, 3, and 5 into the B-tree of Figure 4.19c.
    b) Trace successive deletions of the keys 20, 52, 91, 25, 21, and 81 from the B-tree of Figure 4.16c.

c) Trace successive insertions of the keys 30, 5, 70, 80, and 50 into the final AVL tree of Figure 4.14.

48. Complete the AVL insertion Algorithm 4.4 by designing rebalancing algorithms for the LL and LR cases.
49. Implement Algorithm 4.4.
50. Use your insertion procedure of the previous exercise to compare the number of key comparisons for binary tree sort using an AVL tree with sort using an unbalanced binary search tree. What is a fair way to account for the cost of the rebalancing operations?
51. Design and implement a deletion algorithm for AVL trees.
52. Complete the B-tree algorithms of Section 4.8 by writing the low-level pointer manipulation routines.
53. Remove recursion from the AVL and B-tree algorithms of Sections 4.7 and 4.8.
54. Verify that properties 1 through 4 of the definition of a B-tree are retained after insertion or deletion.
55. Show that the inorder successor of any key not in a leaf of a B-tree must be in a leaf.

Union-Find

56. Trace the operations below in a union-find structure of 12 elements:
Find(1), Union(Find(1), Find(2)), Find(1), Find(2), Union(Find(3), Find(1)), Find(1), Find(3), Union(Find(4), Find(5)), Union(Find(6), Find(7)), Union(3,8), Union(Find(8), Find(9)), Union(9,2), Union (Find(3), Find(2)), Union(Find(3), Find(7)), Union(Find(10), Find(11)), Union(Find(11), Find(5)), Find(10), Find(8), Union (Find(5), Find(2))

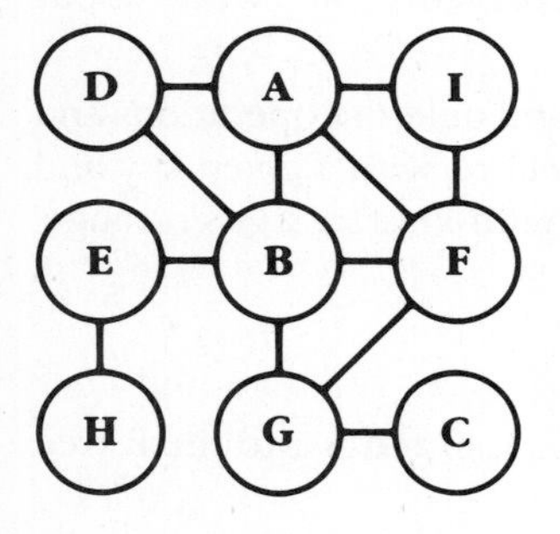

Figure 4.58 An Undirected Graph

*57. Determine the average time required for a successful Find operation for each of the structures of Figures 4.21 and 4.22.

58. Design and implement an algorithm to initialize a union-find structure. Implement the Union and Find algorithms of the text.

*59. Show that the average successful search time is always less in a union-find structure if the root of the larger tree becomes the root of the union than if the root of the smaller tree does.

*60. Show that the worst-case time complexity of the Find operation for union-find structures is $O(\log n)$ if there is no path compression. What if path compression is done?

Search and Traversal

61. Consider the ZOK instance in which $M = 13$, $(w_i) = (5, 7, 3, 4, 2, 1, 6)$ and $(p_i) = (16, 22, 9, 11, 5, 1, 5)$. On this instance, trace
    a) the pure backtracking algorithm,
    b) the backtracking algorithm that prunes nodes from which no feasible solution can be found,
    c) pure breadth-first search,
    d) A*.

    Can you avoid saving hopeless nodes?

62. Repeat the previous problem for the TSP instance with vertices numbered 1 through 5 in which the cost of the edge (i, j) is
    a) ij
    b) $10i + j$
    c) $\lceil 100|\sin i + \cos j| \rceil$

63. Trace the backtracking algorithm on the a) Boolean satisfiability instance

$$(x_1' \vee x_2') \wedge (x_2' \vee x_3) \wedge x_3' \wedge x_4 \wedge (x_4' \vee x_1')$$

Can you do any pruning?

b) Repeat the previous exercise for the Hamiltonian circuit instances

1) consisting of vertices 1 through 10 with the edge (i, j) present iff $(i, j) > 1$ or $i = 1$ or $j = 1$,

2) as (1), except that the vertex 7 is deleted.

64. Do you get the claimed 80 nodes and 34 leaves for the ZOK example of Section 4.11 when pure backtracking is used?

65. Use depth-first search starting at vertex A to find the articulation points of the graph of Figure 4.58.

*66. An old problem involves missionaries and cannibals. There are two missionaries and two cannibals with a canoe on an island. They want to get to the mainland. Only the missionaries can row. If the missionaries are outnumbered by the cannibals on either side of the water, they will be eaten. (If you don't like this formulation of the problem, feel free to switch the roles of missionaries and cannibals.) Construct and search a state-space tree to determine whether or not all four can reach the mainland without disaster. Should you use breadth-first or depth-first search?

67. Give an example of a problem for which the A* algorithm won't terminate.

68. For the ZOK problem, make a reasonable definition for random input and determine the average number $f(n)$ of nodes generated by your favorite search strategy as a function of n. Does $f(n)$ appear to be polynomial in n? Replace ZOK with another problem.

**69. Implement the A* algorithm with a function that takes as parameters only a) a description of the start state, b) a function that will test a state to see whether or not it is a goal, c) a function that will return a list of the successors of a given state, and d) a function that evaluates a given state. For goal states in optimization problems, this last function should agree with the function on which optimization is based.

70. Design an algorithm to determine whether a graph is a tree. What representation would you use? Implement your algorithm.

**71. Can you find a condition on the depth-first numbers of nodes in a graph that will determine which nodes are articulation points? If so, design and implement an algorithm to find the articulation points of a given graph.

72. Design and implement a garbage collection algorithm under the Lisp-like assumptions: a) all nodes have size 1 or 2; b) the nodes of size 1 may not contain pointers when in use, but either field of a larger node may contain a pointer; c) named pointer variables are stored in a list, and all nodes not accessible from named pointers are garbage; and d) nodes not in use are stored on two lists, one for each size. Document any other assumptions you make.

73. Design and implement an algorithm to get free space that works in a similar way to our **Get operation. Assume that it takes requests for nodes of a given size and must return a pointer to a node large enough to meet the request. Consider as an underlying data structure a) a singly linked list for nodes of all sizes, b) different linked lists for nodes of different sizes, c) a binary search tree whose nodes head lists for a given size or a given range of sizes, and d) any other data structure that seems appropriate to you.

How do you initialize your data structure? Can you return nodes to free storage?

74. Modify Algorithm 4.15 so that it doesn't insert nodes that cannot lead to optimal solutions into the set of open nodes.

75. Modify Algorithm 4.15 so that it handles general state space graphs.

*76. The state space model may be applied to deduction. Here facts are represented by letters such as p, q, and r. A *clause* is a disjunction of facts or their negations, for example, $p' \vee q' \vee r$. States consist of sets of clauses where each clause is assumed to be true. A successor state may be obtained by the addition of a new clause obtained from two member clauses by *resolution*. Resolving two clauses means taking the disjunction of the new clauses and eliminating one complementary pair of conjuncts. For example, the state $\{q', r \vee p, p \vee q\}$ would have as a successor the state $\{q', r \vee p, p \vee q, p\}$, where p was obtained by taking the disjunction of q' and $p \vee q$ and eliminating the pair q and q'. A goal state for a desired conclusion is merely one containing that conclusion; the state just constructed would be a goal state for p (i.e., p follows from q', $r \vee p$, and $p \vee q$).

Use breadth-first search with start state $\{s, p' \vee q, p, q' \vee r\}$ and see if you can derive the conclusions

a) q
b) $p \vee r$
c) $p' \vee r$

*77. In the previous exercise, how would you represent the information $p \wedge q$? What about $p \rightarrow q$? What about $p \vee (q \wedge r)$? Show that the new fact added by resolution actually follows from the two clauses resolved.

78. Show that if the estimator function h for the A algorithm is monotonic, then nodes are expanded in order of their f values.

79. Show formally that depth-first search and breadth-first search each generate spanning trees for connected graphs.

**80. Let $A^{(k)}$ be a matrix whose $[i, j]$ entry is 1 iff $i = j$ or there is a path of length at most k between vertices i and j and is 0 otherwise. Show that $A^{(k)}$ is equal to

a. A, if $k = 1$, where A is the adjacency matrix with all entries either 0 or 1, except that $A[i, i] = 1$ for all vertices;
b. $A^{(k-1)} \alpha A$ if $k > 1$, where α is the operator defined exactly as for matrix multiplication except that "and" replaces multiplication and "or" replaces addition.

81. Show that for the α operator above, $A^{(p)} \alpha A^{(q)} = A^{(p+q)}$. Design an efficient algorithm for calculating $A^{(p)}$ when p is a power of 2. What if p is not a power of 2? What is the time complexity of your algorithm?

*82. Use the algorithm of the previous problem to find all connected components of the undirected graph of Figure 4.59.

83. Can you use the approach of the previous three problems to design an all-pairs shortest paths algorithm for a directed graph? If not, why not? If so, what is its time complexity?

```
0 1 1 0 0 0 0 0
1 0 0 1 0 0 0 0
1 0 0 0 0 0 0 0
0 1 0 0 0 1 0 0
0 0 0 0 0 0 1 0
0 0 0 1 0 0 0 1
0 0 0 0 1 0 0 0
0 0 0 0 0 1 0 0
```

Figure 4.59 An Undirected Graph

Game Trees and And-Or Trees

*84. Evaluate the and-or tree of Figure 4.38 if the nodes on the lowest level alternate values between "true" and "false" and all other nodes have value "true" a) beginning with "true," and b) beginning with "false."

85. Find a solution for the and-or tree of Figure 4.38 if all OR nodes become AND nodes and all AND nodes become OR nodes.

*86. Find the value of the game tree of Figure 4.39 to the first player a) if he is a minimizer and the second player is a maximizer, and b) if the values for leaves are given from the point of view of the current player instead of the first player.

*87. Suppose that your probabilities of passing the classes of each of the leaves on the lowest level of Figure 4.38 are respectively 6/7, 7/8, 8/9, 9/10, 5/6, 3/4, 7/8, 5/6, 9/10, 2/3, 5/6, 2/3, 5/6, 2/3, 8/9, and 2/3, and the other leaves have value 1. Assume you are not allowed to fail any courses. Is there a better-than-even chance you will graduate? What is your best strategy? What is the probability of success if you can fail a course but can't retake a course?

*88. Suppose that all of the leaves of the and-or tree of Figure 4.40a were feasible, but had the following respective costs in dollars: 250, 80, 50, 70, 150, 125, 100, 125, 100. Is it possible to get from Los Angeles to New York for $200 or less? What is the cost of the cheapest trip?

89. Trace the alpha-beta algorithm to the game tree of Figure 4.36a assuming a) that the first player is a maximizer and that the values in the leaves are the values to the first player, b) that the first player is a minimizer and that the values in the leaves are the values to the first player, and c) both players are maximizers and the values in the leaves are those to the current player.

*90. In the game of nim, there are initially three piles of sticks. At one turn, a player may remove any number of sticks from one of the piles. The last player to remove a stick wins. Which player should win the game of nim if the initial piles have sizes 1, 2, and 3? What is the winning strategy? What if the last player to remove a stick loses?

*91. Consider the following game. Given a sequence of integers, players A and B construct a subsequence as follows: A decides whether or not to include the first element, B the second, A the third, and so on alternately. Let S be the sum of the elements in the sequence and T the sum of elements in the subsequence. If T is within 1 of $S/2$, then B wins; otherwise A wins. Who should win this game for the sequence (1 2 3 4)? What is the winning strategy?

*92. Consider the following game involving a pile of n sticks. Players A and B alternately remove 1, 2, or 3 sticks. The player who removes the last stick wins. Who should win if the pile originally contains 8 sticks? What if it originally contains 9 sticks? What is the winning strategy? What if the player who removes the last stick loses?

Topological Sort

93. Apply the topological sort algorithm to the prerequisite graphs in which
    a) the vertices are the integers from 1 through 9 and there is an edge from p to q iff $10p + q$ is a perfect square;
    b) the vertices are the integers from 1 through 10 and there is an edge from p to q iff p divides q;
    c) the vertices are the integers from 2 through 9 and there is an edge from p to q iff $p^2 = q \bmod 11$.

    In each case, if you assume that it takes time p for task p, find the critical tasks.

*94. Consider the project of getting to your office in the morning. Assume the following time requirements for each task: 2 minutes to brush your teeth, 3 minutes to walk from the office parking lot to the office, 5 minutes to shine your shoes (why do you always forget to do it at night?), 10 minutes each to shower and to dress, 15 minutes to fix breakfast, 20 minutes to listen to the news on the radio, and 30 minutes each to eat and to drive from home to the office parking lot. Someone else in the family may shine your shoes and fix your breakfast. It's OK to listen to the radio in the shower, while dressing, at breakfast, or in the car. Assume everyone in the family gets up at once. Assume the following prerequisites: you must eat before brushing your teeth, dress and brush your teeth before leaving for work, shower and shine your shoes before dressing, fix breakfast and dress before eating, and listen to the radio before reaching the office parking lot. How much time is needed to get to work?

*95. Find the value of LT for each task in the previous problem. Which tasks are critical?

96. What is the time complexity of topological sort if an adjacency matrix is used instead of adjacency lists?

97. Prove or disprove: The critical path algorithm will always construct ET and LT functions, which are nondecreasing when the tasks are considered in the topological order used to construct them.

98. Show that in general all critical tasks lie on one or more critical paths from the first task to the last.

Algebraic Algorithms

*99. Find the GCD of

a) 109 and 67
b) 143 and 77
c) 1984 and 1776

100. For each case of the previous problem, express the GCD as the sum of integer multiples of the two GCD arguments.

101. Find the representation used by the fast Fourier transform algorithm when $n = 4$ for the polynomial $6x^3 - 3x^2 - 5x - 1$. Check your answer by using the inverse fast Fourier transform.

102. Use the fast Fourier transform algorithm to multiply $x^3 + x - 3$ by $2x^2 + 2$.

103. Show that if $\omega = \cos(2 \times \pi / n) + i \times \sin(2 \times \pi / n)$ then $\omega^k = \cos(2 \times \pi \times k / n) + i \times \sin(2 \times \pi \times k / n)$. What is ω^k if k is a multiple of n? What does this tell you about the nth roots of 1?

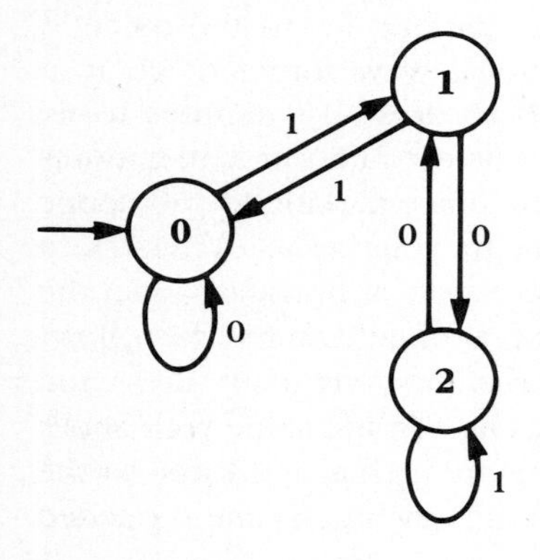

Figure 4.60 A DFA for Calculating Remainders mod 3

104. Show that the quotient of two complex numbers is always a complex number, except for division by zero.

*105. Show that the GCD algorithm for calculating (n, m) has time complexity $O(\log n)$.

106. Count the number of comparisons of each type in the example at the end of Section 4.18. Do you get the claimed total?

107. Use an adversary argument to find a lower bound for the time complexity of searching an unsorted array.

*108. Use an adversary argument to show that the early stages of the tournament sort algorithm give the minimum number of comparisons for finding the largest and second-largest elements of a sequence.

Languages and Finite Automata

109. If P = "ananama" is a pattern and S = "anananananaman" is a subject string, find the first occurrence of P in S by

a) constructing an NFA for P,
b) constructing a DFA for P,
c) using the KMP algorithm.

110. Trace the computation of the TM of Figure 4.54 on input

a) 0101
b) 00111
c) 0110

111. The DFA of Figure 4.60 calculates the remainder when a binary number is divided by 3. Test the DFA on the input 10, 11, 1000, and 1001.

112. How would you simulate an optimization algorithm by a decision algorithm if all numbers in the optimization problem were given as fractions with integral numerators and denominators?

*113. Can the DFA model be applied to the problem of Exercise 66? How many states are there? How many are reachable from the start state? Is the final state reachable from the start state? What sequence of moves will solve the problem?

114. Show that the KMP algorithm is correct (given that the TryNext table is constructed correctly).

115. Show that the TM of Figure 4.54 accepts the claimed language.

*116. Prove that the DFA of Exercise 111 calculates correctly the value mod 3 for any binary input.

117. Show that the two definitions for NDTMs described in Section 4.20 are equivalent.

Additional Reading—Chapter 4

The state-space paradigm is covered in most texts on algorithms and also in most texts on artificial intelligence. Examples of the latter are Charniak and McDermott (1985), Nilsson (1980), and Rich (1983). Pearl (1984) specializes in heuristics and search strategies. The Charniak and McDermott and Rich texts cover deduction as search; Nilsson's treatment of deduction and logic is greatly expanded in Genesereth and Nilsson (1987).

Memory management and garbage collection are covered in most books on data structures, particularly in Standish (1980).

Finite automata and their relation to languages and grammars are the topics of Hopcroft and Ullman (1979) and Lewis and Papadimitriou (1981). Natural languages and computation are the subject of an increasing number of texts, such as Grishman (1986).

Intractable Problems

5.1 NP-Completeness

In previous chapters, we have seen a number of intractable problems (e.g., TSP; Hamiltonian circuit, abbreviated HC; Zero-One Knapsack; and Boolean satisfiability).

On the other hand each of these problems, at least in their decision problem versions, has a simple nondeterministic polynomial-time algorithm for its solution. For example, in ZOK with n items, only n guesses are necessary—one guess about whether or not to include each item. The TSP and HC problems involve choosing a permutation of the vertices, so we need only choose the kth element as k ranges from 1 to n. A solution to the Boolean satisfiability problem requires only guessing an assignment for each variable.

Note that once guesses are made for HC, we can easily check that all vertices chosen are distinct, whether or not each vertex is adjacent to the next vertex guessed, and whether or not the last vertex is adjacent to the first. In fact, this verification can be performed in $O(n)$ time if the graph is represented as an adjacency matrix and in $O(n^2)$ time if it is represented as an adjacency list. Consequently, the Hamiltonian circuit problem has a nondeterministic polynomial-time algorithm for its solution.

In the TSP, the same sort of verification can be performed and the cost of the tour calculated and compared with the target cost. Similarly for ZOK, we may verify that the capacity of the knapsack is not exceeded and compare the total profit with the target. In both cases the verification can be done in polynomial time.

These observations suggest that the following definition may be useful:

Def 5.1 A decision problem is in the class *P* iff it is solvable by a deterministic algorithm of polynomial time complexity. A decision problem is in the set *NP* iff it is solvable by a nondeterministic algorithm of polynomial time complexity.

Note that "NP" stands for "nondeterministic polynomial," not "nonpolynomial." Also note that any problem in P is automatically in NP, since we may consider the deterministic algorithm to be a nondeterministic algorithm with an empty guessing stage. So $P \subseteq NP$.

Intuitively, we think of P as being the class of good problems, although a time complexity of, say, $O(n^{1000})$ may not seem very good. In practice, however, as suggested by earlier chapters, interesting problems in P tend to have fairly low exponents.

We have seen that ZOK, TSP, HC, and Boolean satisfiability are in NP, although none is known to be in P. In fact, it is not known whether there is *any* problem in NP but not in P. There are many problems known to be in NP, but not known to be in P. The question of whether P = NP is perhaps the outstanding unsolved problem of computer science; in the rest of this section we suggest why it is an important question.

The most obvious possible benefit of analysis of algorithms in general is that we might find a good algorithm to solve some problem. An even better result would be to find a class of good algorithms to solve a large class of problems. A less enjoyable but still valuable result would be to show that no good algorithm is possible for one or more problems.

Recall that we are thinking of P as the class of good problems (i.e., problems with good solutions). A strategy for obtaining results like those previously suggested would be to find a class of problems known to contain P and for which membership in the class is easy to determine. The class NP qualifies on both counts. The easiest problems in NP, among those not known to be in P, would be candidates for new good problems. The hardest problems in NP would be candidates for problems that can be shown to be not good.

This strategy assumes that we have a way of telling whether one problem is easier or harder than another. One simple way is to say that one problem is easier than another if a solution to the second problem gives a solution to the first. In other words, we say that one problem is easier than another if the second problem is more general than the first, or if the first problem reduces to the second.

We must be careful about this process of reduction. The reduction algorithm should not be so complicated that it is harder than one of the problems. For example, we do not want to say that multiplication is easier than addition because we can calculate $a \times b$ as $((a + b)^2 - a^2 - b^2) / 2$. Here the reduction algorithm would involve squaring, division by 2, and subtraction.

Fortunately, we have an easy test of simplicity of a reduction algorithm. We assume that the reduction algorithm must be in P. Since we are dealing only with decision problems, the complexity of the reduction algorithm can lie only in transforming the instance of the easy problem into an instance of the hard problem. The output of the hard problem is simply "yes" or "no"; this answer does not have to be interpreted.

We are led to the following definition:

Def 5.2 A problem A is ***polynomially reducible*** to a problem B (notation $A \propto B$) iff there is a polynomial-time algorithm that transforms instances of A into instances of B in such a way that the instance of A and the transformed instance always have the same solution.

It is not difficult to see that if $A \propto B$ and $B \propto C$, then $A \propto C$. In other words, the $\propto$ relation is transitive.

You may already think intuitively that TSP is in some sense a harder problem than HC. We now formalize that intuition in a theorem.

Theorem HC $\propto$ TSP.

Proof An instance of HC consists of a directed graph. We construct an instance of the TSP decision problem, consisting of a target cost T and a directed graph for which every edge has a weight, as follows:

Let the set of vertices in the new instance be the same as in the HC instance. Let the weight of the edge (a, b) be 1 if the directed edge (a, b) is present in the HC graph and 2 otherwise. Let the target cost T equal n.

Since all tours contain n edges, the existence of a tour of cost n implies that each edge included has cost 1, that is, each edge included in the tour appears in the HC instance. So a tour of cost n implies a solution to the HC instance. Conversely, if there is an HC solution, then each edge appearing in the solution has cost 1 in the TSP instance, and there is thus a TSP solution of cost n. Figure 5.1b shows the TSP instance corresponding to a HC instance of Figure 5.1a.

The construction of the TSP instance can clearly be done in polynomial time, so we have shown HC $\propto$ TSP. ∎

We now consider the class of hardest problems in NP.

Def 5.3 A problem X is *NP-complete* iff X is in NP and for every problem A in NP, $A \propto X$. We use *NPC* to stand for the class of NP-complete problems.

It is not immediately obvious that there are any NP-complete problems. Indeed, before we had a model of nondeterministic computation it would have been extremely difficult to find any property that is true of all problems in NP. However, in hindsight it is not tremendously difficult to see that a computation of an NDTM can be modeled by a long Boolean expression. Thus a special case of Boolean satisfiability was the first NP-complete problem found. The foresight belongs to Cook (1971). Some definitions are necessary to define precisely this special case.

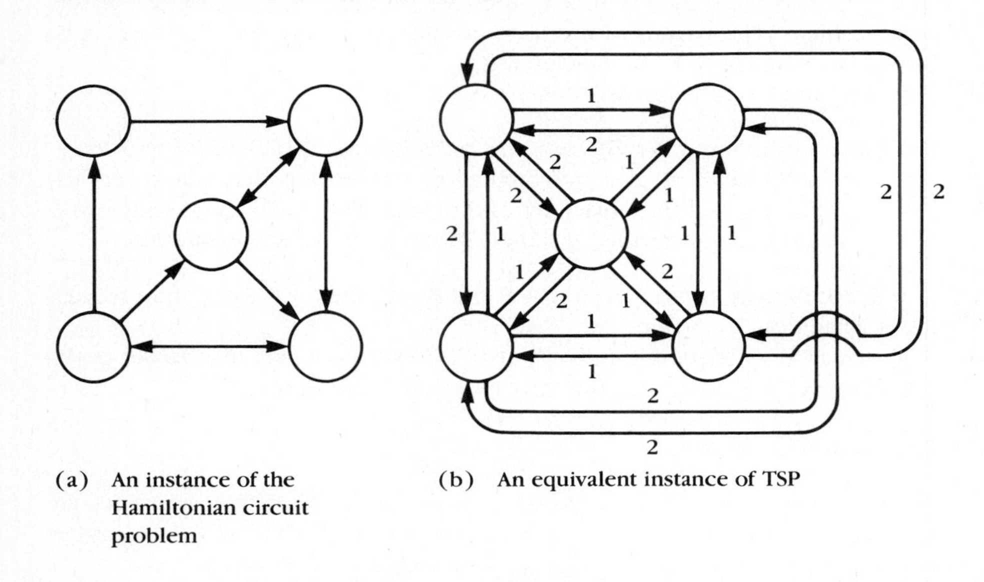

(a) An instance of the Hamiltonian circuit problem

(b) An equivalent instance of TSP

Figure 5.1 Polynomial Reduction of the Hamiltonian Circuit Problem to TSP

Def 5.4 A *literal* is a Boolean variable or its negation. A disjunction of literals is called a *clause.* A formula is in *conjunctive normal form,* or *CNF,* iff it is a conjunction of disjunctions of literals.

The *CNF satisfiability problem* has as an instance a Boolean formula in CNF and asks whether or not there is a satisfying assignment. We prove Cook's Theorem—that this problem is NP-complete—in the next section.

Once one NP-complete problem has been found, it is relatively easy to find others. If A is NP-complete and $A \propto B$, then B is NP-complete—assuming it is in NP. The proof follows from the transitivity of $\propto$; if X is any problem in NP, then $X \propto A$ and $A \propto B$, so $X \propto B$.

Since all NP-complete problems are of the same relative difficulty, if one NP-complete problem is easy, they all should be easy—as should all easier problems. (Recall that the NP-complete problems are the *hardest* in NP.) More precisely, suppose that A is NP-complete and also in P. Then for any X in NP, $X \propto A$, so X can be solved in polynomial time by reducing it to A in polynomial time and then solving A in polynomial time. In other words, if any NP-complete problem is in P, then P = NP.

It turns out that hundreds if not thousands of interesting problems can be shown to be NP-complete. Many of these, such as TSP and ZOK, have been under investigation for many years without yielding a polynomial-time algorithm. By the previous paragraph, we know one of the following is true:

a) all these problems are in P, despite all the effort spent without finding that even one is in P or b) none of them is in P. Most people find (a) extremely difficult to believe.

5.2 NP-Completeness Proofs

The issues involved in showing problems to be NP-complete are large enough to deserve a book in itself [e.g., Garey and Johnson (1979)], so we only hint at the techniques here. In this section our main goal is to show that ZOK and TSP are NP-complete.

We begin with Cook's Theorem. Once it is established, then as suggested in the previous section, further NP-completeness results will be easier to come by.

Theorem (Cook) CNF Boolean satisfiability is NP-complete.

Proof Call the problem CNF. CNF is in NP, by the same argument as for the general Boolean satisfiability problem. Therefore we must show that every problem X in NP is polynomially reducible to CNF. In other words, there is a transformation T of polynomial time complexity such that for every instance I of X $\mathrm{T}(I)$ is an instance of CNF with the same truth value.

Since X is in NP, it has a nondeterministic polynomial-time decision algorithm. We are using TMs as our models of computation, which means that there is a NDTM M and a polynomial p such that M halts in $\mathrm{p}(n)$ steps for all input of size n and accepts exactly those strings corresponding to solutions of X. In this proof we identify instances I of X with input strings of M.

For a given instance I of length n, M uses at most those tape squares numbered between $-\mathrm{p}(n)$ and $\mathrm{p}(n)$, since M moves no more than one square at a time. The computation of M involves $\mathrm{p}(n) + 1$ IDs; if M halts before time $\mathrm{p}(n)$, we simply repeat the final ID until time $\mathrm{p}(n)$. In other words, we can let $\mathrm{d}(\mathrm{Y}, c) = (\mathrm{Y}, c, 0)$ for all c and $\mathrm{d}(\mathrm{N}, c) = (\mathrm{N}, c, 0)$ for all c, where Y is the accepting state and N the rejecting state. We may assume M has states numbered 0 through f and tape symbols numbered 0 through s, where 0 is the blank symbol.

To say that M accepts I is to say that there is a sequence of IDs for M such that

1. Each ID is legal, meaning
   a. it is in a unique state
   b. there is a unique current tape square
   c. each tape square contains a unique symbol
2. The 0th ID is the correct initial ID for I
3. The $\mathrm{p}(n)$th ID corresponds to the accepting state Y

4. Each ID follows from the previous ID by a legal move of M, meaning
   a. the state changes to that given by the move
   b. the tape contents of the current tape square changes to that given by the move
   c. no other tape square has its contents changed
   d. the current tape square changes to that given by the move.

It is enough to construct a Boolean expression in CNF for each of the preceding conditions. Their conjunction will also be in CNF, and its satisfiability will be equivalent to acceptance of I. It will be clear from the constructions that there are polynomially many of them and that each can be constructed in polynomial time.

We will need the following Boolean variables: $S[i, t]$ means that the computation is in state i at time t. $Q[j, t]$ means that the current tape square is the jth at time t. $T[j, k, t]$ means that the contents of tape square j is symbol k at time t. The ranges of the variables are as given previously.

We represent each clause in the following as the set of its literals.

For a particular time t, condition 1a is equivalent to the conjunction of the single clause $\{S[i, t]\}_{i=0}^{s}$ and the conjunction of all $\{S[i, t]', S[i', t]'\}$ for $0 \leq i < i' \leq s$. This is a CNF expression. The first clause means that the computation is in some state i at time t, and the conjunction of the others prevents being in as many as two states at time t. The conjunction of these expressions for all t is still in CNF.

The arguments for 1b and 1c are similar. For a particular t, condition 1b is equivalent to the conjunction of $\{Q[j, t]\}_{j=-\mathrm{p}(n)}^{\mathrm{p}(n)}$ and the conjunction of all $\{Q[j, t]', Q[j', t]'\}$ where $-\mathrm{p}(n) \leq j < j' \leq \mathrm{p}(n)$. We need the conjunction of these expressions for all t.

For particular t and j condition 1c is equivalent to the conjunction of $\{T[j, k, t]\}_{k=0}^{s}$ and the conjunction of all $\{T[j, k, t]', T[j, k', t]'\}$ for all $0 \leq k < k' \leq s$. We need the conjunction of these expressions over all t and j.

Let I_j be the jth symbol of I. Then condition 2 is equivalent to the conjunction of $\{S[\mathrm{g}, 0]\}$, $\{Q[0, 0]\}$, the conjunction of all $\{T[j, I_{j+1}, 0]\}$ over all $0 \leq j < |I|$, and the conjunction of all $\{T[k, 0, 0]\}$ over all $-\mathrm{p}(n) \leq j < 0$ and all $|I| \leq j \leq \mathrm{p}(n)$. This just says that at time 0, the current state is the start state (state g, for an NDTM), the current square is square 0, the input is written one character at a time beginning in square 0, and all other tape symbols are blanks.

Condition 3 is equivalent to $\{S[\mathrm{Y}, \mathrm{p}(n)]\}$.

The components of condition 4 are perhaps easier to write first as implications. For example,

$$(4c) \text{ becomes } Q[j, t]' \wedge T[j, k, t] \rightarrow T[j, k, t+1].$$

Recall that in our version of NDTMs the only nondeterminism is in state g. So for $i \neq \mathrm{g}$, and $\mathrm{d}(i, k) = (i', k', e)$,

$$(4a) \text{ becomes } S[i, t] \wedge Q[j, t] \wedge T[j, k, t] \rightarrow S[i', t+1]$$

(4d) becomes $S[i, t] \wedge Q[j, t] \wedge T[j, k, t] \rightarrow Q[j + e, t + 1]$

(4b) becomes $S[i, t] \wedge Q[j, t] \wedge T[j, k, t] \rightarrow T[j, k', t + 1]$

Recall that if i is the accepting state Y or the rejecting state N, then $d(i, j) = (i, j, 0)$.

From the guessing state g, the current tape square must contain a blank. Any legal guess may be the new contents of the current tape square. The direction of any move must be -1 and the new state may either be g or r. Thus we have

(4a′) $S[g, t] \wedge Q[j, t] \wedge T[j, 0, t] \rightarrow S[g, t + 1] \vee S[r, t + 1]$

(4b′) $S[g, t] \wedge Q[j, t] \wedge T[j, 0, t] \rightarrow Q[j - 1, t + 1]$

(4c′) $S[g, t] \wedge Q[j, t] \wedge T[j, 0, t] \rightarrow$ the disjunction of $T[j, c, t + 1]$ over all legal guesses c.

It is enough to show that implications of the form

$$(p \wedge q \wedge r) \rightarrow d$$

where d is a disjunction of literals, are equivalent to clauses. If so, then we need only add for all i, j, k, and t clauses equivalent to (4a), (4b), (4c) and for all j and t clauses equivalent to (4a′), (4b′), (4c′). But any expression of the form

$$(p \wedge q \wedge r) \rightarrow d \text{ is equivalent to}$$

$$(p \wedge q \wedge r)' \vee d \text{ and so to}$$

$$p' \vee q' \vee r' \vee d$$

which is a clause since d is.

Altogether, there are only a polynomial number of clauses, each of which may be constructed in polynomial time in the size $n = |I|$ of the input. ■

Cliques

Recall that once we have shown one problem X to be NP-complete, another problem Y in NP may be shown to be NP-complete simply by showing that $X \propto Y$.

Our first example is related to the following definition:

Def 5.5 A *clique* of size n in a graph G is a complete subgraph with n vertices.

An instance of the clique decision problem is a graph G and an integer k. The question is whether or not G has a clique of size k. This problem is in NP; a nondeterministic algorithm would guess the k vertices and verify the existence of all the necessary edges in polynomial time. We can show that the problem is NP-complete by reducing CNF satisfiability to it. The original argument is from Karp (1972).

Theorem The clique decision problem is NP-complete.

Proof We observed previously that the problem is in NP. Consider an instance of CNF satisfiability with variables $\{x_i\}$ and clause set $\{C_j\}_{j=1}^r$. The instance is satisfiable iff there is a set $\{z_j\}_{j=1}^r$ of literals such that z_j occurs in C_j and all z_j can be simultaneously true. This latter condition holds iff no z_j is the negation of some z_k.

Consider the set $S_j = \{(w_i, j) \mid w_i \text{ is a literal in } C_j \text{ and } 1 \leq j \leq r\}$. Let this set be the set of vertices of a graph G, and let (w_i, j) and (w_m, k) be adjacent unless $w_i = w_m'$ or $j = k$. Note that G may be constructed in polynomial time.

This definition of adjacency means that if the preceding CNF expression is satisfiable, then $\{(z_j, j)\}$ as previously defined is a clique of size r for G.

On the other hand, suppose that G has a clique of size r. By definition of adjacency in G, the second component of each vertex must be different. Since the clique has size r, for each j its vertex set contains exactly one vertex of the form (w, j) with w in C_j. Assign this w the value "true." Since w is in C_j, this makes C_j true. These assignments can be inconsistent iff there are (w, j) and (z, k) both in the clique with $w = z'$. But by the definition of adjacency this can't happen. ■

Vertex Cover

One other graph-theoretic problem can be shown to be NP-complete by reduction from the clique problem. We need the following definition:

Def 5.6 If G is a graph, a *vertex cover* (or *node cover*) of size k is a subset W of the set of vertices such that for every edge $\{v, w\}$ of G, either v or w is in W.

Theorem (Karp) The vertex cover problem is NP-complete.

Proof by reduction from the clique problem. Let G be a graph of n vertices. Let G' be the complementary graph, that is, G' has the same set V of vertices as G and $\{v, w\}$ is an edge in G' iff it is not an edge in G.

The claim is that G has a clique of size k iff G' has a vertex cover of size $n - k$. Clearly G' and $n - k$ may be constructed from G and k in polynomial time.

Suppose that G has a clique with vertex set X. Let $\{v, w\}$ be an edge of G', that is, an edge not in G. However, both v and w cannot be in X. So one of them is in $V - X$, which is therefore a vertex cover. The set $V - X$ has size $n - k$.

Conversely if G' has a vertex cover W of size k, then it is easy to show that $V - W$ is a clique of size $n - k$. Certainly $V - W$ has size $n - k$. Let $\{v, w\}$ be an edge with v and w in $V - W$. Since W is a vertex cover and neither v nor w is in W, the edge $\{v, w\}$ is not in G'; hence it is in G. Since v and w are arbitrary in $V - W$, we can see $V - W$ is a clique. ■

It turns out that the Hamiltonian circuit problem can be shown to be NP-complete by reduction from the vertex cover problem. The argument is somewhat more complicated than either of the previous two. The construction is given in Exercise 15; the proof that the construction works is given in the solution to that exercise in Appendix C. Since we have already shown HC $\propto$ TSP, and TSP is in NP, this proof would also show TSP is NP-complete.

We will go through all the steps of showing that ZOK is NP-complete. On the way we introduce several classic problems.

3-Satisfiability

A special case of the Boolean satisfiability problem is usually called *3-satisfiability.* We restrict the Boolean expressions in the instance of the CNF satisfiability problem to contain only those for which all clauses have exactly 3 literals. Since it is a special case of a problem in NP, 3-satisfiability is also in NP.

Theorem [Cook (1971)] The 3-satisfiability problem is NP-complete.

Proof by reduction from CNF Boolean satisfiability. We shall construct a 3-satisfiability instance from a general instance clause by clause. In each case the resulting conjunction of clauses will be true under exactly the same conditions as the original clause, so that the new instance is satisfiable iff the old one is.

Let $\{x_i\}_{i=1}^n$ be the disjuncts in a general clause. If $n = 3$, there is nothing to do. If $n = 2$, then introduce a new variable y, distinct from those added to other clauses, and replace $(x_1 \vee x_2)$ by $(x_1 \vee x_2 \vee y) \wedge (x_1 \vee x_2 \vee y')$. If a variable assignment makes the original clause true, then one of $\{x_1, x_2\}$ must get the value "true." In this case, both of the new clauses are satisfied. If a variable assignment makes the original clause false, then neither x_1 nor x_2 gets the value "true." Since y and y' cannot both be assigned the value of "true," the conjunction of the new clauses cannot have the value "true."

If $n = 1$, then add the new variables y and z and replace the old clause by the new clause

$$(x_1 \vee y \vee z) \wedge (x_1 \vee y' \vee z) \wedge (x_1 \vee y \vee z') \wedge (x_1 \vee y' \vee z')$$

An argument similar to the previous paragraph shows that a variable assignment makes the new clause true iff the same assignment made the old clause true.

If $n > 3$, then introduce $n - 3$ new variables $\{y_1\}_{i=1}^{n-3}$ and replace the original clause by the conjunction of the following $n - 2$ clauses:

$$(x_1 \vee x_2 \vee y_1)$$

$$(y_i' \vee x_{i+2} \vee y_{i+1}) \text{ for } 1 \leq i \leq n - 4$$

$$(y_{n-3}' \vee x_{n-1} \vee x_n)$$

Suppose that a variable assignment makes the original clause true. Then some individual x_i must be true. If $i = 1$ or $i = 2$, then the assignment of "false" to all y_i makes all y'_i and hence all clauses true. If $i = n - 1$ or $i = n$, then the assignment of "true" to all y_i makes all the new clauses true. If $2 < i < n - 1$, then we may safely assign "false" to the literals in the same clause as x_i, namely y'_{i-2} and y_{i-1}. By looking at neighboring clauses we see that we may assign "false" to y_j, for $j \geq i - 1$, and "true" to y_j, for $j \leq i - 2$. We can easily check that each new clause is satisfied by this assignment.

If a variable assignment makes the original clause false, then all x_i must have been assigned the value "false." So no x_i, and thus at most $n - 3$ of the literals $\{x_i\} \cup \{y_i\} \cup \{y'_i\}$, can be assigned the value "true." Since no literal appears more than once in the set of $n - 2$ clauses, not all clauses can contain a literal with the value "true." Thus the new clause also cannot be satisfied.

Clearly this reduction can be done for all clauses in polynomial time; so we have shown that 3-satisfiability is NP-complete. ∎

Colorability

Our next example is related to a famous problem in cartography. Mapmakers have long wondered how many colors are necessary to color a map in such a way that adjacent countries are not assigned the same color. Here we need to rule out the case of adjacency at a point; otherwise, the map of a pie with n slices would require n colors. In the case of a map in the plane, it has only recently been shown [see Appel and Haken (1977)] that four colors are always sufficient.

We may define a more abstract instance of the related decision problem using the following definition:

Def 5.7 Suppose G is an undirected graph and k a positive integer. Then G has a *k-coloring* iff there is a function c from G to the first k positive integers such that $c(u) \neq c(v)$ whenever $\{u, v\}$ is an edge in G.

For any k we may easily construct a graph for which k colors are insufficient; consider the complete graph with $k + 1$ vertices.

An instance of the *k-colorability* problem is just a graph G and an integer k.[1] The question is whether or not G has a k-coloring. We can easily see that this problem is in NP; we simply need to guess the value of c for each vertex and verify the desired inequality for each edge of G.

Theorem The k-colorability problem is NP-complete.

Proof by reduction from 3-satisfiability.

Consider a 3-satisfiability instance with a set $\{C_j\}$ of clauses. If there are n variables, the reduction will be to an $(n + 1)$-colorability instance, where the first n colors represent "true" and the last color represents "false."

---

[1] You may see the related minimization problem called the *chromatic number* problem.

We construct a graph as follows: Begin with a complete graph with vertices $\{c_i\}_{i=1}^{n+1}$. As previously suggested, for this graph to be $(n + 1)$-colorable we need a different color for each vertex; we may assume in this case that c_i has color i.

Associate additional vertices x_i and x'_i with each of the n variables in the 3-satisfiability instance. Add edges $\{x_i, x'_i\}$ for each i and edges $\{x_i, c_j\}$ and $\{x'_i, c_j\}$ for all $i \neq j$ and $j \neq n + 1$. This means that if the graph is $(n + 1)$-colorable, either x_i has color i and x'_i has color $n + 1$, or vice versa. According to our interpretation of the colors, x is true if it is assigned color i and false if it is assigned color $n + 1$.

Finally we associate a vertex C_j with each clause in the 3-satisfiability instance. For each j and each literal z in $\{x_i\} \cup \{x'_i\}$, include an edge $\{z, C_j\}$ if z does not appear in C_j. Also add edges $\{C_j, c_{n+1}\}$ for all j.

Clearly the construction of the graph can be done in polynomial time.

Now suppose the graph is $(n + 1)$-colorable. Then each C_j has a color that is not shared by any adjacent vertex. Because of the last set of edges described, this color cannot be $n + 1$. If it has color i for $i \neq n + 1$, we have seen that one of $\{x_i, x'_i\}$ has color i. This vertex is not adjacent to C_j and the other has color $n + 1$. By the construction of the edges, this means that the nonadjacent literal, the one that appears in C_j, is true; so C_j is true. Since j is arbitrary, all C_j are true if there is an $(n + 1)$ coloring.

Finally, we need to show that if there is a satisfying assignment, the graph is $(n + 1)$-colorable. We may reverse the preceding argument. If there is a satisfying assignment, let x_i have color i and x'_i color $n + 1$ if x_i is assigned "true." Let x_i have color $n + 1$ and x'_i color i otherwise. Assign c_i color i. Then all edges of the form $\{x_i, x'_i\}$, $\{x_i, c_j\}$, and $\{c_i, c_j\}$ (from the complete graph) have endpoints with different colors. Since C_j is true, there must be a literal z with color $i < n + 1$ that is both true and a member of C_j. Let C_j have color i. Since no other literal w has color i, all edges of the form $\{w, C_j\}$ have endpoints with different colors, as do all edges of the form $\{C_j, c_{n+1}\}$, and we are done. ■

Exact Cover

Our next problem comes from set theory and is called the *exact cover* problem. An instance of the exact cover problem is a finite family $\{S_j\}$ of finite sets. The question is whether or not there is a subfamily $\{T_k\}$ such that any two T_k are disjoint and the union of all the T_k equals the union of all the S_j.

Clearly this problem is in NP, since we need only guess for each index j whether or not S_j is included in the subfamily. Verification of the properties of the subfamily can then be done in polynomial time.

We now show this problem is NP-complete by reducing k-colorability to it. Given a graph, we need to construct an instance of the exact cover problem. Since the appropriateness of the construction [from Karp (1972)] is not apparent at first glance, we first provide an example.

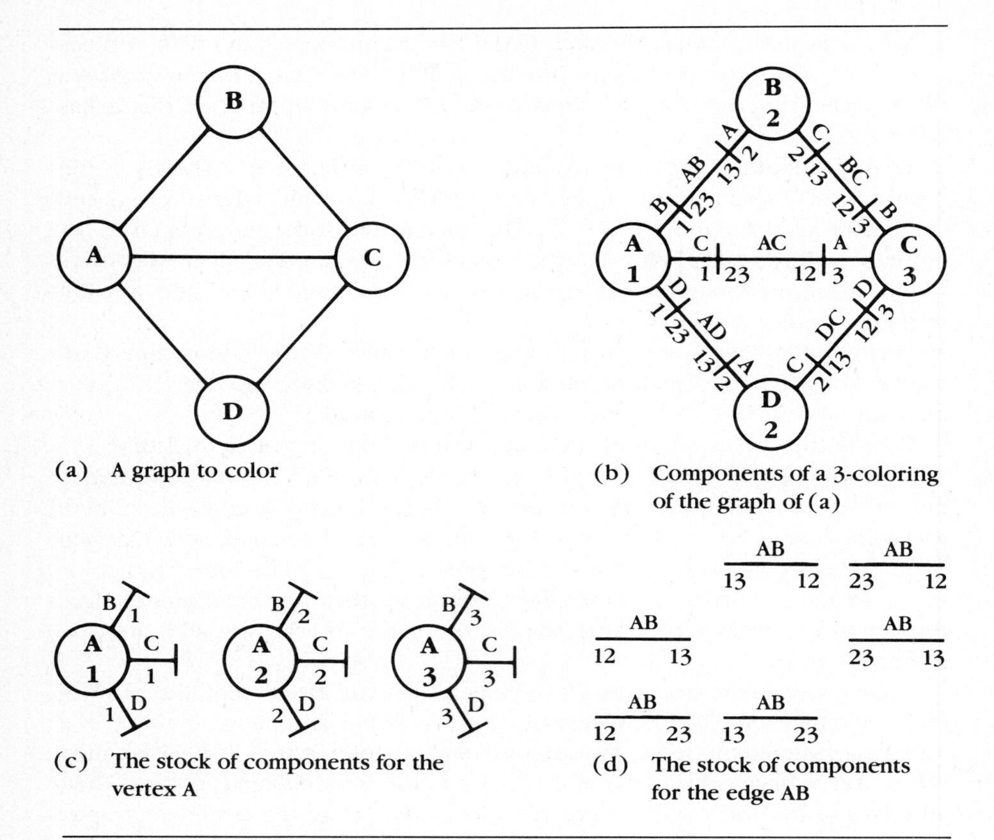

(a) A graph to color

(b) Components of a 3-coloring of the graph of (a)

(c) The stock of components for the vertex A

(d) The stock of components for the edge AB

Figure 5.2 Colorability and a Set for Exact Cover

Consider the graph of Figure 5.2a. We may think of a k-coloring (here $k = 3$) of this graph as being constructed from components as in (b). In this construction one component corresponds to each vertex and one to each edge. Each vertex component must be labeled with the correct vertex name and a color, and outgoing edges have to be labeled correctly. Each edge component must be labeled on each end with the correct vertex name and $k - 1$ colors. The remaining color must match the color assigned to that vertex. The vertex name must match that of the adjacent vertex.

Each vertex component and each edge component has to be chosen from a stock particular to that vertex or edge. The stock of vertex components for vertex A is shown in (c). The stock of edge components for edge $\{A, B\}$ is shown in (d). Note that no two endpoints correspond to the same color. There is a coloring iff a consistently labeled graph like that of Figure 5.2b can be constructed from the components in stock.

Our construction will, given a graph, construct the stock of components for each edge and each vertex.

Theorem The exact cover problem is NP-complete.

Proof by reduction from k-colorability. The general construction is as follows: given a graph $G = (V, E)$, let $S_{v,i} = \{v\} \cup \{(v, w, i) \mid \{v, w\}$ is in $E\}$ for v in V and $1 \leq i \leq k$. Let $S_{v,w,i,j} = \{\{v, w\}\} \cup \{(v, w, m) \mid m \neq i\} \cup \{(w, v, m) \mid m \neq j\}$ for $\{v, w\}$ in E and $1 \leq i \neq j \leq k$. These sets correspond respectively to the possible vertex and edge components. For example, the first vertex component of Figure 5.2c corresponds to $S_{A,1}$ and the first component of Figure 5.2d to $S_{A,B,2,3}$.

There are $k|V|$ sets of the first type, each with at most $|V| + 1$ elements. There are fewer than $k^2|E|$ sets of the second type, each with $2k - 1$ elements. Thus the new instance may be constructed in polynomial time.

Now suppose that there is a coloring. Include in the proposed exact cover the set $S_{v,i}$ iff i is the color assigned to v. Include the set $S_{v,w,i,j}$ iff i is the color assigned to v and j is assigned to w. Each element $\{v\}$ and $\{v, w\}$ is thus included in a unique element of the cover. Consider an element of the form (v, w, i). By construction it can only be a) included in $S_{v,i}$ or b) included in $S_{v,w,i,j}$ for some j. If v has color i, then (a) is true but not (b). Otherwise (b) is true for exactly one choice of j, namely the color assigned to w. In this case (a) is not true. Thus the proposed cover is indeed exact.

The converse is similar. Suppose there is an exact cover. Then for each v in V, only one set of the cover may contain $\{v\}$. This set must be $S_{v,i}$ for some i. Let this i be the color of v.

Now suppose that $\{v, w\}$ is in E and v and w have the same color. In this case there can be no element of the cover containing $\{v, w\}$, a contradiction. Thus the graph has a k-coloring, and we have proved NP-completeness of the exact cover problem. ■

Next we show that exact cover $\propto$ ZOK. Since we have already shown that ZOK is in NP, this reducibility result will imply that ZOK is NP-complete.

Theorem ZOK is NP-complete.

Proof by reduction from exact cover. Given an instance $\{S_j\}$ of the exact cover problem, let S be the union of all the S_j and let $n = 1 + |\{S_j\}|$. Impose an arbitrary order on the elements of S such that each s in S has the form s_i, where 0 is the smallest index. For each j, let $w_j = p_j = \Sigma\, n^i$ over all i with s_i in S_j. This sum is just the interpretation of the bit string representing S_j as an integer in base n [see Figure 5.3a for $S_j = \{s_1, s_3, s_4, s_7\}$]. Let the capacity M and the target profit T both equal $\Sigma\, n^i$ over all i. Note that this sum corresponds to the base n integer consisting entirely of 1's. It is enough to show that an exact cover exists iff the knapsack instance with the given weights w_i, profits p_i, capacity M, and target profit T has a solution, since the instance can clearly be constructed in polynomial time.

First suppose that there is an exact cover $\{T_k\}$. Let $x_j = 1$ if S_j is in the cover and 0 otherwise. Considering each w_j as an integer in base n, exactness implies that the digit in the ith place of $\sum_j x_j w_j$ is 1 for all i [see Figure 5.3(b) and (c)].

$$
\begin{array}{lcccccccc}
 & s_7 & s_6 & s_5 & s_4 & s_3 & s_2 & s_1 & s_0 \\
S_j \leftrightarrow & 1 & 0 & 0 & 1 & 1 & 0 & 1 & 0_n
\end{array}
$$

$$S_j \leftrightarrow n^7 + n^4 + n^3 + n^1$$

(a) Representation of a set S_j as an integer

$$
\begin{aligned}
s_1 &\leftrightarrow 00011_4 = 4^1 + 4^0 = 5 \quad = w_1 = p_1 \\
s_2 &\leftrightarrow 01100_4 = 4^3 + 4^2 = 80 \quad = w_2 = p_2 \\
s_3 &\leftrightarrow 10100_4 = 4^4 + 4^2 = 272 = w_3 = p_3 \\
s_4 &\leftrightarrow 01000_4 = 4^3 + \quad = 64 \quad = w_4 = p_4 \\
M = T &\leftrightarrow 11111_4 = 4^4 + 4^3 + 4^2 + 4^1 + 4^0 = 341
\end{aligned}
$$

(b) Instance of ZOK corresponding to an exact cover instance

$$
\begin{array}{rcrcr}
w_1 & = & 00011_4 & = & 5 \\
w_3 & = & 10100_4 & = & 272 \\
+\, w_4 & = & 01000_4 & = & 64 \\
\hline
 & & 11111_4 & = & 341
\end{array}
$$

(c) An exact cover for the instance of (b)

Figure 5.3 Exact Cover and the Zero-One Knapsack Problem

Thus the total weight included in the knapsack is $\sum_i n^i = M$. Since each $w_i = p_i$, the total profit included is also M, which equals T, the target profit. Thus there is a knapsack solution.

Conversely, if there is a knapsack solution, the total profit must equal the total weight. Since

$$\text{the total profit} \geq T = M \geq \text{the total weight},$$

both totals must equal $M = \sum_i n^i$. But the total weight is just the sum of w_j over all j included in the knapsack solution. By the definition of n, this sum has at most $n - 1$ terms, and each term in base n consists of 0's and 1's. Thus there can be no carrying, and by definition of M, a 1 must occur in the ith place exactly once for each i (again, see Figure 5.3b and c). This implies that a cover consisting of those S_j for which item j is included in the knapsack is exact, and we are done with the proof of NP-completeness for ZOK. ∎

For more examples of NP-completeness proofs involving polynomial reduction, see Garey and Johnson (1979). This standard work also provides an extensive set of examples of NP-complete problems complete with commentary and bibliographic references. More traditional algorithms textbooks with easy-to-read sections on NP-completeness include Aho, Hopcroft, and Ullman (1974) and Baase (1988).

5.3 Strategies for NP-Complete Problems

In the previous sections, we saw that the discovery that a particular problem is NP-complete very strongly suggests that no efficient algorithm can be found for a general solution of the problem. In this section, we consider what to do after discovering that a problem is NP-complete. We shall see that the outlook is not completely bleak.

In this section we make some simple observations about dealing with NP-complete problems. The remaining sections of this text deal with particular strategies for getting around NP-completeness.

One useful observation is that NP-completeness is essentially a worst-case phenomenon. In other words, if P $\neq$ NP, we can only conclude that an NP-complete problem fails to have a polynomial-time worst-case complexity. Its average-case behavior may be polynomial. This is a large part of the motivation behind the tree and graph searching strategies of Chapter 4. For example, although ZOK is NP-complete, it is likely to have a relatively efficient solution using branch-and-bound techniques.

Another example deals with a very important problem called the *linear programming* problem. In this problem one tries to maximize a linear function of n variables subject to a set of m linear constraints. More precisely, an instance consists of a set $\{a_{i,j}\}_{1 \le j \le m, 0 \le i \le n}$ of integer coefficients, a set $\{c_j\}_{j=1}^{m}$ of integer constraints, and a target integer T. The question is whether or not there is a sequence $\{x_i\}_{i=1}^{n}$ of rational numbers with $\sum_{i=1}^{n} x_i a_{i,j} \le c_j$ for $1 \le j \le m$ and $\sum_{i=1}^{n} x_i a_{i,0} \ge T$.

The classical simplex algorithm for solving this problem is not a polynomial-time algorithm, but seemed to work well in the average case. Indeed, for a long time it was not known whether or not the problem was in P. Finally, Khachian (1979) found a polynomial-time algorithm, although it did not seem to work so well in the average case. A still later algorithm by Karmarkar (1984) seems to give the best of both worlds.

A second possibility is that the problem in question is only a special case of an NP-complete problem. The special case may be in P.

For example, although the general k-colorability problem is NP-complete, 2-colorability can be determined in polynomial time. We may assume the graph is connected. The simplest approach is to find a breadth-first spanning tree for the graph. Nodes on odd-numbered levels must have one color and those on even-numbered levels the other color. Except for swapping colors, this is the only possible coloring; whether or not it works may be verified in polynomial time.

The color assignments and checks may be done in parallel with the construction of the spanning tree; design of such an algorithm is left as an exercise.

The general 3-colorability problem is NP-complete. The 4-colorability problem for graphs lying in the plane is trivially in P; all such graphs are 4-colorable.

Sometimes fixing one of the parameters to a problem will turn an NP-complete problem into a tractable problem—although it doesn't help in general to fix k in the k-colorability problem. For example, finding a clique of size 2 is the same as asking whether there is any edge in the graph at all. In general for a fixed k the number of subgraphs on k vertices is $O(n^k)$, where n is the number of vertices in the graph.

Finally there are cases when problems restricted to instances of a reasonable size will allow a polynomial-time algorithm. For example, there is a natural dynamic programming algorithm for ZOK. Consider a particular set S of n items with their weights w_i and profits p_i and let S_k be the subset consisting of the first k items. In particular $S = S_n$. Then finding the solution $f(n, M)$ of the ZOK optimization problem with set S_n and capacity M just means finding the larger of $f(n - 1, M)$ and $p_i + f(n - 1, M - w_i)$. The first expression corresponds to excluding item n and the second corresponds to including n. If M is bounded by a polynomial function $p(n)$ of n, then a table may be constructed for all values of f in time $O(np(n))$, which is polynomial in n.

5.4 Approximation Algorithms

Given an optimization problem whose related decision problem is NP-complete (e.g., ZOK, TSP), it is often possible to find a polynomial-time algorithm whose solution is approximately optimal. In fact, sometimes there is an entire family of approximation algorithms for a given problem, in which the better approximations require more time. Since the subject of polynomial-time approximations to intractable optimization problems is a rather large subject, we can only hint at it here; nevertheless, we do cover a few simple examples.

In this section, we are concerned with the error made by the approximating algorithm only as a fraction of the optimal solution. If the optimal solution is large, a large absolute error can be tolerated as long as the relative error is small. In order to handle both maximization problems and minimization problems in the same formalism, we make the following definition.

Def 5.8 An algorithm A is an ε-*approximation* algorithm for a problem P iff there is a number ε such that for every instance I of P, the approximate solution $S_A(I)$ given by A is related to the exact solution on $S_P(I)$ by

$$|(S_P(I) - S_A(I))/S_P(I)| < \varepsilon$$

This is just a formal way of saying that the relative error is always bounded by ε.

To show examples of ε-approximation algorithms, we introduce two new problems, both related to the knapsack problems. The first is called the *bin-*

packing problem. An instance of the bin-packing problem consists of a sequence $(w_i)_{i=1}^n$ of positive weights. Intuitively, we want to know how many bins of capacity M are necessary to contain all the weights without overflow. More precisely, we want the smallest value of c such that there is a function $f: W \to C$, where W is the set of weights, and C is a set of size c, such that $\sum_{f(w)=c} w \le M$ for all c. Naturally we assume that each $w_i \le M$.

It is not too difficult to construct an ε-approximation algorithm for $\varepsilon = 1$. First sort the w_i into nonincreasing order. Then assign each weight in order to the first bin capable of accepting it. If no existing bin will accept the current weight, then allocate a new bin. Showing that this *first-fit* algorithm has polynomial time complexity has been left as an exercise.

As an example, let $M = 100$ and consider a sequence (40, 40, 30, 30, 30, 30, 25, 25, 25) of weights. The first-fit algorithm would allocate one bin for the first two weights, another for the next three, another for the next three, and a fourth bin for the final weight. In no case would there be enough room to reuse a bin already rejected. One optimal packing would fill two bins each with weights of size 40, 30, and 30 and allocate a single bin for the last three weights. For this instance the first-fit algorithm has relative error $(4 - 3)/3 = 1/3$.

Theorem The first-fit algorithm for the bin-packing problem is a 1-approximation algorithm.

Proof by minimal counterexample. Suppose there is an instance with a relative error of more than 1. Consider a minimal counterexample, that is, an instance I with the smallest number n of weights. We claim that first-fit would have assigned the nth weight a new bin. If not, compare the current instance with the instance obtained by deleting the nth weight. In this new instance, the same number of bins are used by the algorithm, and no more bins are needed in the optimal solution. Thus the relative error could not have decreased and must still be greater than 1, violating minimality of n.

By dividing all the weights by the capacity M, we may assume the capacity is 1 and no weight is greater than 1.

Let c be the number of bins required by the algorithm for instance I. The fact that when w_n is considered, none of the existing $c - 1$ bins has sufficient room implies that each bin contains at least $(1 - w_n)$ weight. Thus the sum of the first n weights is at least $(c - 1)(1 - w_n) + w_n = (c - 2)(1 - w_n) + 1$, which is a lower bound on the optimal solution (since each bin has capacity 1).

On the other hand, we can find a lower bound on $1 - w_n$. If $w_n > 1/2$, then by sortedness all weights would be greater than 1/2, and n bins would be required in both the optimal and first-fit solutions. So $w_n \le 1/2$, and $1 - w_n \ge 1/2$.

Thus the ratio

$$\left| \frac{S_P(I) - S_A(I)}{S_P(I)} \right| =$$

$$\frac{(S_A(I) - S_P(I))}{S_P(I)} =$$

$$\frac{S_A(I)}{S_P(I)} - 1 \leq$$

$$\frac{c}{[(c-2)(1-w_n)+1]} - 1 \leq$$

$$\frac{c}{[(c-2)\times(1/2)+1]} - 1 =$$

$$\frac{2c}{[(c-2)+2]} - 1 =$$

$$\frac{2c}{c} - 1 = 1,$$ violating the assumption of a counterexample. ∎

This bound for the relative error is not especially good, partly because we did not use a particularly good estimate for the optimum. For example, if all weights are just greater than $1/2$, then the first-fit algorithm will require n bins, while the total weight is just over $n/2$. However even the optimal solution requires n bins, since no two weights may be combined in a bin.

It has been shown in Johnson and others (1974) that the preceding *first-fit* algorithm has an asymptotic relative error of $2/9$. More precisely, if A is the first-fit algorithm, then $S_A(I) \leq (11/9)S_P(I) + 4$ for all I.

A dual problem to the bin-packing problem is the *processor-scheduling problem.* Here instead of considering an unbounded number of bins of fixed capacity, we consider a fixed number of bins of unbounded capacity. The name of the problem comes from the model of a fixed number m of processors; each of a set of n tasks must be assigned to one of those processors. Each task has a time t_i required for its completion, and each processor can work on only one task at a time; the question is, how much time is required? We assume that any task can be scheduled at any time as long as there is an available processor. A more general problem allows for a prerequisite structure on the set of tasks along the lines of topological sort.

Algebraically, we have a sequence $(t_i)_{i=1}^{n}$ of tasks, and we want the smallest T such that there is a function f: $W \rightarrow M$, where W is the set of tasks and M the set of m processors, which we may identify with integers from 1 through m, and $\sum_{f(t)=i} t \leq T$ for all i.

An analog of the first-fit algorithm would sort the tasks in nonincreasing order and assign the tasks in this order, so that each task was assigned to the processor with the smallest total time assigned so far.

For example, given tasks with times 50, 46, 44, 40, 37, 34, and 25, the first-fit algorithm would require running times of $50 + 34 = 84$, $46 + 37 + 25 = 108$, and $44 + 40 = 84$ for the three processors. Even the best assignment would require some processor to have three tasks. The smallest time required for three tasks is $37 + 34 + 25 = 96$, allowing times of $50 + 40 = 90$ and $46 + 44 = 90$ for the other two processors. For this instance the relative error would be $(108 - 96)/96 = 1/8$.

It is not difficult to show that this version of first-fit is an ε-approximation algorithm for $\varepsilon = (m - 1)/(m + 1)$. Note that $\varepsilon < 1$. We may mimic the previous proof.

Theorem The first-fit algorithm for the processor-scheduling problem is an ε-approximation algorithm for $\varepsilon = (m - 1)/(m + 1)$.

Proof by minimal counterexample. In such a counterexample the last task t_n—here we identify the tasks with their times—must have the latest finishing time. If not, then removing this task will give a smaller counterexample, since the algorithm's finishing time will not change and the optimal finishing time can only decrease.

Let $s = S_A(I)$ be the finishing time for this last task. Its starting time is thus $s - t_n$. All processors must be continuously occupied until this time, or t_n would have been assigned to a different processor. So the total time required by all n tasks is at least $m(s - t_n) + t_n$, and a lower bound for the optimum solution is therefore $[m(s - t_n) + t_n]/m = (s - t_n) + t_n/m$.

In particular, the total time $s - t_n$ assigned to the first processor must be at least t_1, which is at least t_n, so $s - t_n \geq t_n$, or $2t_n \leq s$, or $t_n \leq s/2$.

The ratio $\left|\dfrac{S_A(I) - S_P(I)}{S_P(I)}\right|$ is then equal to

$$\frac{S_A(I)}{S_P(I)} - 1 \leq$$

$$\frac{s}{[(s - t_n) + t_n/m]} - 1 =$$

$$\frac{s}{[s - t_n(m - 1)/m]} - 1 =$$

$$\frac{ms}{[ms - (m - 1)t_n]} - 1 \leq$$

$$\frac{ms}{[ms - (m - 1)s/2]} - 1 =$$

$$\frac{2ms}{[2ms - (m - 1)s]} - 1 =$$

$$\frac{2ms}{[(m + 1)s]} - 1 =$$

$$\frac{2m}{(m + 1)} - 1 = \frac{(m - 1)}{(m + 1)}$$

■

A more involved argument from Graham (1969) actually shows that this algorithm is an ε-approximation algorithm for $\varepsilon = 1/3 - 1/(3m)$. The details are left as an exercise.

Our final approximation algorithm involves the TSP. We assume an undirected graph whose edge-cost function c satisfies the *triangle inequality*—that $c(\{x, y\}) \le c(\{x, z\}) + c(\{z, y\})$ for all vertices x, y, and z. The triangle inequality is satisfied by the ordinary distance function in the plane.

The approximation algorithm fist constructs a minimum-cost spanning tree. Since any TSP solution becomes a spanning tree if a vertex is removed, the minimum cost of a spanning tree is at most that of the optimal TSP solution.

Then the approximation algorithm chooses a starting vertex (i.e., a root) and constructs a recursive traversal of the tree. If we think of following edges toward the root as well as away from the root, we get a path containing all edges twice. Its cost is just twice that of the minimum-cost spanning tree.

For example, the instance of Figure 5.4a consists of 9 points in the plane. Edge costs are the ordinary Euclidean distances. A minimum-cost spanning tree must have 8 edges, each of cost at least 1. Since the spanning tree of (b) of the figure has cost 8, it must be minimal. A traversal of it beginning at vertex 1 proceeds through the sequence (1 2 3 2 1 4 5 6 5 4 7 8 9 8 7 4 1) of vertices. The cost of the corresponding path is $16 = 2 \times 8$.

To turn this path into a Hamiltonian cycle, we simply skip the second and subsequent occurrences of vertices. The triangle inequality says that skipping a vertex cannot lengthen the path. Consequently, if the original path had cost at most twice the cost of the minimum-cost spanning tree, so will the new cycle. And since the cost of this spanning tree is less than the cost of the optimal Hamiltonian circuit, the new cycle has cost less than twice the optimal cost.

In our example, when duplicate vertices are omitted, we get the sequence (1 2 3 4 5 6 7 8 9 1) of vertices, corresponding to the cycle of Figure 5.4c. It has cost $1 + 1 + \sqrt{5} + 1 + 1 + \sqrt{5} + 1 + 1 + \sqrt{8} \approx 13.3$. The optimal TSP solution is at least 9, since all edge costs are at least 1. The exact value of 9 is not achievable, but the next best possibility is achieved. Here all edge costs except one are 1 and the other has the next best value of $\sqrt{2}$. An example is given in (d) of the figure. The cost of this cycle is $8 \times 1 + \sqrt{2} \approx 9.4$. The relative error for this example is $(6 + 2 \times \sqrt{5} + \sqrt{8})/(8 + \sqrt{2}) - 1$ or about 41%.

We suggested at the beginning of this section that successively better approximations could be obtained at the cost of greater time complexity. Naturally we want the time required by our successive algorithms to remain polynomial in the problem size and not to grow too quickly as ε decreases. One way to impose this latter constraint is to require the time to be polynomial in $1/\varepsilon$. This sort of *polynomial-time approximation scheme* is feasible in a number of cases.

It is beyond the scope of this text to describe in detail a polynomial-time approximation scheme and prove that it has the desired properties [see Horowitz and Sahni (1978)]. However, we can describe informally a useful technique for the construction. We may begin by dividing the problem in-

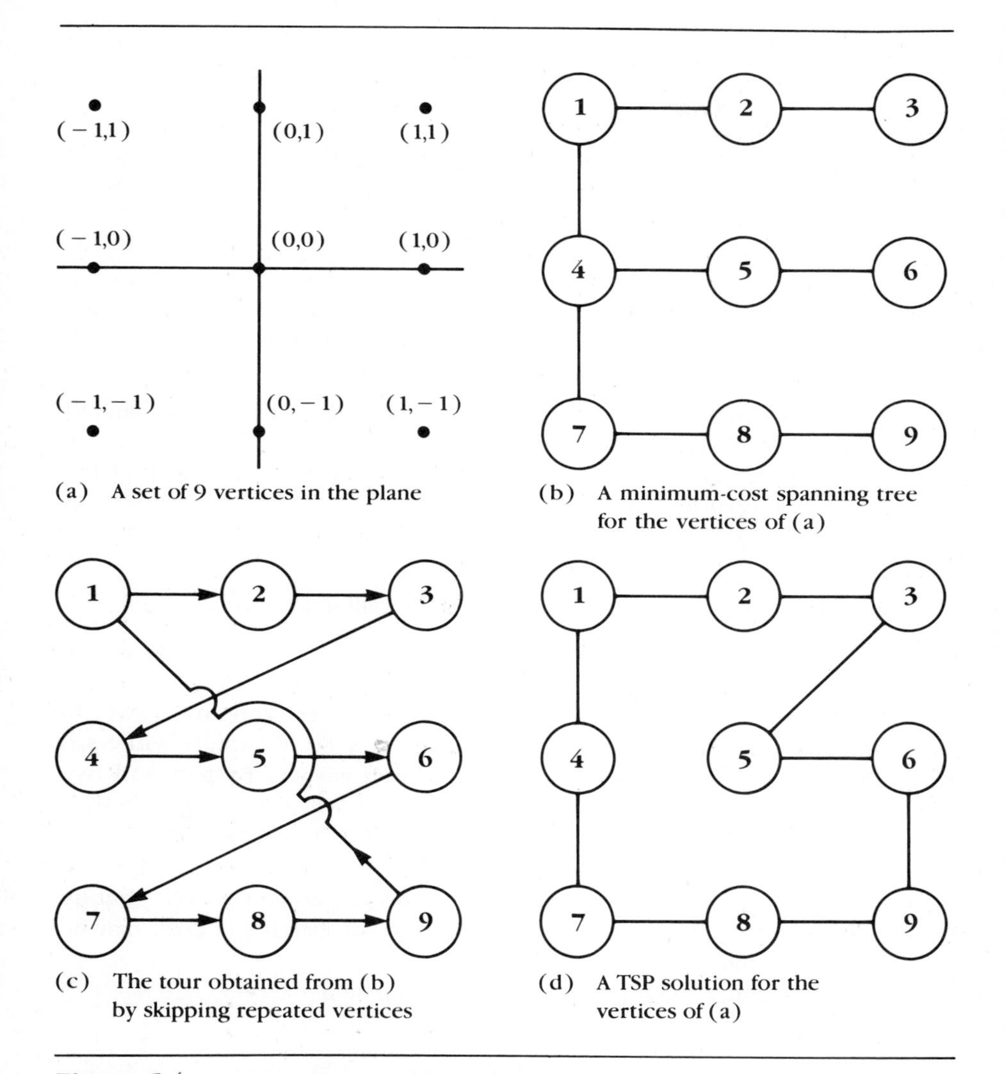

Figure 5.4 Approximating a TSP Solution

stance into bad items and good items, solve exactly the instance consisting only of the bad items, and then finish the solution approximately with the good items. For example in the processor-scheduling problem, the bad items would be those tasks requiring a long time. For this problem, a polynomial-time approximation scheme is possible.

The exact solution for the bad items could not in general be done in polynomial time, but if there are few enough bad items—how few depending on the choice of ε—the overall algorithm could still be polynomial. Of course if there are too few bad items, then the approximate solution for the large number of good items might not be sufficiently close.

5.5 Probabilistic Algorithms

In this section we assume some elementary probability.

Algorithms that are likely but not certain to find the right answer can be very useful under certain circumstances. We first need to define what we mean by such an algorithm.

Def 5.9 Suppose there is a decision problem P and an algorithm A, which takes as input an instance I of P and an extra parameter w. $A = A(I, w)$ is a *probabilistic algorithm* for P iff

1. Whenever the answer to I is "yes," $A(I, w)$ gives the answer "yes";
2. If the answer to I is "no," then $A(I, w)$ answers "no" with probability at least $1/2$;
3. For a given I, the events "$A(I, w)$ answers 'yes' " and "$A(I, x)$ answers 'yes' " are independent.

Each element w used as input to A is called a *witness*.

By parts 1 and 2, any answer of "no" must be correct and any answer of "yes" must be correct with probability at least $1/2$. So if a single witness is used, the algorithm will probably give the correct answer. By part 3, if two witnesses are used, the probability that the algorithm will answer incorrectly is at most $(1/2)^2 = 1/4$. A simple induction shows that if n witnesses are used, the probability of error is at most $1/2^n$.

There is no point at which we can be sure that the answer is "yes," but by randomly choosing sufficiently many witnesses the probability of error may be made as small as we like.

We may construct the following algorithm schema:

```
1. SolveProbably(I, A, α)            {Assume that A is a probabilistic
2.                                   algorithm for P. Assume that an error
3.                                        of at most α can be tolerated.}
4.    find n such that (1/2)^n < α
5.    for each of n randomly chosen witnesses w_i
6.       find A(I, w_i)
7.       if A(I, w_i) is "no," then return "no"
8.    end for
9.    return "yes"
```

Algorithm 5.10 Finding a Probable Solution

In Algorithm 5.10, if any witness answers "no," then the algorithm returns the correct answer of "no." If all n witnesses answer "yes," then the algorithm returns "yes," an answer that is probably correct. Overall, the probability of error is less than the given bound α.

Note that there is nothing special about the number $1/2$ in the definition of a probabilistic algorithm or Algorithm 5.10. Any positive probability would be as good in the sense that enough witnesses would give an answer with any desired degree of reliability. This probability must be a constant; however, it cannot vary from witness to witness (see Exercise 27).

If the problem P is NP-complete, then the algorithm A may still be polynomial. The number n of witnesses (i.e., of repetitions of A) is logarithmic in $1/\alpha$, the reciprocal of the tolerable error. So the preceding schema may give a reasonably efficient algorithm for P, although an infrequent incorrect answer is possible.

Probabilistic notions may be combined with other approaches, including those described previously. For example, we may have an efficient optimization algorithm for which the probability of correctness is very large. Or we may have an efficient approximation algorithm for which the probability of a good approximation is large. Or we may have an algorithm that for most witnesses is polynomial. In these cases it may be possible to make the probability of error arbitrarily small by accepting a greater time complexity, as for polynomial-time approximation schemes.

The statistical arguments necessary to handle or even measure the probabilities for a particular problem A are in general beyond the scope of this text. By way of example, we construct a probabilistic algorithm for a simple problem not known to be NP-complete, but not known to be in P either. This problem also has a very interesting application to encryption. The problem is that of primality.

An instance for this problem is just an integer n. The question is whether n is prime. We may assume that n is odd. At first glance the use of witnesses may be better suited to the complementary problem. Any integer r less than n can be used as a witness. If r divides n evenly, then r is a witness to the fact that n is composite. Unfortunately, we have no analog of property 2 in this case.

The complete proof that the probabilistic algorithm of the next theorem has the desired properties relies heavily on number theory. We sketch its outline, although there are a few sections in which the number theory is a little more than we can cover here. The investigation of this algorithm for small values of n, both prime and composite, is left as another exercise.

For each witness w, the algorithm determines whether two functions on w and n have the same value. The first function is just $w^{(n-1)/2} \bmod n$. Powers of $w \bmod n$ may be found efficiently by observing $w^k = (w^2)^{(k/2)}$ if k is even and $w(w^2)^{(n-1)/2}$ if k is odd. Since residues mod n are of bounded size, the time complexity of this computation for $w^{(k-1)/2}$ is just $\Theta(\log n)$.

The second function we define below.

Def 5.11 The *Jacobi symbol* [notation: $J(w, n)$] has a value in $\{+1, -1\}$ given by

1. $J(1, n) = +1$

1′. $J(2, n) = -1$ if $n \bmod 8 = 3$ or 5
$+1$ if $n \bmod 8 = 1$ or 7
2. If $w = 2^a b$ where b is odd, then $J(w, n) = J(2, n)^a J(b, n)$
2′. For b odd, $J(b, n) = J(n \bmod b, b) \times (-1)^{(n-1)(b-1)/4}$

In (2′) we need $(b, n) = 1$, but since this is true for (w, n) it will be true after each application of (3d), since $(n \bmod b, b) = (b, n)$.

Note that the calculation of J is very similar to that of the GCD algorithm 4.24. Its time complexity is also $O(\log n)$.

All witnesses w must be in the interval $(0, n - 1)$ and satisfy $(w, n) = 1$. This latter condition can be tested in time $O(\log n)$ by the GCD algorithm; if it fails, then n is composite with factor (w, n). Given w, the algorithm returns "yes" iff $w^{(n-1)/2} = J(w, n)$.

Now we sketch the correctness proof.

Theorem The probabilistic algorithm for primality with $A(I, w) = (w^{(n-1)/2} = J(w, n))$ is correct.

Proof sketch If n is prime, then it turns out that there is some residue a mod n such that $a^{n-1} = 1$ and all other residues b mod n are powers of a (mod n). A typical witness w thus has the form $w = a^s$. Consider $w^{(n-1)/2} = a^{s(n-1)/2}$. If s is even, then $s = 2r$ and we get $a^{(n-1)r} = 1^r = 1$. If s is odd, then $s = 2r + 1$, and we get $a^{(n-1)r+(n-1)/2} = a^{(n-1)r}a^{(n-1)/2} = 1^r a^{(n-1)/2} = a^{(n-1)/2}$. This last value is a square root of $a^{(n-1)} = 1$. Since the first $n - 1$ powers of a are all different, the square root can't be 1, so it must be -1. Thus $w^{(n-1)/2}$ is equal to 1 iff w is an even power of a, hence a square, and equal to -1 otherwise.

On the other hand, it turns out that $J(w, n)$ is equal to $+1$ iff w is a square mod n. If s is even, say $s = 2r$, then w^s is the square of both w^r and $-w^r$. Accordingly $J(w, n) = +1$. There are $(n - 1)/2$ such choices for w, accounting for the squares of $2 \times (n - 1)/2$ residues mod n. Since this is all the nonzero residues mod n, there can be no other squares mod n. Thus if w is an odd power of a, w cannot be a square mod n and $J(w, n) = -1$. Therefore $J(w, n)$ always has the same value as $w^{(n-1)/2}$ if n is prime.

If n is not prime, then there is no particular reason that $w^{(n-1)/2}$ must have one of the values $+1$ or -1. We leave as an exercise the proof that these values fail to be attained for at least half the residues mod n. ■

For example, let $n = 15$ and suppose 2 is the witness. Then $2^{(n-1)/2} = 2^7 = 2 \times 4^3 = 2 \times 4 \times 16^1 = 2 \times 4 \times 1^1 = 2 \times 4 \times 1 = 8 \bmod 15$. Since this value is not $+1$, 15 cannot be prime.

If $n = 17$, then if 2 is a witness, $2^{(17-1)/2} = 2^8 = 4^4 = 16^2 = 256 = 1 \bmod 17$. $J(2, 17) = 1$ by part c of the definition of J, so the witness 2 gives answer "yes." If 7 is a witness, then $7^8 = 49^4 = 15^4 = 225^2 = 4^2 = 16 = -1 \bmod 17$. $J(7, 17) = J(3, 7)(-1)^{16\times6/4} = J(3, 7)(-1)^{24} = J(3, 7) = J(1, 3)(-1)^{2\times6/4} = J(1, 3)(-1)^3 = -J(1, 3) = -1$. For witness 7 we have $7^{(17-1)/2} = J(7, 17)$, so this witness also gives an answer of "yes."

A new and efficient method of encryption, which seems to be but has not been shown to be as hard as factorization, based on the existence of good probabilistic algorithms for primality has been developed by Rivest, Shamir, and Adleman (1978) from a suggestion by Diffie and Hellman (1976).

In encryption, we want to find an encrypting function E and a decrypting function D such that $D(E(x)) = x$ for all text strings x.[2] As in hashing, we may assume that x is an integer. Application of both E and D should have low time complexity. It should not be easy to find x given $E(x)$. In fact if it is not easy to find D given E, then E may be published in the equivalent of a telephone directory. Anyone desiring to send a secret message x to the possessor of D need only apply E to x. (In the exercises we suggest that this same technique can give the digital equivalent of a signature that cannot be forged.) This latter scheme is called *public-key encryption*.

If also $E(D(y)) = y$ in general [this is certainly true for y of the form $D(x)$], then the possessor of D may send $D(y)$ as a message. If the receiver can decode it with E, it must have been encoded with D. D is called a *digital signature*. If two different encrypting functions are available, they may be composed to get the benefits of both public-key encryption and digital signatures.

It is not easy to find good functions D and E for public-key encryption. Usually if E has a reasonable inverse function D, then D may be obtained from a knowledge of E. But if p and q are large primes and $n = pq$, then we may proceed as follows.

For $x < n = pq$, choose e such that $(e, (p-1)(q-1)) = 1$. This condition is easy to check knowing p and q. It's automatically true if e is prime. Now let $E(x) = x^e \bmod n$. Then we may easily find, knowing p and q, a value d such that $x^{ed} = x \bmod n$. In fact, we can show that $y^{(p-1)(q-1)} = y \bmod n$, so that d is just the inverse of $e \bmod (p-1)(q-1)$. This is easy to do by Algorithm 4.24 if p and q are known. In general no way of doing so is known if the factorization of n is not known.[3]

So if $D(y) = y^d \bmod n$, then E and D are inverses. We have seen that exponentiation mod n can be done efficiently. It is also relatively easy to find new encryption functions E; all we need to do is find new large (probable) primes p and q, and suitable exponents e. In the unlikely event that the probabilistic primality-testing algorithm has falsely stated that p and q are prime, the

---

[2] In practice there will be an upper bound on the size of x, with long texts divided into shorter strings, each of which is encrypted separately.

[3] The value of having a secret algorithm for inverting an encryption function E of the given form has created a situation perhaps unique in the history of mathematics—someone may have a long-sought mathematical result (and efficient factorization algorithms have been sought for centuries) and a very powerful reason for keeping it private. Thus it is possible that such an algorithm is known but is being kept secret. Ironically, number theory has long attracted those interested in a discipline with the absolute minimum of useful applications.

decryption function is extremely unlikely to work properly, so that this sort of error is easy to discover.

In practice the primes p and q are taken to be several hundred bits long. For an example involving much smaller numbers, let $p = 5$ and $q = 11$, so $n = 55$. Let $e = 17$, a prime. The inverse of e mod $(p - 1)(q - 1) = e$ mod 40 was computed to be 33 in Section 4.16.

Suppose the message is 2. Then, with all arithmetic done mod 55, the encrypted message is

$$2^{17} = 2 \times 4^8 = 2 \times 16^4 = 2 \times 256^2 = 2 \times 36^2 = 2 \times 1296 = 2 \times 31 = 62 = 7.$$

The decrypted message is

$7^{33} = 7 \times 49^{16} = 7 \times 2401^8 = 7 \times 36^8 = 7 \times 1296^4 = 7 \times 31^4 = 7 \times 961^2 = 7 \times 26^2 = 7 \times 676 = 7 \times 16 = 112 = 2$, as expected.

5.6 Parallel Algorithms

One approach to speeding up any algorithm, whether or not it has polynomial time complexity, is to consider assigning more than one processor to the algorithm. Ideally the processors will be able to complete much of their work in parallel.

It is natural to approach a divide-and-conquer algorithm this way. The two or more pieces produced by the "divide" phase can be handled by different processors.

Another way of looking at the matter is through decision trees and nondeterminism. Different subtrees or different possible nondeterministic guesses can be assigned to different processors.

The use of parallel algorithms is not a panacea. First of all, the use of p processors in parallel can always be simulated sequentially with a time complexity penalty factor of $O(p)$. Thus the best possible (worst-case) speedup to hope for is $O(p)$. So for example, reducing the time complexity of an algorithm by $O(n)$ will require at least $O(n)$ processors.

For nonpolynomial algorithms, the ramifications discussed in Chapter 1 still apply. Suppose that an algorithm taking time $c \times 2^n$ could be speeded up by a factor of 100 by the use of parallelism (note that at least 100 processors would be required). A simple calculation similar to those of Chapter 1 would imply that increasing n by 7 would more than offset the speedup due to parallelism. On the other hand, we might be comforted to know that the same problem that used to take, say, a day, would now take just 15 minutes.

In many cases, the use of p processors will not yield anything like a p-fold speedup. Some problems are not well suited to parallelization. Even tree searches are not as parallelizable as they may appear. For example, if 100 processors are used to search a binary tree without the binary search property, there will be idle processors until level 7, and most processors will be idle before level 6.

On the other hand, some apparently inherently serial operations, such as finding the end of a linked list, can be effectively parallelized.

Communication between processors is also not trivial. If processors do not share memory, then there must be some way of communicating from one processor to another. It is not feasible, if there are more than a very few processors, to have every processor connected directly to every other processor. This means there may be communication delays between processors.

However if processors do share memory, then time-consuming safeguards are necessary to make sure one processor doesn't overwrite or monopolize access to a memory location desired by another processor. In some cases, these problems can dissipate most of the speedup that parallelism might be expected to achieve.

The study of parallel algorithms is a large field in itself, and even a complete textbook on the subject, such as that by Quinn (1987), can barely scratch the surface. One complication suggested by the previous paragraph is that the form and efficiency of an algorithm will depend on the architecture of the underlying machine. Both introduction of the different feasible architectures and the details of parallel algorithms for particular NP-complete problems are beyond the scope of this text. Instead, we provide several examples in an attempt to give the flavor of parallel algorithms. For each algorithm, we comment on the architectures for which it is appropriate.

Summation

For our first example, we assume that we have an n-element array A distributed over n processors numbered from 0 through $n - 1$ such that $A[i]$ is stored in the ith processor. The goal is to find the sum of all n elements. We assume that either n is a power of 2 or that all array elements between n and $2^{\lceil \log_2 n \rceil}$ are zero.

Algorithm 5.12 corresponds to a bottom-up processing of the tree of Figure 5.5. The initial value of $A[i]$ is $2i + 1$. At each level the sums of the original elements from each box of the figure are stored in the leftmost positions of the box. All additions on a given level are to be performed in parallel. Since there are $\lceil \log_2 n \rceil$ levels, the parallel time complexity is $O(\log n)$ using n processors, as compared to $O(n)$ for the straightforward serial algorithm.

We can easily show that after iteration k of the outer loop, $A[j]$ contains the sum of $A[j]$ through $A[j + 2^k - 1]$ whenever j is a multiple of 2^k. The proof of this claim and of the correctness of the algorithm based on this invariant is left as an exercise. The claim implies that the final answer is in $A[0]$.

In Algorithm 5.12 we assume that each processor knows its own number.

At the kth level from the bottom, if $A[p]$ is added to $A[q]$, then p and q differ by 2^k. In particular, p and q are composed of the same bits except that they differ at position k. The *hypercube* model assumes that processors p and q are

```
 1. SumArray(A)                 {sums elements of an array A in parallel.
 2.                                        See documentation in text.
 3.                             Assumes array size available via size(A)}
 4. for k := 1 to ⌈log₂ n⌉                   {for levels 1 through k}
 5.    for all j in parallel             {this loop executed in parallel}
 6.       if j is a multiple of 2^k then
 7.          A[j] := A[j] + A[j + 2^(k-1)]         {find sums on kth level}
 8.       end if
 9.    end for
10. end for
11. return A[0]
```

Algorithm 5.12 Summing the Elements of a Vector in Parallel

connected directly iff their binary representations differ by a single bit. If there are $n = 2^b$ processors, then this assumes that each processor can communicate efficiently with $k = \log_2 n$ different processors. Note that this may be too many direct connections for massive parallelism; for $n = 1024$ each processor must be directly connected to 10 neighbors. Other architectures are capable of simulating these direct connections indirectly.

It is most natural to assume for this algorithm that each processor has its own separate memory, but note that at no stage in the parallel loop is the same array element accessed by more than one processor. Thus shared memory is also possible.

Note that if each processor can add only 2 numbers at a time, after level k it can contain the sum of only 2^k elements. Thus $\lceil \log_2 n \rceil$ is a lower bound for the number of parallel iterations as well as an upper bound.

The algorithm can be modified so that it terminates with $A[j]$ containing $\sum_{k=0}^{j} A[k]$ for all j; the construction is left as an exercise.

Finding the End of a Linked List

It is also possible in parallel time $O(\log n)$ to find the last element of a linked list of length n. The algorithm we describe [from Hillis and Steele (1986)] will actually handle multiple lists in parallel.

We assume that each list element is assigned to a different processor. In the same processor as each element is a field containing the address of its successor and a field to contain the end of the list beginning at the given element. We call these fields **next** and **last** respectively. For elements with no successors the **last** field will end up containing NIL.

The approach is not so different from that of the previous algorithm—at the end of iteration k each element will contain in its **last** field the location 2^k

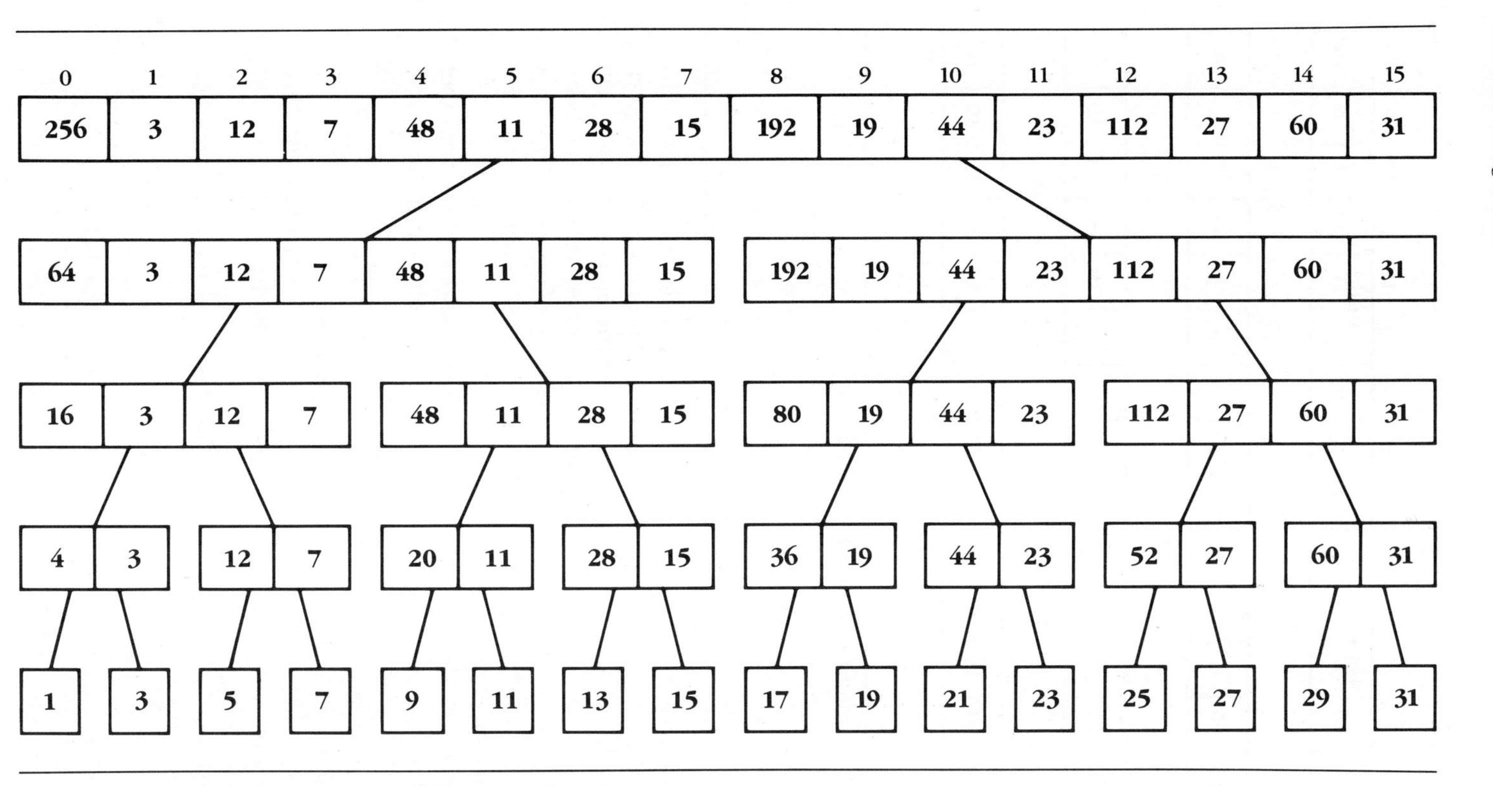

Figure 5.5 Parallel Summation of Elements of an Array

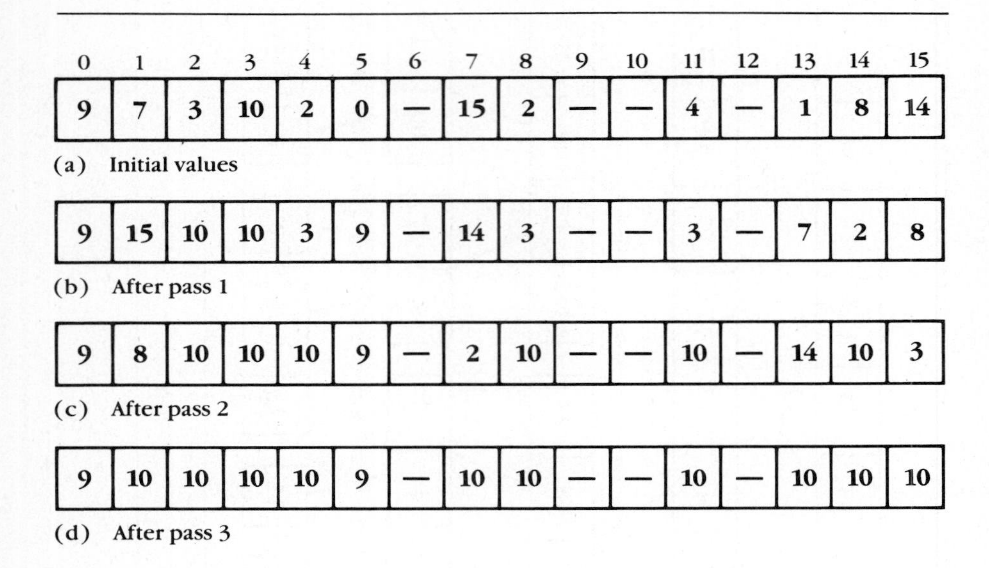

	0	1	2	3	4	5	6	7	8	9	10	11	12	13	14	15
(a) Initial values	9	7	3	10	2	0	—	15	2	—	—	4	—	1	8	14
(b) After pass 1	9	15	10	10	3	9	—	14	3	—	—	3	—	7	2	8
(c) After pass 2	9	8	10	10	10	9	—	2	10	—	—	10	—	14	10	3
(d) After pass 3	9	10	10	10	10	9	—	10	10	—	—	10	—	10	10	10

Figure 5.6 Finding the Ends of Linked Lists

```
 1. FindLast(j)                 {returns the last element in the linked list
 2.                               beginning at processor j. If the given
 3.                               element is at the end of a list, NIL will
 4.                               be returned. List nodes are allocated one
 5.                               per processor. Each processor has a last
 6.                               and a next field}
 7. for all k in parallel
 8.    last[k] := next[k]                  {each next field contains the}
 9. end for                                {immediate successor}
10. while last[j] ≠ nil and last[last[j]] ≠ nil
11.    for all k in parallel
12.       if last[k] ≠ nil and last[last[k]] ≠ nil then
13.          last[k] := last[last[k]]                 {replace the 2^i th}
14.       end if                   {successor by the 2^(i+1) st successor}
15.    end for
16. end while                   {stop when last gives the end of the list}
17. return last[j]
```

Algorithm 5.13 Find the Last Element of a List in Parallel

positions down the list. This will naturally give logarithmic time complexity. We need only make sure we don't fall off the end of the list.

Figure 5.6 gives an example for 16 processors. Identifying the list elements with their addresses, we assume linked lists corresponding to the sequences (13 1 7 15 14 8 2 3 10), (12), (6), and (5 0 9). We also assume that the sequence beginning (11 4 2) shares the suffix (2 3 10) with the preceding longest sequence, which causes no problem for the algorithm. Successive rows in the figure give the value of the **last** field of each processor. After the third iteration, the values don't change.

Algorithm 5.13 assumes that only the end of a particular list is wanted. If the last element of all lists is wanted, then we need to continue until last(N) or last(last(N)) is empty for all nodes N, or until we have executed $\log_2 P$ iterations, where P is the number of processors.

Here we assume that every processor can send a value to every other processor. Hillis and Steele (1986) describe one architecture implementing this.

Matrix Multiplication

We next consider multiplying two n-by-n matrices using the straightforward $O(n^3)$ algorithm. Here we assume n^2 processors indexed by i and j, where $1 \le i, j \le n$. Processor P$[i, j]$ is responsible for computing the (i, j) entry of the product matrix. Assume that A and B are the two matrices to be multiplied. The algorithm is responsible for ensuring that $A[i, k]$and $B[k, j]$ reach processor P$[i, j]$ simultaneously so that their product may be added into the total.

We assume that initially P$[i, j]$ contains $A[i, j]$ and $B[i, j]$. We would prefer a configuration like that of Figure 5.7a. With this initial configuration, a leftward cyclic permutation of each row and an upward cyclic permutation of each column would provide each processor with two of the factors that make up the sum it is computing. This is illustrated in the rest of the figure. The two outer loops of Algorithm 5.14, essentially from Quinn (1987), respectively obtain a configuration like that of Figure 5.7a and repeatedly perform the cyclic permutations.

More precisely, at the beginning of the kth iteration of the second loop, P$[i, j]$ contains $A[i, i + (j - 1) + (k - 1)]$ and $B[j + (i - 1) + (k - 1), j]$, where all index computations are done mod n. Note that the first subscript of A and the second of B agree with those of P, and hence of the product. The first subscript of A and the second of B agree with each other as required. When $k = 1$, we get the appropriate generalization of the configuration of Figure 5.7a. Also notice that as k ranges from 1 to n, all possible values of this common subscript are obtained. Thus the sum being formed in $C[i, j]$ contains all its terms.

$j = 1$	$j = 2$	$j = 3$	
$A[1, 1]$ $B[1, 1]$	$A[1, 2]$ $B[2, 2]$	$A[1, 3]$ $B[3, 3]$	$i = 1$
$A[2, 2]$ $B[2, 1]$	$A[2, 3]$ $B[3, 2]$	$A[2, 1]$ $B[1, 3]$	$i = 2$
$A[3, 3]$ $B[3, 1]$	$A[3, 1]$ $B[1, 2]$	$A[3, 2]$ $B[2, 3]$	$i = 3$

(a) Contents of $P[i, j]$ before first iteration of lines 12–18

$j = 1$	$j = 2$	$j = 3$	
$A[1, 2]$ $B[2, 1]$	$A[1, 3]$ $B[3, 2]$	$A[1, 1]$ $B[1, 3]$	$i = 1$
$A[2, 3]$ $B[3, 1]$	$A[2, 1]$ $B[1, 2]$	$A[2, 2]$ $B[2, 3]$	$i = 2$
$A[3, 1]$ $B[1, 1]$	$A[3, 2]$ $B[2, 2]$	$A[3, 3]$ $B[3, 3]$	$i = 3$

(b) Contents of $P[i, j]$ before the second iteration

$j = 1$	$j = 2$	$j = 3$	
$A[1, 3]$ $B[3, 1]$	$A[1, 1]$ $B[1, 2]$	$A[1, 2]$ $B[2, 3]$	$i = 1$
$A[2, 1]$ $B[1, 1]$	$A[2, 2]$ $B[2, 2]$	$A[2, 3]$ $B[3, 3]$	$i = 2$
$A[3, 2]$ $B[2, 1]$	$A[3, 3]$ $B[3, 2]$	$A[3, 1]$ $B[1, 3]$	$i = 3$

(c) Contents of $P[i, j]$ before the third iteration

Figure 5.7 Assignments of Elements to Processors in Algorithm 5.14

Both outer loops have time complexity $O(n)$, which is therefore the time complexity of the entire algorithm. This time complexity is the best possible for n^2 processors, at least when implementing the straightforward multiplication algorithm, since n^3 multiplications are required overall.

The only communication involved among processors is that between $P[i, j]$ and $P[i + 1, j + 1]$, where again the differences are taken mod n. The *two-dimensional mesh* architecture uses only these connections (between $P[i, j]$ and $P[i \pm 1, j \pm 1]$) as direct connections.

```
 1. Multiply(A, B)         {The matrix product of A and B is returned.}
 2.                 {A matrix C is used locally. See documentation in text}
 3. for k from 1 to n − 1
 4.    for all P[i, j] in parallel
 5.       if i > k then A[i,j] := A[i,j + 1]     {cycle row i, i − 1 times}
 6.       if j > k then B[i,j] := B[i + 1,j] {cycle column j, j − 1 times}
 7.    end for
 8. end for
 9. for all P[i, j] in parallel
10.    C[i,j] := 0                                  {initialize product}
11. end for
12. for k from 1 to n
13.    for all P[i, j] in parallel
14.       C[i,j] := C[i,j] + A[i,j] × B[i,j]{update entry (i,j) of product}
15.       A[i,j] := A[i,j + 1]                            {permute row}
16.       B[i,j] := B[i + 1,j]                           {and column}
17.    end for
18. end for
19. return C
```

Algorithm 5.14 Parallel Matrix Multiplication

Bitonic Merge

Our final example will be a sorting algorithm related to mergesort. The problem with mergesort is that the merging phase doesn't lend itself well to parallelism. In the following algorithm, attributed to Batcher (1968), we simulate a merge of two runs of size 2^{k-1} by k passes that can be parallelized. Here we assume the size n of the sequence to be merged is 2^r.

From the point of view of the merge tree, the main difference between Batcher's algorithm and mergesort is the reversal of a number of the runs. An example is shown in Figure 5.8a for the sequence (4 3 12 9 2 11 16 6 1 7 10 13 5 14 8 15). The corresponding merge tree is shown in (b) of the figure. Note that any two runs that are merged go in opposite directions and that the resulting run goes in the same direction as the first input run.

(To simplify the following discussion we use "increasing" for "nondecreasing" and "decreasing" for "nonincreasing.")

In general, if the locations are labeled from 0 through $2^r - 1$ during the kth merge (i.e., in the kth level from the bottom), a simple induction shows that location j is part of an increasing run iff bit (j, k) = bit $(j, k + 1)$ where bit

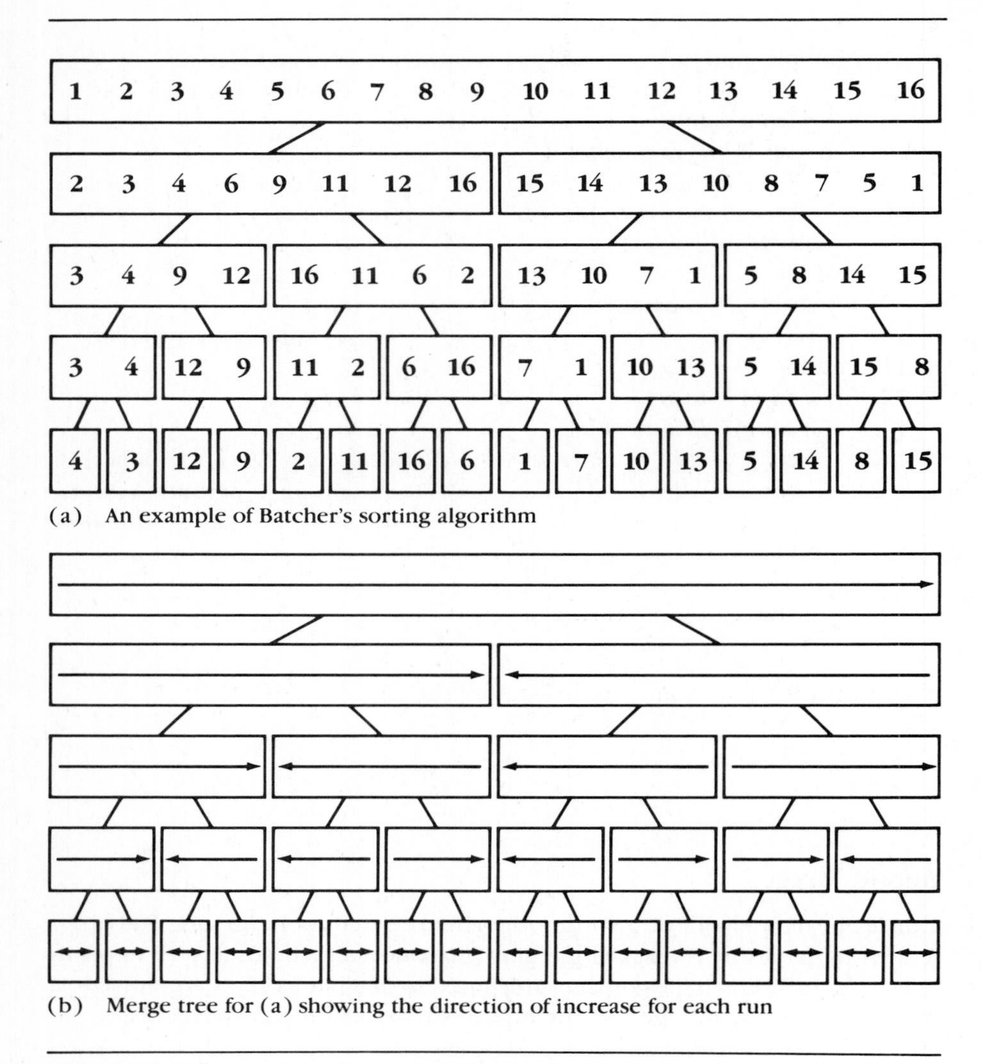

(a) An example of Batcher's sorting algorithm

(b) Merge tree for (a) showing the direction of increase for each run

Figure 5.8 Batcher's Sorting Algorithm and Merge Trees

(j, k) is the kth bit of the integer j, counting from the right starting at 0. At the top level, corresponding to the rth merge, both bits are always 0.

The counterpart of the merge algorithm is called *bitonic merge,* which assumes that one input run is increasing and the other is decreasing. By symmetry we may assume that the first run is increasing and that the result is to be increasing; we do so in the analysis that follows.

We claim that the kth merge of Batcher's algorithm, which merges runs of size 2^{k-1}, has parallel time complexity $\Theta(k)$. If so, then for a sequence of size

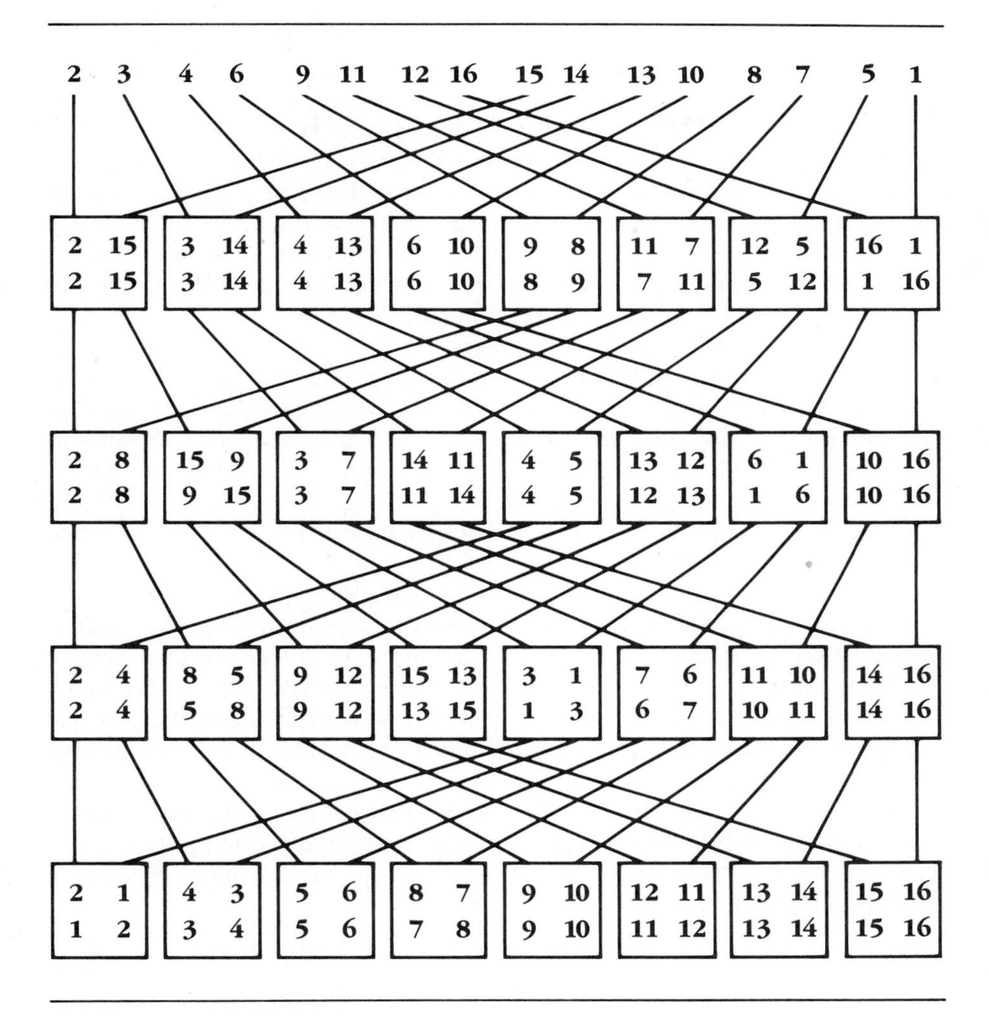

Figure 5.9 A Bitonic Merge

$n = 2^r$, the entire algorithm has for its r passes a total parallel time complexity of $\Theta\left(\sum_{k=1}^{r} k\right)$, which is $\Theta(r^2)$ or $\Theta(\log^2 n)$.

The bitonic merge algorithm itself is divided into passes. Each pass works by comparing elements from one of a set of disjoint pairs of locations, putting the smaller in one location and the larger in the next location—or vice versa if the output run is to be decreasing. Since both the input pairs and the output pairs are all disjoint, these comparisons may all be done in parallel.

We assume in the following analysis that the input sequence has size $n = 2^r$

and is stored one element per processor in processors indexed from 0 to $n - 1$. The element in processor i will be denoted a_i.

The result after each pass of the last bitonic merge of Figure 5.8 is given in Figure 5.9. We have drawn the figure to suggest that pairs of elements are first moved to a box responsible for comparing the two elements, which are then swapped if they are out of order. Note that the assignment of pairs to boxes is the same for each pass. More precisely, the input pair (a_{s+k}, a_{s+k+2^p}) is associated with output pair, (a_{s+2k}, a_{s+2k+1}), where s is the first location of the first run and we are in the pth merge of the overall algorithm. Here we assume that bit p of a_{s+k} is equal to 0 so that bit p of a_{s+k+2^p} is 1.

In Figure 5.10 the passes of the merge of Figure 5.9 are interpreted as creating shorter subsequences (each composed of an increasing and a decreasing segment) from long ones. The subsequences are in rows; the array may be recovered by reading downward in the columns. No subsequence contains an element smaller than that of any previous subsequence. When there are n such subsequences, the merge is complete.

The sequence after each pass of Batcher's algorithm for the input of Figure 5.8 is given in Figure 5.11.

Note that the permutations of lines 9–13 of Algorithm 5.15 (see Figure 5.9) correspond to a perfect shuffle of a deck of cards. In both cases the input is divided exactly in half, and the halves are perfectly interweaved. A *perfect shuffle* architecture has been suggested in which processor i is connected to processor $2i \bmod n - 1$, with processor $n - 1$ connected to itself. We have left as an exercise that the permutations of lines 9–13 may be described this way for all values of n.

We will prove correctness of Algorithm 5.15 by showing that each pass of bitonic merge does create the appropriate generalization of an increasing run followed by a decreasing run. In order to do so, we first must define this appropriate generalization.

Def 5.16 A sequence $(a_j)_{j=0}^{2q-1}$ is a ***bitonic sequence*** iff

a. there is some m such that $a_{j-1} \leq a_j$ for $j \leq m$ and $a_{j-1} \geq a_j$ for $j > m$ (i.e., the sequence increases up to a_m and decreases thereafter) or,
b. for some c the sequence $(a_{j+c \bmod 2q})_{j=0}^{2q-1}$ is bitonic in the sense of (a).

Note that in (b) adding $c \bmod 2q$ to the index j permutes the residues mod $2q$. This type of permutation is called a *cyclic permutation.*

The rows of Figure 5.10 are bitonic sequences. Note that sequences consisting of decreasing runs followed by increasing runs qualify as bitonic by (b).

Theorem Algorithm 5.15 sorts correctly.

Proof We first show that each pass of bitonic merge divides a bitonic sequence into two bitonic sequences of half the length. We also show that each

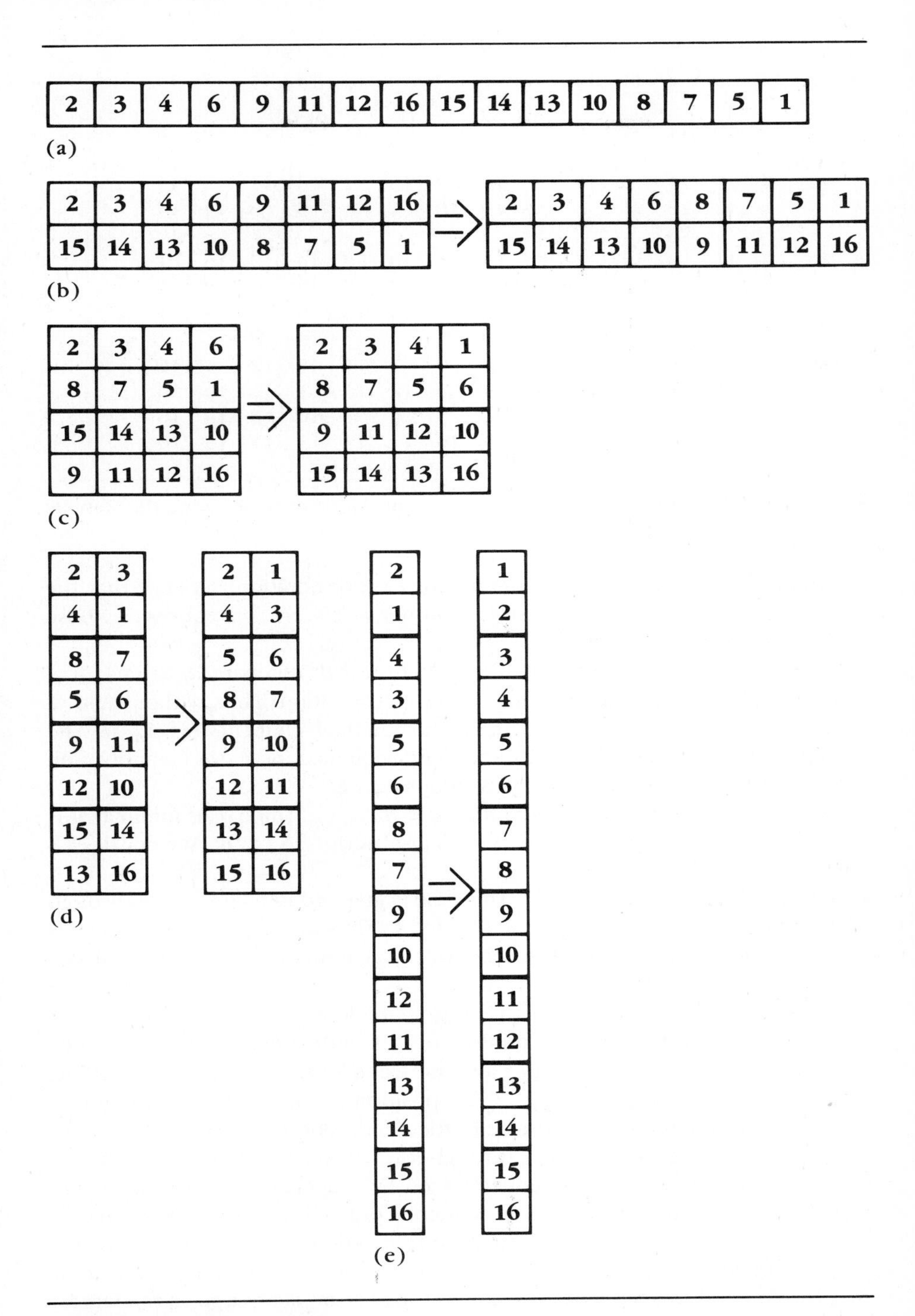

Figure 5.10 Bitonic Merge

4	3	12	9	2	11	16	6	1	7	10	13	5	14	8	15
3	4	12	9	11	2	6	16	7	1	10	13	5	14	15	8
3	12	4	9	11	6	16	2	10	7	13	1	5	15	8	14
3	4	9	12	16	11	6	2	13	10	7	1	5	8	14	15
3	16	4	11	6	9	2	12	13	5	10	8	14	7	15	1
3	6	9	16	2	4	11	12	14	13	7	5	15	10	8	1
2	3	4	6	9	11	12	16	15	14	13	10	8	7	5	1
2	15	3	14	4	13	6	10	8	9	7	11	5	12	1	16
2	8	9	15	3	7	11	14	4	5	12	13	1	6	10	16
2	4	5	8	9	12	13	15	1	3	6	7	10	11	14	16
1	2	3	4	5	6	7	8	9	10	11	12	13	14	15	16

Figure 5.11 An Example of Batcher's Sorting Algorithm, Showing the Results After Each Pass

minimum is less than each maximum. For ease of notation, suppose that the bitonic sequence is $\{a_j\}_{j=0}^{2k-1}$. Our argument will also work for sequences whose lowest subscript is nonzero.

For a sequence of length $2k$, bitonic merge will compare a_j to $a_{j+k \bmod 2k}$. Note that this remains true for any cyclic permutation. Thus without loss of generality we may assume that a_0 is the minimal element. Let a_m be the maximal element. In other words, the input sequence increases from position 0 to position m, and then decreases to position $2k - 1$.

Let j be the least subscript such that $a_j \geq a_{j+k \bmod 2k}$. Since a_0 is minimal, k is such a subscript, so $j \leq k$. Also m is such a subscript, so $j \leq m$. We consider 3 cases:

Case 1: $j = k$. Here $k = j \leq m$. This case is graphed in Figure 5.12a through d. By definition of j, $a_j < a_{j+k}$ for all $0 \leq j < k$. Thus the sequence of minima is $\{a_j\}_{j=0}^{k-1}$, which is increasing. The sequence of maxima is $\{a_j\}_{j=k}^{2k-1}$. It increases from k to m, then decreases.

Case 2: $j < k$ and $k \leq m$. This case is graphed in Figure 5.12e through h. By definition of j the sequence of minima begins with the subsequence of elements a_0 through a_{j-1}. Here and also between positions j and $k - 1$, the sequence is increasing. Since $a_{j+k} < a_j$, position $j + k$ and thus the segment from $j + k$ through $2k - 1$ must be in the decreasing segment. So the rest of the minima are the elements a_{j+k} through a_{2k-1}. Thus the sequence of minima consists of an increasing segment followed by a decreasing segment, so the sequence of minima is bitonic. The sequence of maxima may be permuted to run from a_j through a_{j+k-1}, increasing up to position a_m and then decreasing.

The largest minimum is a_{j-1} or a_{j+k}. The smallest maximum is either a_j or a_{j+k-1}. We need to show that the smallest maximum is as large as the largest minimum. By the definition of j, $a_{j+k} \leq a_j$ and $a_{j-1} \leq a_{j+k-1}$. We have $a_{j+k} <$

```
 1. ParallelSort(a, n)   {Sorts in parallel an input sequence a stored in
 2.                            processors 0 through n − 1 where n is a
 3.                         power of 2. The kth element is accessed as a[k].
 4.                            Variables pass and size are used locally}
 5. pass := 1                              {initialize the pass number}
 6. size := 1                        {initialize the size of an input run}
 7. while size < n − 1                       {until there is just one run}
 8.    merge(a, n, size, pass)                 {perform a bitonic merge}
 9.    pass := pass + 1               {update pass and size values—we}
10.    size := 2 × size                   {always have size := 2^(pass−1)}
11. end while
```

```
 1. Merge(a, n, size, pass)      {Merges pairs of adjacent runs each of
 2.                           size size. The array a (implemented as above) is
 3.                          passed by value-result. Parameters size and pass
 4.                                are as above. A temporary array b is used
 5.                       locally as in ordinary Merge. Assumes bit as in text}
 6. for j := 1 to pass do               {merging takes pass local passes}
 7.    for i := 0 to n − 1 in parallel
 8.       start := i − i mod (2 × size)         {find start of pair of runs}
 9.       if bit(i, pass − 1) = 0 then         {interweave two input runs}
10.          b[start + 2 × (i − start)] := a[i]              {into new array}
11.       else                                           {in preparation for}
12.          b[start + 2 × (i − start − size) + 1] := a[i] {comparisons}
13.       end if
14.    end for
15.    for i := 0 to n − 1 in parallel
16.       start := i − i mod (2 × size)         {find start of pair of runs}
17.       dir := (bit(start, pass + 1) = bit(start, pass)) {find direction}
18.       if (bit(i, 0) = 0) and ((b[i + 1] < b[i]) = dir) then
19.          swap(b[i + 1], b[i])       {of sort—if in wrong order, swap}
20.       end if
21.    end for
22.    for i := 0 to n − 1 do in parallel
23.       a[i] := b[i]                       {copy back to original array}
24.    end for
25. end for
```

Algorithm 5.15 Batcher's Parallel Sort

a_{j+k-1}, since both are in the decreasing segment. Also $a_{j-1} \le a_j$, since both are in the increasing segment; so we are done with case 2.

Case 3: $j < k$ and $m < k$. Recall $j \le m$. This case is graphed in Figure 5.12i through l. Again the sequence of minima begins with the elements a_0 through

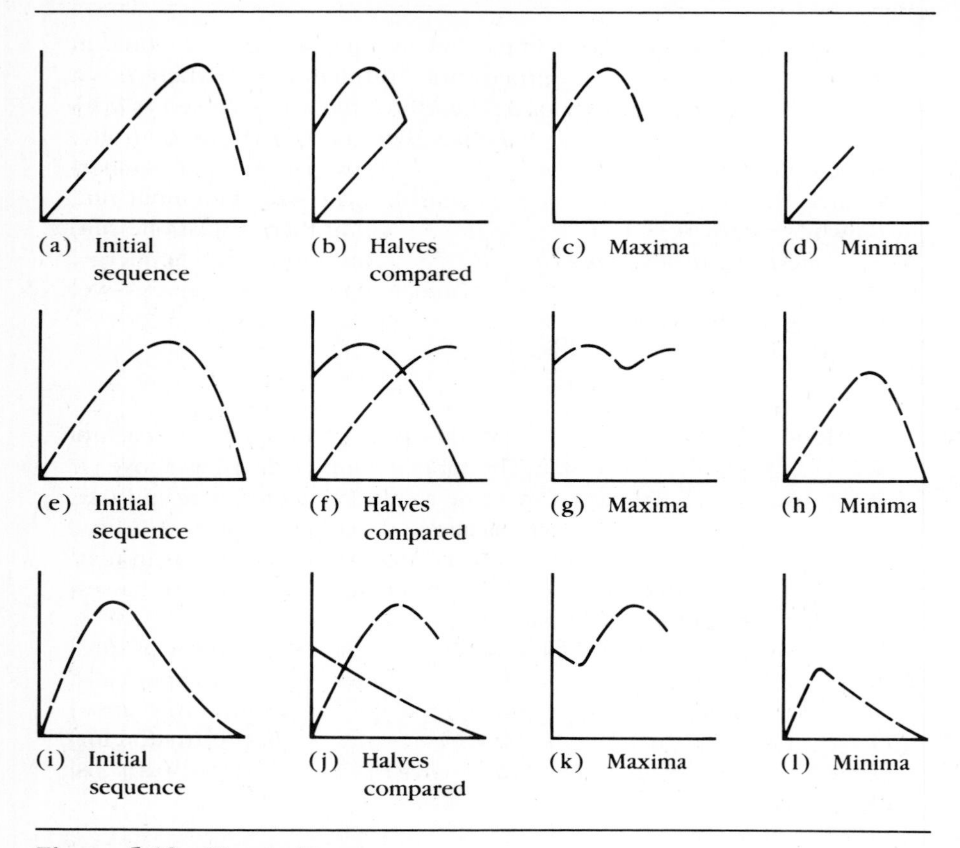

Figure 5.12 Bitonic Merge

a_{j-1}. For i between j and m, $a_i \geq a_j \geq a_{j+k} \geq a_{i+k}$. For i between m and k, $a_i \geq a_{i+k}$ since both elements are in the decreasing segment. So the rest of the sequence of minima consists of the elements a_{j+k} through a_{k-1}. The sequence of maxima may be permuted to run from a_j through a_{j+k-1}. The proof that no minimum can be larger than a maximum is the same as in case 2.

We next claim that after pass k of bitonic merge we have 2^k bitonic sequences $\{S_i\}$ indexed from 0 through 2^{k-1} such that no element in S_i is greater than any element in S_j if $i < j$.[4] This is trivial after pass 0. After pass k, let the bitonic sequences created from S_m be S_{2m} (containing the minima) and S_{2m+1} (containing the maxima). If an element from S_i is greater than one from S_j and $i < j$, then the first element is in $S\lfloor i/2 \rfloor$ at the end of the pass $k - 1$ and the second is in $S\lfloor j/2 \rfloor$. By induction, $\lfloor i/2 \rfloor \geq \lfloor j/2 \rfloor$. A strict inequality would contradict $i < j$. The only possibility is that $\lfloor i/2 \rfloor = \lfloor j/2 \rfloor = m$. Since $i < j$

[4] S_i is stored in locations $\{a_i + p2^k\}$.

we have $i = 2m$ and $j = 2m + 1$. But pass k cannot leave a larger element in S_{2m} and a smaller one in S_{2m+1}.

Now after the final pass, we have n bitonic sequences $\{S_i\}$ of size 1. Since no element in S_i is greater than any of S_j for $i < j$, the bitonic merge algorithm has created a sorted sequence. Thus Batcher's algorithm sorts correctly. ■

5.7 Exercises

Solutions of exercises marked with an asterisk appear in Appendix C; references are given in that appendix for those marked with a double asterisk. You may assume the given problems are not trivial.

NP-Completeness

*1. Suppose that A and B are problems in P, that C and D are problems in NP, that E is NP-complete, and that F is not in P. For each of the following questions, answer either "false" (i.e., not necessarily true), "true if P = NP," or "true" (i.e., whether or not P = NP). For a problem X, notation X' refers to the complementary problem of X.

a) is $A \propto B$?
b) is $A \propto C$?
c) is $C \propto A$?
d) is $F \propto A$?
e) is $A \propto E$?
f) is $E \propto A$?
g) is $C \propto E$?
h) is $C \propto D$?
i) is E in P?
j) is A' in P?
k) is 3SAT $\propto C$?
l) is $C \propto$ 3SAT?
m) if $X \propto E$, is X NP-complete?
n) is A in NP?
o) if $E \propto X$, is X NP-complete?
p) is E in NP?
q) is 3SAT $\propto E$?
r) is $E \propto$ 3SAT?
s) if $X \propto C$, is X in NP?
t) if $C \propto X$, is X in NP?
u) does A require only polynomial space?
v) if $X \propto A$, is X in P?
w) is A NP-complete?

2. Show that the bin-packing problem is in NP. Show that the processor-scheduling problem is in NP.

*3. Find

a) the clique of maximum size,
b) the vertex cover of minimum size, and
c) the minimum k for which the graph is k-colorable for the graphs of Figures 4.41a and 5.1a.

*4. Why can't we show HC $\propto$ 3DM, by, given a graph G, letting X and Y be the set of the n vertices in the graph, Z be the set of the first n integers, and M be $\{(x, y, n) \mid (x, y)$ is an edge of the graph$\}$?

5. A graph is *bipartite* iff its vertex set V is the disjoint union of sets W and X and each edge of the graph connects a vertex of W and a vertex of X. Show that deciding whether or not a graph is bipartite is polynomially reducible to the 2-colorability decision problem.

6. Show that the 2-colorability decision problem is in the class P. What does this say about the bipartite decision problem?

*7. In the proof that 3-satisfiability $\propto$ k-colorability in Section 5.2, why can't both x_i and x_i' be nonadjacent to C_j?

*8. What's wrong with the following argument, which claims to show that 3-colorability is polynomially reducible to 4-colorability?

Let $G = (V, E)$ be an instance of 3-colorability. Use the same G as the instance of 4-colorability. Clearly this is a polynomial-time transformation. If G is 3-colorable, there is a function $c: V \rightarrow \{1, 2, 3\}$ such that $c(v) = c(w)$ implies $\{v, w\}$ is not in E. Let $c': V \rightarrow \{1, 2, 3, 4\}$ be defined by $c'(v) = c(v)$. Since $c'(v) = c'(w)$ iff $c(v) = c(w)$, which implies that $\{v, w\}$ is not in E, c' is a 4-coloring.

9. Design and implement a polynomial-time algorithm for the 2-colorability decision problem that assigns colors to vertices in parallel with a breadth-first search.

10. Show that the bin-packing problem is NP-complete.(Hint: reduce partition to it.)

*11. An instance of the *minimum penalty job sequencing* problem is an integer n, sequences $\{t_i\}$, $\{d_i\}$, $\{p_i\}$ indexed from 1 through n, and a target k. For jobs numbered 1 through n, t_i is interpreted as the time required for job i, d_i as the deadline for job i, and p_i the penalty to be incurred if the deadline for job i is not met. The question is whether or not there is a schedule of jobs for a single processor so the total penalty can be held under the target value k. Show this problem is NP-complete (hint: reduce ZOK to it).

12. An instance of the *sum of subsets* problem is a set S of integers and a target sum T. The question is whether or not there is a subset U of S such that the sum of the elements in U equals T. Show that

a) the sum of subsets problem is NP-complete.(Hint: look at the proof of NP-completeness of ZOK.)
b) sum of subsets $\propto$ ZOK.

13. An instance of the *partition* problem is a set S of integers. The question is whether or not there is a subset T of S such that the sum of the elements of T equals the sum of the elements in $S - T$. Show that
    a) partition $\propto$ sum of subsets and
    b) sum of subsets $\propto$ partition (so that partition is NP-complete).
14. Show that the partition problem is polynomially reducible to ZOK.
*15. Show that vertex cover $\propto$ the Hamiltonian circuit problem for directed graphs by using the following construction from Karp (1972). Given an instance (G, k) of vertex cover, assume that the edges of G are numbered. Replace each edge $i = \{v, w\}$ of G by a structure G_i like that of Figure 5.13. Add additional vertices 1 through k. For each G_i, if $i = \{v, w\}$ then there is an edge from its copy of v_1 to the v_0 of G_j where j is the next-highest-numbered edge connected to v. If there is no such edge, then the edge from v_1 goes to all the vertices 1 through k. Also for each G_i there is an edge from its copy of w_i to the w_0 of G_r where r is the next-highest-numbered edge connected to w. If no such edge exists, then the edge from v_1 goes to all the vertices 1 through k. Also for all v, if m is the lowest-numbered edge connected to v, then there are edges from all vertices 0 through k to G_m's copy of v_0.
16. An instance of *3-dimensional matching* consists of three sets X, Y, and Z with the same number of elements and a subset M of $X \times Y \times Z$. The question is whether M has a subset S such that each element of $X \times Y \times Z$ appears exactly once as a component of a member of S.
    a) Show that 3-dimensional matching is NP-complete. (Hint: reduce 3SAT to it.)
    b) Show that 3-dimensional matching $\propto$ partition.
17. An instance of the *exact cover by 3-sets* problem is a set X and a family F of subsets of size 3 of X. The question is whether or not there is a subfamily of F containing each element of X exactly once. Note that we may assume that the size of X is a multiple of 3.

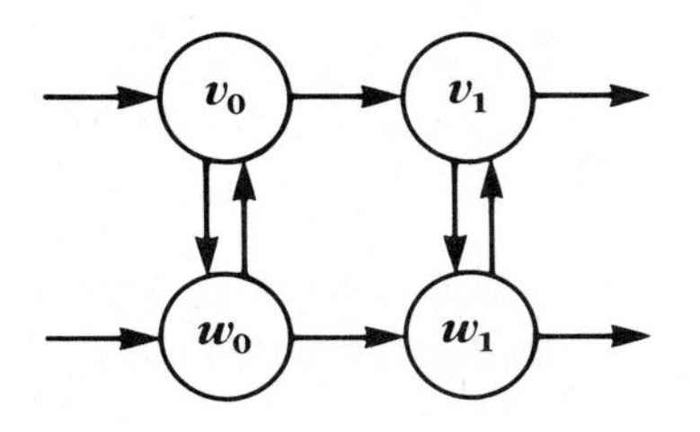

Figure 5.13 Edge Replacement in Reducing from Vertex Cover to Hamiltonian Circuit

a. Show that this problem is NP-complete. (Hint: reduce 3-dimensional matching to it.)
b. Show that exact cover by 3-sets $\propto$ exact cover.

**18. An instance of *zero-one integer programming* consists of a set of m variables $(x_j)_{j=1}^{m}$, a finite set of constraints where the ith constraint has the form $\sum_{j=1}^{m} a_{i,j}x_j \leq b_i$, and an additional constraint $\sum_{j=1}^{m} c_j x_j \geq b$. Here b may be considered a target profit. The question is whether or not there is an assignment of values from $\{0, 1\}$ to the variables $\{y_j\}$ such that all constraints are met. Show that this problem is NP-complete.

19. Show that TSP with the triangle inequality is still NP-complete.

20. A problem A is in the class *co-NP* iff A′ is in NP. Show that either of the following conditions is enough to imply NP = co-NP:

a) P = NP or
b) there is an NP-complete problem in co-NP.

*21. Show that the set of all Hamiltonian circuits of a graph with n vertices cannot be listed in polynomial time.

22. Show that the set of parse trees of a string of length n cannot be listed in polynomial time.

23. Which of the problems defined in these exercises are in NP?

Probabilistic Algorithms

24. Trace the probabilistic primality testing algorithm on the instances 133 and 137. Assume that four witnesses are sufficient.

*25. Trace the public-key encryption algorithm with $p = 23$, $q = 29$, $d = 17$. After you have found the appropriate e, find E(2) and D(E(2)). Verify that D(E(x)) = x and E(D(y)) = y for other messages.

26. Show that the number-theoretical statements made in the proof sketch of the probabilistic primality testing algorithm are correct.

*27. Can you still get a probability arbitrarily close to 1 of a correct answer with a probabilistic algorithm in which the kth witness may answer "yes" when it should answer "no"

a) with probability $1 - 1/(k + 1)$?
b) with probability $1 - 1/2^k$?

28. Is hash insertion using a second hash function a probabilistic algorithm?

29. Could you implement a probabilistic algorithm for testing equality of large data objects in terms of the hash signatures of Exercise 196, Chapter 2?

30. Implement a public-key encryption system.

Approximation Algorithms

**31. Show that the algorithm given for the processor scheduling problem (Section 5.4) is an ε-approximation algorithm for $\varepsilon = 1/3 - 1/(3m)$.

*32. Show that 2 / 9 is the best possible bound for the relative error in the first-fit algorithm for bin packing.

*33. In the approximation algorithm for TSP with triangle inequality is the upper bound achieved? arbitrarily closely approximated?

*34. Show that the greedy approximation algorithm for ZOK used in Section 2.11 in connection with best-first search is not an ε-approximation algorithm for any ε.

35. Implement the two first-fit algorithms of Section 5.4. Verify that they have polynomial time complexity.

*36. Consider the bin-packing strategy that works like first fit except that once a bin is rejected, it is never reconsidered. Show this strategy leads to a 1-approximation algorithm.

Parallel Algorithms

*37. Trace the Batcher parallel sort Algorithm 5.16 on the input sequence (2 12 1 8 7 16 13 14 9 4 3 15 11 6 10 5).

38. Use Algorithm 5.13 to find the end of the linked list representing the sequence (14 3 15 4 1 11 12 7 6 9). Assume there are 16 processors.

39. Trace the matrix multiplication Algorithm 5.14 on input of your choice.

40. Trace the parallel sum Algorithm 5.12 for $n = 16$ if $a[i] = i$ initially.

41. How can you use Algorithm 5.12 if the number of elements to be summed is not a power of 2? What about the other algorithms of Section 5.6?

*42. Where does the hypercube model get its name?

**43. Construct a parallel algorithm that, given a sequence $(a_k)_{k=1}^{n}$, will for each k return $\sum_{j=0}^{k} a_k$. Document any assumptions that you make.

**44. Design a parallel radix sort algorithm.

45. Design a parallel algorithm to convert a sorted linked list to a balanced binary search tree with the same nodes. What assumptions do you need to make about connections among processors?

**46. Design a parallel algorithm to solve the all-pairs shortest-path problem given n^3 processors. You may assume that n is a power of 2. (Hint: see Exercise 83 of Chapter 4 and the matrix multiplication algorithm of Section 5.6.) What assumptions do you need to make about connections among processors?

*47. Show that parallel search in a sorted array of n elements with p processors requires $\lceil \log(n+1)/\log(p+1) \rceil$ comparisons. You may assume that $n + 1$ is a power of $p + 1$. Note the speedup compared to the use of a single processor is only $\Theta(\log p)$.

**48. Find an algorithm that achieves the lower bound of the previous problem.

49. Show that Algorithm 5.12 is correct.

50. Show that Algorithm 5.13 is correct.

Additional Reading—Chapter 5

The papers of Cook (1971) and Karp (1972) are fundamental to the history of NP-completeness. A standard text is Garey and Johnson (1979); Johnson also has contributed a regular column on NP-completeness to the *Journal of Algorithms.* Pratt (1975) showed that determining whether or not a number is prime is in NP; this proof is not quite so easy as that for being composite.

Approximation algorithms are covered in Garey and Johnson as well as in some algorithms texts such as Baase (1988) and Horowitz and Sahni (1978).

There are many texts on number theory, such as Edgar (1988). Most will cover quadratic reciprocity, which is essentially the special case of the Jacobi symbol $\left(\frac{m}{n}\right)$ for prime n.

A quickly growing field is that of parallel algorithms. Quinn (1987) has a text devoted to the subject. The December 1986 issue of the *Communications of the ACM* was devoted to parallel computation; it contains other interesting articles besides the Hillis and Steele article previously cited.

Appendix

A

Proofs of Results on O-Notation

Before proving the results of Section 1.6, we first justify the remark of Section 1.5 that the definition of the O-relation simplifies when f and g have positive values.

Theorem If f and g are positive-valued functions defined on the positive integers, then $f(n)$ is $O(g(n))$ iff there is a $c > 0$ such that $f(n) < cg(n)$ for all positive integers n.

Proof The "if" part is clear. For the "only if" part, assume that $f(n)$ is $O(g(n))$ and let c and N be the positive numbers given by the definition of the O relation. By assumption, $f(n)/g(n)$ exists and is positive for all $n < N$. Let d be $\max_{0 \le n < N} \{f(n)/g(n)\}$. Then $f(n) \le d \times g(n)$ for all $n < N$. Now let $c' = \max\{c, d\}$. By the previous inequality, $f(n) \le c'g(n)$ for all $n < N$. By the definition of c and N, $f(n) \le cg(n) \le c'g(n)$ for all $n \ge N$. So the desired inequality is proven for all positive integers n. ■

Now we prove the properties of O-notation of Section 1.6.

Theorem 1 $(kf)(n)$ is $O(f(n))$ for any real number k.

Proof Let $c = k$ and $N = 1$. Clearly if $n \ge N = 1$, then $|(kf)(n)| \le k|f(n)|$, so $(kf)(n)$ is $O(f(n))$. ■

Corollary k is $O(1)$.

Proof Let $f(n) = 1$ above. ■

Corollary If $k \ne 0$ then $(kf)(n)$ is $\Theta(f(n))$.

Proof By Theorem 1 and the definition of Θ, all we need to show is that $(kf)(n)$ is $\Omega(f(n))$, that is, that $f(n)$ is $O((kf)(n))$. However, this is immediate from Theorem 1, since $f = (1/k)(kf)$. ■

From now on we will often omit the argument from functions (i.e., we may write f is $O(g)$ instead of $f(n)$ is $O(g(n))$.

Theorem 2 If f is $O(g)$ and g is $O(h)$, then f is $O(h)$.

Proof By hypothesis, there are c, d, N_1, and N_2 such that whenever $n \geq N_1$, $|f(n)| < c|g(n)|$ and whenever $n \geq N_2$, $|g(n)| < d|h(n)|$. Let $N = \max\{N_1, N_2\}$. Then if $n \geq N$, $|f(n)| < c|g(n)|$ and $|g(n)| < d|h(n)|$. So $|f(n)| < c|g(n)| < cd|h(n)|$ whenever $n \geq N$, which is enough to show that f is O(h). ■

Theorem 3 If f is O(h) and g is O(h), then f + g is O(h).

Proof By hypothesis, there are c, d, N_1, and N_2 such that whenever $n \geq N_1$, $|f(n)| < c|h(n)|$ and whenever $n \geq N_2$, $|g(n)| < d|h(n)|$. Let $N = \max\{N_1, N_2\}$. Then if $n \geq N$, $|f(n)| < c|h(n)|$ and $|g(n)| < d|h(n)|$. Thus $|(f + g)(n)| = |f(n) + g(n)| \leq |f(n)| + |g(n)| < c|h(n)| + d|h(n)| = (c + d)|h(n)|$. This completes the proof, since we have shown that whenever $n \geq N$, $|f(n)| \leq (c + d)|h(n)|$.

■

Corollary If f is O(g) then (f + g) is O(g).

Proof Let g = h in Theorem 3. ■

Theorem 4 If $0 \leq r < s$, then n^r is $O(n^s)$ and n^s is not $O(n^r)$.

Proof The first part is easy since $|n^r| = n^r \leq n^s = |n^s|$ if $0 \leq r < s$. In other words, we can let $N = c = 1$ in the definition. For the second part, assume that n^s is $O(n^r)$. With this assumption, there must be a c and an N such that whenever $n \geq N$, $n^s = |n^s| \leq c|n^r| = cn^r$. We may divide this inequality by n^r to conclude that for $n \geq N$, $n^{s-r} \leq c$, that is, that n^{s-r} is bounded. But n^{s-r} cannot be bounded for $s - r > 0$. This contradiction establishes the theorem. ■

Theorem 5 If p(n) is a polynomial of degree d, then p(n) is $\Theta(n^d)$.

Proof by strong induction on d. If $d = 0$, then p is constant, so is O(1). For $d > 0$, we may write p(n) as $a_d n^d + q(n)$, where q has degree $b < d$. By induction, q(n) is $O(n^b)$. By Theorems 4 and 2, q(n) is $O(n^d)$. By Theorem 1, $a_d n^d$ is $O(n^d)$. Finally by Theorem 3, $p(n) = a_d n^d + q(n)$ is $O(n^d)$. ■

Theorem 6 If f is O(h) and g is O(k), then fg is O(hk).

Proof By hypothesis there are N_1, N_2, c, and d such that for $n \geq N_1$, $|f(n)| \leq c|h(n)|$ and for $n \geq N_2$, $|g(n)| \leq d|k(n)|$. Let $N = \max\{N_1, N_2\}$. Then for $n \geq N$, $|(fg)(n)| = |f(n)|\,|g(n)| \leq cd|h(n)|\,|k(n)| = cd|(hk)(n)|$. This is enough to show that fg is O(hk). ■

Theorem 7 If $b > 1$ then n^k is $O(b^n)$ and b^n is not $O(n^k)$.

Proof One simple proof would use L'Hopital's rule from calculus; however, in case you haven't mastered calculus, we present another proof. Intuitively, we see that for fixed k, the ratio $(n + 1)^k/n^k$ of consecutive kth powers is $(1 + 1/n)^k$, which gets arbitrarily close to 1, while the ratio b^{n+1}/b^n is always b. Thus for large n, the ratio b^n/n^k gets larger by a factor near b every time n

increases by 1. Eventually this will make up for any head start the sequence n^k may have. We formalize this argument as follows:

1. Find M such that $(n + 1)^k/n^k$ is small for $n > M$. More precisely, by "small" we mean less than $\sqrt{b} = b^{1/2}$. To make $(M + 1)^k/M^k = b^{1/2}$, a little algebra shows that M should equal $1/(b^{1/2k} - 1)$. Since $(n + 1)^k/n^k$ is a decreasing function, this M will work.

2. Show that $(M + s)^k/M^k \le b^{s/2}$ for $s \ge 1$. Proof by induction on s. We have shown this inequality for $s = 1$. For $s > 1$, we have $(M + s)^k/(M + s - 1)^k \le b^{1/2}$ by (1). By induction, $(M + s - 1)^k/M^k \le b^{(s-1)/2}$. Multiplying these two inequalities gives the inequality we want.

3. After the Mth term of the sequence, find how long the sequence must run to make $b^n > cn^k$ for a given c. We want s such that $c \le b^{M+s}/(M + s)^k = b^M/M^k \times b^s \times M^k/(M + s)^k$. By (2), it's enough to find an s such that $c \le b^M/M^k \times b^s \times 1/b^{s/2} = b^M/M^k \times b^{s/2}$. So it's enough to choose s such that $b^{s/2} \ge cM^k/b^M$. Then for $n \ge M + s$, $b^n \ge cn^k$. ∎

Theorem 8 If $k > 0$ then $\ln n$ is $O(n^k)$.

Proof For $x \ge 1$, $\ln x \le x$. Thus for $x \ge 1$, $\ln x^{1/k} = (1/k) \ln x \le (1/k)x$. Letting $x = n^k$, we have $\ln n \le (1/k)n^k$ for $n^k \ge 1$. But for $k > 0$, $n^k \ge 1$ iff $n \ge 1$. Letting $c = 1/k$ and $N = 1$ in the definition of the O-relation, we have that $\ln n$ is $O(n^k)$ whenever $k > 0$. ∎

Theorem 9 If $b > 1$ and $c > 1$ then $\log_b n$ is $\Theta(\log_c n)$.

Proof This result follows from the change of base formula $\log_b n = \log_b c \times \log_c n$ and the corollaries to Theorem 1. ∎

Theorem 10 $\sum_{k=1}^{n} k^r$ is $\Theta(n^{r+1})$.

Proof Done in Example 5 of Section 2.1. ∎

Appendix

B

◆

Character	Decimal	Octal	Hex	Character	Decimal	Octal	Hex
(null)	0	0	0	(space)	32	40	20
(start heading)	1	1	1	!	33	41	21
(start text)	2	2	2	"	34	42	22
(end text)	3	3	3	#	35	43	23
(end transm.)	4	4	4	$	36	44	24
(enquiry)	5	5	5	%	37	45	25
(acknowledge)	6	6	6	&	38	46	26
(bell)	7	7	7	'	39	47	27
(back space)	8	10	8	(	40	50	28
(horiz. tab)	9	11	9	)	41	51	29
(line feed)	10	12	A	*	42	52	2A
(vert. tab)	11	13	B	+	43	53	2B
(form feed)	12	14	C	,	44	54	2C
(carr. return)	13	15	D	-	45	55	2D
(shift out)	14	16	E	.	46	56	2E
(shift in)	15	17	F	/	47	57	2F
(data line esc.)	16	20	10	0	48	60	30
(dev. ctl. 1)	17	21	11	1	49	61	31
(dev. ctl. 2)	18	22	12	2	50	62	32
(dev. ctl. 3)	19	23	13	3	51	63	33
(dev. ctl. 4)	20	24	14	4	52	64	34
(neg. acknowl.)	21	25	15	5	53	65	35
(synchr. file)	22	26	16	6	54	66	36
(end block)	23	27	17	7	55	67	37
(cancel)	24	30	18	8	56	70	38
(end medium)	25	31	19	9	57	71	39
(substitute)	26	32	1A	:	58	72	3A
(escape)	27	33	1B	;	59	73	3B
(file sep.)	28	34	1C	<	60	74	3C
(group sep.)	29	35	1D	=	61	75	3D
(record sep.)	30	36	1E	>	62	76	3E
(unit sep.)	31	37	1F	?	63	77	3F

The American Standard Code for Information Interchange (ASCII) is a 7-bit character code allowing for 128 characters. Of these, the first 32 are not true characters but instead instructions to input/output devices. These have been described in an abbreviated manner in parentheses. Some of these instructions, such as (backspace) and (horizontal tab), correspond to keys on the keyboard; others do not. The numerical codes corresponding to each character have been given in decimal, octal (base 8), and hexadecimal (base 16).

Character	Decimal	Octal	Hex	Character	Decimal	Octal	Hex
@	64	100	40	`	96	140	60
A	65	101	41	a	97	141	61
B	66	102	42	b	98	142	62
C	67	103	43	c	99	143	63
D	68	104	44	d	100	144	64
E	69	105	45	e	101	145	65
F	70	106	46	f	102	146	66
G	71	107	47	g	103	147	67
H	72	110	48	h	104	150	68
I	73	111	49	i	105	151	69
J	74	112	4A	j	106	152	6A
K	75	113	4B	k	107	153	6B
L	76	114	4C	l	108	154	6C
M	77	115	4D	m	109	155	6D
N	78	116	4E	n	110	156	6E
O	79	117	4F	o	111	157	6F
P	80	120	50	p	112	160	70
Q	81	121	51	q	113	161	71
R	82	122	52	r	114	162	72
S	83	123	53	s	115	163	73
T	84	124	54	t	116	164	74
U	85	125	55	u	117	165	75
V	86	126	56	v	118	166	76
W	87	127	57	w	119	167	77
X	88	130	58	x	120	170	78
Y	89	131	59	y	121	171	79
Z	90	132	5A	z	122	172	7A
[	91	133	5B	{	123	173	7B
\	92	134	5C			124	174	7C
]	93	135	5D	}	125	175	7D
^	94	136	5E	~	126	176	7E
_	95	137	5F	(rub out)	127	177	7F

Appendix

C

Solutions to Selected Exercises

Chapter 1

1. The values are 4, 1, 45, 0, and 1.
2. The values are 4, 3, 3, 3, 0, -1, -5, and -5.
6. a) n^3
   b) n^5
   c) n^2
   d) n
   e) n
   f) $\sqrt{n}$
   g) 1
   h) $\log n$
   i) n
   j) n^3
   k) $n \log n$
7. Since for $n > 1$ we have $1 < \log_2 \log_2 n < \log_2 n$, we have 1 is $O(\log_2 \log_2 n)$ and $\log_2 \log_2 n$ is $O(\log n)$. Our function is $O(\log n)$, so it necessarily is in the O-relation to the faster-growing functions n, $n \log n$, n^2, n^3, and 2^n.

   For $n!$, note that $n! > 2^n$ for $n = 4$, and that the quotient $n!/2^n$ increases by $n/2$ as n increases. Thus the quotient gets arbitrarily large, so 2^n is $O(n!)$, but $n!$ is not $O(2^n)$.
8. a) $\Theta(n)$
   b) $\Theta(n^2)$
   c) $\Theta(n^2)$
   d) $\Theta(n^3)$
9. One example would be:

```
for i := 1 to n
  j := 1;
  While j < n do
    j := 2 × j;
    . . .
  end while
  end for
```

10. a) $O(n)$, since each integer must be inspected,
    b) $O(n^2)$ if we use the usual algorithm, since there are n^2 pairs of terms that we need to multiply,
    c) $O(n^2)$, since n^2 products need to be printed.

14. The program whose running time is asymptotically less may require more space to run or store or may be more difficult for humans to deal with. For small values of n (for example, $n < 1{,}000{,}000{,}000$) it may actually be slower than the other program. If the programs calculate approximations, the second program may find a better approximation. The asymptotic notation only deals with the worst case; the second program may be faster in the average case. See also Exercise 26.

17. 20. See Appendix A.

23. By definition, $x \bmod y = m$ implies that $x = qy + m$ for some q. By hypothesis, x and y are both multiples of a, say $x = ba$ and $y = ca$. Thus $ba = qca + m$, and therefore a divides $m = a(b - qc)$.

24. As above, $x = qy + m$ for some q. Also $m = pz + n$ for some p. Combining the two equations, we have $x = qy + pz + n$. By hypothesis, $y = sz$ for some s, so $x = qsz + pz + n$. Thus $x = (qs + p)z + n$. Since n is a residue mod z, we must have $n < z$, and we are done.

26. One possibility is to let $g(n)$ be $1/2$ for all n and $f(n)$ be a function with values in the interval $(0, 1)$ that get arbitrarily close to one. For example, $f(n)$ may be $\sin(n)$ or $f(n)$ may be $1/n$ if n is a perfect square and 1 otherwise.

Chapter 2

1. a) 15 b) 91 c) 77160 d) 10

2. $1! = 1$, $2! = 2$, $3! = 6$, $4! = 24$, $5! = 120$, $6! = 720$, $7! = 5040$, $8! = 40{,}320$, $9! = 362{,}880$, $10! = 3{,}628{,}800$

3. Starting with a_0, we get the sequence (1 3 7 15 31 . . .). If 2^n is replaced by 3^n, we get (1 4 13 40 121 . . .).

4. Of size 0: $\varnothing$. Of size 1: $\{1\}, \{2\}, \{3\}, \{4\}$. Of size 2: $\{1, 2\}, \{1, 3\}, \{1, 4\}, \{2, 3\}, \{2, 4\}, \{3, 4\}$. Of size 3: $\{1, 2, 3\}, \{1, 2, 4\}, \{1, 3, 4\}, \{2, 3, 4\}$. Of size 4: $\{1, 2, 3, 4\}$. This gives a total of $1 + 4 + 6 + 4 + 1 = 16$ subsets.

5. (1 2 3 4), (1 2 4 3), (1 3 2 4), (1 3 4 2), (1 4 2 3), (1 4 3 2), (2 1 3 4), (2 1 4 3), (2 3 1 4), (2 3 4 1), (2 4 1 3), (2 4 3 1), (3 1 2 4), (3 1 4 2), (3 2 1 4), (3 2 4 1), (3 4 1 2), (3 4 2 1), (4 1 2 3), (4 1 3 2), (4 2 1 3), (4 2 3 1), (4 3 1 2), (4 3 2 1).

6. 1, 1, 2, 3, 5, 8, 13, 21, 34, 55, 89, 144, 233, 377, 610.

7. See Figure C.1.

9. For $r \leq 3$, $A(r, 0) = 0$, $A(r, 1) = 2$, and $A(r, 2) = 4$. $A(0, 3) = 6$, $A(1, 3) = A(0, 4) = 8$, $A(2, 3) = A(1, 4) = 16$, $A(3, 3) = A(2, 4) = 65{,}536$. $A(3, 4)$ is an extremely large number; see Exercise 35.

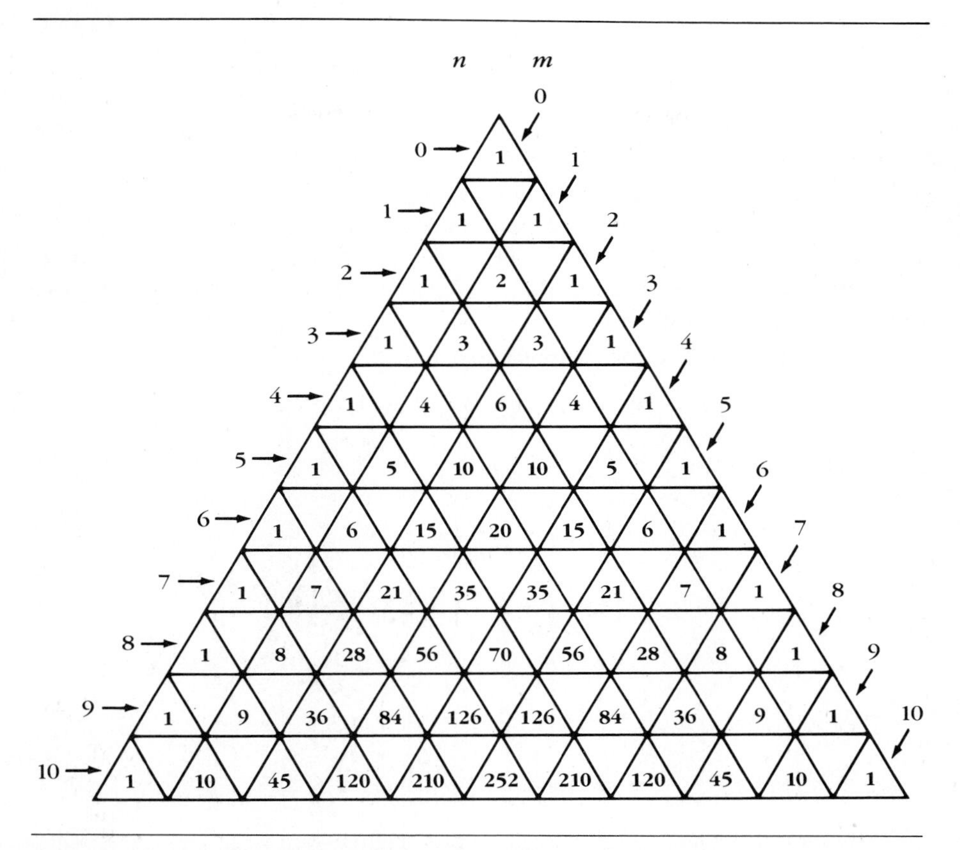

Figure C.1 Binomial Coefficients (Pascal's Triangle) Through $n = 10$

18. Proof by induction. The induction hypothesis is the statement of the theorem.
Basis: $p^1 = p < q = q^1$.
Induction step: By induction $p^{n-1} < q^{n-1}$. We may multiply this inequality by the inequality $p < q$ to get $pp^{n-1} < qq^{n-1}$, or $p^n < q^n$.

21. a) Proof by induction, beginning at $n = 1$.
Basis: The sum is just $f_1 = 1$, while $f_3 - 1 = 2 - 1 = 1$.
Induction step: $\sum_{k=1}^{n} f_k = f_n + \sum_{k=1}^{n-1} f_k$. By induction, this latter sum is $f_{n-1+2} - 1$, so the right-hand side becomes $f_n + f_{n+1} - 1$. By the recurrence defining the Fibonacci sequence, this is simply $f_{n+2} - 1$.

b) Proof by induction, beginning at $n = 1$.
Basis: The sum is just $f_1^2 = 1$. Also, $f_1 f_{1+1} = 1$.
Induction step: The sum is just $f_n^2 + \sum_{k=1}^{n-1} f_k^2$. By induction, this is just $f_n^2 + f_{n-1}f_n$, or $f_n(f_n + f_{n-1})$, or $f_n f_{n+1}$.

c) Proof by induction, beginning at $n = 1$.
Basis: $f_{1+1}f_{1+2} = 1 \times 2$. Also, $f_n f_{n+3} = 1 \times 3 = 3$, and $2 = 3 + (-1)^1$.
Induction step: $f_{n+1}f_{n+2} = (f_{n-1} + f_n)f_{n+2} = f_{n-1}f_{n+2} + f_n f_{n+2}$. By induction, this sum is $f_n f_{n+1} + f_n f_{n+2} - (-1)^{n-1}$, or $f_n(f_{n+1} + f_{n+2}) + (-1)^n$, or $f_n f_{n+3} + (-1)^n$.

d) Proof by induction, beginning at $n = 2$.
Basis: $f_{2-1}f_{2+1} = 1 \times 2 = 2$. Also, $f_2^2 + (-1)^2 = 1 + 1 = 2$.
Induction step: $f_{n-1}f_{n+1} = f_{n-1}(f_n + f_{n-1}) = f_{n-1}f_n + f_{n-1}^2$. By induction, the right-hand side is $f_{n-1}f_n + f_{n-2}f_n - (-1)^{n-1}$, or $(f_{n-1} + f_{n-2})f_n + (-1)^n$, or $f_n^2 + (-1)^n$.

22. a) Proof by induction, beginning at $n = 0$.
Basis: The sum is just $\binom{0}{0} = 1$. Also, $2^0 = 1$.

Induction step: $\sum_{m=0}^{n} \binom{n}{m}$

$$= 1 + 1 + \sum_{m=1}^{n-1} \binom{n}{m}$$

$$= 1 + \sum_{m=1}^{n-1} \binom{n-1}{m-1} + 1 + \sum_{m=1}^{n-1} \binom{n-1}{m}$$

$$= 1 + \sum_{m=0}^{n-2} \binom{n-1}{m} + 1 + \sum_{m=1}^{n-1} \binom{n-1}{m}$$

$$= \sum_{m=0}^{n-1} \binom{n-1}{m} + \sum_{m=0}^{n-1} \binom{n-1}{m}$$

By induction, each sum is 2^{n-1}, so the original sum is $2^{n-1} + 2^{n-1} = 2^n$.

b) Proof by induction on n, beginning with $n = 0$.
Basis: For $n = 0$, we need only consider $m = 0$. Here

$n!/[m!\,(n - m)!] = 0!/[0!\,0!] = 1/[1 \times 1] = 1 = \binom{n}{m}$.

Induction step: For $m = 0$, we have $n!/[0!\,(n - 0)!] = n!/[1 \times n!] = 1$ $= \binom{n}{0}$. Similarly, for $m = \text{n}$ we have $n![n!\,(n - n)!] = n!/[n! \times 1] = 1$ $= \binom{n}{n}$. Otherwise, we have $\binom{n}{m} = \binom{n-1}{m} + \binom{n-1}{m-1}$

$= (n - 1)!/[m!\,(n - 1 - m)!] + (n - 1)!/[(m - 1)!\,(n - m)!]$

$= [(n - 1)!\,(n - m)]/[m!\,(n - m)!] + [(n - 1)!\,m]/[m!\,(n - m)!]$

$= [(n - 1)!\,n]/[m!\,(n - m)]! = n!/[m!\,(n - m)!]$

c) Proof by induction on n, beginning with $n = 0$.
Basis: $(a + b)^0 = 1 = 1a^0b^0$, so the coefficient of a^0b^{0-0} is just $1 = \binom{0}{0}$.

The case $m = 0$ is the only one to consider when $n = 0$.
Induction step: $(a + b)^n = (a + b)^{n-1}(a + b)$. Using induction to obtain the coefficients in $(a + b)^{n-1}$, we see that for $0 < m < n$, the $a^m b^{n-m}$

term of the product is the sum of $\binom{n-1}{m-1} a^{m-1}b^{n-m} \times a$ and $\binom{n-1}{m}$ $a^m b^{n-1-m} \times b$. This is just $\left[\binom{n-1}{m-1} + \binom{n-1}{m}\right] a^m b^{n-m}$ or $\binom{n}{m} a^m b^{n-m}$.
If $m = 0$, then the only contribution to the b^n term of the product is from $\binom{n-1}{0} a^0 b^{n-1-0} \times b = b^n$. The coefficient here is just $1 = \binom{n}{0}$. Similarly, the only contribution to the a^n term of the product is from $\binom{n-1}{n-1}$ $a^{n-1}b^0 \times a = a^n$. The coefficient here is also $1 = \binom{n}{n}$.
d) Using (b), we see that the desired result is true for $k \neq n/2$ iff

$$n!/[(n/2)!\,(n/2)!] > n!/[k!\,(n-k)!]$$

$$\text{iff } [(n/2)!\,(n/2)!] < [k!\,(n-k)!]$$

By the symmetry of k and $n - k$, we need only show this for $k > n/2$. This proof will be by induction on k, beginning with $k = 1 + n/2$.
Basis: $[(n/2)!\,(n/2)!] < [(1 + n/2)!\,(n - (1 + n/2))!]$

$$\text{iff } [(n/2)!\,(n/2)!] < [(1 + n/2)!\,(-1 + n/2)!]$$

$$\text{iff } [1 \times (n/2)] < [(1 + n/2) \times 1]$$

which is true.
Induction step: By induction, it's enough to show that

$$[(k-1)!\,(n-(k-1))!] < [k!\,(n-k)!],$$

since the left-hand side is greater than $[(n/2)!\,(n/2)!]$. But the inequality is true iff

$$[1 \times (n - k + 1)] < [k \times 1]$$

iff $n - k + 1 < k$, or $n + 1 < 2k$. Since by assumption $k > (n/2)$ and k is an integer, we are done.
e) By (d), the term $\binom{n}{n/2}$ is the largest term in the sum $\sum_{m=0}^{n} \binom{n}{m}$. By (a), this sum is just 2^n. So $2^n = \sum_{m=0}^{n} \binom{n}{m} \leq \sum_{m=0}^{n} \binom{n}{n/2} = (n+1)\binom{n}{n/2}$ or $\binom{n}{n/2} \geq 2^n/(n+1)$. The result follows.

23. Proof of existence by induction. The induction hypothesis is the statement of the theorem.
Basis: If $0 \leq n \leq 9$, then the single digit n is its own decimal representation.
Induction step: For $n \geq 10$, there are integers p and q such that $n = 10p + q$, p and q are less than n, and $q < 10$. By induction, p has a representation. The representation of p followed by the digit q is a representation for n.
Proof of uniqueness by induction. Again the induction hypothesis is the statement of the theorem.
Basis: If $0 \leq n \leq 9$, then the representation must have just one digit. There is no choice other than n itself.
Induction step: Suppose that n has two representations. Let q and s be

the respective final digits. Then $q = n \bmod 10 = s$. When q and s are deleted from the two representations, we get two representations for $(n - q)/10$. Since this value is less than n, by induction these representations are the same. So the representations are the same except for their last digit, and they have the same last digit.

Nothing in either proof used any special properties of the number 10, so the proofs do generalize to other bases.

26. Proof by induction on n beginning with $n = a$ in both cases. The induction hypothesis is just the statement of the theorem.

    a) Basis: The left-hand sum is $f_a + g_a$, as is the right-hand sum.
    Induction step: The left-hand side is $f_n + g_n + \sum_{k=a}^{n-1} (f_k + g_k)$. The right-hand sum is $f_n + \left(\sum_{k=a}^{n-1} f_k\right) + g_n + \sum_{k=a}^{n-1} g_k$. By induction the Σ-expression on the left is equal to the sum of the two Σ-expressions on the right, so we are done.

    b) Basis: Both sides equal cf_a.
    Induction step: The left-hand side is $cf_n + \sum_{k=a}^{n-1} cf_k$. The right-hand side is $c(f_n + \sum_{k=a}^{n-1} f_k) = cf_n + c \times \sum_{k=1}^{n-1} f_k$. By induction this last Σ-expression is equal to that of the left-hand side, so we are done.

29. Proof by induction on the population n. The induction hypothesis is just the statement of the theorem.
    Basis: The population can't be 1, since that person would have to have zero hairs. The population could be 2. In that case both must have exactly 1 hair, and we have the desired duplication.
    Induction: If more than one person has $n - 1$ hairs, we are done. If exactly one person has $n - 1$ hairs, then if that person is removed, the remaining population $n - 1$ is greater than the number of hairs on any remaining person's head. By induction there is a duplication in the remaining population, hence in the original population. If no one has exactly $n - 1$ hairs, then remove one person at random from the population. As in the previous case the induction hypothesis is satisfied, so there is a duplication in the reduced population, hence in the original population.

30. The argument for the induction step of the previous exercise still works. The problem is the basis. In this case there may be just one person in the world who has zero hairs. This makes the entire induction proof fail. For example, there may be 100 people in the world, one each with every number of hairs from 0 through 99.

31. Proof by minimal counterexample. If there were any uninteresting positive integers, there would have to be a smallest one. But that's a very interesting property!

38. Yes. Let's use "Sum" instead of "Sigma" (Σ) for the new definition. We need to show that $\sum_{k=a}^{n} f_k$ equals $\operatorname*{Sum}_{k=a}^{n} f_k$. The proof is by strong induction on the number of terms in the sum, beginning at 0. The induction hypothesis is just the statement of the theorem.
Basis: The empty sum is 0 by both definitions. The sum from k = a to a by the both definitions is just f_a.
Induction step: If there are more than 2 terms,

$$\operatorname*{Sum}_{k=a}^{n} f_k = \text{(by definition)}$$

$$f_a + \operatorname*{Sum}_{k=a+1}^{n} f_k = \text{(by induction)}$$

$$f_a + \sum_{k=a+1}^{n} f_k = \text{(by definition)}$$

$$f_a + \left(\sum_{k=a+1}^{n-1} f_k\right) + f_n = \text{(by induction)}$$

$$f_a + \left(\operatorname*{Sum}_{k=a+1}^{n-1} f_k\right) + f_n = \text{(by definition)}$$

$$\left(\operatorname*{Sum}_{k=a}^{n-1} f_k\right) + f_n = \text{(by induction)}$$

$$\left(\sum_{k=a}^{n-1} f_k\right) + f_n = \text{(by definition)}$$

$$\sum_{k=a}^{n} f_k$$

39. We want to show that if there is no minimal counterexample to P, then P is true for all nonnegative integers. The proof will be by strong induction—we assume P is true for all k less than n and then show it is true for n.
Basis: For n = 0, if P(0) were false, it would be the minimal counterexample. So it must be true.
Induction step: Suppose that P(k) is true for all k less than n. If P(n) is false, then n would be the minimal counterexample. So P(n) must be true.

40. Let P be an arbitrary property and define Q(n) to be true iff P(m) holds for each nonnegative integer m less than or equal to n. We want to show that if a) P(0) and b) P(n) whenever P(k) for all $0 \leq k \leq n$ hold for P, then P(n) is true for each nonnegative n. But b) is equivalent to the assertion that P(n) is true whenever Q($n - 1$) is. Also, P(n) and Q($n - 1$) together imply Q(n), so Q(n) is true whenever Q($n - 1$) is. Since P(0) and Q(0) are equivalent, we have

a′) Q(0)

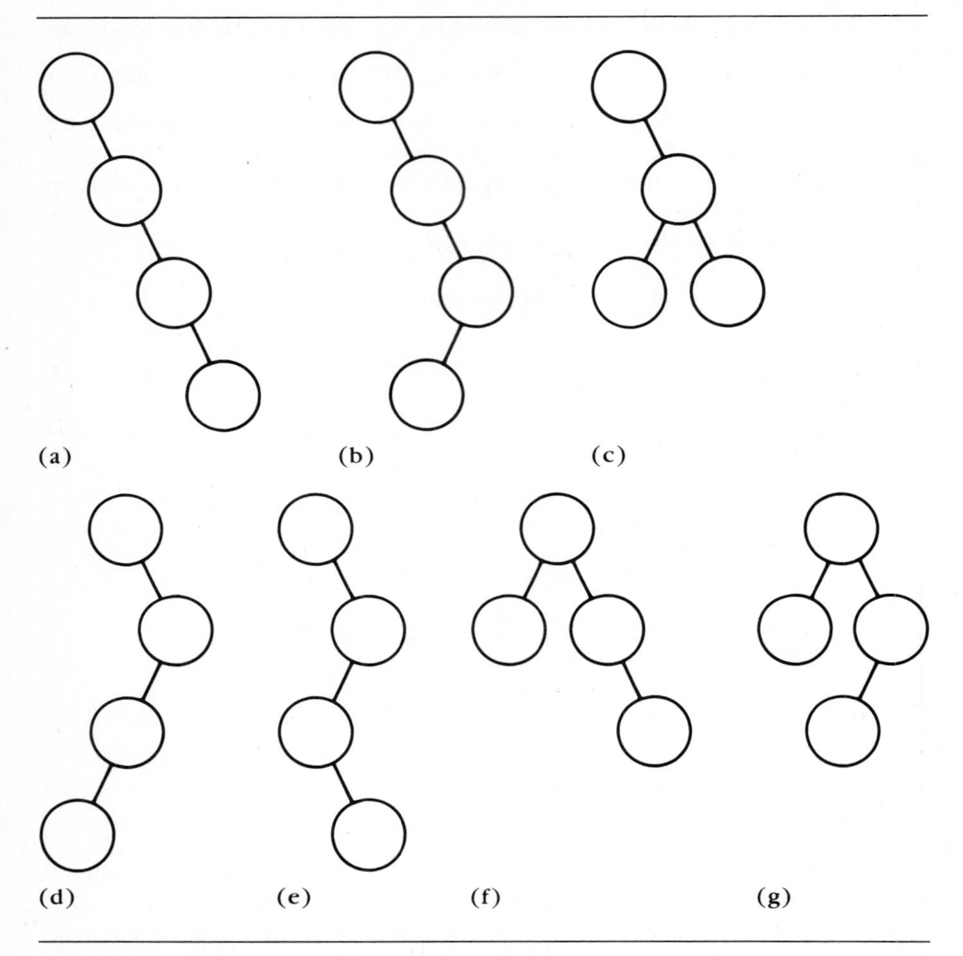

Figure C.2 Seven of the Fourteen Binary Trees on Four Nodes

and b′) $Q(n)$ whenever $Q(n-1)$. Thus by ordinary induction $Q(n)$ is true for every nonnegative integer. Since $Q(n)$ implies $P(n)$, we are done.

41. Define $Q(n)$ to be equivalent to $P(n+a)$. Then showing $P(a)$ is the same as showing $Q(0)$, and the induction step is the same for P and Q. So Q follows by induction, and the desired property for P follows.

65. No, if the input sequence is sorted then only time $\Theta(n)$ is necessary.

66. There are two different binary trees with 2 nodes, one with a root and its left child and one with a root and its right child. Each of these may occur as a left subtree or a right subtree of a root in a binary tree with 3 nodes. In addition a binary tree with 3 nodes may have a root, a right child, and a left child. This gives 5 possible trees on 3 nodes.

In a tree with 4 nodes, the left subtree may have 0, 1, 2, or 3 nodes. If it has 0 nodes, then there are 5 possibilities for the right subtree and hence for the entire tree. If it has one node then there are two possibilities for the right subtree and hence for the tree. The last two cases give another 7 subtrees, which can be seen by reversing the roles of the left and right subtrees in the above argument. There are therefore 14 different binary trees with 7 nodes. The 7 with a larger left subtree than right subtree are shown in Figure C.2; the other 7 may be obtained by reflection. The average height of the 14 trees is $36/14 = 2.57$.

67. Proof by induction on the number of nodes in the tree. The induction hypothesis is simply the statement of the theorem applied to preorder and inorder (the case of postorder and inorder is symmetric) and sufficient to prove the theorem.
Basis: If $n = 0$, then there is nothing to prove.
Induction step: If the tree is not empty, then preorder will first visit the root, which is not a leaf. So we can ignore this step. Both traversals will next traverse the left subtree; by induction leaves will be visited in the same order. Inorder will next visit the root, and again we may ignore this step. Finally both traversals will visit the right subtree; by induction the leaves will be visited in the same order.

71. Figure C.3 provides a counterexample. The key 20 should not be in the right subtree of the tree rooted at 50.

72. In certain cases the given algorithm will not terminate. For example, let $low = 1$, $upp = 2$, and $x = a[2]$. The first time through the **while** loop, *mid* has the value 1, and the values of *low* and *upp* will not change.

80. The binary search trees are (a) and (b). The only AVL tree is (a), although (c) satisfies the balance condition for AVL trees.

81. The four assignments of keys are shown in Figure C.4.

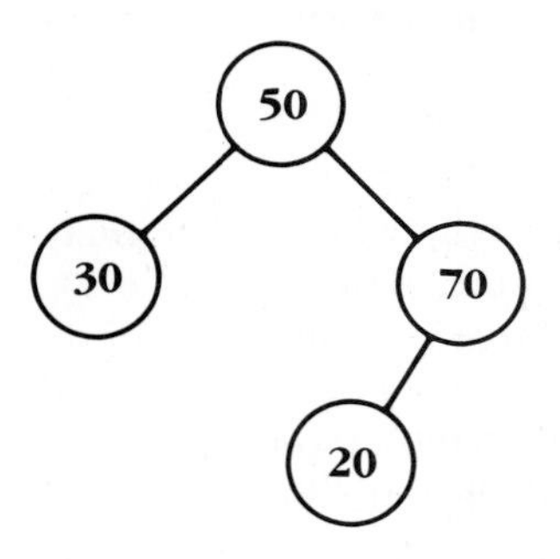

Figure C.3 A Tree That Is Not a Binary Search Tree

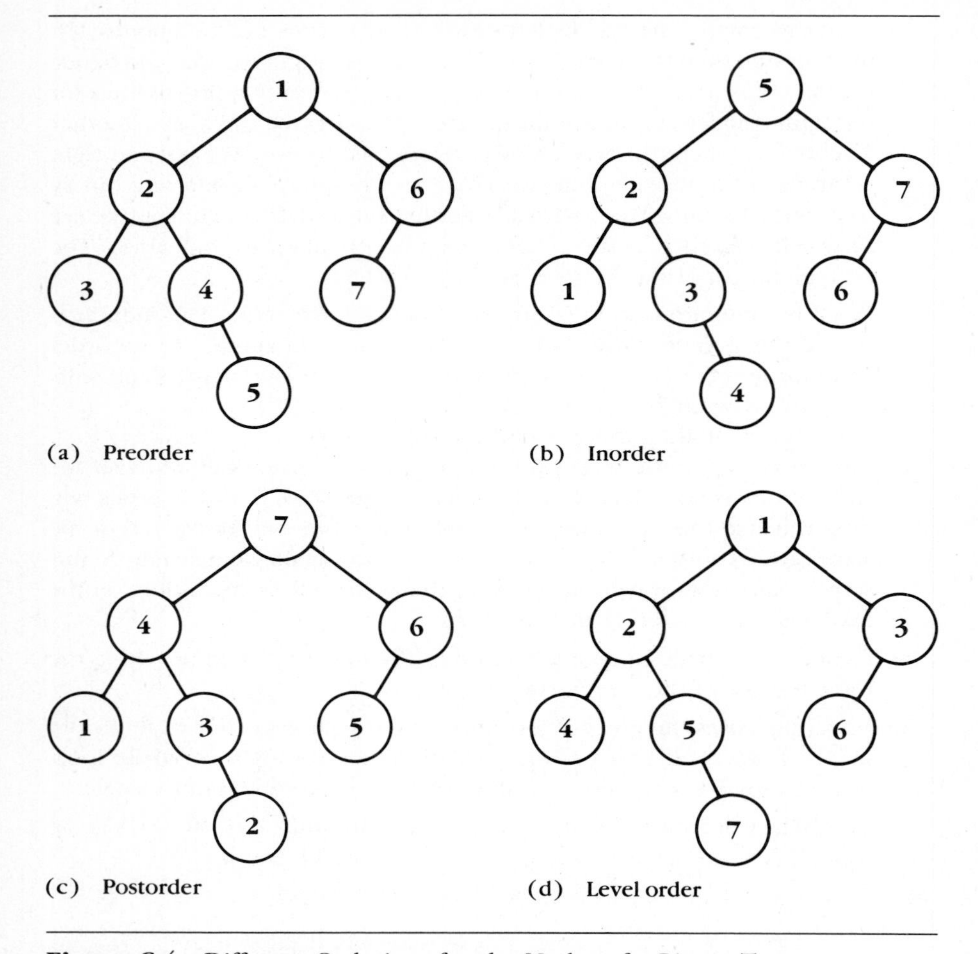

Figure C.4 Different Orderings for the Nodes of a Binary Tree

84. The successive trees are shown in Figure C.5. The inorder traversal of the final binary search tree gives

$$2, 3, 4, 7, 10, 14, 16, 18, 20, 21$$

87. Tree (a) in the discussion of Exercise 66 occurs only when the original input is (1 2 3 4). Tree (b) occurs only when the input is (1 2 4 3); tree (c) only with input (1 3 2 4) or (1 3 4 2); tree (d) only with input (1 4 3 2); tree (e) only with input (1 4 2 3); tree (f) only with input (2 1 3 4), (2 3 1 4), or (2 3 4 1); and tree (g) only with input (2 1 4 3), (2 4 1 3), or (2 4 3 1). Here the better balanced trees are more likely to occur. The average height for these 12 trees is 28/12 or 2.33. By symmetry we get the same answer if the input begins with 3 or 4, and thus 2.33 is the average height under the assumptions of the problem. Note that

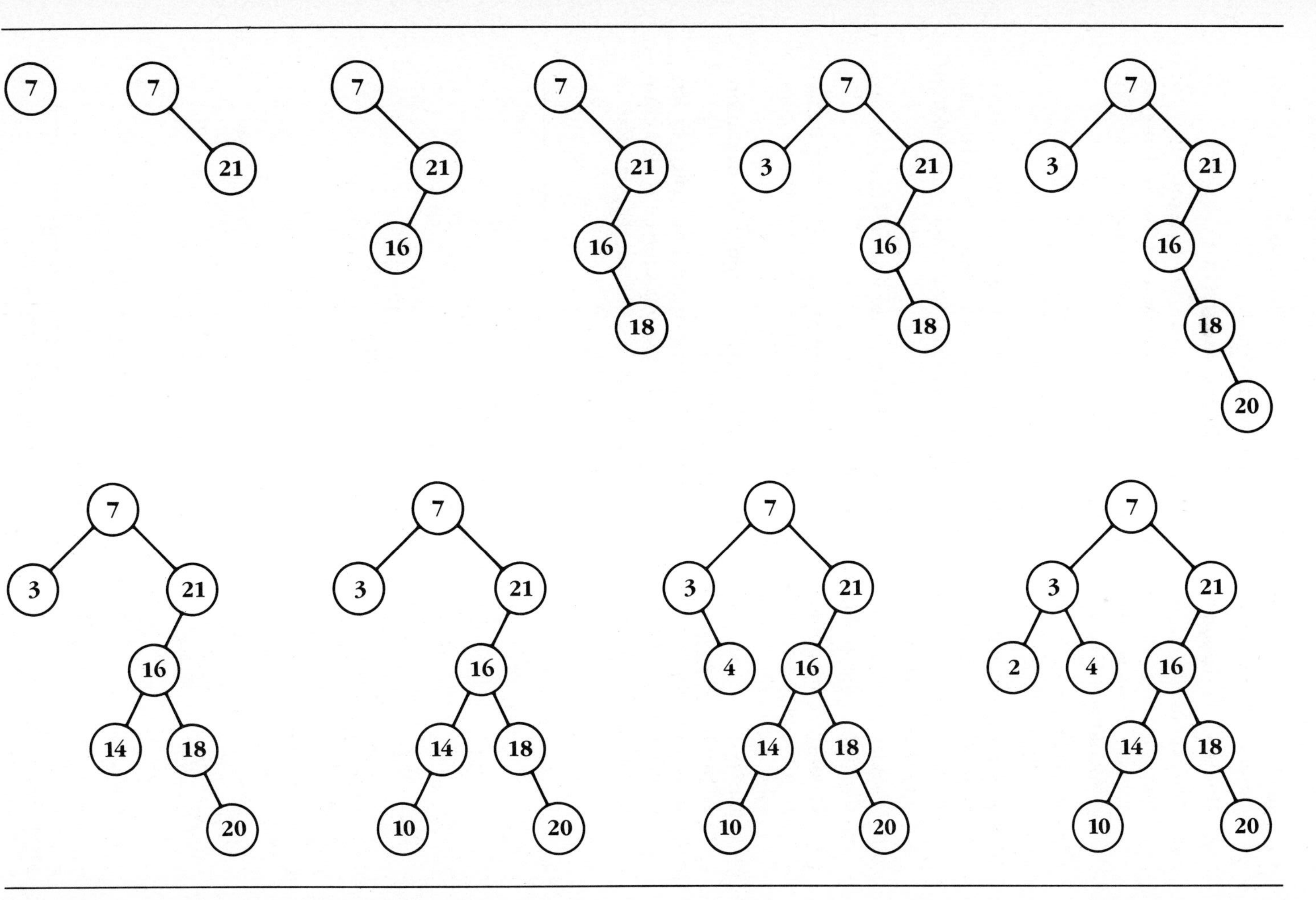

Figure C.5 Insertion into a Binary Search Tree

since shorter trees are more common than in Exercise 66, we would expect the average height to be shorter and it is.

88. No. Since nodes may have a single child, condition 2(d) may fail.

98. The average search requires about $2 \ln n$ comparisons. In terms of $\log_2 n$, this expression is $2 \log_e 2 \log_2 n$, or approximately $1.38 \log_2 n$. Thus the average case requires about an extra 38% more comparisons. [See Reingold and Hansen (1983), 298–301.]

105. See Figure C.6.

106. See Figure C.7.

111. Algorithm C.1 will traverse a tree in level order. It assumes queue operations Create, Destroy, Insert, Empty, and Delete, and functions NumberOfChildren and Child that return respectively the number of children and a given child of a node.

116. The parent would be at location 6; the left child at location 26, if it exists; the right child at location 27, if it exists.

123. Proof by induction on the height h, beginning with $h = -1$. The induction hypothesis will be that a completely full tree of height h is a complete tree of height h.
Basis: If $h = -1$, then the completely full tree is empty. By definition, empty trees are complete trees of height -1.
Induction step: Let T be a completely full tree of height h. Then its left subtree is completely full of height $h - 1$. Its right subtree is also completely full of height $h - 1$; by induction this subtree is complete of

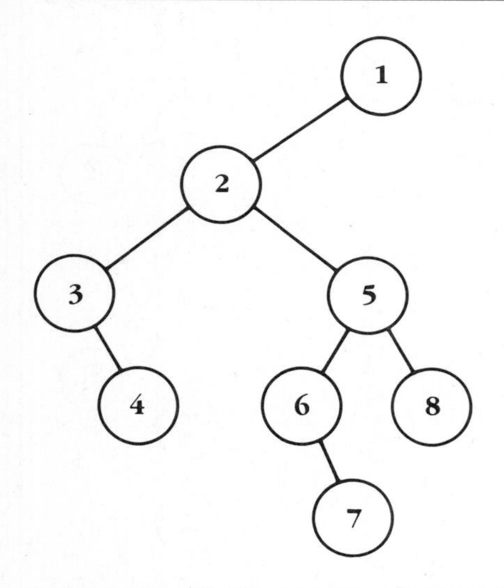

Figure C.6 Binary Tree Representation of an Ordered Tree

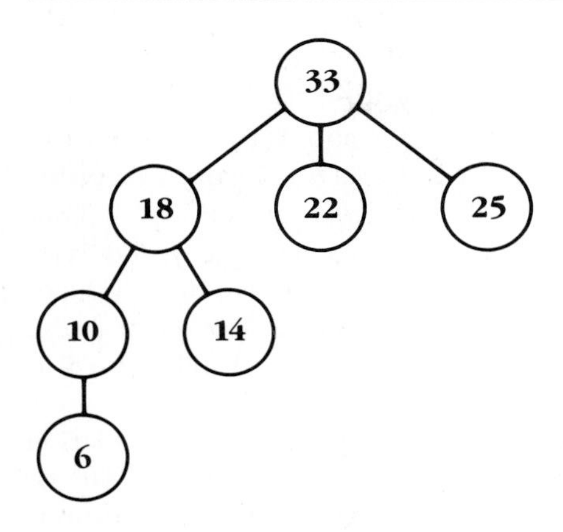

Figure C.7 Ordered Tree Represented by a Binary Tree

height $h - 1$. Therefore T is complete by $(2')$ of Definition 2.50 of a complete tree.

131. Suppose that E − 1 annihilates the sequence (t_n). Then $t_{n+1} - t_n = 0$ for all n, that is, $t_{n+1} = t_n$ for all n. Thus all the terms of the sequence are identical, and the sequence is a constant sequence.

132. Suppose that $(\text{E} - 1)\langle t_n \rangle = \langle b \rangle$. Then we have that $t_{n+1} - t_n = b$ for all $n \geq 0$. A simple induction shows that $t_{n+k} - t_n = bk$ for all k and n. In particular, if $t_0 = c$, then $t_n = t_{0+n} = bn + c$. Thus the sequences $\langle t_n \rangle$ that are converted to constant sequences by E − 1 are those sequences of the form $\langle bn + c \rangle$, where b is the constant of the constant sequence.

133. Let $\langle t_n \rangle$ be annihilated by $(\text{E} - 2)$. Let $s_n = t_n/2^n$. Then since $t_{n+1} = 2t_n$, $s_{n+1} = t_{n+1}/2^{n+1} = 2t_n/2^{n+1} = t_n/2^n = s_n$. In other words, the sequence

```
 1. LevelOrder(tree)                            {see documentation above}
 2. Create(q)                                   {Create a queue called q}
 3. Insert(tree, q)                             {put the root into the queue}
 4. while not Empty(q)                {as long as there are nodes in the queue}
 5.    current := Delete(q)                     {get a node from the queue}
 6.    visit(current)                                           {visit it}
 7.    for k := 1 to NumberOfChildren(current)
 8.       Insert(Child(current, k))             {and insert all its children}
 9.    end for
10. end while
11. Destroy(q)
```

Algorithm C.1 Level Order Traversal (Using a Queue)

$\langle s_n \rangle = \langle t_n/2^n \rangle$ is constant. If the constant is b, then $t_n/2^n = b$, so $t_n = b2^n$. We can easily check that all such sequences are annihilated by E $-$ 2, so if $\langle t_n \rangle$ is annihilated by E $-$ 2, it has the form $\langle b2^n \rangle$ for some b.

134. Suppose that x has b bits. If n is a power of 2 and $T(n)$ is the cost of finding x^n in terms of $x^{n/2}$, then $T(n) = T(n/2) + (bn/2)(bn/2)$. Letting $n = 2^u$, we get $(E - 1)\langle t_u \rangle = \langle 2^{u-1}2^{u-1}b^2 \rangle = \langle 4^{u-1}b^2 \rangle$. Thus $(E - 4)(E - 1)\langle t_u \rangle = \langle 0 \rangle$, so $t_u = p4^u + q1^u$ for some p and q, and $T(n) = pn^2 + q$. Since $T(1) = 0$ and $T(2) = 0 + b \times b = b^2$, we may evaluate the constants to get $T(n) = b^2(n^2 - 1)/3$. If n is not a power of 2, the asymptotic result still holds, since at most one extra multiplication will be required at each stage.

    Finding x^n directly from the definition leads to the recurrence relation $T(n) = T(n - 1) + b \times (n - 1)b$ or $(E - 1)\langle t_n \rangle = \langle (n - 1)b^2 \rangle$ or $(E - 1)^3\langle t_n \rangle = \langle 0 \rangle$. So $T(n) = pn^2 + qn + r$. Using the initial conditions $T(1) = 0$, $T(2) = 0 + b \times 1 \times b = b^2$, $T(3) = b^2 + b \times 2 \times b = 3b^2$, we can evaluate the constants to get $T(n) = b^2(n^2 + n)/2$, which is about 50% more bit multiplications than the previous method for n a power of 2. You might want to see what happens for small values of n that are not powers of 2.

135. In this case, the powers of x (mod n) stay bounded in size. Let m be the size of M in bits. For the first method, we have $T(n) = T(n/2) + m^2$ or $(E - 1)\langle t_u \rangle = \langle m^2 \rangle$, where $n = 2^u$. Thus $(E - 1)^2\langle t_u \rangle = \langle 0 \rangle$, so $t_u = pu + q = q + p \log_2 n = O(\log n)$. For the second method, we have $T(n) = T(n - 1) + bm$, so $(E - 1)\langle t_n \rangle = \langle bm \rangle$ and $(E - 1)^2\langle t_n \rangle = \langle 0 \rangle$. Thus $t_n = rn + s = O(n)$. So for finding powers mod n, the first method is asymptotically superior to the second.

137. Assuming that time $O(n^2)$ is required to add two n-by-n matrices, a divide-and-conquer algorithm for matrix multiplication using the observation of Exercise 136 would have time complexity given by $T(n) = 7T(n/2) + cn^2$.

    To solve this recurrence, let $n = 2^u$ and $t_n = s_u$, so that

$$t_n = 7t_{n/2} + cn^2 \text{ or}$$

$$s_u = 7s_{u-1} + c(2^u)^2 \text{ or}$$

$$(E - 7)\langle s_u \rangle = c2^{2u} = c4^u \text{ or}$$

$$(E - 7)(E - 4)\langle s_u \rangle = \langle 0 \rangle$$

    We therefore get that $s_u = a7^u + b4^u$, so that s_u is $O(7^u)$. In terms of n, we have $t_n = s_u = O(7^u) = O(n^d)$, where $d = \log_2 7$. Note that this time complexity is somewhat better than $O(n^3)$.

141. Yes, by the assignment of "false" to x_1 and x_2 and "true" to x_3.

142. The minimum cost is 37. The tour (1 5 2 3 4 1) or its reversal (1 4 3 2 5 1) or any cyclic permutation of these will give the minimum cost.

144. The tables for *C*, *S*, and *Root*, together with the resulting optimal tree, are shown in Figure C.8. The weights are multiplied by 100 in the tables.

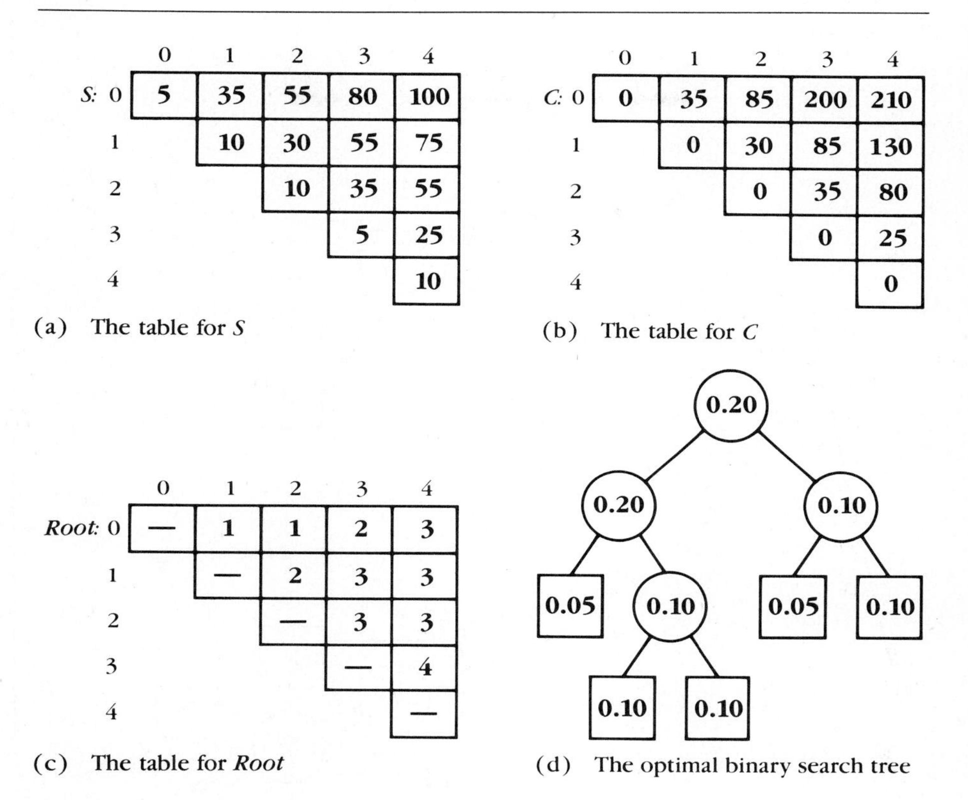

S:	0	1	2	3	4
0	5	35	55	80	100
1		10	30	55	75
2			10	35	55
3				5	25
4					10

(a) The table for *S*

C:	0	1	2	3	4
0	0	35	85	200	210
1		0	30	85	130
2			0	35	80
3				0	25
4					0

(b) The table for *C*

Root:	0	1	2	3	4
0	—	1	1	2	3
1		—	2	3	3
2			—	3	3
3				—	4
4					—

(c) The table for *Root*

(d) The optimal binary search tree

Figure C.8 Construction of an Optimal Binary Search Tree

147. In Figure C.9, leaves are labeled with the appropriate letter; internal nodes with the weight of the subtree rooted there (where the weights are the probabilities multiplied by 1000).

148. See Figure C.10.

149. The edges of weights 1, 2, 3, and 4 may be included. Those of weights 5 and 6 complete cycles. The edge of weight 7 may be included. The final spanning tree is shown in Figure C.11; its total cost is 17.

152. Yes. Applying a single-source algorithm n times is enough to solve the all-pairs problem. Dijkstra's Algorithm 2.78 for the single-source problem is easily seen to have time complexity $O(n^2)$, so this approach to the all-pairs problem has time complexity $O(n^3)$, just as the original all-pairs algorithm did. You might try implementing both and see which runs faster in practice.

158. See Brassard and Bratley (1988, 159–62) or Horowitz and Sahni (1978, 231–34).

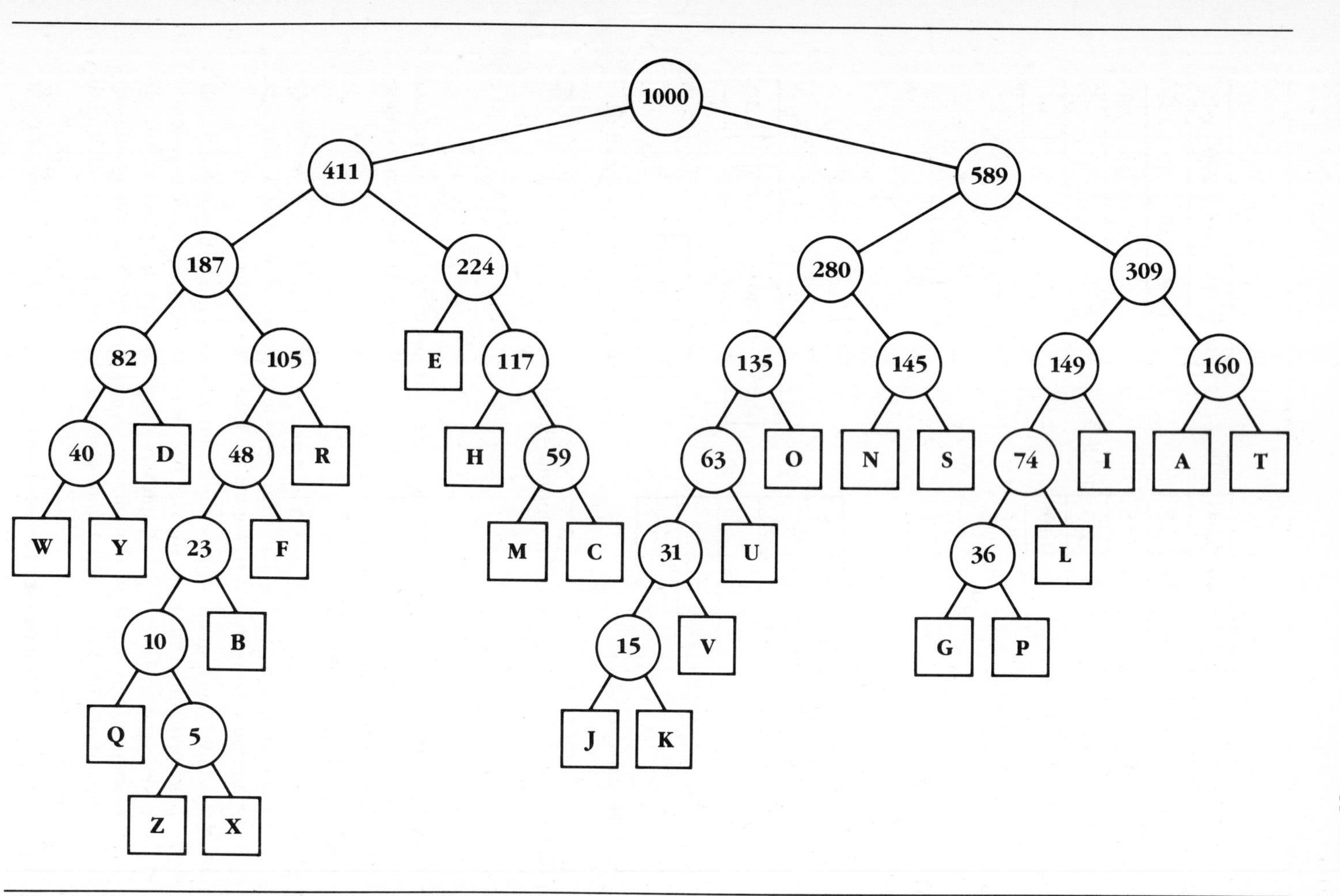

Figure C.9 Tree Created by the Huffman Algorithm

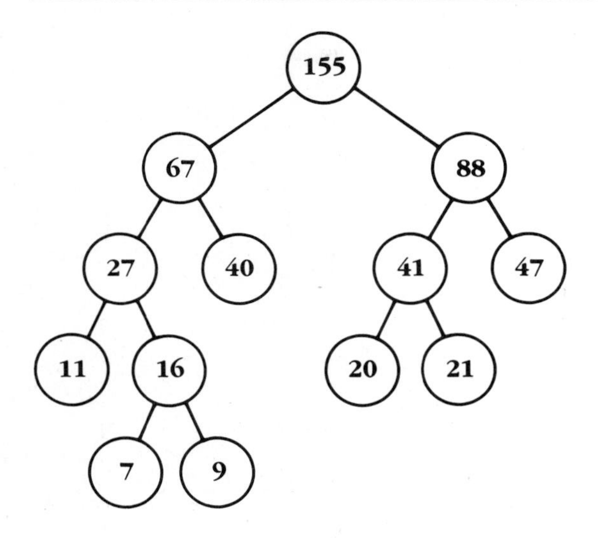

Figure C.10 Tree for Optimal File Merging

160. No. The key 38 (in position 3) is less than its left child 39 (in position 6).
161. See Figure C.12.
162. See Figure C.13.
163. See Figure C.14.
164. See Figure C.15. The result of moving each successive key down is shown.

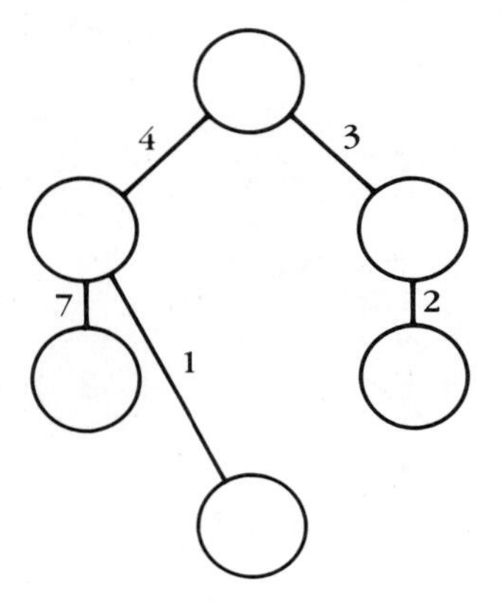

Figure C.11 A Minimum-Cost Spanning Tree

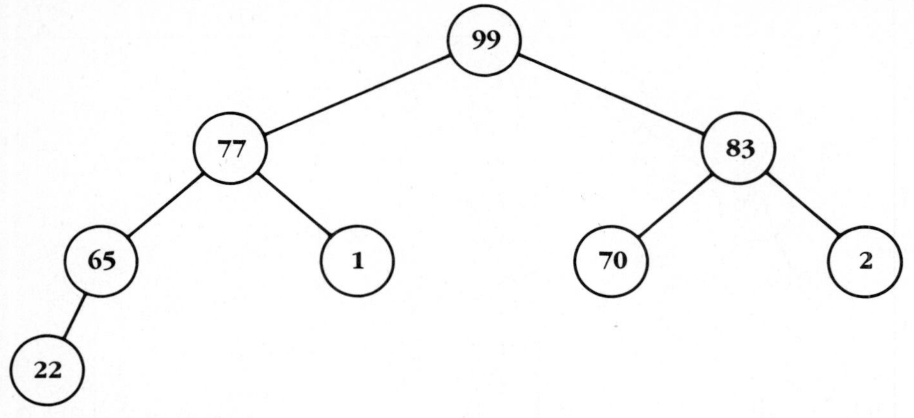

(a) The original heap

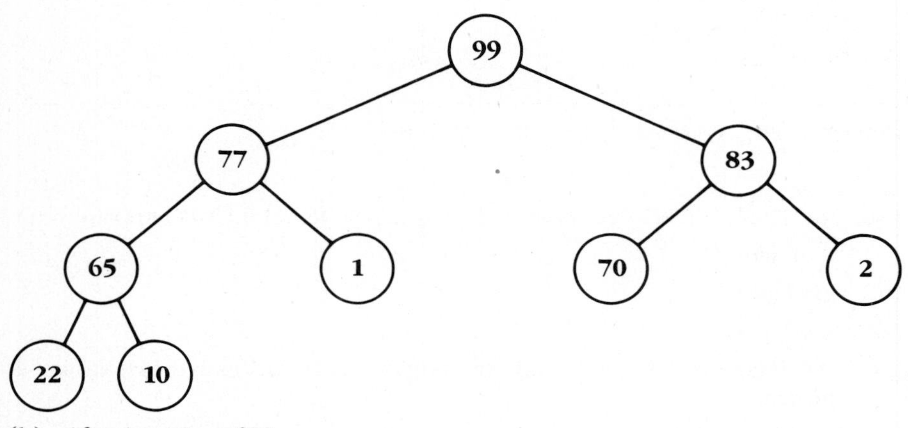

(b) After insertion of 10

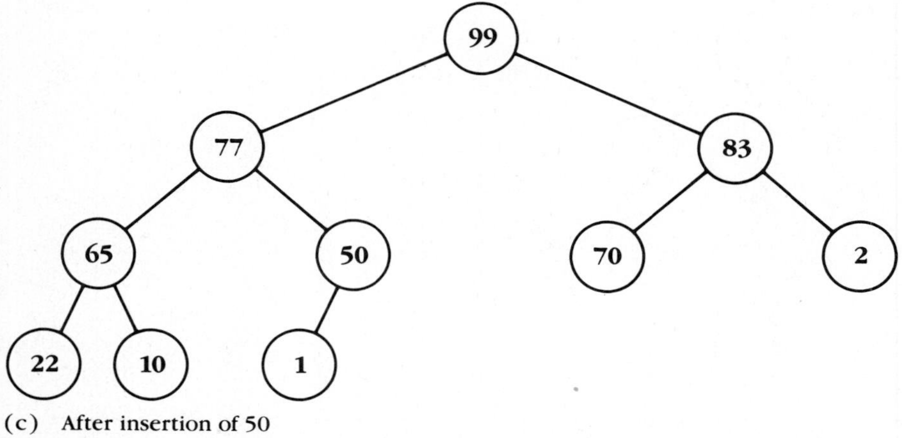

(c) After insertion of 50

Figure C.12 Insertions into a Heap

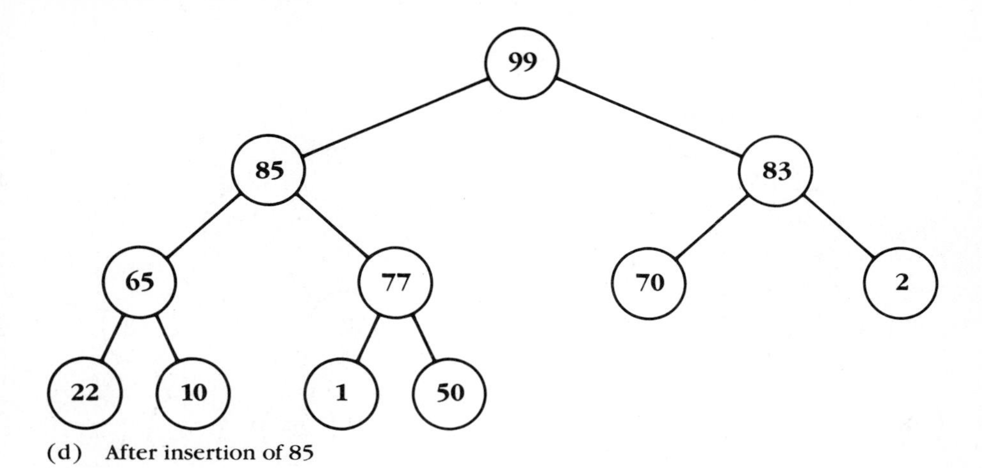

(d) After insertion of 85

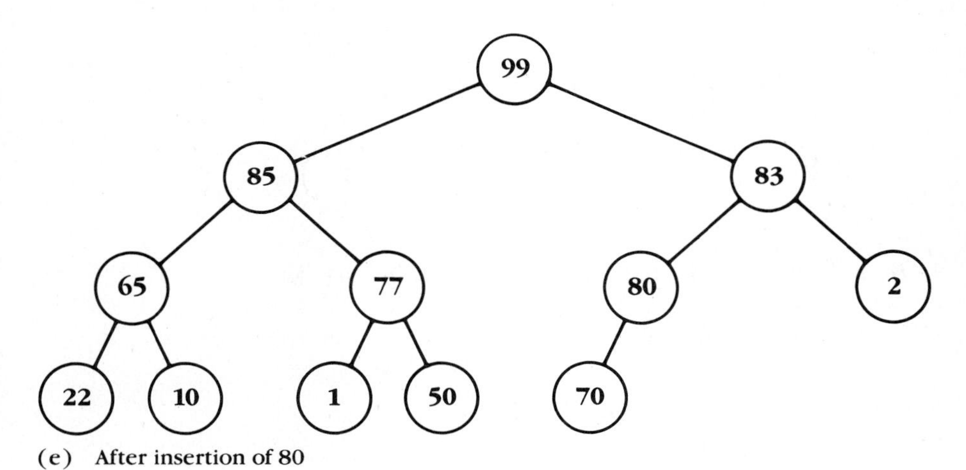

(e) After insertion of 80

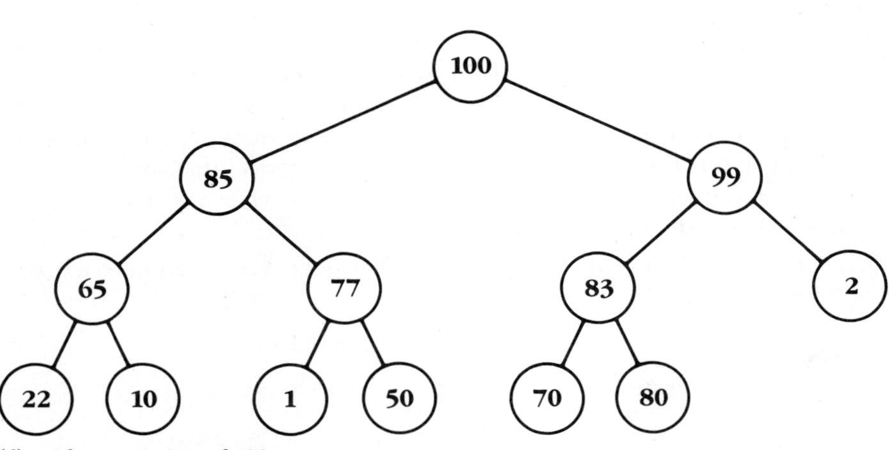

(f) After insertion of 100

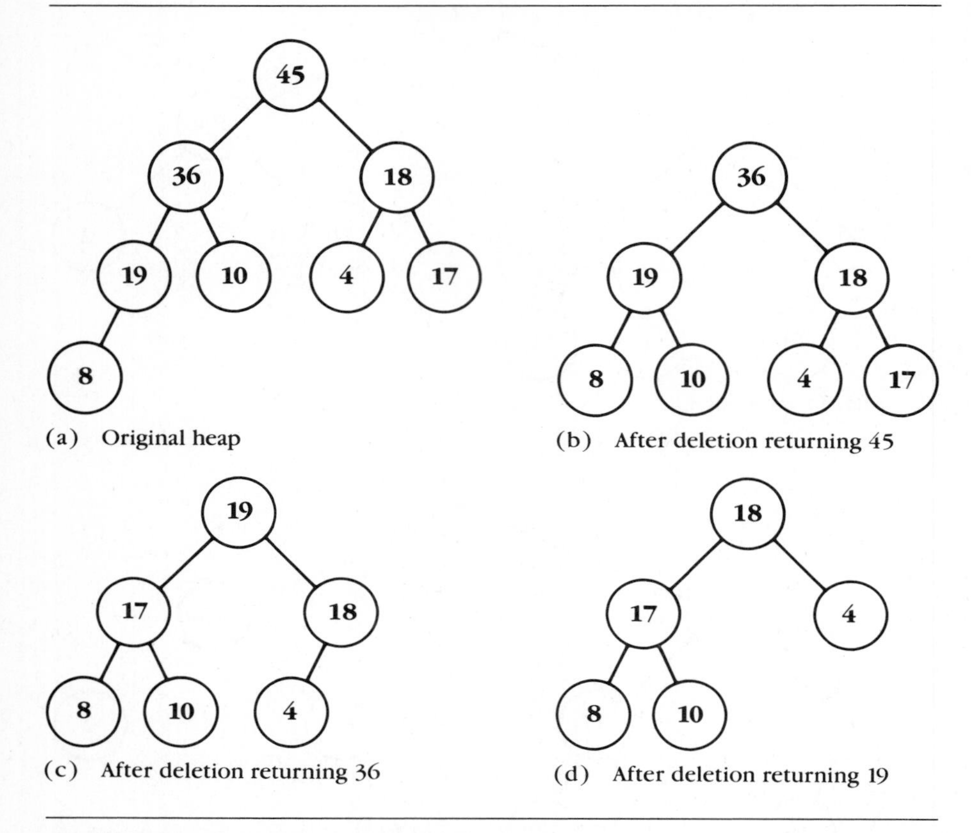

Figure C.13 Deletions from a Heap

165. See Figure C.16. The result of each deletion is shown.
166. The original heap is shown in Figure C.17a. After moving 77 up from position 2, we get the heap of Figure C.17b. The key 70 is in its correct location in position 3; after moving 22 up from position 4 we are in the situation of Figure C.17c. The key 99 moves up from position 5 to give the position of part (d) of the figure; the key 83 moves up from position 6 to give the heap of part (e). The remaining keys are in their proper position, so the result is the heap of Figure C.17e.
167. If MoveDown is used, the example of Figure C.17a requires for the element originally in

    position 4, 1 key comparison and 1 exchange,

    position 3, 1 key comparison and 1 exchange,

    position 2, 1 key comparison and 1 exchange,

    position 1, 2 key comparisons and 2 exchanges,

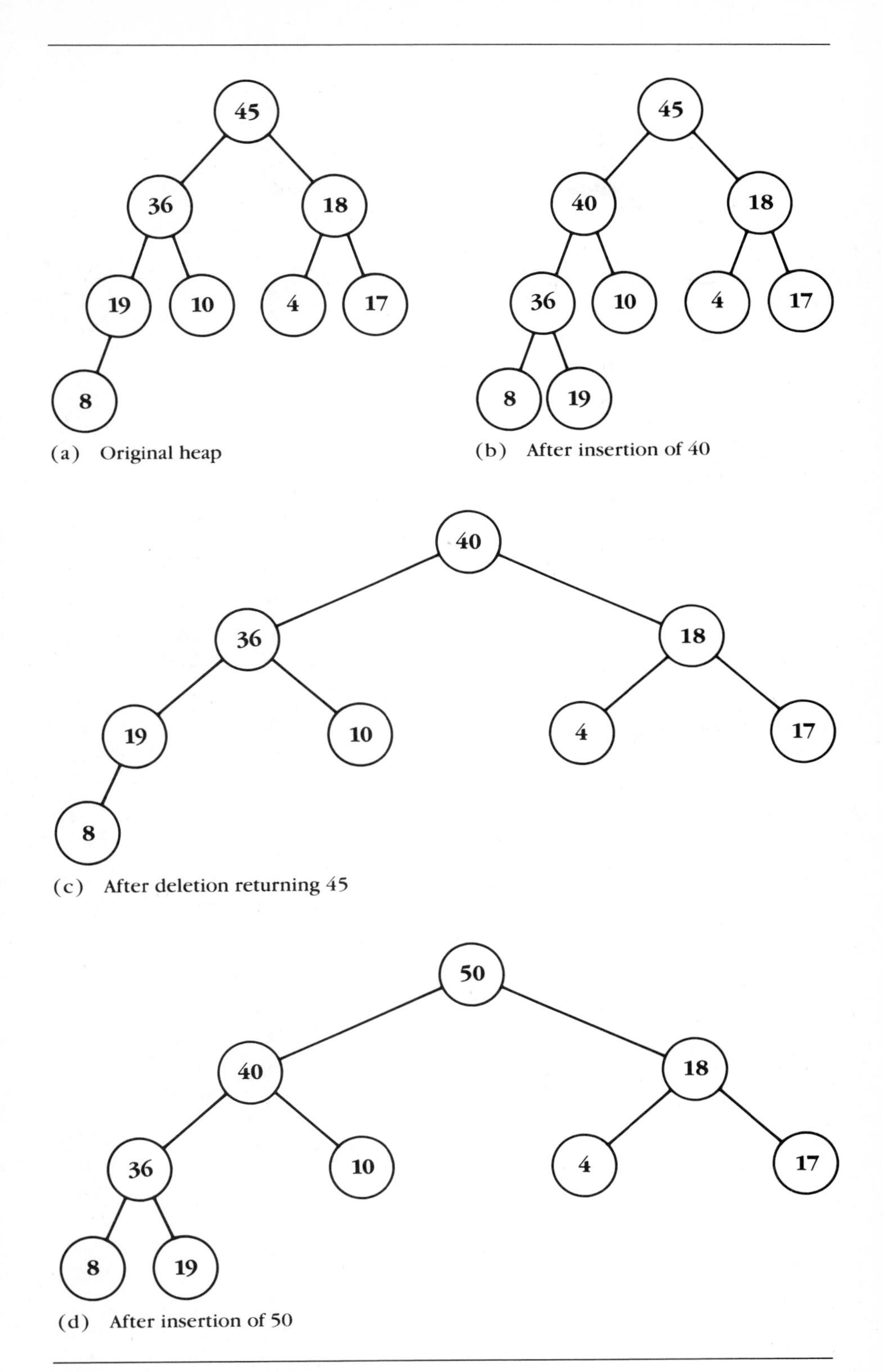

Figure C.14 Mixed Operations in a Heap (*continued on next page*)

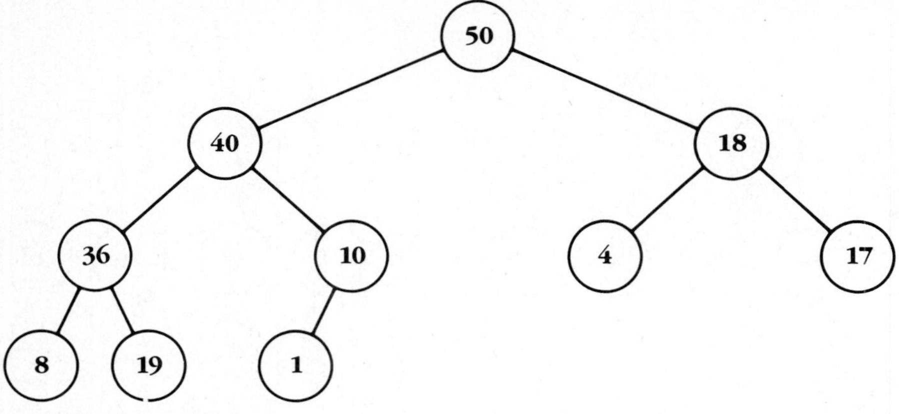

(e) After insertion of 1

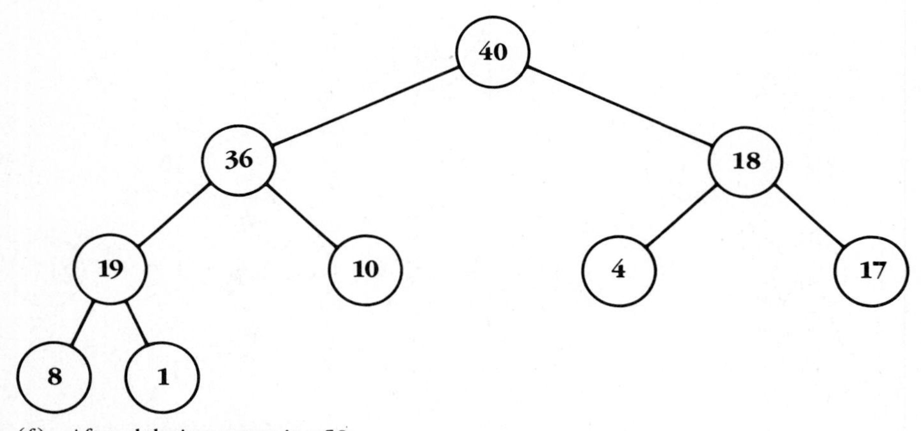

(f) After deletion returning 50

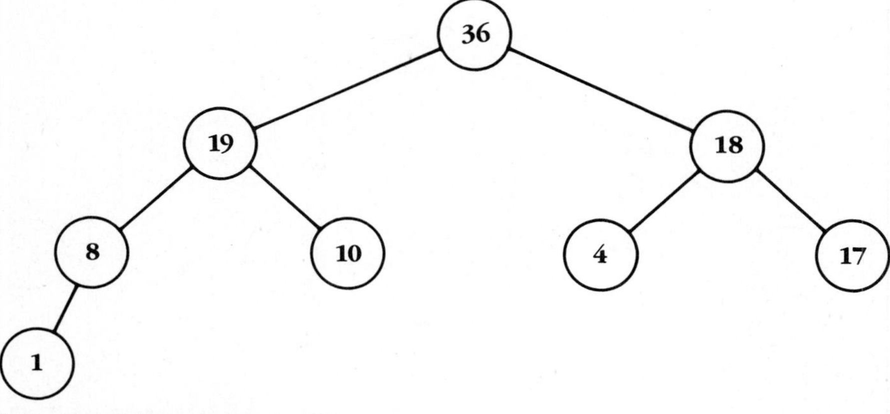

(g) After deletion returning 40

Figure C.14 (*continued*)

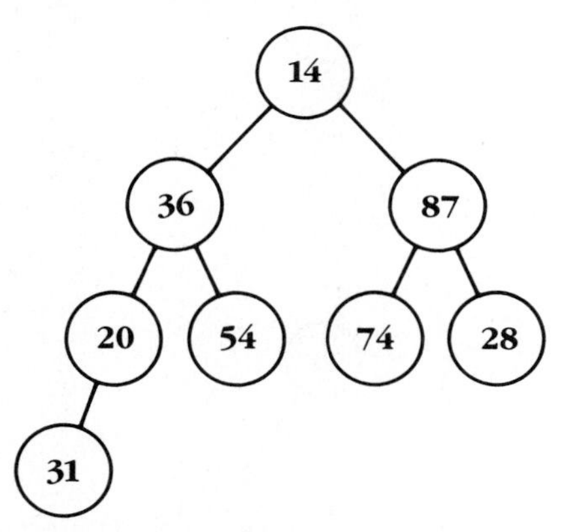

(a) The original heap

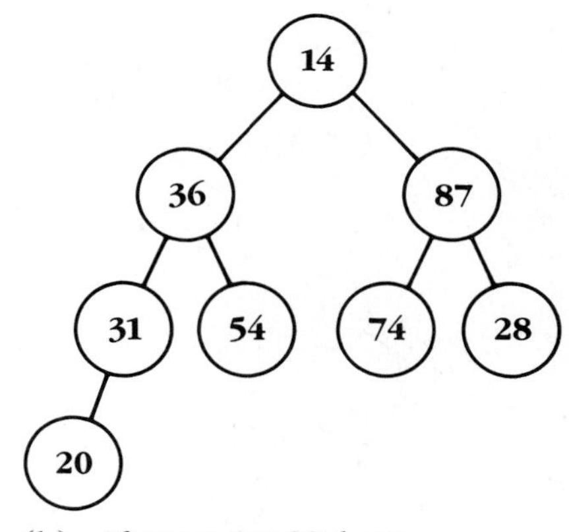

(b) After moving 20 down

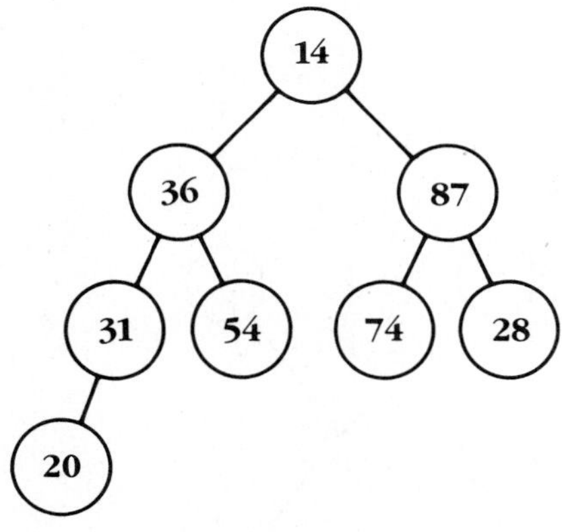

(c) After moving 87 down

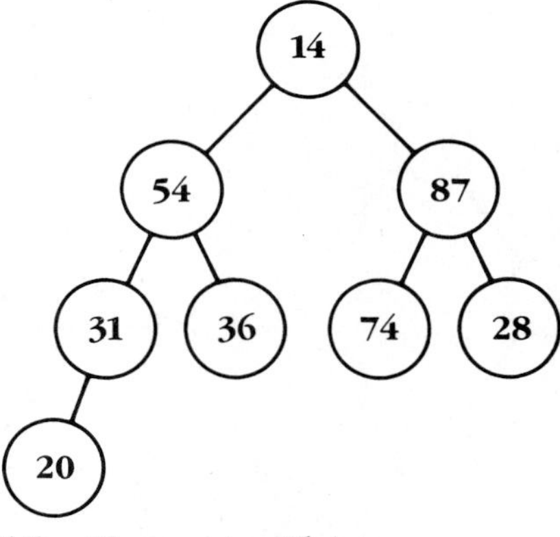

(d) After moving 36 down

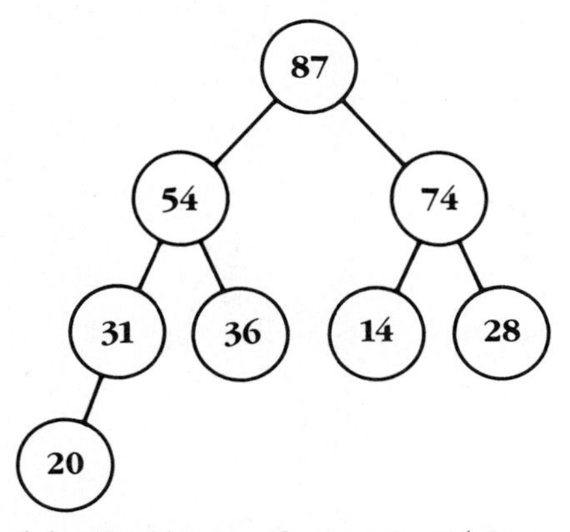

(e) Final heap – after moving 14 down

Figure C.15 Stages in the Heapify Algorithm

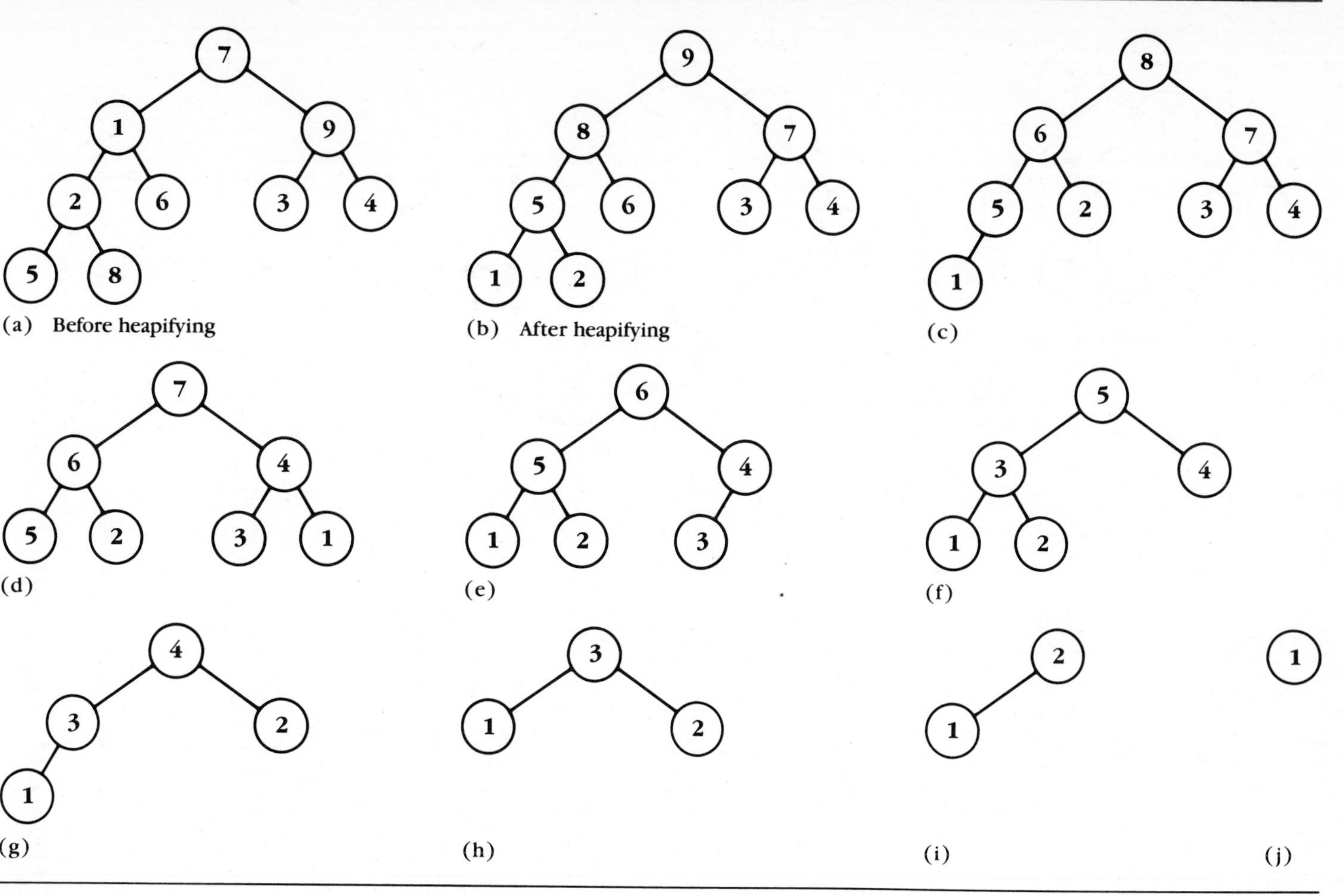

Figure C.16 Stages in the Heapsort Algorithm

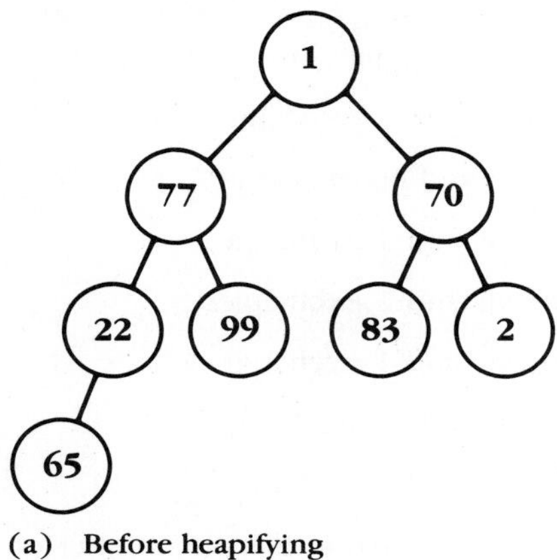

(a) Before heapifying

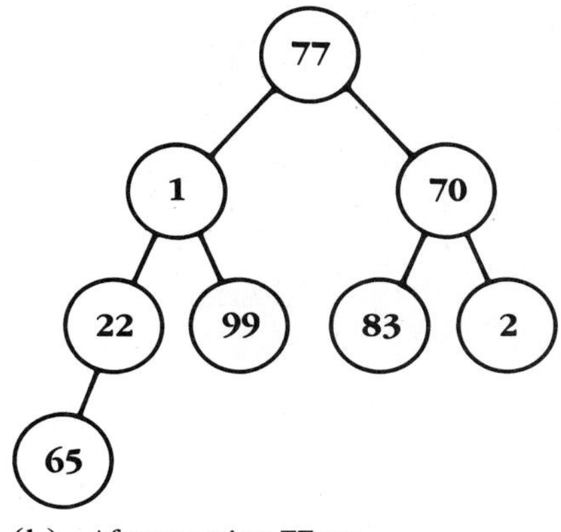

(b) After moving 77 up

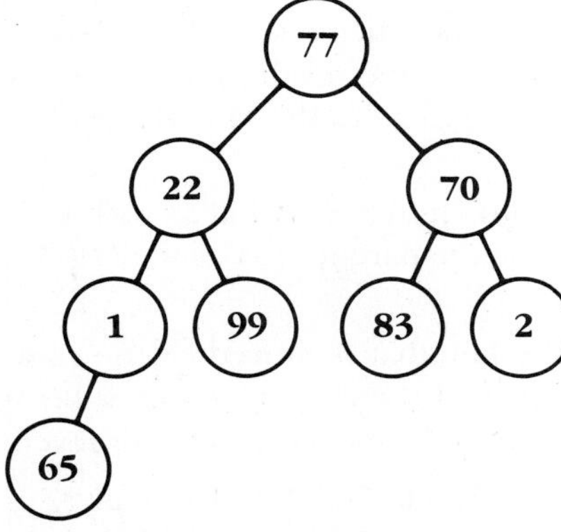

(c) After moving 1 and 70 up

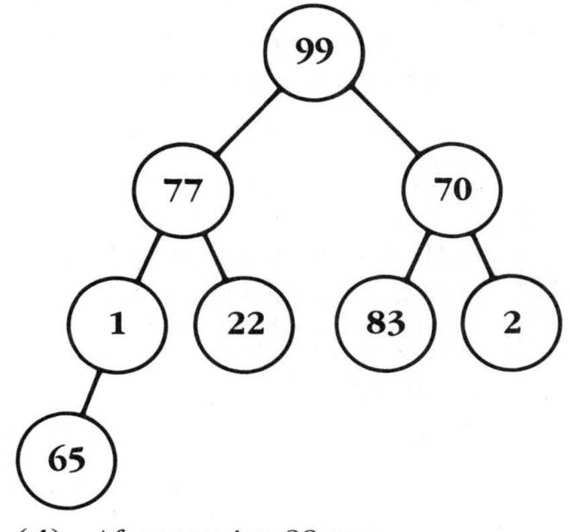

(d) After moving 99 up

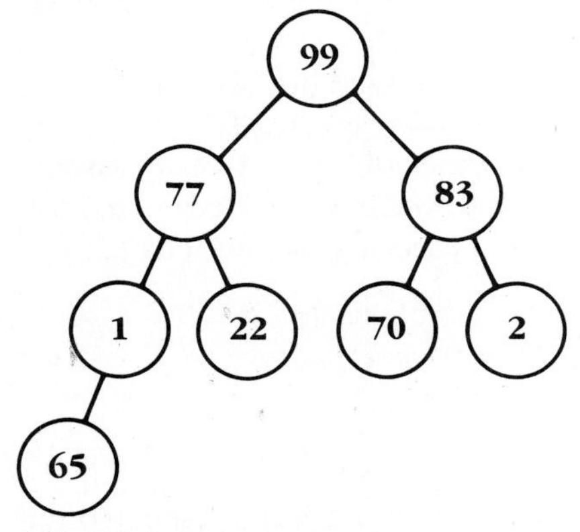

(e) After moving 83 up

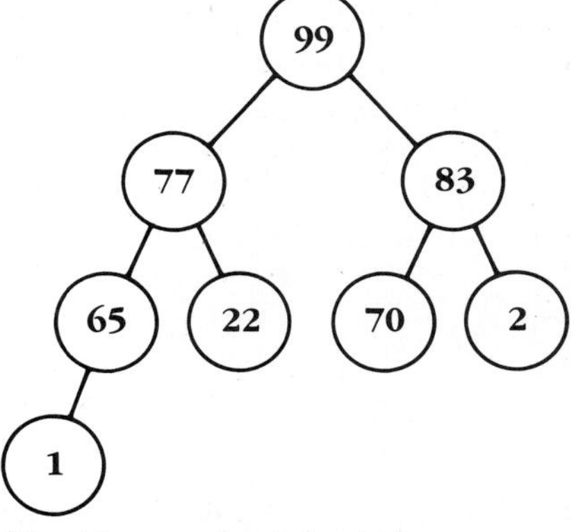

(f) After moving 2 (and 1) up

Figure C.17 Using Heapify with MoveUp

for a total of 5 of each. If we use MoveUp, then as in the previous exercise, we start at location 2 and let the index increase from 2 to 8 (see Figure C.17). We require for the element originally in

position 2, 1 key comparison and 1 exchange,

position 3, 1 key comparison and no exchange,

position 4, 2 key comparisons and 1 exchange,

position 5, 2 key comparisons and 2 exchanges,

position 6, 2 key comparisons and 1 exchange,

position 7, 1 key comparison and no exchange,

position 8, 2 key comparisons and 1 exchange,

for a total of 11 comparisons and 6 exchanges. Note that the number of key comparisons is misleading here, since MoveDown determines when to stop by making an index comparison rather than a key comparison.

Inserting elements one at a time gives the same result as using MoveUp, since single insertions are performed by a call to the MoveUp procedure.

168. Given a choice of exchanging a key K with child keys L and M with $L < M$, if we exchange K with L, then L will be the parent of M. Since $L < M$, this will violate the heap property.

172. When a key is deleted from a heap, the key that is moved to the root comes from the lowest level. Thus it tends to be among the smallest keys in the heap and is more likely than average to be moved to a lower level.

174. It requires 1 comparison to find the key 043, 2 to find 066, 1 to find 221, 1 to find 446, 1 to find 717, 2 to find 700, 3 to find 726, and 1 to find 883. This gives a total of 12 and an average of $12/n = 12/8 = 1.5$. Note that the formula $1 + (n - 1)/2b$ predicts 1.35.

175. The number of comparisons to reach the end of each bucket 0 through 9 (including the comparison with the null key) is respectively 3, 1, 2, 1, 2, 1, 1, 4, 2, and 1. This gives an average case number of comparisons of $18/10 = 1.8$. If buckets 7, 8, and 9 were requested twice as often, the average case number of comparisons would change to $(18 + 4 + 2 + 1)/13 = 25/13 \approx 1.923$.

176. a) 1 4 7 10 13
b) 5 14 7 0 9
c) 12 13 14 15 0
d) 0 7 14 5 12

177. a) The keys would be inserted into locations 7, 9, 12, and 3 respectively.
b) The keys would be inserted into locations 2, 7, 12, and 3 respectively.

178. The terms in each of the following sums refer to the number of comparisons for hash function values of 0 through 7 respectively.

a) $(1 + 3 + 2 + 1 + 2 + 1 + 2 + 1)/8 = 1.625$
b) $(1 + 2 + 1 + 2 + 1 + 2 + 1 + 2)/8 = 1.5$
c) $(1 + 4 + 3 + 2 + 1 + 1 + 2 + 1)/8 = 1.875$
d) $(1 + 5 + 4 + 3 + 2 + 1 + 1 + 1)/8 = 2.25$

179. We need

26 for the 1-character identifiers,

260 for the 2-character identifiers,

2,600 for the 3-character identifiers,

26,000 for the 4-character identifiers,

260,000 for the 5-character identifiers,

and 2,600,000 for the 6-character identifiers

Thus we would need 2,888,886 table locations altogether. In most languages, of course, more than one letter is allowed, so we would need many more table locations.

180. The AND operation cannot be used because of its bias toward producing zeros. It will produce a 0 result if the inputs are 0 and 0, 1 and 0, or 0 and 1, and a 1 result only if both results are 1. Similarly the OR operation is biased toward producing ones. The exclusive OR operation is not so biased and will produce ones and zeros with equal probability if the input is random. Therefore the exclusive OR operation can be used in folding.

186. See Reingold and Hansen (1983, 355–59) or Standish (1980, 149–53) for a discussion of "ordered hashing."

187. a) The squares of the binary numbers from 0 to 1111 are 0, 1, 100, 1001, 10000, 11001, 100100, 110001, 1000000, 1010001, 1100100, 1111001, 10010000, 10101001, 11000100, and 11100001. Taken together, these squares have 4, 6, 6, 6, 4, 4, 0, and 8 ones in the 8-bit positions from left to right. Since there are 16 numbers, we should use the rightmost bit and one of the bit positions giving 6 ones. Of the three possibilities, the second bit from the left gives the most even distribution with 5 function values 00, 5 of 01, 3 of 10, and 3 of 11.

b) Counting from left to right, the 6-bit positions contain respectively 5, 6, 2, 4, 1, and 6 ones. Since 4 is optimal, we should use the bit positions with 5 and 4 ones.

c) The ASCII codes only differ in the 5 rightmost bit positions, where they have expectations of containing a 1 bit counting from left to right of .328, .384, .553, .413, and .568. The closest 2 bits to the optimal .5 are thus the rightmost and the third from the right.

190. Assume $(b, c) = 1$. We want to show that none of the first b terms of the probe sequence have the same value. Suppose the jth term equals the kth term. We may assume $j > k$. Then $h + jc \equiv h + kc \bmod b$, so $jc - kc$

$= (j - k)c$ is a multiple of b, and $(j - k)c/b$ is an integer. Since the entire denominator disappears when this fraction is reduced, and since by assumption we cannot reduce c/b, it must be that $j - k$ is divisible by b. But $j - k < b$, so $j - k = 0$. In other words $j = k$; we have shown that no two of the first b terms can have the same value unless they are the same term, as desired.

If $(b, c) = d > 1$, then the probe sequence will repeat after b/d terms, thus failing to visit all b table locations (note $b/d < b$). To show this, we note first that if the ith term is r_i, then the $(i + b/d)$th term is $\equiv r_i + (b/d)c \equiv r_i + bc/d \bmod b$. Since $d|c$, bc/d is divisible by b, so $r_i + bc/d \equiv r_i \bmod b$, QED.

191. In successful searches, we are assuming that each node, not each bucket, is equally likely to be requested. So the average length of a chain must be multiplied by a weight proportional to its length (i.e., by the probability we will be searching within it). So our estimate is too low.

192. The argument of the previous exercise works for U_k to give a value of $1 + k/b$. The number of comparisons required to search for the kth key inserted is just U_{k-1}, so S_n is the average of U_k as k ranges from 0 through $n - 1$. This average is just $(1/n) \sum_{k=0}^{n-1} (1 + k/b)$

$$= 1 + (1/nb) \sum_{k=0}^{n-1} k$$
$$= 1 + (1/nb)(n - 1)n/2$$
$$= 1 + (n - 1)/2b$$

This expression is similar but not equal to the incorrect expression of the previous exercise.

193. Let $\alpha = n/b$. The probability that the probe sequence will require a second comparison is α, the probability that it will require a third comparison is α^2, and in general the probability that it will require a kth comparison is α^{k-1}. Thus the expected number of comparisons is $\sum_{k=1}^{\infty} \alpha^{k-1}$, or $1/(1 - \alpha)$. Since our assumptions ignore the fact that a well-designed probe sequence will not repeat terms, this result is only a good approximation to the behavior of actual probe sequences.

As in the previous problem, S_n is the average value of U_k. More precisely, $S_n = (1/n) \sum_{k=0}^{n-1} 1/(1 - k/b)$. Approximating this sum by $\int_1^n 1/(1 - k/b)$, we get a good estimate for S_n of (b/n) $[-\ln(b - n) + \ln b] = (b/n) \ln(b/(b - n)) = (b/n) \ln(1/(1 - \alpha))$.

196. Before comparing two sets for equality, their hash signatures may be compared. If these differ, we needn't perform a costly set comparison. We can also use hash signatures for character strings to avoid string comparisons in case they are long or stored externally.

197. A reasonable guess would be that if the desired element *key* is $r\%$ of the way from $A[low]$ to $A[upp]$, then the index m of *key* should be $r\%$ of the way from *low* to *upp*. This gives an equation

$$(m - low)/(upp - low) = (key - A[low])/(A[upp] - A[low]) \text{ or}$$

$$m = low + (upp - low)(key - A[low])/(A[upp] - A[low])$$

This estimate is likely to require fewer iterations than simply letting m be halfway between *low* and *upp*, especially if the distance between consecutive array elements is roughly constant. On the other hand, the multiplication and division used in this approach, called *interpolation search*, are more time-consuming than the simple division by 2 used in ordinary binary search.

202. For $n = 1$ we may use the sequence $(0\ 1)$. For $n > 1$, assume we have a Gray code for $n - 1$. A Gray code for n may be constructed by concatenating two subsequences. The first is obtained by prefixing 0 to each term of the Gray code for $n - 1$. The second is obtained by reversing the Gray code for $n - 1$ and prefixing 1 to each element. By induction, consecutive terms within each subsequence differ by exactly one bit. The last term of the first subsequence and the first term of the second differ only in their leftmost bit.

203. To initialize A we need to find the minima of r groups of size s. We can find each minimum in time $\Theta(s)$, so initialization takes time $r \times \Theta(s) = \Theta(rs) = \Theta(n)$. For each key we output, we take time $\Theta(s)$ to replace it in A and time $\Theta(r)$ to find the new smallest element in A. To do this for each of the n keys we are to output takes time $n(\Theta(r) + \Theta(s)) = \Theta(nr + ns)$. Our total time requirement is thus $\Theta(n) + \Theta(n(r + s)) = \Theta(n(r + s))$.

    To minimize $n(r + s) = n(r + n/r)$ we use the usual methods of calculus. Here n is fixed, so we minimize $r + n/r$ by setting its derivative $1 - n/r^2$ to zero. This means that $1 = n/r^2$ or $n = r^2$ or $r = \sqrt{n}$. Our algorithm then takes time $\Theta(n(r + n/r)) = \Theta(n(\sqrt{n} + n/\sqrt{n})) = \Theta(n(2 \times \sqrt{n})) = \Theta(n \times \sqrt{n}) = \Theta(n^{3/2})$.

    Because of the way that r is chosen, this algorithm is sometimes called *quadratic selection sort.*

205. Yes, by adding a preliminary pass to check if the input sequence is already sorted.

208. If the key is in position i in the unsorted sequence then, assuming it is equally likely to be in any of the positions of the unsorted sequence, its average displacement from the correct position is

$$(1/n) \sum_{j=1}^{n} |i - j|$$

$$= (1/n) \left(\sum_{j=1}^{i-1} (i - j) + \sum_{j=i+1}^{n} (j - i) \right)$$

$$= (1/n)\left(\sum_{k=1}^{i-1} k + \sum_{k=1}^{n-i} k\right)$$
$$= (1/n)((i-1)i/2 + (n-i)(n-i+1)/2)$$
$$= (1/2n)(i^2 - i + n^2 - 2ni + i^2 + n - i)$$
$$= (1/2n)(2i^2 - 2(n+1)i + (n^2+n))$$

The average of this value as i ranges from 1 to n is

$$(1/n)\sum_{i=1}^{n}(1/2n)(2i^2 - 2(n+1)i + (n^2+n))$$
$$= (1/2n^2)\left(\sum_{i=1}^{n}(2i^2) - 2(n+1)\sum_{i=1}^{n}(i) + \sum_{i=1}^{n}(n^2+n)\right)$$
$$= (1/2n^2)[2n(n+1)(2n+1)/6 - (n+1)n(n+1) + n(n^2+n)/2]$$
$$= [n(n+1)/2n^2][(4n+2)/6 - (n+1) + n]$$
$$= [(n+1)/2n][(4n+2-6n-6+6n)/6]$$
$$= (n+1)(4n-4)/12n = (n+1)(n-1)/3 = (n^2-1)/3$$

Using the above solution, there are n keys moved an average distance of $\Theta(n)$. In bubblesort it takes one exchange to move a key by 1 location, so each key must be involved in $\Theta(n)$ exchanges in the average case. This means a total number of exchanges of $1/2 \times n \times \Theta(n) = \Theta(n^2)$. The average-case behavior of bubblesort can therefore be no better than $\Theta(n^2)$, which is also the worst-case time required.

For another average-case analysis see Aho, Hopcroft, and Ullman (1983, 259).

209. Shellsort is discussed in Baase (1988, 80–83); Wirth (1976, 68–70); and Knuth (1973, vol. 3, 84–95). The original paper is Shell (1959).

210. Although bubblesort requires time $\Theta(n^2)$ in the average case, it runs well when either n is small or the file is nearly sorted. In the early passes of shellsort, bubblesort is being applied to small subfiles. By the time the files have become large, previous passes have nearly sorted them, so bubblesort will work well for these passes as well.

The formal analysis of shellsort is difficult and depends on the values of the increments. For example with increments of the form $2^k - 1$, shellsort takes time $O(n^{3/2})$. Peterson and Russell found empirically that the average case for several families of increments is about $\Theta(n^{5/4})$. See Knuth (1973, vol. 3, 84–95) where the Peterson and Russell result is cited or Augenstein and Tenenbaum (1979, 488–91).

Chapter 3

8. If the doubling operation requires constant time, then each bit requires $O(1)$ time, giving an overall $O(n)$ time complexity. If doubling a k-bit number has time complexity $O(k)$, then $O(n^2)$ time will be required. In

either case, if n represents the number and not its length, the time complexity will be $O(\log n)$.

11.

b	number of passes $\log_b m$	time per pass $5 \times b + 10 \times n$	time	time if $n = 25$
2^1	20	$10 + 10n$	$200 + 200n$	5200
2^2	10	$20 + 10n$	$200 + 100n$	2700
2^4	5	$80 + 10n$	$400 + 50n$	1650
2^5	4	$160 + 10n$	$640 + 40n$	1640
2^{10}	2	$5120 + 10n$	$10240 + 20n$	10740

The choice $b = 2^5$ is slightly better than 2^4.

12. The time T is $(\log_b m)(5b + 250) = [(\ln m)/(\ln b)][5b + 250]$ so $dT/db = 5(\ln m)/(\ln b) + (5b + 250)[(-\ln m)/(\ln b)^2][1/b]$. This equals 0 iff $5 = (5b + 250)/(b \ln b)$ or $b \ln b = b + 50$. The solution to this equation lies between 23 and 24, so that the answer to Exercise 11 is sensible.

25. Algorithm C.2 behaves as does the earlier Partition Algorithm 3.4 except that it returns the new location of the pivot element. When called from Quicksort, a sentinel of $+\infty$ is necessary just after the end of the array so that the loop of lines 6–8 will terminate. A sentinel is not necessary for subarrays.

26. For row-major order, the offset of the element in position (i, j) from the first location of an m by n array is given by $(i - 1)n + j - 1$. For column-major order, the offset is $(j - 1)m + i - 1$.

```
 1. Partition(A, low, upp)                      {see documentation in text}
 2. pivot := A[low]                 {the first element is used as the pivot}
 3. i := low − 1                     {i moves right looking for large elements}
 4. j := upp + 1                     {j moves left looking for small elements}
 5. repeat                                  {as long as i and j haven't crossed}
 6.    repeat                        {move i right until it finds a large element}
 7.       i := i + 1
 8.    end repeat if A[i] ≥ pivot
 9.    repeat                        {move j left until it finds a small element}
10.       j := j − 1
11.    end repeat if A[j] ≤ pivot
12.    if i < j then swap(A[i], A[j])        {swap large and small elements}
13. end repeat if i > j                          {but if i and j cross then quit}
14. swap(A[1], A[j])                         {put pivot element in right place}
15. return j
```

Algorithm C.2 Partition

27. The natural generalization of the previous exercise gives an offset for element $(i_1, \ldots, i_r)$ in an array with dimensions of $(n_1, \ldots, n_r)$ of

$$(i_1 - l_1) \ldots n_r + (i_2 - l_2)n_3 \ldots n_r + \ldots + (i_{r-1} - l_{r-1})n_r + (i_r - l_r)$$

The terms involving the l_i are constants that may be evaluated once and for all for a given array, to give

$$i_1 n_2 \ldots n_r + i_2 n_3 \ldots n_r + \ldots + i_{r-1} n_r + i_r + K \text{ where}$$

$$K = -l_1 n_2 \ldots n_r - l_2 n_3 \ldots n_r - \ldots - l_{r-1} n_r - l_r$$

28. The number of multiplications can be reduced by factoring out the n_i to get

$$(\ldots (i_1 n_2) + i_2)n_3 + \ldots)n_{r-1} + i_{r-1})n_r + i_r + K$$

35. See Lorin (1975, 144–48).

Chapter 4

1. (a), (b), (c), (e), and (f)
2. (a), (b), (e), (f), and (g)
3. a) value: 45
   prefix: * + 3 6 − 7 / 4 2
   postfix: 3 6 + 7 4 2 / − *
   b) value: 43
   prefix: + 3 − * 6 7 / 4 2
   postfix: 3 6 7 * 4 2 / − +
   c) value: 61
   prefix: − * + 3 6 7 / 4 2
   postfix: 3 6 + 7 * 4 2 / −
4. See Figure C.18.
6. $n + 1$. See Exercise 77 of Chapter 2.
7. Functions, including user-defined functions, are written in infix notation in APL. More than two arguments could lead to ambiguity.
11. Inserting l_3 into the sequence of length 3 containing l_1, w_1, and w_2 requires 2 comparisons. Inserting l_5 into the sequence of length 7 containing l_1 through l_3 and w_1 through w_4 requires 3 comparisons. Inserting l_{11} into the sequence of length 15 containing l_1 through l_5 and w_1 through w_{10} requires 4 comparisons. In general we want the insertion of t_i to require consideration of $2^i - 1$ elements, and t_{i-1} of these are of the form l_k. This leaves $2^i - 1 - t_{i-1}$ elements, which must be the w_k for $k < t_i$. Thus $t_{i-1} - 1 = 2^i - 1 - t_{i-1}$. This is equivalent to the recurrence of Exercise 125b of Chapter 2; its solution is $t_i = [2^{i+1} + (-1)^i]/3$.

    So by definition, if $k = t_i$, $2^i - 1$ elements must be considered when inserting l_k. When inserting l_{10}, say, w_{10} need not be considered, but l_{11} must, so the size of the sequence is the same as for l_{11}. In general, if $t_{i-1} <$

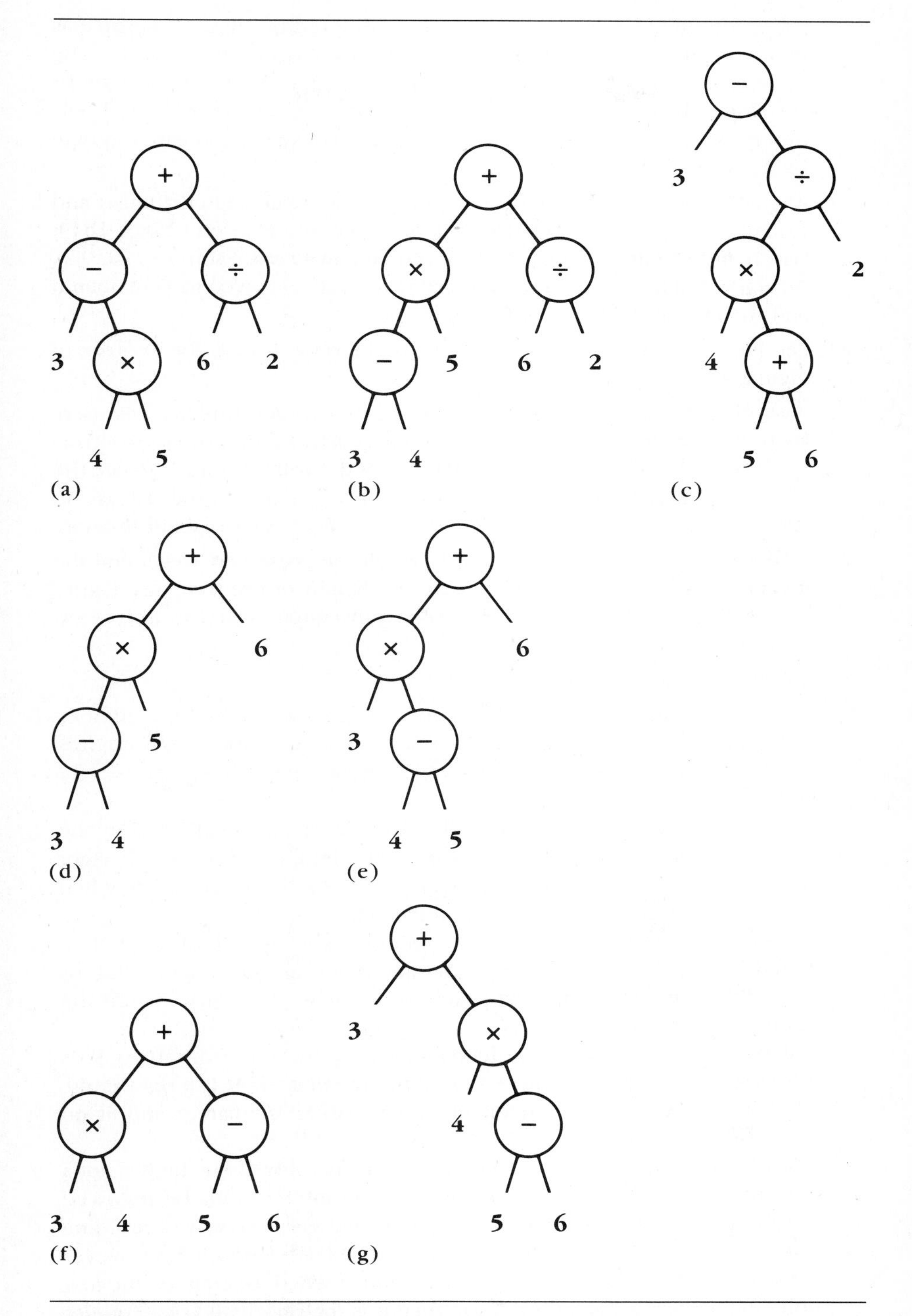

Figure C.18 Trees Representing Algebraic Expressions

$j \le t_i$, the number of elements to be considered is still $2^i - 1$, since as many of the w_k can be omitted as there are extra l_k to consider. The numbers of comparisons required for sequences of sizes 1 through 13 are respectively 0, 1, 3, 5, 7, 10, 13, 16, 19, 22, 26, 30, and 34. These numbers are the best possible except for the last two, which are too high by 1.

18. Algorithms for converting from infix to postfix are given in Brillinger and Cohen [1972, 522 (in FORTRAN)]; Reingold and Hansen (1983, 181); Baron and Shapiro (1980, 58–60); and Horowitz and Sahni (1976, 93). An algorithm for converting from prefix to postfix is given in Tenenbaum and Augenstein (1981, 129ff).

25. See Figure C.19. Figure 4.11a shows the inorder threads for the tree of Figure 4.9.

26. Assuming both predecessor and successor inorder threads are used, there will be $2n$ pointer fields in the usual representation of a binary tree. Of these, $n - 1$ contain ordinary pointers and 2 will be empty—since the first node in inorder has no predecessor or left child and the last has no successor or right child. So there are $n - 1$ fields that contain threads.

27. Although the binary search property would be preserved, in general the given method could greatly increase the height of the tree (see Figure C.20). This method would slow down subsequent searches and insertions in the tree.

34. See Reingold and Hansen (1983, 228).

36., 37., 38. The time for traversal can be measured in terms of the number $f(n)$ of links (pointers or threads) followed. Since Exercise 36 implies that $f(n) < 2n$, its solution will give an $O(n)$ solution for Exercises 37 and 38.

For Exercise 36, we first note that threads are only followed in the case when the node last visited has no right child. In this case only right threads are followed, and a given right thread is followed precisely when we have just visited the node it points from.

In the cases when the node last visited has a right child, we follow exactly one right pointer, namely the one pointing to the right child. So each right pointer is followed precisely when we have just visited the node it points from.

Finally, suppose that some left pointer, say the one from M to N, is followed at two different times during the traversal. If M is a right child, we follow this pointer immediately after visiting M's parent and at no other time.

If M is a left child, then let S be the lowest ancestor of M which is not a left child. If S is the root, then the given pointer cannot be followed during any inorder successor operation, since there is no way of reaching M by following a right pointer and then a series of left pointers. However we will reach M and follow the given pointer when looking for the first node in inorder. If S is not the root, then it is the right child of a node U,

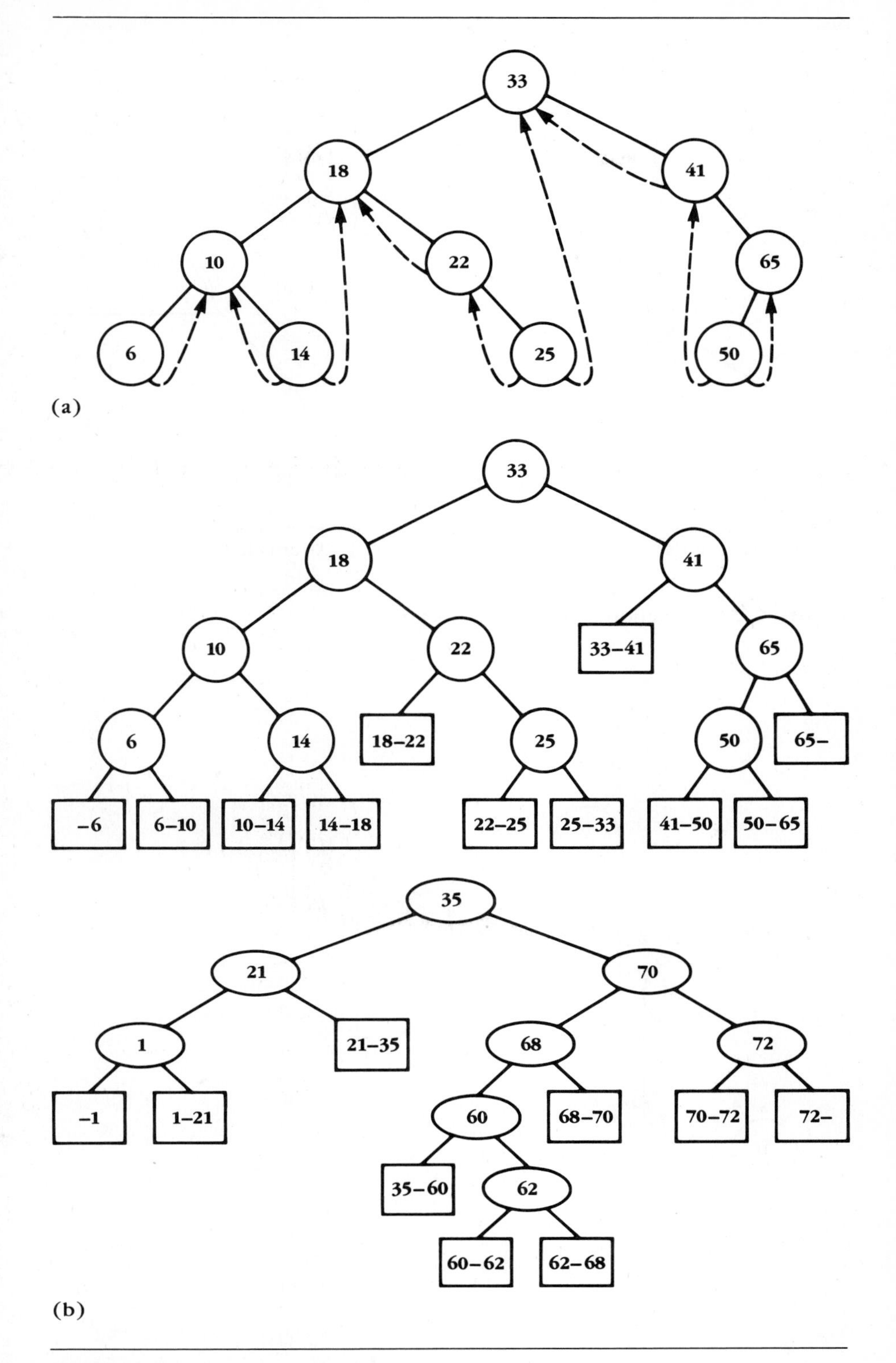

Figure C.19 Exercises on Binary Trees

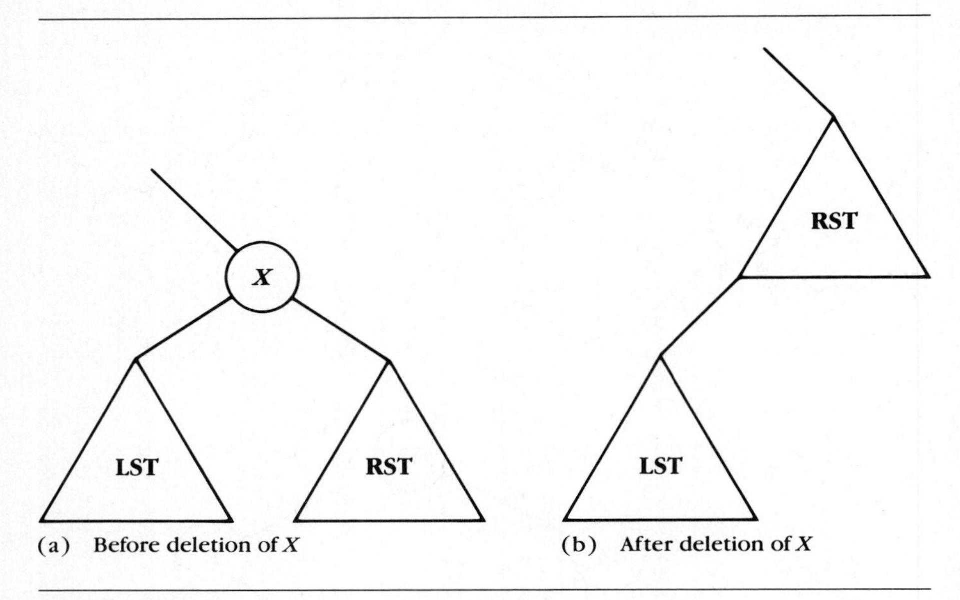

(a) Before deletion of *X* (b) After deletion of *X*

Figure C.20 Deletion from a Binary Search Tree

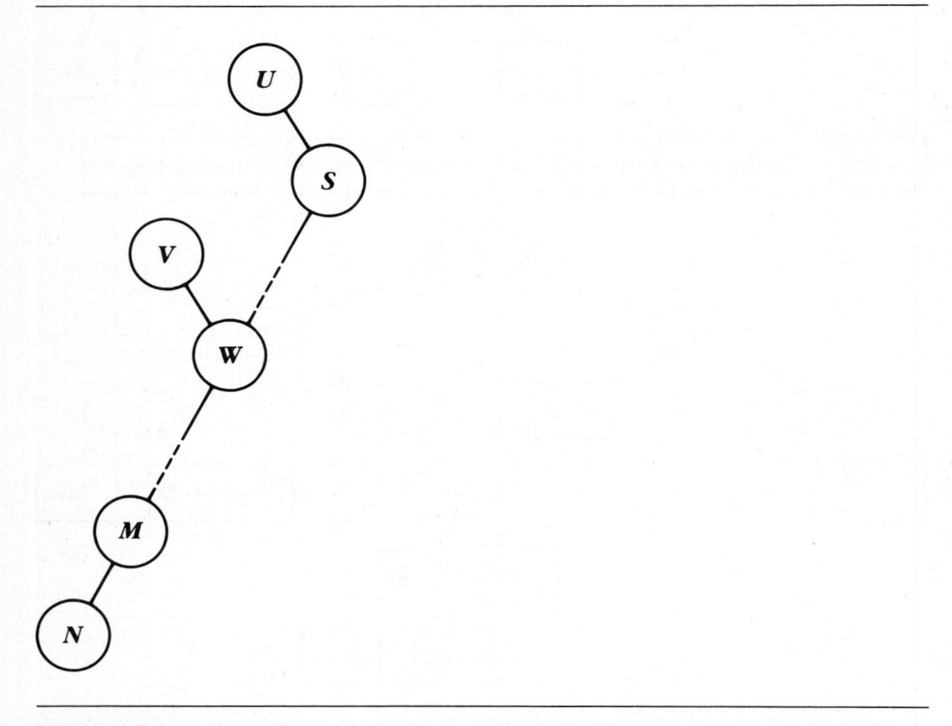

Figure C.21 Hypothetical Counterexample in Threaded Tree Traversal

and we will follow the given pointer when looking for the inorder successor of U. We need only show that it is followed at no other time.

If the given pointer is followed when looking for the inorder successor of the node V, and $V \neq U$, then let W be the right child of V (see Figure C.21). $W \neq S$, since they have different parents. Also, W must be an ancestor of M. S cannot be on the path from V to N, since W is the only right child on that path. But W is then the lowest ancestor of M that is not a left child, contradicting the choice of S.

41. Suppose a node S has a left subtree. Then its inorder predecessor will be in this subtree and will have a right thread (it can't have a right pointer) that points to S. Similarly if S has a right subtree, there is a left thread pointing to S. So there are at least as many threads pointing to each node as pointers pointing from the node. But by Exercise 26 (see preceding solution), there are exactly $n - 1$ nonempty threads. Since there are only $n - 1$ pointers, each thread has been accounted for, and there exists a 1-1 correspondence between pointers from a node and threads to a node.

44. By symmetry it's enough to consider the case of right threads. Suppose a right thread from S points to T. Consider the path from the root to S. If this path follows only right pointers, then S is the last node in inorder (since it has no right child) and its right thread is empty. This contradiction shows that some node in the path is a left child.

Let U be the lowest node on the path that is a left child of its parent V. Note that U may equal S. S is the last node in inorder in the tree rooted at U (since only right pointers are followed from U to S, and S has no right child). The inorder traversal of the subtree that is rooted at V begins by traversing its left subtree (rooted at U) and then visits V. Since S is the last node in inorder in this left subtree, V is the inorder successor of S. Thus $T = V$. So T is the parent of U and U is an ancestor of S, making T an ancestor of S.

57. For Figure 4.21a, the average number of comparisons is $(1 + 2 + 2 + 3 + 3)/5 = 2.0$. For Figure 4.21b, the average number is $(1 + 2 + 3 + 3 + 4)/5 = 2.6$, a larger value.

For Figure 4.22a, the average number is $(1 + 2 + 2 + 2 + 3 + 3 + 3 + 4)/8 = 2.5$. For Figure 4.22b, the average is $(1 + 2 + 2 + 2 + 2 + 2 + 3 + 3)/8 = 2.125$.

59. Suppose the arguments to the Union operation are structures A and B with respective sizes $|A|$ and $|B|$. Since the way to minimize the cost of the Find operation is to minimize the internal path length of the tree, we need to ask how the internal path length of the union relates to the individual internal path lengths. If the root of A is made to point to the root of B, then the height of each element in B is unchanged, while the height of each element of A increases by 1. This gives an increase of $|A|$ in the internal path length of the union. Similarly pointing the root of B to the root of A will increase the internal path length by $|B|$. To get the

smaller increment, we should change the root pointer in the smaller structure.

60. A little experimenting should lead to the following claim: In a tree of height h, there are at least 2^h nodes in the tree. The claim may be proved by induction; it is certainly true for $h = 0$. Suppose the claim is true for $h - 1$. Let the number of nodes in the tree be n. The only way the height can increase is via the Union operation, and then only if the other operand to Union has more than n nodes. So the resulting tree will have height h and at least $2n$ nodes. By induction $n \geq 2^{h-1}$, so $2n \geq 2^h$, and the claim has been shown. Thus the height of the tree is logarithmic in the number of nodes, and we are done.

66. The information in each node in Figure C.22 consists of the number of missionaries, the number of cannibals, and the number of boats on the original side of the river. Since there are infinite paths in the tree, we use breadth-first search. The last node visited is therefore in the rightmost position at the lowest level. We indicate with an "X" those nodes that do not represent legal positions. Note that many nodes are repeated; since the number of possible nodes is small, it would be feasible to determine which nodes have already been visited and avoid inspecting them again.

69. See Charniak and McDermott (1985, 269).

71. See Baase (1988, 184); Horowitz and Sahni (1978, 302); or Aho, Hopcroft, and Ullman (1983, 244).

73. Free space routines and techniques are discussed in Standish (1980, chap. 5); Reingold and Hansen (1983, chap. 6); and Smith (1987, chap. 11).

76. At level 1 we may add those clauses obtainable by resolution from the original clauses. This would add the clause q (from $p' \vee q$ and p) and the clause $p' \vee r$ (from $p' \vee q$ and $q' \vee r$). At level 2 we may add those clauses obtainable by resolving clauses from level 1 with other clauses. This gives only the clause r (from either p and $p' \vee r$ or $q' \vee r$ and q). At level 3 there are no new clauses, so as we have defined resolution, the clause $p \vee r$ cannot be derived, although it follows from the given clauses.

    The definition of resolution may be modified to allow $x \vee y$ to be derived from the 2 clauses x and y, but this would mean that every pair of clauses could be resolved, which would make proof by resolution a very slow process. More commonly, we use resolution not to generate all consequences of a particular set of assumptions, but to determine whether a single given formula can be derived from the assumptions. In this case, $x \vee y$ would be handled by first attempting to prove x, then attempting to prove y.

77. The set $\{p, q\}$ may represent $p \wedge q$. The set $\{p' \vee q\}$ may represent $p \rightarrow q$. The set $\{p \vee q, p \vee r\}$ may represent $p \vee (q \wedge r)$.

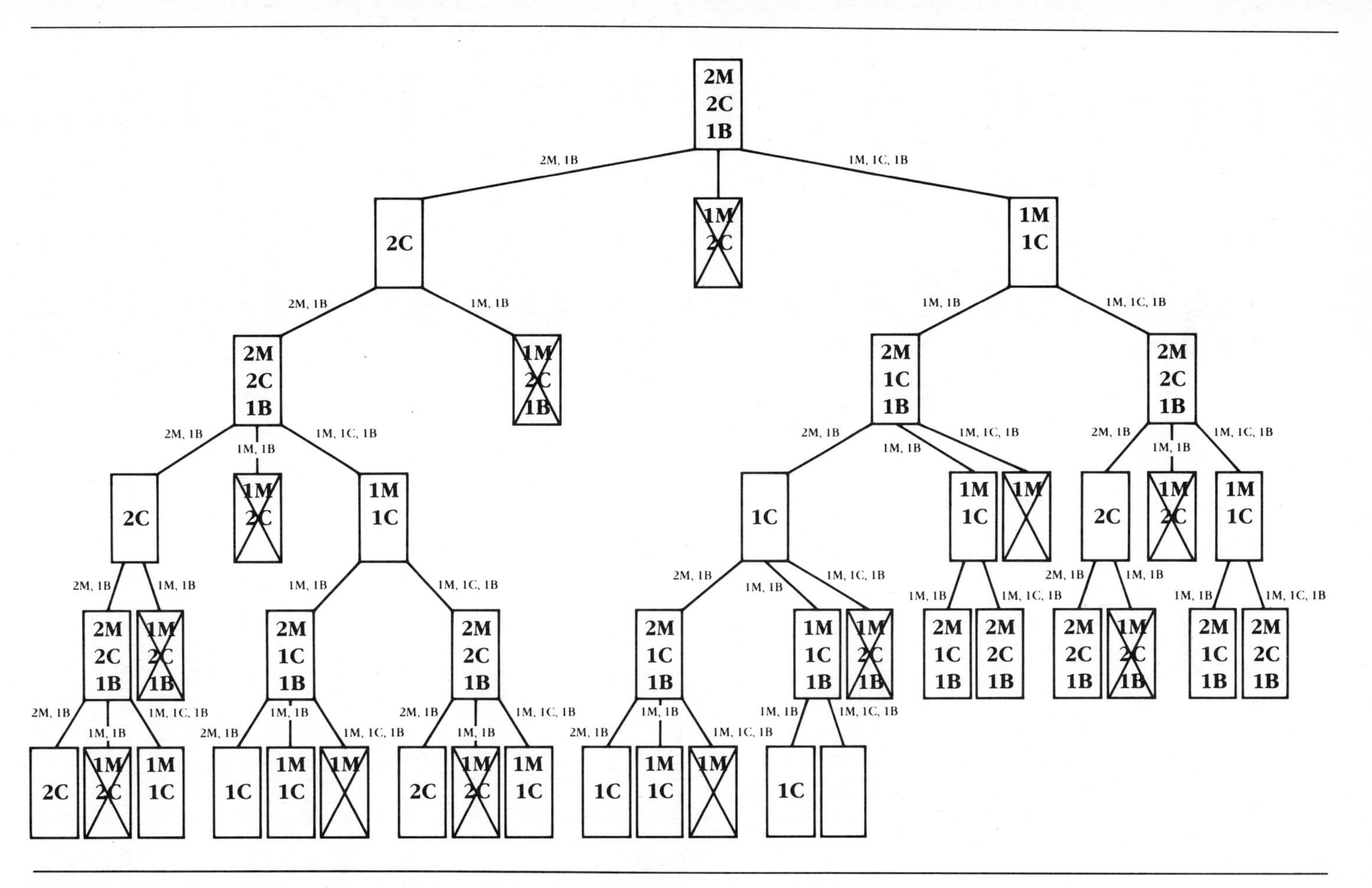

Figure C.22 Tree for the Missionaries-and-Cannibals Problem

For the second part, assume that $x \vee p$ and $y \vee p'$ are true, where x and y may themselves be disjunctions. We want to show that $x \vee y$ is true. If either x or y is true, we are done, so suppose not. In this case since $x \vee p$ is true and x is not, p must be true. Similarly since $y \vee p'$ is true and y is not, p' must be true. But p and p' can't both be true, so we are done.

78. Suppose that the conclusion is false, and let M and N be the first pair of nodes such that M is expanded immediately before N while $\mathrm{f}(M) > \mathrm{f}(N)$. Then N must be a successor of M, since otherwise N would have been open when M was chosen to have minimum f value among open nodes. We have

by monotonicity	$\mathrm{h}(M) \leq \mathrm{c}(M, N) + \mathrm{h}(N)$
by definition of g*	$\mathrm{h}(M) \leq \mathrm{g}^*(N) - \mathrm{g}^*(M) + \mathrm{h}(N)$
by the last theorem of Sec. 4.11	$\mathrm{h}(M) \leq \mathrm{g}(N) - \mathrm{g}(M) + \mathrm{h}(N)$
so	$\mathrm{h}(M) + \mathrm{g}(M) \leq \mathrm{g}(N) + \mathrm{h}(N)$
or	$\mathrm{f}(M) \leq \mathrm{f}(N)$

a contradiction.

80. See Horowitz and Sahni (1976, 300–301).

82. The matrices $A^{(1)}$ through $A^{(5)}$ are given together with the underlying graph in Figure C.23. Since $A^{(6)} = A^{(5)}$, $A^{(5)}$ gives the final answer, showing the two components of the graph to be {5, 7} and {1, 2, 3, 4, 6, 8}. Note that $A^{(4)} = A^{(2)} \,\alpha\, A^{(2)}$, $A^{(8)} = A^{(4)} \,\alpha\, A^{(4)}$, and so on, as suggested by Exercise 81. Also note that the claims of Exercise 80 are true for this example. Finally, comparing this algorithm to the all-pairs shortest-paths algorithm is instructive. The general approach to these algorithms is attributed to Warshall (1962).

84. See Figure C.24 for the answer to (a). For (b), the values of all internal nodes are as in (a).

86. See Figure C.25.

87. See Figure C.26.

88. See Figure C.27. The value of OR nodes is the minimum of the values of the children, and the value of AND nodes is the sum of the value of the children. There is a flight for exactly \$200 going through Denver and St. Louis and another through Dallas, if the second flight from there to New York is taken.

90. The second player should win. This can be seen by sketching the game tree or by first noting that any player who leaves just two nonempty piles of equal size may win simply by mimicking the other player's move. Representing states by a triple giving the sizes of each pile, the states that may be left by the first player are (0, 2, 3), (1, 1, 3), (1, 0, 3), (1, 2, 2), (1, 2, 1), and (1, 2, 0). The second player may in response leave the

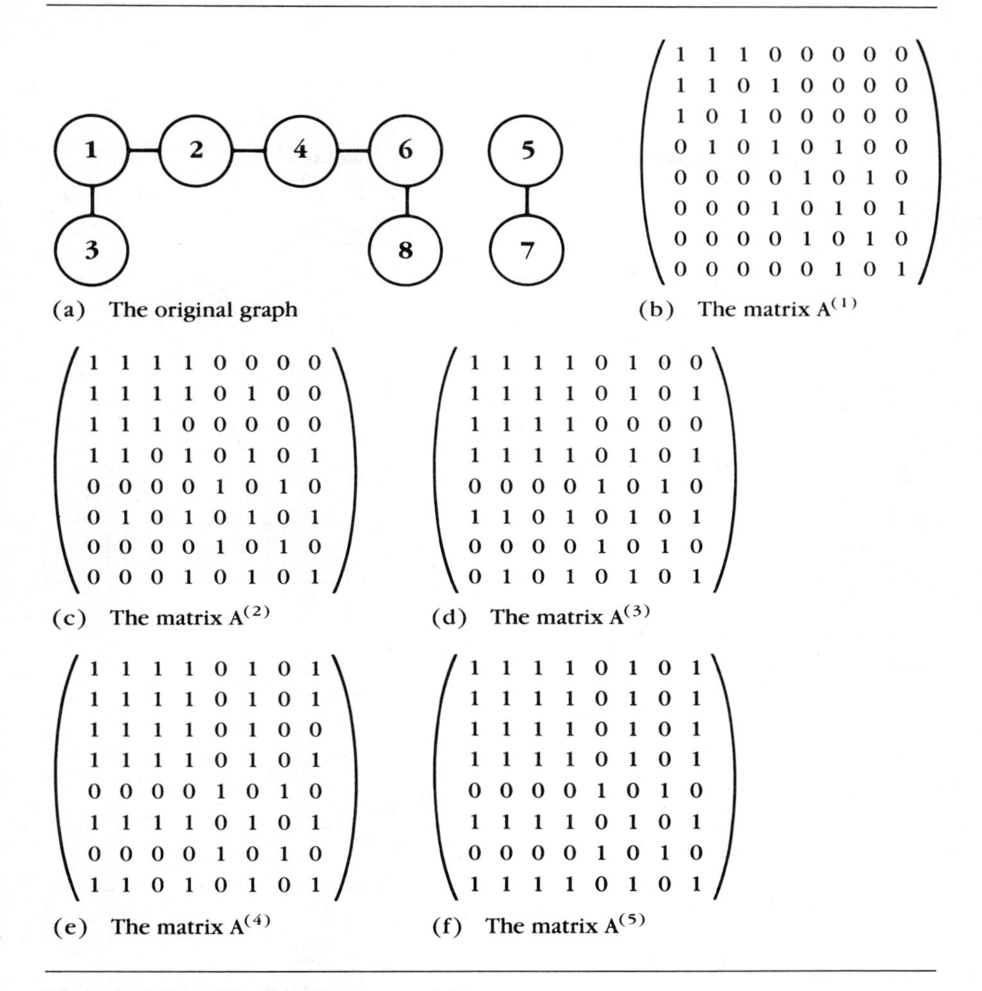

Figure C.23 Finding Connected Components

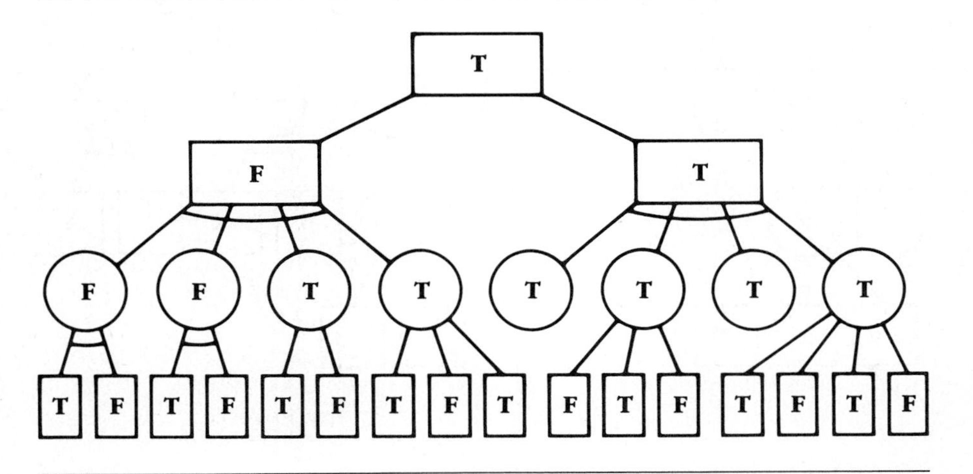

Figure C.24 Evaluation of an And-Or Tree

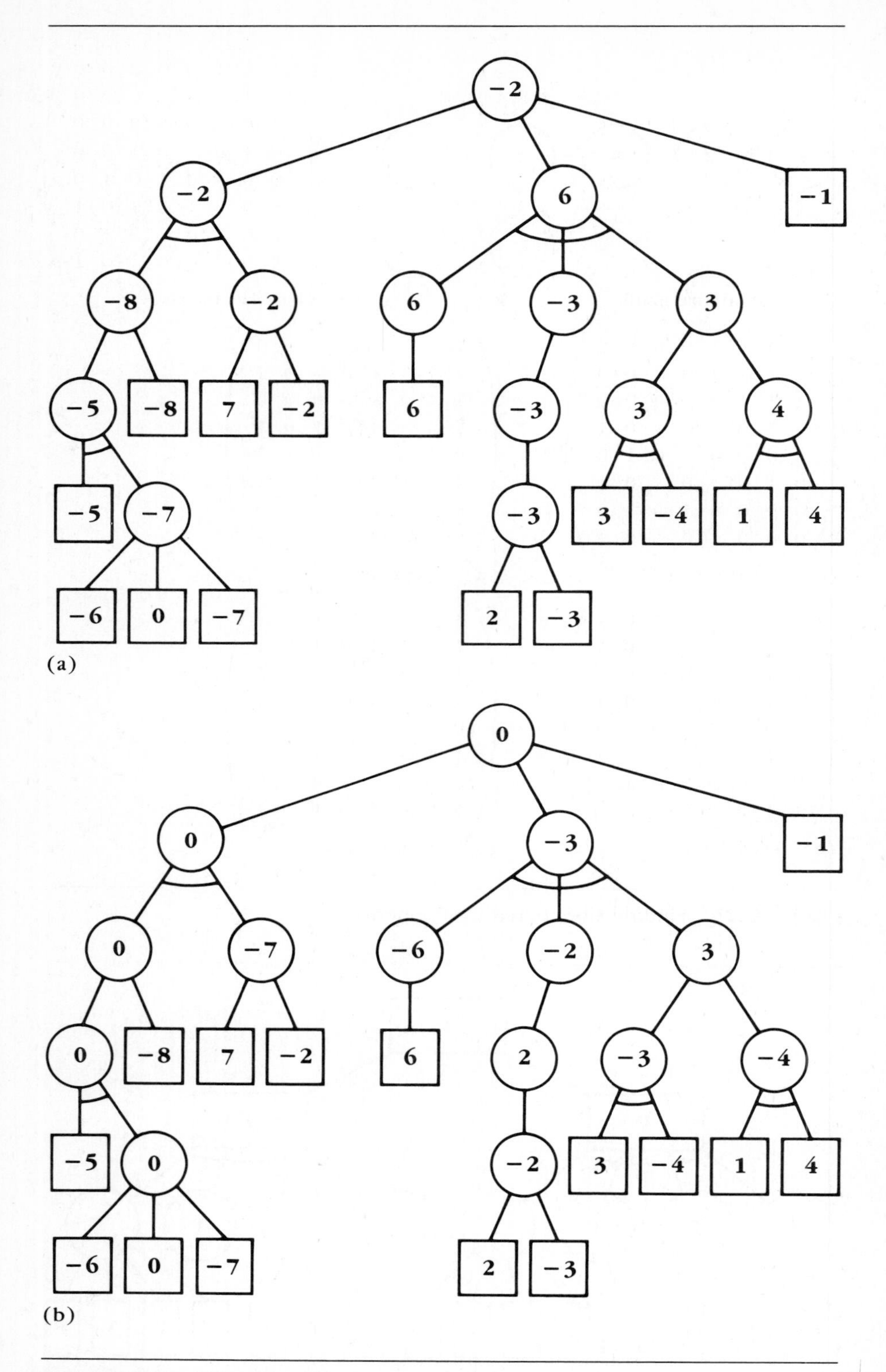

Figure C.25 Evaluating a Game Tree

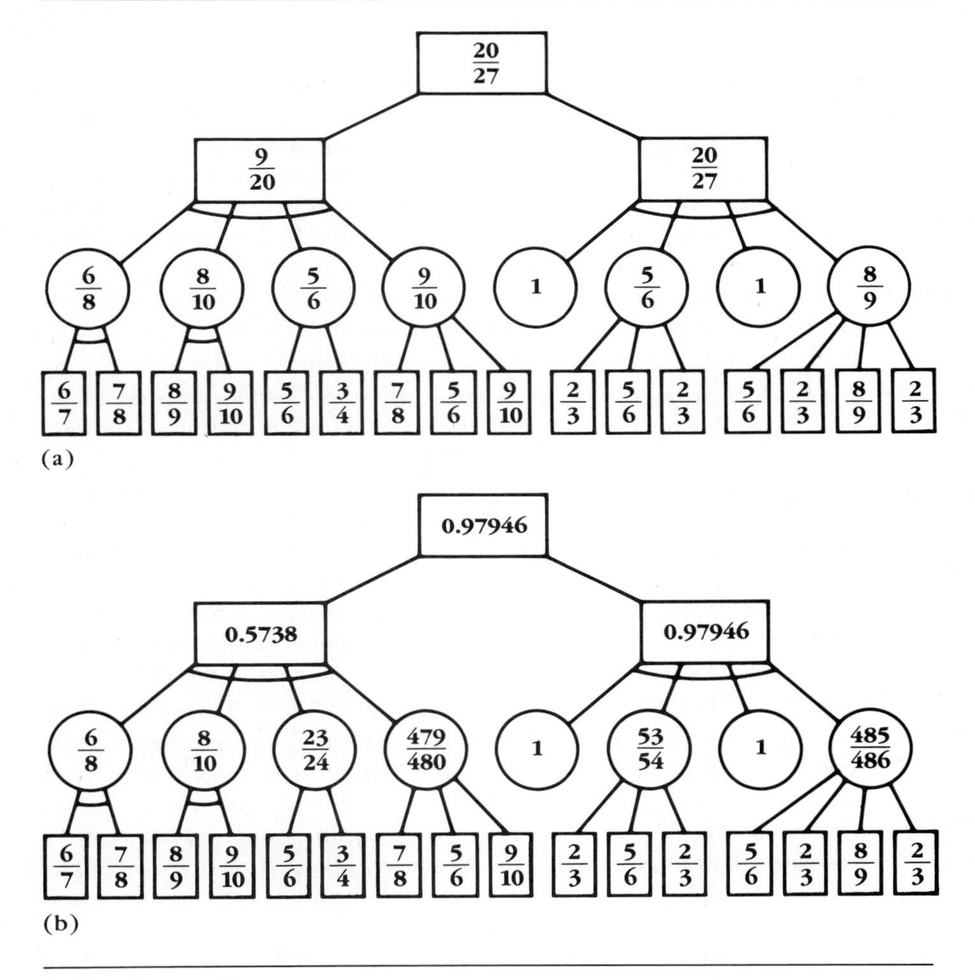

Figure C.26 Two Evaluations of an And-Or Tree

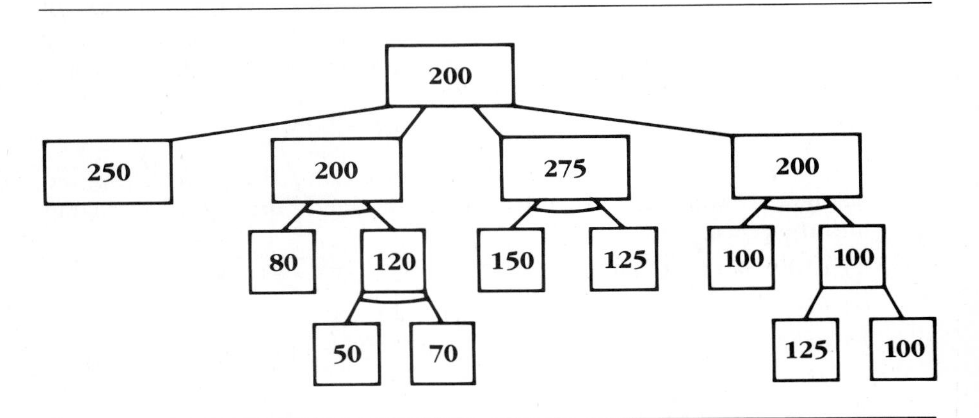

Figure C.27 Evaluation of the And-Or Tree of Figure 4.40a

respective states (0, 2, 2), (1, 1, 0), (1, 0, 1), (0, 2, 2), (1, 0, 1), and (1, 1, 0). You may want to show that in general a) if the three sizes are considered as binary numbers, any player can win who leaves a state where each bit position contains among its three bits an even number of 1 bits, b) any state not of this form can be converted by a legal move to one of this form, and c) no state of this form may be converted by a legal move to one of this form.

If the player taking the last stick loses, the given position is still a win for the second player. The respective responses in this case are (0, 2, 2), (1, 1, 1), (1, 0, 0), (0, 2, 2), (1, 1, 1), and (1, 0, 0).

91. We label each node with the sum of the items included so far in the subset (see Figure C.28). Below each leaf we have indicated its value, that is, whether it is a win or a loss for A. Alongside each internal node we have indicated its value, where $+1$ means a win for the first player and -1 a win for the second player. The recursive postorder traversal used in applying the definition corresponds to evaluating nodes from the bottom level up. The value of the root is -1, so B should win.

    B's strategy, shown by heavy lines in the figure, is as follows: If in the previous move A has included a number, B should not include the next number, and vice versa. This will include one number of the subset $\{1, 2\}$ and one number of $\{3, 4\}$. The sum will therefore be from the set $\{4, 5, 6\}$.

92. Each state may be uniquely described by indicating the number of sticks remaining and the player to move next. If 1, 2, or 3 sticks remain, then the player to move wins. Otherwise we use the algorithm of an earlier problem and get the values of Figure C.29, again assuming a value of $+1$ if the first player A wins and -1 if the second player B wins. For example, $v(4, A) = \max\{v(3, B), v(2, B), v(1, B)\} = \max\{-1, -1, -1\} = -1$ and $v(5, B) = \min\{-1, 1, 1\} = -1$.

    If the original set has size 9, then A should win. If it has size 8, then B should win. If a player can leave $4k$ sticks, then that player can remove $4 - s$ sticks whenever the other removes s sticks and thus again leave a multiple of 4. Thus B can only win if the original number is a multiple of 4 or if A blunders. If the player who takes the last stick loses, then leaving 1 stick will win, and a player may adapt the preceding strategy to leave always one more than a multiple of 4.

94. The answer is 85 minutes. See Figure C.30.

95. The values of ET and LT are given in Figure C.30. The critical tasks are taking a shower, dressing, eating, brushing teeth, driving, and walking into the office.

99. a) $(109, 67) = (67, 42) = (42, 25) = (25, 17) = (17, 8) = (8, 1) = 1$

    b) $(143, 77) = (77, 66) = (66, 11) = 11$

    c) $(1984, 1776) = (1776, 208) = (208, 112) = (112, 96) = (96, 16) = 16$

105. We want to show that the values of a and b in Algorithm 4.24 grow small quickly. Since the new value of a is just the old value of b, the value of a

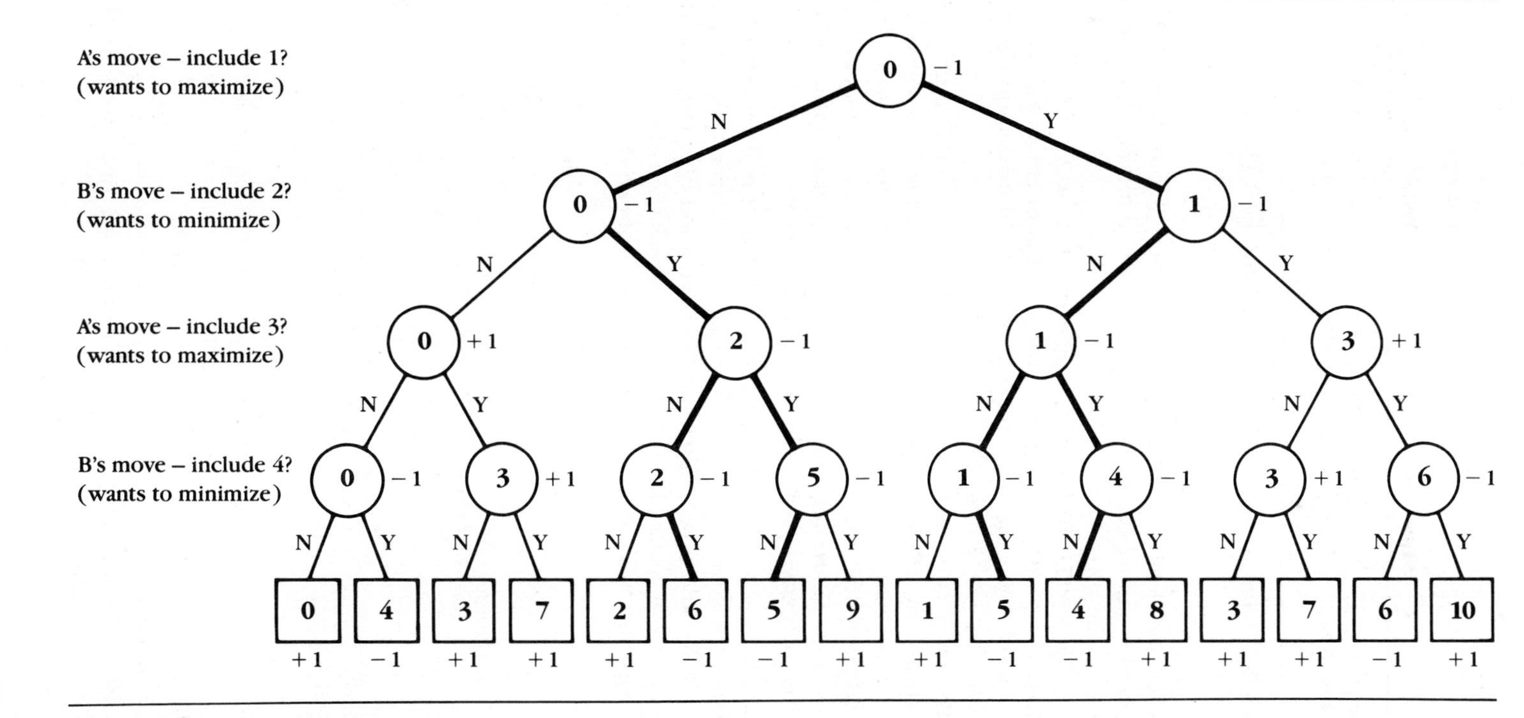

Figure C.28 Evaluation of a Game Tree

To Play	9	8	7	6	5	4	3	2	1
	Number of Sticks								
A (first player)	+1	−1	+1	+1	+1	−1	+1	+1	+1
B (second player)	−1	+1	−1	−1	−1	+1	−1	−1	−1

Figure C.29 Value of the Game of Sticks to the First Player

may not shrink very fast after just one iteration. But the second value of a is a mod b, and we can show that this value is at most $a/2$. There are two cases:

Case 1: $a \geq 2b$. In this case $a \bmod b < b \leq a/2$.

Case 2: $a < 2b$. In this case the truncated quotient q of a/b must be 1, so $a \bmod b = a - qb = a - b$. Since $b > a/2$ by assumption, $a \bmod b = a - b < a - a/2 = a/2$.

Therefore after each two iterations the value of the larger argument has shrunk by a factor of at least 2. A simple induction shows that the time complexity is therefore logarithmic. A more careful argument would show that the worst-case behavior is given by two consecutive terms in the Fibonacci sequence, and that the nonasymptotic worst-case time complexity is very close to $\log_r a$, where $r = (1 + \sqrt{5})/2$.

108. Unlike the case of finding just the maximum, it matters what happens when two elements that have lost once are compared. In this case the adversary could give a minimum of information by saying that the element that seems most likely to be the second largest is the winner of the comparison. One way to give minimal information is to keep track for all

Task	*ET* (Earliest time)	*LT* (Latest time)
Shower	10	10
Fix breakfast	15	20
Shine shoes	5	15
Listen to radio	20	82
Dress	20	20
Eat	50	50
Brush teeth	52	52
Drive	82	82
Walk to office	85	85

Figure C.30 Earliest and Latest Time for Completing Tasks to Get to Work in the Morning (see Exercises 59 and 60)

elements x of an upper bound for the number of elements known to be smaller than x. In any comparison, x beats y if BEAT(x) $\geq$ BEAT(y), with ties broken arbitrarily. If x beats y, BEAT(x) gets the new value BEAT(x) + BEAT(y) + 1. Initially, BEAT(x) = 0 for all x.

The tournament sort algorithm requires $n - 1$ comparisons to find the maximum and another $\lceil \log_2 n \rceil - 1$ comparisons to find the second largest. To be ruled out as the maximum, each nonmaximal element must lose a comparison, for $n - 1$ comparisons in all. Except for the second-largest element, each element that loses to the maximum must lose an additional comparison to be ruled out as the second largest. So it's enough to show that given the preceding strategy for the adversary, the maximum element must be involved in $\lceil \log_2 n \rceil$ comparisons.

By analogy with Exercise 60, we may make the following claim: If an element x has won c comparisons, BEAT(x) $\leq 2^c - 1$, which is clearly true for $c = 0$. If it is true for $c - 1$, then when x wins another comparison, say against y, the adversary's strategy implies that BEAT(y) $\leq$ BEAT(x). So the new value of BEAT(x) is at most 1 more than twice the old value, or $1 + 2(2^{c-1} - 1) = 2^c - 1$, and we have proved the claim.

Now for BEAT(x) to be at least $n - 1$, we must have $n - 1 \leq$ BEAT(x) $\leq 2^c - 1$ or $n \leq 2^c$ or $c \geq \log_2 n$. Since c is an integer, $c \geq \lceil \log_2 n \rceil$, and we are done.

113. Using the notation of Exercise 66, the only legal states correspond to 2M − 2C, 2M − 1C, 1M − 1C, 2M − 0C, 0M − 2C, 0M − 1C, and 0M − 0C. If 2 missionaries are on the same side, the boat must be with them, so only 8 states are accessible. The initial and final states are again 2M − 2C − 1B and 0M − 0C − 0B. There are at most 3 legal transitions from every state, corresponding to 1 missionary, 2 missionaries, and 1 missionary and 1 cannibal. Sometimes these transitions lead to illegal states. (See Figure C.31.) Note that the solution found for Exercise 66 is the unique minimal solution.

116. Identifying the empty string with 0, we will show by induction that for any number n, s(n) = n mod 3, where s is the function of Definition 4.29 and n is interpreted as a binary number. The induction will be on the length of the string.

Basis: Any string of length 0 will be empty. We have identified this string with 0, and its s value is 0, the number of the start state.

Induction: If x is a string (including the empty string) and b a bit, then the concatenation xb has value $2x + b$. The value of $2x + b$ mod 3 depends only on b and (x mod 3). We can easily verify that $2r + b$ = d(r, b) mod 3 for all residues r mod 3 and bits b, so we are done.

Chapter 5

1. Parts (a), (b), (e), (g), (j), (l), (n), (p), (q), (r), (s), (u), and (v) are true. Parts (d), (o), and (t) are false. Parts (c), (f), (h), (i), (k), (m), and (w) are true if P = NP.

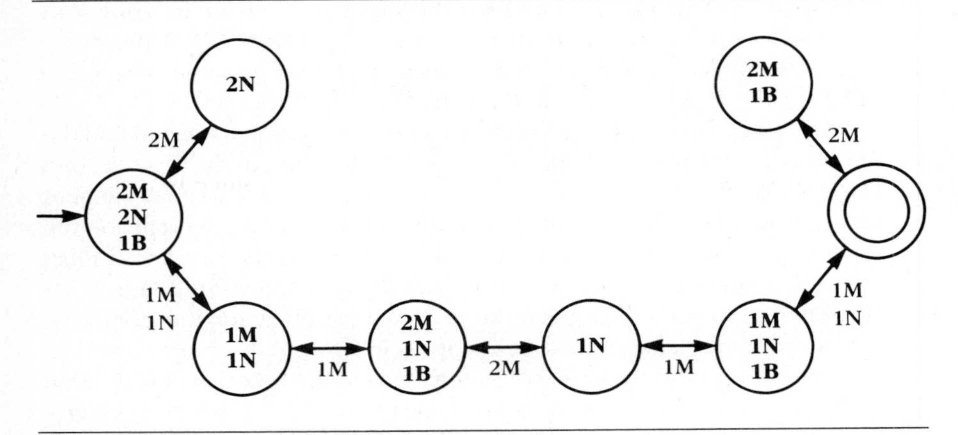

Figure C.31 A DFA for the Missionaries-and-Cannibals Problem

In (a), the reduction would simply solve A in polynomial time and return a true instance of B if A was true and a false instance otherwise (assuming that B is nontrivial). A similar proof works for (b), and for (c) and (f) if P = NP. The truth of (d) would imply that F is in P. The truth of (e), (g), (l), (q), and (r) follows from the definition of NP-complete. The answers to (h), (i), (k), (m), and (w) follow from the observation that if P = NP, then by (b) all nontrivial problems in P = NP are NP-complete. Part (o) is true if X is in NP. For (u), as in the proof of Cook's theorem, we observe that an algorithm can only use polynomial space in polynomial time.

3. For the graph of Figure 4.41a, the maximum clique size is 3 (e.g., {A, G, H}); the minimum vertex cover size is 4 (e.g., {A, B, C, G}); and the minimum k for k-colorability is 3. (An example for minimum k is {A, B, F}, which may have color 1; {C, G} color 2; and {D, E, H} color 3.) For the graph of Figure 5.1a, the maximum clique size, minimum vertex cover size, and minimum k for k-colorability are all 3.
4. Consider the graph on 6 vertices whose only edges are {1, 2}, {2, 3}, {1, 3}, {4, 5}, {5, 6} and {4, 6}. This has no Hamiltonian circuit even though the corresponding instance of 3-dimensional matching has a solution.
7. If neither is adjacent, then both x_i and x_i' appear in C_j, so C_j is trivially satisfiable.
8. Although a solution for 3-colorability implies a solution for 4-colorability, that a 4-colorability solution implies a 3-colorability solution has not been shown—and is not true.
11. Given an instance of ZOK with n items, profits (p_i), weights (w_i), capacity M, and target T, let the corresponding instance of the job-sequencing problem be defined by $t_i = w_i$ for all i, $d_i = M$ for all i, $c_i = p_i$, and the

target penalty $k = \left(\sum_{i=1}^{n} p_i\right) - T$. Clearly this instance may be constructed in polynomial time.

If there is a solution (x_i) for ZOK, let $S = \{i \mid x_i = 1\}$. Intuitively, this is the set of jobs that may be completed by their deadlines. Formally this is so since if we perform the jobs indexed by S first, the last of them will be completed by time $\sum_{i \text{ in } S} t_i = \sum_{i \text{ in } S} x_i t_i = \sum_{i \text{ in } S} x_i w_i \leq M$, since the knapsack capacity is not exceeded. The only possible penalties are those from jobs not in S; the total of these penalties is $\sum_{i \text{ not in } S} c_i$

$$= \left(\sum_i c_i\right) - \left(\sum_{i \text{ in } S} x_i c_i\right)$$

$$= \left(\sum_i p_i\right) - \left(\sum_i x_i p_i\right)$$

$$\leq \left(\sum_i p_i\right) - T = k$$

where the last inequality follows since (x_i) is a solution to the ZOK instance.

If there is a solution to the job-sequencing problem, letting S be the set of jobs met by their deadlines and letting x_i equal 1 iff i is in S gives a solution for ZOK, as may be seen by reversing the argument above.

15. Suppose there is a vertex cover of size k. We may call the ith vertex of the cover v_i. Construct a Hamiltonian circuit as follows: After visiting $v = v_i$, process in numerical order the edges corresponding to v. Here processing an edge $\{v, w\}$ means visiting the vertices v_0, w_0, w_1, and v_1 in that order for $\{v, w\}$ if w is not in the cover and visiting only v_0 and v_1 in that order if w is in the cover. Since vertex i is connected to the first v_0, each v_1 is connected to the v_0 for the next edge, and the last v_1 is connected to vertex $i + 1$ (or vertex 1 if $i = k$), this is a cycle in the graph. No vertices are repeated and since each original edge is assigned to an original vertex, each vertex in the new graph is visited. So the cycle is a Hamiltonian circuit. An example of such a Hamiltonian circuit is shown in Figure C.32. In this figure the original graph is shown in part (a) and the new directed graph is superimposed on it in (b). Edge 1 is assigned to A, edge 2 to A and B, and edges 3 and 4 to B. The part of corresponding Hamiltonian circuit from vertex 1 to vertex 2 is shown in (c); the part from vertex 2 back to vertex 1 is shown in (d).

Conversely, if there is a Hamiltonian circuit, it must include the vertices 1 through k. For each original edge, the edge's copy of v_0 is visited between two of these vertices, say after vertex j. If v_0 is visited after the edge's copy of w_0, then assign the edge to w; otherwise assign it to v. In either case, by construction of the new graph the vertex assigned to the edge must be an endpoint of the edge. Since each original edge must be represented in the Hamiltonian circuit, each original edge is assigned to an endpoint, and we are done.

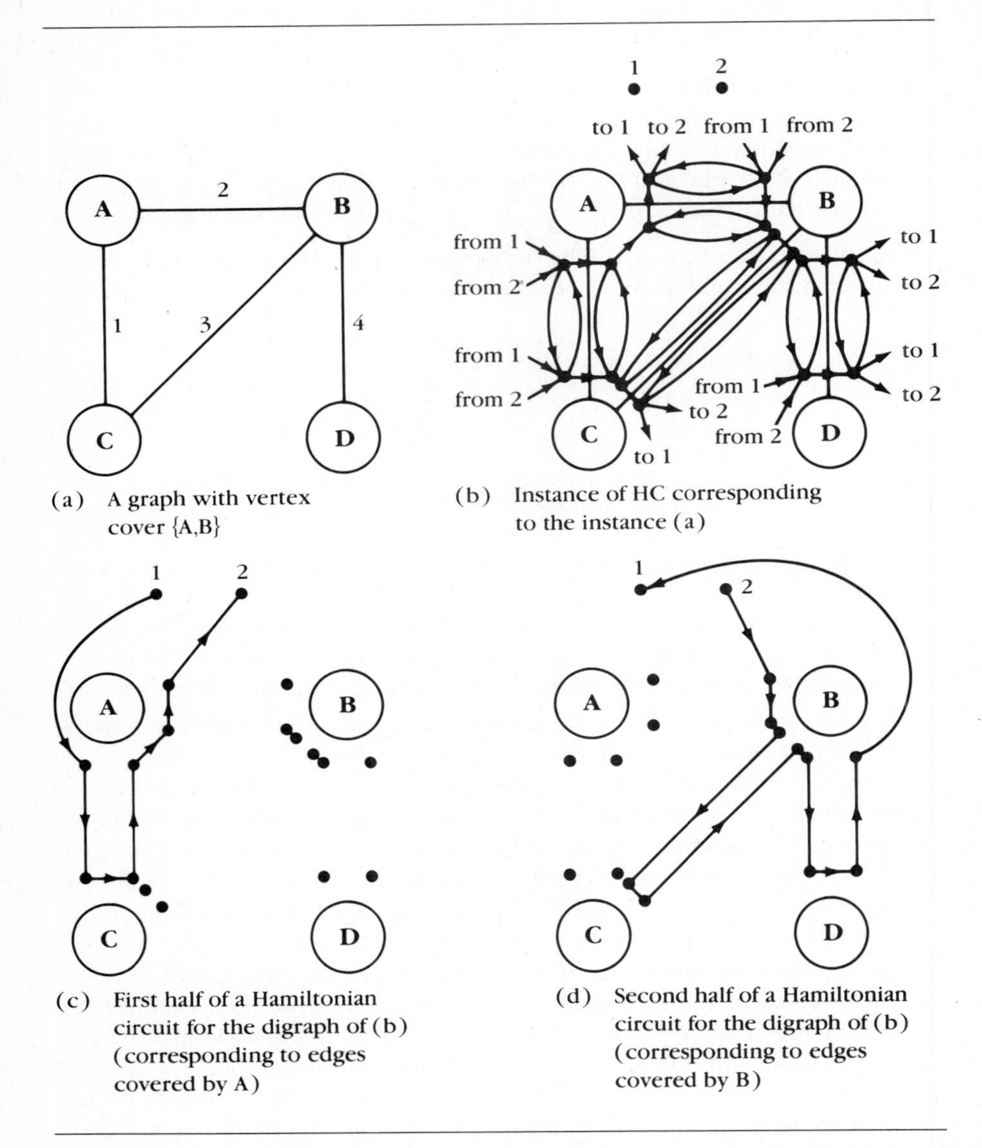

(a) A graph with vertex cover {A,B}

(b) Instance of HC corresponding to the instance (a)

(c) First half of a Hamiltonian circuit for the digraph of (b) (corresponding to edges covered by A)

(d) Second half of a Hamiltonian circuit for the digraph of (b) (corresponding to edges covered by B)

Figure C.32 An Instance of Vertex Cover and a Corresponding Instance of the Hamiltonian Circuit Problem

18. See Karp (1972).

21. In the complete graph on n vertices, each of the $n!$ possible permutations of the vertices gives a different Hamiltonian circuit. Note that $n!$ is not polynomial in n, nor is $n!$ polynomial in n^2, the size of the adjacency matrix.

25. To find the inverse of 17 mod $22 \times 28 = 616$, we use Algorithm 4.24 and see that $616 = 36 \times 17 + 4$, so that $4 = 1 \times 616 - 36 \times 17$ and that $17 =$

$4 \times 4 + 1$, so that $1 = -4 \times 616 + 145 \times 17$. Thus $e = 145$. Now $23 \times 29 = 667$, so if all arithmetic is done mod 667, $2^{145} = 2 \times 4^{72} = 2 \times 256^{18} = 2 \times 170^9 = 2 \times 170 \times 219^4 = 340 \times 604^2 = 340 \times 634 = 119$. To decode, evaluate $119^{17} = 119 \times 154^8 = 119 \times 371^4 = 119 \times 239^2 = 119 \times 426 = 2$.

27. a) Yes. The probability that the kth witness will answer "yes" incorrectly is 1/2, 2/3, 3/4, 4/5 for the first few values of k. It is easy to guess and verify by induction that the product of the first k terms of this sequence is just $1/(k + 1)$. This probability may be made arbitrarily small, so the probability of its complement, that some answer will be "no," may be made arbitrarily close to 1.

    b) Here inspection, perhaps using a calculator, suggests that the product of the terms 1/2, 3/4, 7/8, 15/16, . . . does not approach 0. One way to prove this is to note that the probability that some answer will be "no" is bounded above by the sum of the probabilities of "no" answers. For the original sequence this sum is $\sum_{k=1}^{\infty} 1/2^n = 1$, which doesn't help; but for the terms after the first term, the sum is $\sum_{k=2}^{\infty} 1/2^n = 1/2$. Thus the probability that no answer after the first will be "no" is at least 1/2, and thus the probability that no answer at all will be "no" is at least $(1/2)(1/2) = 1/4$. Thus the probability of a correct answer for the probabilistic algorithm is at most 3/4.

31. See Horowitz and Sahni (1978, 578–81).

32. Given m processors, consider $m - 1$ jobs requiring 5 units of time, $m - 1$ jobs requiring 4 units, and three requiring 3 units. The total time required is $5(m - 1) + 4(m - 1) + 3 \times 3 = 9m$; an optimal solution is obtainable by assigning the jobs requiring 3 units to the same processor and giving each remaining processor a job requiring 5 units and another job requiring 4 units. The first-fit algorithm would begin by assigning two jobs to each processor, giving the first two jobs with 3 units to processors already scheduled for 5 units. The last job would have to go to a processor already scheduled for 8 units, giving a total of 11 units of time. The relative error is $(11 - 9)/9 = 2/9$.

33. Consider a graph with vertices numbered from 1 to n with edge cost function given by $c(\{1, k\}) = 1$ for $k > 1$, $c(\{k, k + 1\}) = 2$ for k from 1 to $n - 1$, and $c(\{j, k\}) = 1 + \varepsilon$ otherwise. This cost function satisfies the triangle inequality. The minimum-cost spanning tree includes exactly the $n - 1$ edges of cost 1, so has cost $n - 1$. When this is converted to a TSP tour, it will have 2 edges of cost 1 and $n - 2$ of cost 2, for a total cost of $2 + 2(n - 2) = 2n - 2$. For $n \geq 6$, there is a tour containing the same 2 edges of cost 1 and $n - 2$ edges of cost $1 + \varepsilon$, for a total cost of $2 + (n - 2)(1 + \varepsilon)$ or $n + (n - 2)\varepsilon$. The ratio $(2n - 2)/(n + (n - 2)\varepsilon)$ gets arbitrarily close to 2 as ε gets small and n gets large, so the relative error gets arbitrarily close to 100%.

34. Given $N > 2$, consider the ZOK instance with capacity $N + 2$ and two items, one with weight 1 and profit 1 and the other with weight $N + 2$ and profit $N + 1$. The greedy algorithm will include only the first item; the optimal solution is to include only the second. The relative error is $(N + 1 - 1)/1 = N$.

36. The optimal strategy must use at least $\lceil W/M \rceil$ bins, where W is the sum of all the weights. We claim that the given strategy will use at most $2W/M$ bins; it's enough to establish this claim, which is intuitively true because two adjacent bins must contain at least weight M. More precisely, consider a counterexample to the claim using a minimum number of bins. This minimum number must be at least 2, so consider the instance resulting from deleting the items in the last two bins. The given strategy will give the same result as the original strategy up to these last two bins, so it uses exactly 2 fewer bins. If w_i is the first item included in the original last bin, and X is the total weight included in the original next-to-last bin, then $X + w_i > n$, or w_i would have been included in the original next-to-last bin. Thus the sum of the weights in the new instance is at most $W - X - w_i < W - M$, and for the new instance, $\left(\sum_j w_j\right)/M < (W - M)/M = W/M - 1$, so the number of bins in the optimal solution decreases by at least 1. Thus the new instance is also a counterexample, contradicting minimality. Additional exercise: Why won't this argument work if in the claim we replace the theoretical optimum $\lceil W/M \rceil$ by the actual optimum?

37. For an original sequence of elements

(2 12 1 8 7 16 13 14 9 4 3 15 11 6 10 5)

the results after the first merge are

(2 12 8 1 16 7 13 14 4 9 15 3 11 6 5 10)

the results after each pass of the second merge are

(2 8 1 12 16 13 14 7 15 4 9 3 5 11 6 10) and

(1 2 8 12 16 14 13 7 15 9 4 3 5 6 10 11)

the results after each pass of the third merge are

(1 16 2 14 8 13 7 12 15 5 9 6 10 4 11 3)

(1 8 13 16 2 7 12 14 15 10 5 4 11 9 6 3) and

(1 2 7 8 12 13 14 16 15 11 10 9 6 5 4 3)

and the results after each pass of the final merge are

(1 15 2 11 7 10 8 9 6 12 5 13 4 14 3 16)

(1 6 12 15 2 5 11 13 4 7 10 14 3 8 9 16)

(1 4 6 7 10 12 14 15 2 3 5 8 9 11 13 16) and

(1 2 3 4 5 6 7 8 9 10 11 12 13 14 15 16)

42. In 3 dimensions, eight processors may be stationed at the corners (0, 0, 0), (0, 0, 1), (0, 1, 0), (0, 1, 1), (1, 0, 0), (1, 0, 1), (1, 1, 0), and (1, 1, 1) of a cube. In this case the processors are directly connected iff the vertices are adjacent in the cube. The hypercube mode simply generalizes this to higher dimensions.
43. See Quinn (1987, 138–39) or Hillis and Steele (1986, 1173).
44. See Hillis and Steele (1986, 1175).
46. See Quinn (1987, 165).
47. Let $n + 1 = (p + 1)^k$. The basis, $k = 0$, is trivial. For $k > 0$, a single step can compare with at most p elements. This leaves $(p + 1)^k - 1 - p$ elements uncompared. The p elements used for comparison define $p + 1$ intervals; at least one of these intervals must have the average number $[(p + 1)^k - 1 - p]/(p + 1) = (p + 1)^{k-1} - 1$ of uncompared keys. By induction, $k - 1$ steps are necessary to search this interval, so k steps are required altogether.
48. An algorithm achieving the lower bound of the previous problem would need to give each interval the average number of keys; see Quinn (1987, 108–9).

Appendix

D

Elementary Data Types and Their Implementation

This text assumes that the reader has had at least as much of an introduction to the most common types of data structures as is given in a second programming course. This appendix contains a *brief* introduction to these data types and their implementation in three classes of languages:

1. languages with lists as a primitive data type, such as Lisp,
2. languages with pointer variables, such as Pascal,
3. languages with neither, such as FORTRAN and assembly languages.

To use linked lists to represent sequences, we need to a) find the first element or *head* of a sequence, b) find the rest or *tail* of a sequence, c) construct a new list with a given head and tail, and d) represent the empty list. In Lisp, this is particularly easy, since the three operations correspond to the primitive functions CAR, CDR, and CONS respectively. The empty list is represented by NIL.

In Pascal, one thinks of data as stored in a *field,* or generally in several fields, of a node implemented as a *record.* In another field of the node is stored a pointer to the successor. A *type* statement is used to define the record type and the type of pointer that may point to it. In this statement, the record is declared to contain one or more fields of the appropriate type for the data and a field of the correct pointer type. A list is represented by a pointer to its first node.

This representation may be pictured as in Figure D.1. An entire list representing the sequence (G E T Q) is shown in Figure D.2a.

If L is the name of the list pointer, and the fields are called **data** and **next,** then the head is accessed by $L^\wedge$.data and the tail by $L^\wedge$.next. An empty list has value NIL. A *null* pointer is one whose value is NIL. To construct a new list M with head H and tail T, the following statements may be used:

$$\text{new}(M); M^\wedge.\text{head} := H; M^\wedge.\text{tail} := T$$

The **new** statement points its argument M to a new node of the appropriate type (i.e., the type to which M may point). It is similar to the **Get** function of this text.

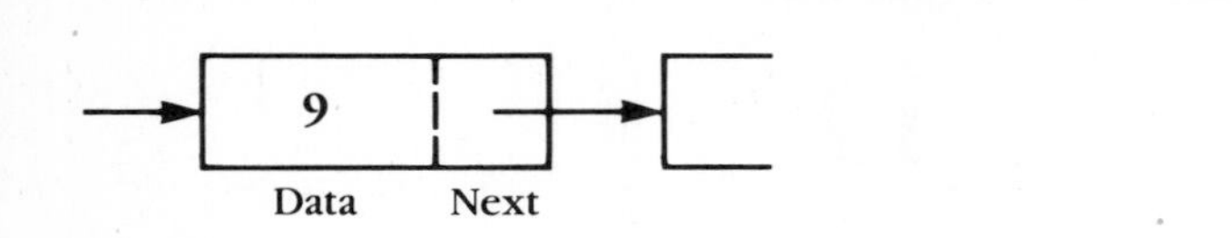

Figure D.1 Representation of a List Node Using Pointer Variables

Insertion may be accomplished by the process suggested by Figure D.2. Here it is sufficient to specify the element to be inserted and a pointer to the node that is to be the element's predecessor. Deletion of a node, given a pointer to that node, is more complicated, since the pointer field of that node's predecessor must be changed, but the predecessor is not accessible from the given pointer. One solution is to provide as an additional parameter the list from which the node is to be deleted. Another solution is to use a *variable parameter,* corresponding to call by value-result, to allow an assignment by the deletion routine to be reflected in the calling routine. Or we may allocate and maintain a second pointer field for each node that is to point to the node's predecessor. This structure is called a *doubly linked list,* as opposed to the *singly linked list* we have been discussing, and is pictured in Figure D.3.

Insertion into a doubly linked list requires getting a new node, initializing its three fields, and pointing two pointers to it. Deletion of a given node from a doubly linked list requires pointing two pointers around it. Operations at the beginning or end of the list require special care. One way to avoid complications in, for example, deleting from the beginning of a list is to maintain a special dummy node called the *header,* which is never deleted. The header is the natural place to maintain extra or global information about the list, such as its size or a pointer to its last element.

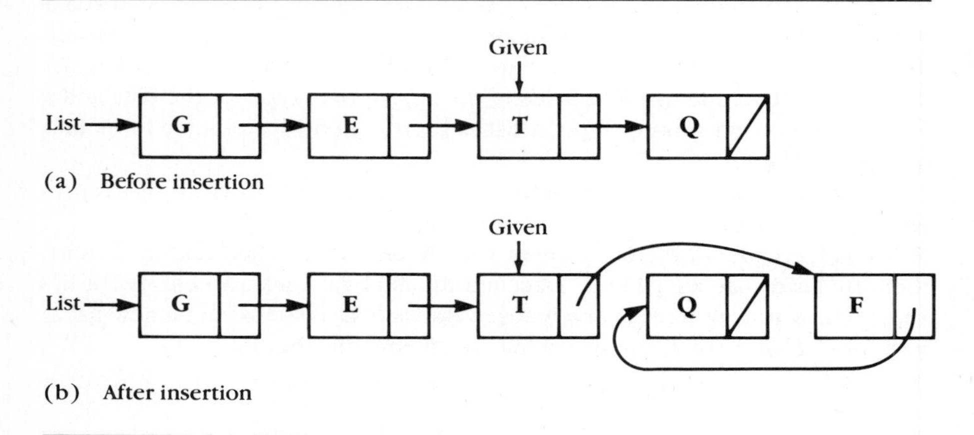

Figure D.2 Insertion of the Key F After the Key T in the Linked List Representing (G E T Q)

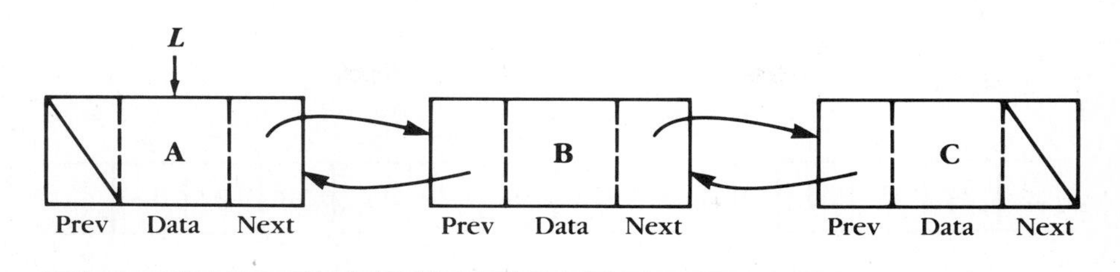

Figure D.3 A Doubly Linked List (Representing the Sequence (A B C))

Binary trees may also be represented with two pointer fields and one or more data fields as in Figure D.4. The tree is accessed by way of a named pointer to the root node. Finding the left and right subtrees simply requires accessing a pointer field; a new tree with given subtrees may be constructed by getting a new node and initializing its fields. Insertion of a child where previously there was no child, as for binary search trees, is a little trickier. An insertion algorithm follows pointers down the tree, either recursively or iteratively, until the empty pointer field is reached. Then the new node is created and initialized. But at this point if only the null value of the pointer field is available, there may be no way to point to the new node from its new parent. Again the use of a variable parameter in Pascal—or in general, a call by value-result—solves the problem by allowing assignment to this pointer, with original value NIL, to be reflected in a new value for the parameter (i.e., the parent's pointer field) in the calling routine. Another possibility is to maintain a parent pointer as an extra parameter.

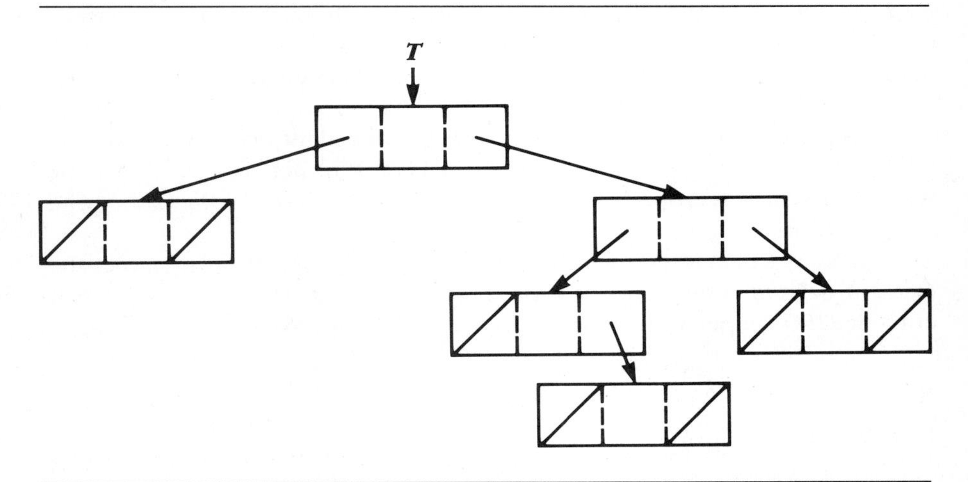

Figure D.4 Linked Representation of a Binary Tree

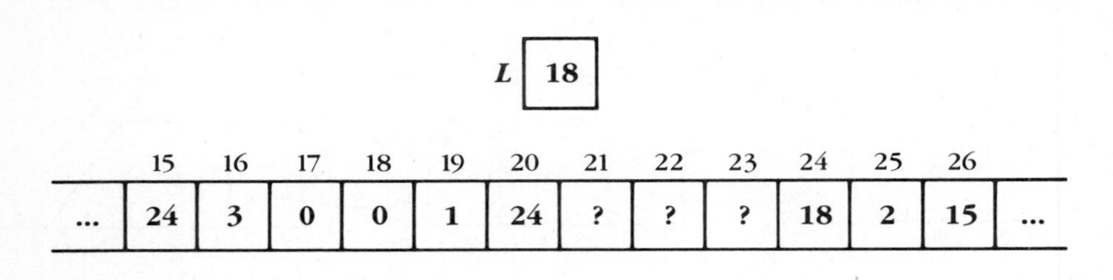

Figure D.5 Sample Doubly Linked Representation in a Language Without Pointer Variables

In languages of type (c), the general solution is to implement pointer values as integers. These integers are analogous to array indices, and available memory may be thought of as a large array or as several parallel arrays. For example, one doubly linked representation of the sequence (1 2 3) is that of Figure D.5. In that figure we assume that a pointer value of 0 represents the empty pointer NIL.

Getting new storage may be accomplished by keeping track of the first unused location in the array that represents all of memory and by returning and updating this value after each request. Unfortunately, restoring storage that is no longer in use to this list is extremely awkward. This issue of memory management arises in all languages; however, in Pascal it is hidden from the user by the **new** function. In Lisp, memory management is also hidden from the user, but since all nodes have the same size and shape, they may simply be linked together on a *free space list.*[1]

The operations needed for a Lisp free space list are the same as those for a *stack.* A stack has operations *insert* or *push* and *delete* or *pop* such that the element popped is the stack element most recently pushed. This property, called the "last in, first out" or *LIFO* property, is the property needed in implementing subroutine calls to save old values of parameters and other local variables.

A push operation can be implemented as the linked list operation (c) previously defined, corresponding to CONS in Lisp. The pop operation can be implemented as operation (b), returning the tail of the list, as long as the head is also returned.

A queue is a data structure with insert and delete operations such that the element deleted is the element least recently inserted. This is a "first in, first out," or *FIFO* property, as opposed to the LIFO property of a stack. A queue

[1] A particular implementation may have nodes of several sizes and shapes, but in any case the number of possibilities is small.

may be implemented as a linked list if elements are inserted at the end and deleted from the beginning. Header nodes are useful here to contain a pointer to the end of the list. This pointer should of course be updated by the insert operation.

Stacks and queues may also be implemented in terms of arrays.

Bibliography

Abelson, Harold, and Gerald Jay Sussman with Julie Sussman. *Structure and Interpretation of Computer Programs.* Cambridge MA: MIT Press, 1985.

Adelson-Velskii, G. M., and E. M. Landis. "An algorithm for the organization of information." *Dokl. Akad. Nauk. SSSR Mathemat.* 146(2): 263–66 (1962).

Aho, Alfred V., John E. Hopcroft, and Jeffrey D. Ullman. *The Design and Analysis of Computer Algorithms.* Reading MA: Addison-Wesley, 1974.

———. *Data Structures and Algorithms.* Reading MA: Addison-Wesley, 1983.

Appel, K., and W. Haken. "Every planar map is 4 colourable." *Ill. J. Math.* 21(3): 429–90 (part I) and 491–567 (part II) (1977).

Augenstein, Moshe J., and Aaron M. Tenenbaum. *Data Structures and PL/I Programming.* Englewood Cliffs NJ: Prentice-Hall, 1979.

Baase, Sara. *Computer Algorithms—Introduction to Design and Analysis.* Reading MA: Addison-Wesley, 1978. 2d ed. 1988.

Baron, Robert J., and Linda G. Shapiro. *Data Structures and Their Implementation.* New York: Van Nostrand Reinhold, 1980.

Batcher, K. E. "Sorting networks and their applications." In *Proceedings of the Spring Joint Computer Conference,* vol. 32, 307–14. Reston VA: AFIPS Press, 1968.

Bayer, R., and E. McCreight. "Organization and maintenance of large ordered indexes." *Acta Informatica* 1(4): 290–306 (1972).

Bentley, Jon L. "Programming pearls." *Communications of the ACM* 24(4): 287–91 (April 1984).

Boyer, Robert S., and J. Strother Moore. "A fast string searching algorithm." *Communications of the ACM* 20(10): 762–72 (October 1977).

Brassard, Gilles, and Paul Bratley. *Algorithmics: Theory and Practice.* Englewood Cliffs NJ: Prentice-Hall, 1988.

Brillinger, P. C., and D. J. Cohen. *Introduction to Data Structures and Non-Numeric Computation.* Englewood Cliffs NJ: Prentice-Hall, 1972.

Charniak, Eugene, and Drew McDermott. *Introduction to Artificial Intelligence.* Reading MA: Addison-Wesley, 1985.

Cook, S. A. "The complexity of theorem-proving procedures." In *Proceedings of the Third ACM Symposium on Theory of Computing,* 151–58. New York: Association for Computing Machinery, 1971.

Date, C. J. *An Introduction to Database Systems.* 3d ed. Reading MA: Addison-Wesley, 1981.

Diffie, W., and M. E. Hellman. "New directions in cryptography." *IEEE Trans. on Info. Theory* 226: 644–54 (November 1976).

Dijkstra, E. W. "A note on two problems in connection with graphs." *Numerische Mathematik* 1: 269–71 (1959).

Edgar, Hugh M. *A First Course in Number Theory.* Belmont CA: Wadsworth, 1988.

Floyd, Robert W. "Algorithm 97: Shortest path." *Communications of the ACM* 5(6): 345 (1962).

———. "Algorithm 245: Treesort 3." *Communications of the ACM* 7(12): 701 (1964).

———, and Ronald L. Rivest. "Algorithm 489: Select." *Communications of the ACM* 18(3): 173 (March 1975).

Ford, Lester, Jr., and Selmer M. Johnson. "A tournament problem." *American Mathematical Monthly* 66: 387–89 (1959).

Garey, Michael R., and David S. Johnson. *Computers and Intractability—A Guide to the Theory of NP-completeness.* Paperback. San Francisco: W. H. Freeman, 1979.

Gelperin, David. "On the optimality of A*." *Artificial Intelligence* 8(1): 69–76 (February 1977).

Genesereth, Michael R., and Nils J. Nilsson. *Logical Foundations of Artificial Intelligence.* Los Altos CA: Morgan Kaufmann Publishers, 1987.

Graham, R. "Bounds on multiprocessor timing anomalies." *SIAM J. on Applied Math.* 17(2): 416–29 (1969).

Grishman, Ralph. *Computational Linguistics.* Cambridge UK: Cambridge University Press, 1986.

Hart, P. E., N. J. Nilsson, and B. Raphael. "A formal basis for the heuristic determination of minimum cost paths." *IEEE Trans. Syst. Sci. and Cybernetics* 4(2): 100–107 (1968).

———, "Correction to 'A formal basis for the heuristic determination of minimum cost paths.'" *SIGART Newsletter* #37: 28–29 (December 1972).

Hillis, W. Daniel, and Guy L. Steele, Jr. "Data parallel algorithms." *Communications of the ACM* 29(12): 1170–83 (December 1986).

Hopcroft, John E., and Jeffrey D. Ullman. *Introduction to Automata Theory, Languages, and Computation.* Reading MA: Addison-Wesley, 1979.

Horowitz, Ellis, and Sartaj Sahni. *Fundamentals of Data Structures.* Potomac MD: Computer Science Press, 1976.

———. *Fundamentals of Computer Algorithms.* Potomac MD: Computer Science Press, 1978.

Huffman, D. A. "A method for the construction of minimum-redundancy codes." In *Proc. IRE,* vol. 40, 1098–1101. 1952.

Johnson, D., A. Devers, J. Ullman, M. Garey, and R. Graham. "Performance bounds for simple one dimensional bin packing algorithms." *SIAM J. on Computing* 3(4): 299–325 (1974).

Jones, Douglas W. "An empirical comparison of priority queue and event set implementations." *Communications of the ACM* 29(4): 300–311 (April 1986).

Karlton, P. L., S. H. Fuller, R. E. Scroggs, and E. B. Koehler. "Performance of height-balanced trees." *Communications of the ACM* 19(1): 23–28 (January 1976).

Karmarkar, N. "A new polynomial-time algorithm for linear programming." *Combinatorica* 4: 373–95 (1984).

Karp, Richard M., "Reducibility Among Combinatorial Problems." In *Complexity of Computer Computations.* New York: Plenum Press, 1972.

Kasami, T. "An efficient recognition and syntax algorithm for context-free languages." Bedford MA: Scientific Report AFCRL-65-758, Air Force Cambridge Research Lab, 1965.

Khachian, L. G. "A polynomial algorithm in linear programming." *Soviet Math. Dokl.* 20(1): 191–94 (1979).

Knuth, Donald. *The Art of Computer Programming,* 3 vols. Reading MA: Addison-Wesley, Vol. 1, 2d ed. 1973, Vol. 2, 2d ed. 1981, Vol. 3, 1st ed. 1973.

———, James H. Morris, Jr., and Vaughn R. Pratt. "Fast pattern matching in strings." *SIAM J. on Computing* 6(2): 323–50 (June 1977).

Kruskal, J. B., Jr. "On the shortest spanning subtree of a graph and the traveling salesman problem." *Proc. AMS* 7(1): 48–50 (1956).

Lewis, Harry R., and Christos H. Papadimitriou. *Elements of the Theory of Computation.* Englewood Cliffs NJ: Prentice-Hall, 1981.

Lorin, Harold. *Sorting and Sort Systems.* Reading MA: Addison-Wesley, 1975.

Lueker, George S. "Some techniques for solving recurrences." *Computing Surveys* 12(4): 419–36 (December 1980).

MacWilliams, F. J., and N. J. A. Sloane. *Theory of Error Correcting Codes.* Amsterdam: Elsevier, 1978.

Merritt, Susan M. "An inverted taxonomy of sorting algorithms." *Communications of the ACM* 28(1): 96–99 (January 1985).

Nilsson, Nils J. *Principles of Artificial Intelligence.* Palo Alto CA: Tioga Press, 1980.

Padua, David A., and Michael J. Wolfe. "Advanced compiler optimizations for supercomputers." *Communications of the ACM* 29(12): 1184–1201 (December 1986).

Pearl, Judea. *Heuristics—Intelligent Search Strategies for Computer Problem Solving.* Reading MA: Addison-Wesley, 1984.

Pereira, Fernando C. N., and Stuart M. Shieber, *Prolog and Natural Language Analysis,* Stanford CA: Center for the Study of Language and Information, 1987.

Perlis, A. J., and C. Thornton. "Symbol manipulation by threaded lists." *Communications of the ACM* 3(4): 195–204 (April 1960).

Peterson, William Wesley, and E. J. Weldon, Jr. *Error-Correcting Codes,* 2d ed. Cambridge MA: MIT Press, 1972.

Pratt, Vaughan R. "Every prime has a succinct certificate." *SIAM J. on Computing* 4(3): 214–20, September 1975.

Quinn, Michael J. *Designing Efficient Algorithms for Parallel Computers.* New York: McGraw-Hill, 1987.

Reingold, Edward M., and Wilfred J. Hansen. *Data Structures.* Boston: Little, Brown and Co., 1983.

Rich, Elaine. *Artificial Intelligence.* New York: McGraw-Hill, 1983.

Rivest, R. L., A. Shamir, and L. Adleman. "A method for obtaining digital signatures and public-key cryptosystems." *Communications of the ACM* 21(2): 120–26 (February 1978).

Ross, Kenneth A., and Charles R. B. Wright. *Discrete Mathematics.* 2d ed. Englewood Cliffs NJ: Prentice-Hall, 1988.

Sahni, Sartaj. *Concepts in Discrete Mathematics,* 2d ed. Fridley MN: Camelot Publishing Co., 1985.

Shell, D. L. "A high-speed sorting procedure." *Communications of the ACM* 2(7): 30–33 (July 1959).

Smith, Harry F. *Data Structures, Form and Function.* San Diego: Harcourt Brace Jovanovich, 1987.

Standish, Thomas. *Data Structure Techniques.* Reading MA: Addison-Wesley, 1980.

Strassen, Volker. "Gaussian elimination is not optimal." *Numerische Mathematik* 13(4): 354–56 (1969).

Stubbs, Daniel F., and Neil W. Webre. *Data Structures with Abstract Data Types and Pascal.* Belmont CA: Brooks-Cole Publishing Co., 1985.

Tarjan, R. "Efficiency of a good but not linear set union algorithm." *Journal of the ACM* 22(2): 215–25 (April 1975).

Tenenbaum, Aaron M., and Moshe J. Augenstein. *Data Structures Using PASCAL.* Englewood Cliffs NJ: Prentice-Hall, 1981.

Tucker, Alan. *Applied Combinatorics.* New York: John Wiley & Sons, 1980.

Turing, Alan M. "On computable numbers with an application to the Entscheidungsproblem." *Proc. London Math. Soc.* 2(42): 230–65 (1936).

Ullman, Jeffrey D. *Principles of Database Systems.* Rockville MD: Computer Science Press, 1980.

Warshall, S. "A theory on Boolean matrices." *Journal of the ACM* 9(1): 11–12 (1962).

Wiederhold, Gio. *Database Design.* 2d ed. New York: McGraw Hill, 1983.

Wirth, Niklaus. *Algorithms + Data Structures = Programs.* Englewood Cliffs NJ: Prentice-Hall, 1976.

Yao, A. C., and F. F. Yao, "The complexity of searching an ordered random table." In *IEEE Computer Soc. Proc. 17 Ann. Symp. Found. Comp. Sci.,* 173–77 (October 1976).

Younger, D. H. "Recognition and parsing of context free languages in time n^3." *Information and Control* 10(2): 189–208 (1967).

Index

Entries in **boldface** refer to a definition or algorithm. Entries followed by an *n* indicate a footnote.